Emerging Imaging Technologies in Medicine

IMAGING IN MEDICAL DIAGNOSIS AND THERAPY

William R. Hendee, Series Editor

Quality and Safety in Radiotherapy
Todd Pawlicki, Peter B. Dunscombe, Arno J. Mundt, and Pierre Scalliet, Editors
ISBN: 978-1-4398-0436-0

Adaptive Radiation Therapy
X. Allen Li, Editor
ISBN: 978-1-4398-1634-9

Quantitative MRI in Cancer
Thomas E. Yankeelov, David R. Pickens, and Ronald R. Price, Editors
ISBN: 978-1-4398-2057-5

Informatics in Medical Imaging
George C. Kagadis and Steve G. Langer, Editors
ISBN: 978-1-4398-3124-3

Adaptive Motion Compensation in Radiotherapy
Martin J. Murphy, Editor
ISBN: 978-1-4398-2193-0

Image-Guided Radiation Therapy
Daniel J. Bourland, Editor
ISBN: 978-1-4398-0273-1

Targeted Molecular Imaging
Michael J. Welch and William C. Eckelman, Editors
ISBN: 978-1-4398-4195-0

Proton and Carbon Ion Therapy
C.-M. Charlie Ma and Tony Lomax, Editors
ISBN: 978-1-4398-1607-3

Comprehensive Brachytherapy: Physical and Clinical Aspects
Jack Venselaar, Dimos Baltas, Peter J. Hoskin, and Ali Soleimani-Meigooni, Editors
ISBN: 978-1-4398-4498-4

Physics of Mammographic Imaging
Mia K. Markey, Editor
ISBN: 978-1-4398-7544-5

Physics of Thermal Therapy: Fundamentals and Clinical Applications
Eduardo Moros, Editor
ISBN: 978-1-4398-4890-6

Emerging Imaging Technologies in Medicine
Mark A. Anastasio and Patrick La Riviere, Editors
ISBN: 978-1-4398-8041-8

Forthcoming titles in the series

Informatics in Radiation Oncology
Bruce H. Curran and George Starkschall, Editors
ISBN: 978-1-4398-2582-2

Cancer Nanotechnology: Principles and Applications in Radiation Oncology
Sang Hyun Cho and Sunil Krishnan, Editors
ISBN: 978-1-4398-7875-0

Monte Carlo Techniques in Radiation Therapy
Joao Seco and Frank Verhaegen, Editors
ISBN: 978-1-4398-1875-6

Image Processing in Radiation Therapy
Kristy Kay Brock, Editor
ISBN: 978-1-4398-3017-8

Stereotactic Radiosurgery and Radiotherapy
Stanley H. Benedict, Brian D. Kavanagh, and David J. Schlesinger, Editors
ISBN: 978-1-4398-4197-6

Cone Beam Computed Tomography
Chris C. Shaw, Editor
ISBN: 978-1-4398-4626-1

Emerging Imaging Technologies in Medicine

Edited by
Mark A. Anastasio
Patrick La Riviere

CRC Press
Taylor & Francis Group
Boca Raton London New York

CRC Press is an imprint of the
Taylor & Francis Group, an **informa** business

A TAYLOR & FRANCIS BOOK

First published 2013 by Taylor & Francis

Published 2019 by CRC Press
Taylor & Francis Group
6000 Broken Sound Parkway NW, Suite 300
Boca Raton, FL 33487-2742

© 2013 by Taylor & Francis Group, LLC
CRC Press is an imprint of Taylor & Francis Group, an Informa business

First issued in paperback 2019

No claim to original U.S. Government works

ISBN 13: 978-0-367-44589-8 (pbk)
ISBN 13: 978-1-4398-8041-8 (hbk)

Library of Congress Cataloging-in-Publication Data

Emerging imaging technologies in medicine / editors, Mark A. Anastasio, Patrick La Riviere.
    p. ; cm. -- (Imaging in medical diagnosis and therapy)
  Includes bibliographical references and index.
  ISBN 978-1-4398-8041-8 (hardcover : alk. paper)
  I. Anastasio, Mark A. II. La Riviere, Patrick. III. Series: Imaging in medical diagnosis and therapy.
  [DNLM: 1. Diagnostic Imaging--methods. 2. Image Processing, Computer-Assisted--methods. 3. Medical Informatics--methods. WN 180]

616.07'54--dc23
    2012031497

Visit the Taylor & Francis Web site at
http://www.taylorandfrancis.com

and the CRC Press Web site at
http://www.crcpress.com

To my parents and family
—M.A.

To my loving family, Heather, Catherine, and Elizabeth,
and my parents, Andi and Jean
—P.L.

Contents

SECTION I Absorption-Based X-Ray Imaging

SECTION II X-Ray Phase-Contrast Imaging

SECTION III Photoacoustic Imaging and Tomography

SECTION IV Diffuse Optical Imaging

Series Preface

Advances in the science and technology of medical imaging and radiation therapy are more profound and rapid than ever before, since their inception over a century ago. Further, the disciplines are increasingly cross-linked as imaging methods become more widely used to plan, guide, monitor, and assess the treatments in radiation therapy. Today, the technologies of medical imaging and radiation therapy are so complex and so computer-driven that it is difficult for the persons (physicians and technologists) responsible for their clinical use to know exactly what is happening at the point of care, when a patient is being examined or treated. The persons best equipped to understand the technologies and their applications are medical physicists, and these individuals are assuming greater responsibilities in the clinical arena to ensure that what is intended for the patient is actually delivered in a safe and effective manner.

The growing responsibilities of medical physicists in the clinical arenas of medical imaging and radiation therapy are not without their challenges, however. Most medical physicists are knowledgeable in either radiation therapy or medical imaging, and expert in one or a small number of areas within their discipline. They sustain their expertise in these areas by reading scientific articles and attending the scientific talks at the meetings. In contrast, their responsibilities increasingly extend beyond their specific areas of expertise. To meet these responsibilities, medical physicists must periodically refresh their knowledge of advances in medical imaging or radiation therapy, and they must be prepared to function at the intersection of these two fields. How to accomplish these objectives is a challenge.

At the 2007 annual meeting of the American Association of Physicists in Medicine in Minneapolis, this challenge was the topic of conversation during a lunch hosted by Taylor & Francis Publishers and involving a group of senior medical physicists (Arthur L. Boyer, Joseph O. Deasy, C.-M. Charlie Ma, Todd A. Pawlicki, Ervin B. Podgorsak, Elke Reitzel, Anthony B. Wolbarst, and Ellen D. Yorke). The conclusion of this discussion was that a book series should be launched under the Taylor & Francis banner, with each volume in the series addressing a rapidly advancing area of medical imaging or radiation therapy of importance to medical physicists. The aim would be for each volume to provide medical physicists with the information needed to understand technologies driving a rapid advance and their applications to safe and effective delivery of patient care.

Each volume in the series is edited by one or more individuals with recognized expertise in the technological area encompassed by the book. The editors are responsible for selecting the authors of individual chapters and ensuring that the chapters are comprehensive and intelligible to someone without such expertise. The enthusiasm of volume editors and chapter authors has been gratifying and reinforces the conclusion of the Minneapolis luncheon that this series of books addresses a major need of medical physicists.

Imaging in Medical Diagnosis and Therapy would not have been possible without the encouragement and support of the series manager, Luna Han of Taylor & Francis Publishers. The editors and authors, and most of all I, are indebted to her steady guidance of the entire project.

William Hendee
Series Editor
Rochester, Minnesota

Preface

From the discovery of x-rays in 1895, through the emergence of computed tomography (CT) in the 1970s and magnetic resonance imaging (MRI) in the 1980s, noninvasive imaging has revolutionized the practice of medicine. While these technologies have thoroughly penetrated clinical practice, researchers continue to develop novel approaches that promise to push imaging into entirely new clinical realms, while addressing the issues of dose, sensitivity, or specificity that limit the existing imaging approaches.

This volume surveys a number of emerging technologies that have the promise to find routine clinical use in the near- (<5 year), mid- (5–10 year) and long-term (more than 10 year) time frames. It focuses on modalities with clinical potential rather than those, such as bioluminescence imaging, likely to have an impact mainly in preclinical animal imaging. We have chosen to focus primarily on novel imaging physics and/ or hardware configurations rather than algorithmic or software advances such as novel image reconstruction algorithms or MRI pulse sequences. This book also does not survey the remarkable developments in contrast agents and imaging probes that have given rise to the molecular imaging revolution now underway.

The last 10 years have been a period of fervent creativity and progress in imaging technology, with improvements in computational power, nanofabrication, and laser and detector technology leading to major new developments in phase-contrast imaging, photoacoustic imaging, and optical imaging. The success of PET-CT has paved the way for new multimodality systems such as PET-MRI. As some of these modalities are on the cusp of clinical trials, it is an auspicious time for a published volume surveying these emerging imaging modalities.

The target audiences of this book are scientists and clinicians interested in learning the basic technology and potential clinical applications driving a variety of promising new imaging modalities that may find their way into clinical use in the near-, mid-, and longer term. Each chapter begins with an accessible overview explaining the modality to a technically literate but non-expert reader. The chapters then contain a more technical and mathematical overview of the modalities aimed at physicists, engineers, and technically inclined clinicians seeking a self-contained explanation of these emerging modalities. They end with a section on the modality's clinical potential.

This book is organized into seven sections comprising a total of 20 individual chapters. The seven sections are as follows:

I. *Absorption-Based X-Ray Imaging*: The volume begins with x-ray absorption-based modalities, specifically tomosynthesis and spectral CT, which are perhaps closest to clinical use of all the modalities discussed in the volume.

II. *X-Ray Phase-Contrast Imaging*: This section considers x-ray phase-contrast imaging, which has enjoyed a renaissance in the last few years thanks to advances in nanofabrication of diffraction gratings and the computational power needed to enable propagation-based imaging.

III. *Photoacoustic Imaging and Tomography*: Next, we turn to photoacoustic imaging and tomography, with chapters on applications to breast imaging and quantitative imaging of physiology.

IV. *Diffuse Optical Imaging*: The fourth section focuses on diffuse optical tomography, beginning with an overview chapter on imaging physics and image formation, followed by application chapters on breast imaging and brain imaging.

V. *Acoustical Imaging*: The fifth section addresses advances in acoustic imaging, including ultrasound tomography (primarily for breast imaging), ultrasound elastography, and acoustic radiation force imaging, all of which allow for more quantitative measurement of tissue elastic properties.

VI. *Multimodality Imaging*: In the sixth section, we consider four emerging hybrid modalities: (1) PET/MRI, which holds great promise for brain and cardiac imaging; (2) photoacoustic tomography combined with reflection-mode ultrasound, which promises to combine functional photoacoustic images with anatomical ultrasound images; (3) photacoustic tomography combined with diffuse optical tomography, which promises to improve the quality of both pairs of images by further constraining the optical properties of the imaged medium; and (4) diffuse optical tomography and MRI, which will allow MRI anatomical information to help constrain the DOT reconstruction problem as well as allow superposition of functional DOT information on anatomical MRI images.

VII. *Three-Dimensional in vivo Microscopy*: The final section surveys three *in vivo* microscopy modalities: photoacoustic microscopy, with a focus on ophthalmic applications, a generalization of optical coherence tomography called interferometric synthetic aperture microscopy, and Fourier transform infrared spectroscopy, which has potential for label-free imaging of biomolecules.

Overall, the hope is that the book will leave readers with a picture of the state of the art in novel imaging technologies, and also a sense of the roadmap for further development toward clinical use, with an emphasis on the most likely clinical applications of the various modalities and the principal challenges to their deployment.

Editors

Mark A. Anastasio earned his PhD degree at the University of Chicago in 2001 and is currently a professor of biomedical engineering at Washington University in St. Louis (WUSTL). Prior to joining WUSTL, he was a faculty member at the Illinois Institute of Technology. Dr. Anastasio is an internationally recognized expert on tomographic image reconstruction, imaging physics, and the development of novel biomedical imaging systems. He has conducted pioneering research in the fields of diffraction tomography, x-ray phase-contrast imaging, and photoacoustic tomography. He has published extensively on these topics and was a recipient of an NSF CAREER award. He is on the editorial board of the *Journal of Biomedical Optics* and is on the organizing committee for the SPIE Photonics West Photons plus Ultrasound Conference and the SPIE Conference on Image Reconstruction from Incomplete Data. He was also a program chair for the 2009 Optical Society of America Signal Recovery and Synthesis Topical Meeting.

Patrick La Riviere received his AB degree in physics from Harvard University in 1994 and his PhD degree from the Graduate Programs in Medical Physics in the Department of Radiology at the University of Chicago in 2000. In between, he studied the history and philosophy of physics while on the Lionel de Jersey–Harvard scholarship to Cambridge University. He is currently an associate professor in the Department of Radiology at the University of Chicago, where his research interests include tomographic reconstruction in computed tomography, x-ray fluorescence computed tomography, and optoacoustic tomography. In 2005, he received the IEEE Young Investigator Medical Imaging Scientist Award, given to a young investigator within 6 years of the PhD for significant contributions to medical imaging research. He is the author of more than 50 peer-reviewed articles and peer-reviewed conference proceedings and eight book chapters.

Contributors

Steven Adie
Beckman Institute for Advanced
    Science and Technology
University of Illinois at Urbana-Champaign
Urbana, Illinois

Ken Bates
Applied Concepts, LLC
Beaverton, Oregon

Adam Q. Bauer
Department of Radiology
Washington University School of
    Medicine
St. Louis, Missouri

Rohit Bhargava
Department of Bioengineering
and
Beckman Institute for Advanced
    Science and Technology
and
Department of Electrical and Computer
    Engineering
University of Illinois at Urbana-Champaign
Urbana, Illinois

Stephen A. Boppart
Beckman Institute for Advanced
    Science and Technology
University of Illinois at
    Urbana-Champaign
Urbana, Illinois

Jovan G. Brankov
Department of Electrical and
    Computer Engineering
Illinois Institute of Technology
Chicago, Illinois

Jochen Cammin
Department of Radiology and
    Radiological Science
Johns Hopkins University
Baltimore, Maryland

P. Scott Carney
Beckman Institute for Advanced
    Science and Technology
University of Illinois at Urbana-Champaign
Urbana, Illinois

Xiaoyang Cheng
Delphinus Medical Technologies
Plymouth, Michigan

Rinaldo Cubeddu
Dipartimento di Fisica
Politecnico di Milano
and
CNR—Istituto di Fotonica e
    Nanotecnologie
Milan, Italy

Joseph P. Culver
Department of Radiology
and
Department of Biomedical Engineering
Washington University School of
    Medicine
St. Louis, Missouri

Jeremy J. Dahl
Department of Biomedical
    Engineering
Duke University
Durham, North Carolina

Scott C. Davis
Thayer School of Engineering
Dartmouth College
Hanover, New Hampshire

Stephen Del Rio
OptoSonics, Inc.
Oriental, North Carolina

Neb Duric
Barbara Ann Karmanos Cancer Institute
Wayne State University
Detroit, Michigan

and
Delphinus Medical Technologies
Plymouth, Michigan

Todd N. Erpelding
Philips Research North America
Briarcliff Manor, New York

Peter van Es
MIRA-Institute for Biomedical
    Technology and Technical
    Medicine
University of Twente
Enschede, the Netherlands

Georges El Fakhri
Center for Advanced Radiological
    Sciences
Massachusetts General Hospital
Harvard Medical School
Boston, Massachusetts

Christopher Favazza
Department of Biomedical Engineering
Washington University
St. Louis, Missouri

Jeffrey Goll
Delphinus Medical Technologies
Plymouth, Michigan

Jeremy C. Hebden
Department of Medical Physics and
    Bioengineering
University College London
London, United Kingdom

Song Hu
Department of Biomedical
    Engineering
Washington University
St. Louis, Missouri

Jan S. Iwanczyk
DxRay, Inc.
Northridge, California

Roman Janer
Delphinus Medical Technologies
Plymouth, Michigan

Shudong Jiang
Thayer School of Engineering
Dartmouth College
Hanover, New Hampshire

Shuliang Jiao
Department of Ophthalmology
University of Southern California
Los Angeles, California

Soren D. Konecky
Caliper—A PerkinElmer Company
Alameda, Califomia

Robert Kruger
OptoSonics, Inc.
Oriental, North Carolina

Dave Kunz
Delphinus Medical Technologies
Plymouth, Michigan

Richard Lam
OptoSonics, Inc.
Oriental, North Carolina

Ton G. van Leeuwen
MIRA-Institute for Biomedical
    Technology and Technical Medicine
University of Twente
Enschede, the Netherlands
and
Department of Biomedical Engineering
    and Physics
Academic Medical Center
University of Amsterdam
Amsterdam, the Netherlands

Cuiping Li
Delphinus Medical Technologies
Plymouth, Michigan

and

Barbara Ann Karmanos Cancer Institute
Wayne State University
Detroit, Michigan

Quanzheng Li
Center for Advanced Radiological
    Sciences
Massachusetts General Hospital
Harvard Medical School
Boston, Massachusetts

Peter Littrup
Barbara Ann Karmanos Cancer Institute
Wayne State University
Detroit, Michigan

and

Delphinus Medical Technologies
Plymouth, Michigan

Srirang Manohar
MIRA-Institute for Biomedical
    Technology and Technical
    Medicine
University of Twente
Enschede, the Netherlands

Michael A. Mastanduno
Thayer School of Engineering
Dartmouth College
Hanover, New Hampshire

Kelly E. Michaelsen
Thayer School of Engineering
Dartmouth College
Hanover, New Hampshire

Peter Modregger
Paul Scherrer Institute
University of Lausanne
Villigen PSI, Switzerland

Kaye S. Morgan
School of Physics
Monash University
Victoria, Australia

Glenn R. Myers
Department of Applied Mathematics
The Australian National University
ACT, Australia

Ralph E. Nothdurft
Department of Radiology
Washington University School of
    Medicine
St. Louis, Missouri

Jinsong Ouyang
Center for Advanced Radiological
    Sciences
Massachusetts General Hospital
Harvard Medical School
Boston, Massachusetts

Kevin J. Parker
Department of Electrical
    and Computer
    Engineering
University of Rochester
Rochester, New York

Keith D. Paulsen
Thayer School of Engineering
Dartmouth College
Hanover, New Hampshire

Antonio Pifferi
Dipartimento di Fisica
Politecnico di Milano
and
CNR—Istituto di Fotonica e
    Nanotecnologie
Milan, Italy

Bernd Pinzer
Paul Scherrer Institute
Villigen PSI, Switzerland

Brian W. Pogue
Thayer School of Engineering
Dartmouth College
Hanover, New Hampshire

Giovanna Quarto
Dipartimento di Fisica
Politecnico di Milano
Milan, Italy

Rohith Reddy
Department of Bioengineering
and
Beckman Institute for Advanced Science
    and Technology
University of Illinois at
    Urbana-Champaign
Urbana, Illinois

Daniel Reinecke
OptoSonics, Inc.
Oriental, North Carolina

Ingrid Reiser
Department of Radiology
The University of Chicago
Chicago, Illinois

Olivier Roy
Barbara Ann Karmanos Cancer Institute
Wayne State University
Detroit, Michigan

and

Delphinus Medical Technologies
Plymouth, Michigan

Steve Schmidt
Delphinus Medical Technologies
Plymouth, Michigan

Karen K.W. Siu
School of Physics
and
Monash Biomedical Imaging
Monash University
and
Australian Synchrotron
Victoria, Australia

Lorenzo Spinelli
CNR—Istituto di Fotonica e Nanotecnologie
Milan, Italy

Marco Stampanoni
Paul Scherrer Institute
ETH and University of Zürich
Villigen PSI, Switzerland

Wiendelt Steenbergen
MIRA-Institute for Biomedical
    Technology and Technical Medicine
University of Twente
Enschede, the Netherlands

Katsuyuki Taguchi
Department of Radiology and
    Radiological Science
Johns Hopkins University
Baltimore, Maryland

Paola Taroni
Dipartimento di Fisica
Politecnico di Milano
and
CNR—Istituto di Fotonica e
    Nanotecnologie
Milan, Italy

Alessandro Torricelli
Dipartimento di Fisica
Politecnico di Milano
Milan, Italy

Lihong V. Wang
Department of Biomedical Engineering
Washington University
St. Louis, Missouri

Zhentian Wang
Paul Scherrer Institute
Villigen PSI, Switzerland

Kevin Yeh
Department of Bioengineering
and
Beckman Institute for
    Advanced Science
    and Technology
University of Illinois at
    Urbana-Champaign
Urbana, Illinois

Hao F. Zhang
Department of Biomedical
    Engineering
Northwestern University
Evanston, Illinois

Adam M. Zysk
Department of Electrical and
    Computer Engineering
Illinois Institute of Technology
Chicago, Illinois

Absorption-Based X-Ray Imaging

1

X-Ray Tomosynthesis

Ingrid Reiser
The University of Chicago

1.1 Introduction

Planar radiography was the earliest clinical application of x-ray imaging and is still the most widely used imaging tool. In a planar radiograph, x-rays travel through the patient's body, where they are partially absorbed by anatomic tissues, and the transmitted x-rays are detected by film or by a digital detector to form the image. In this image, the complex 3D anatomy is projected onto the 2D plane of the x-ray detector. As a result, normal anatomy can hide anomalies, or summation of structures can mimic an abnormality. Both situations can potentially result in a misdiagnosis. This limitation is addressed by tomosynthesis, which adds some (but not complete) depth resolution to the planar radiograph.

In tomosynthesis, a sequence of x-ray projections is used to synthesize a quasi-3D image. The tomosynthesis projections are acquired over a limited angular range of typically <60°. From such data, it is not possible to reconstruct a fully resolved image volume. Instead, a volume can be synthesized that contains some depth information. The individual tomosynthesis slices are visually similar to a planar radiograph; however, structures at the depth of the given slice are in focus, while structures at other depths appear blurred.

1.1.1 Principles of Tomosynthesis

Figures 1.1 and 1.2 illustrate the principle of tomosynthesis imaging. In Figure 1.1, an object is shown that contains structures at different depths, indicated by planes A and B. Three projection

views are acquired from different x-ray source locations. In each projection view, the relative positions of the structures depend on their depths within the object.

Tomosynthesis slices at the depths of planes A and B are reconstructed as illustrated in Figure 1.2. Each projection view is shifted by an amount that depends on the reconstructed slice depth. In the tomosynthesized planes, structures that are physically located at the slice depth are enhanced, while structures located at other depths appear blurred.

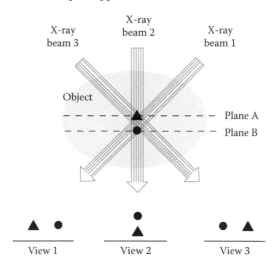

FIGURE 1.1 Tomosynthesis projections of an object containing structures at different depths.

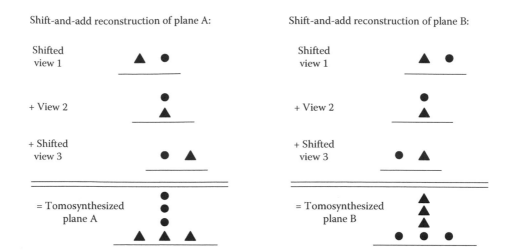

Shift-and-add reconstruction of plane A: Shift-and-add reconstruction of plane B:

FIGURE 1.2   Shift-and-add reconstruction of the tomosynthesis planes A and B.

Conceptually, tomosynthesis can be considered as limited-angle, limited-view computed tomography (CT). However, there are fundamental differences between CT and tomosynthesis. CT is a cross-sectional modality, where images are typically viewed in the plane of x-ray source rotation (Figure 1.3a). Under certain ideal conditions, exact image reconstruction is possible.

On the other hand, tomosynthesis is a planar modality where images are viewed in the plane perpendicular to the x-ray source motion (Figure 1.3b). Because of the limited scan angle (α), exact reconstruction is not possible, even for an ideal system (with the exception of reconstructing a single uniform object). In a tomosynthesis slice at a given depth, in addition to structures physically located at that depth, out-of-focus structures can be perceived that are located at different depths within the object. Typically, a cut through a tomosynthesis volume parallel to the detector is called *in-plane*, while a cut perpendicular to the detector is called *in-depth* or *out-of-plane*.

The principle of x-ray tomosynthesis was pioneered by Ziedses des Plantes as early as 1932 (des Plantes, 1932). Originally, tomosynthesis was based on the use of x-ray film, which was cumbersome and therefore received little attention. Recent advances in large-area digital detector technology sparked renewed interest in tomosynthesis. Digital planar radiography systems can perform tomosynthesis by acquiring projections from different x-ray source positions. These digital projections can then be readily reconstructed into a volume representation.

The development of tomosynthesis is driven by clinical demands. Tomosynthesis maintains the high spatial resolution and low dose of planar radiography, but provides some depth resolution that allows for moderate resolution of tissue structures. The dose delivered to the patient in a tomosynthesis scan is similar to that delivered by a radiograph, while patient dose from a conventional diagnostic CT can be up to two orders of magnitude greater. In terms of clinical workflow, patient positioning for tomosynthesis is the same as that for a planar radiograph, allowing for acquisition of either a projection radiograph or a tomosynthesis scan, or both, on the same unit. Tomosynthesis, thus, fits into the clinical workflow the way planar radiography does, which again differentiates tomosynthesis from CT.

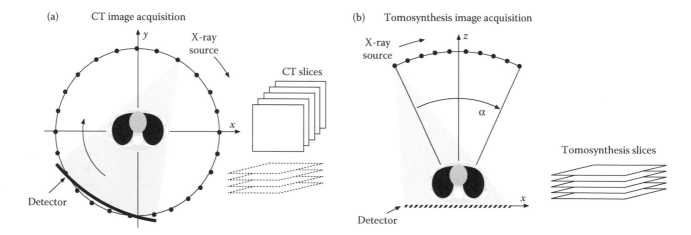

FIGURE 1.3   Schematics of (a) CT imaging and (b) tomosynthesis imaging of the thorax.

1.1.2 Tomosynthesis Image

As mentioned previously, a tomosynthesis image is very different from a CT image, but bears some similarity to a projection radiograph. Figure 1.4 shows cuts through a tomosynthesis breast volume. The appearance of the in-plane cut is similar to a projection radiograph (i.e., a mammogram). However, tissue structures appear sharp (i.e., in-focus) only in the tomosynthesis slice at the structure's true depth, and blurred (i.e., out-of-focus) in slices at other depths. The bottom panel of Figure 1.4 shows a cut perpendicular to the detector surface, and parallel to the direction of the x-ray source path. Anatomic structures are blurred and practically no information is gained from this view.

The depth resolution of tomosynthesis is determined by the scan angle and the in-plane size of the object structures. This is due to the Fourier slice theorem (see Section 1.2.1.1), which states that a projection along a direction θ corresponds to a line along θ in the spatial frequency domain. When projection views are acquired along a limited scan angle, a wedge in the Fourier domain is covered, where the angle of the wedge is equal to the arc of the tomosynthesis scan.

As a result, depth resolution in the tomosynthesis image depends on spatial frequency: Smaller objects (corresponding to higher spatial frequencies) are better resolved than larger objects (corresponding to lower spatial frequencies), and larger scan angles improve depth resolution. Figure 1.5 shows this effect for uniform spheres of different sizes. One row contains two single spheres. In the tomosynthesis reconstruction, the spheres are elongated in the depth dimension, where the elongation corresponds to the shadow cast by the in-plane dimension of the object. The second row contains spheres that have been stacked on top of each other, with their centers spaced by 2 cm. In the 60° tomosynthesis scan, the large spheres are resolved, while in the 20° scan, the stacked large spheres appear as one single sphere.

Thus, resolution in the tomosynthesis image is highly nonuniform. In-plane resolution is similar to image resolution in the projection images, but out-of-plane resolution is about an order of magnitude lower. Since little information is gained from the depth direction, the tomosynthesis image tends to be reconstructed on a highly nonuniform grid. As an example, breast tomosynthesis images are typically reconstructed on a grid with voxel size of about 100 µm in-plane and 1 mm in-depth,

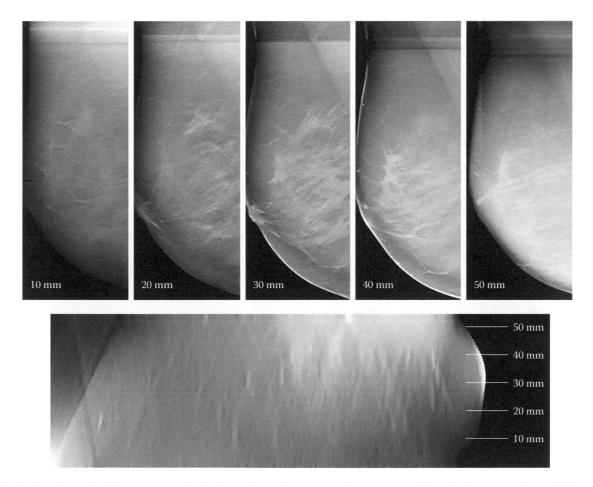

FIGURE 1.4 Cuts through a breast tomosynthesis image volume. The top row shows cuts parallel to the detector surface at different depths. The bottom panel shows a cut through the breast volume perpendicular to the detector, along the x-ray source motion path. The lines indicate the depths at which the in-plane slices are located. This image was reconstructed using the TV algorithm (Section 1.2.3).

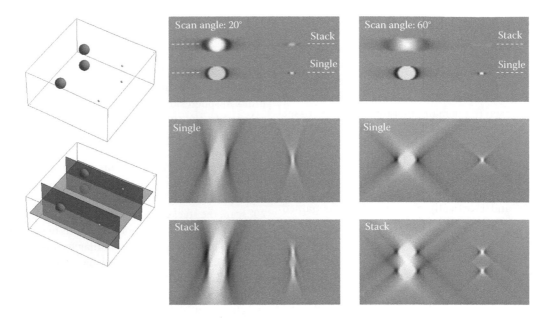

FIGURE 1.5 Tomosynthesis image of a phantom that contains spheres of two different sizes ($r_1 = 0.5$ mm, $\mu_1 = 0.2$ cm^{-1} and $r_2 = 0.1$ mm, $\mu_2 = 0.4$ cm^{-1}). One row contains two *single* spheres, while another row contains *stacked* spheres. For the 20° scan angle, the larger stacked spheres are perceived as a single large high-contrast sphere, whereas the stack is resolved for both sphere diameters when the scan angle is 60°.

resulting in voxels that are 10 times longer than they are wide. Such a reconstruction grid can produce noise aliasing. This is shown in Figure 1.6. The noise power spectra (NPS) in these figures were generated by reconstructing projections that contained uncorrelated Gaussian noise ($\mu = 0$, $\sigma = 1$). The reconstruction algorithm was FBP with a Hanning window (see Section 1.2.1.2). From the reconstructed images, the NPS was computed by

$$\text{NPS} = <|\, FT\{\mathcal{R} - <\mathcal{R}>\} \,|^2>, \qquad (1.1)$$

where $FT\{\cdot\}$ is the Fourier transform, $< \cdot >$ is ensemble average, and $\mathcal{R}$ is a region in the reconstructed image taken in the (x, z) plane (or, the plane containing the x-ray source motion).

1.2 Tomosynthesis Image Reconstruction

1.2.1 Reconstruction Preliminaries

To better understand tomosynthesis image reconstruction, as well as the properties of the resulting tomosynthesis image volume, it is helpful to revisit the Fourier slice theorem, and the basic filtered back-projection algorithm for image reconstruction from parallel projections. The derivations presented here are based on that by Kak and Slaney (1988).

1.2.1.1 Fourier Slice Theorem

Given a spatially varying object $f(x, y)$, the line integral $P_\theta(t)$ along a direction $\hat{\theta}$ is given by

$$P_\theta(t) = \int_{-\infty}^{\infty} \int_{-\infty}^{\infty} (x, y)\delta(x\cos\theta + y\sin\theta - t)\, dx\, dy, \qquad (1.2)$$

where $\delta(\cdot)$ is a delta function and t is a coordinate axis perpendicular to $\hat{\theta}$. The Fourier transform of the line integral ($S_\theta(w)$) is then

$$S_\theta(w) = \int_{-\infty}^{\infty} P_\theta(t)e^{-2\pi iwt}\, dt. \qquad (1.3)$$

The Fourier transform of the object $f(x, y)$ is given by

$$F(u, v) = \int_{-\infty}^{\infty} \int_{-\infty}^{\infty} (x, y)e^{-2\pi i(ux + vy)}\, dx\, dy. \qquad (1.4)$$

One can now look at a single line in the Fourier space. For simplicity, the Fourier transform of the object along the $v = 0$ line is

$$F(u, 0) = \int_{-\infty}^{\infty} \left[\int_{-\infty}^{\infty} f(x, y)\, dy \right] e^{-2\pi i(ux)}\, dx. \qquad (1.5)$$

Comparison of the inner integral with Equation 1.2 reveals that it is equal to $P_{\theta = 0}(t)$. Thus, the parallel projection of an object described by $f(x, y)$ corresponds to a line through the Fourier transform of the object, $F(u, v)$ (Figure 1.7). This relationship is valid for any angle θ and is known as the *Fourier slice theorem*.

In tomosynthesis, projection images are acquired along a limited scan angle. As a result, the tomosynthesis projection data fills a wedge in the Fourier domain. This leads to nonisotropic resolution of the tomosynthesis volume, as discussed in Section 1.1.2.

1.2.1.2 Filtered Back-Projection for Parallel Projections

Filtered back-projection is a reconstruction algorithm that has been developed for recovery of the object $f(x, y)$ from a

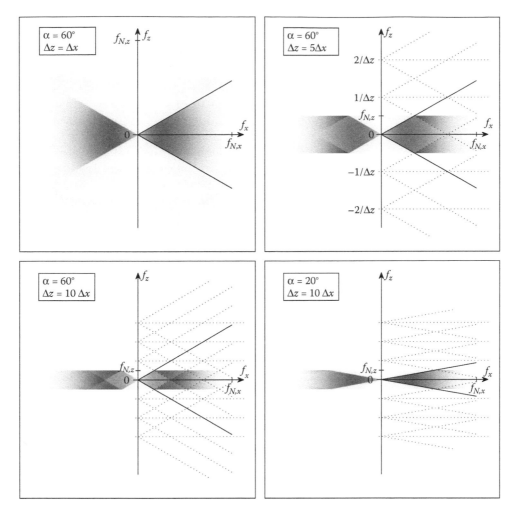

FIGURE 1.6 Noise-power spectra for different sampling conditions and tomosynthesis scan angles, $\alpha = 60°$ and $\alpha = 20°$.

series of projections $P_\theta(t)$. Given here is a derivation of the algorithm for reconstruction of a 2D slice from parallel beam projections (Kak and Slaney, 1988). The basic idea of this algorithm is to make use of the central slice theorem to generate a Fourier domain representation of the object, $F(u, v)$. From this, the object $f(x, y)$ is reconstructed using the inverse Fourier transform

$$f(x, y) = \int_{-\infty}^{\infty} \int_{-\infty}^{\infty} F(u, v) e^{2\pi i(ux+vy)} \, du \, dv. \tag{1.6}$$

Transformation of the integration variable into polar coordinates (w, θ) with $u = w \cos \theta$, $v = w \sin \theta$ results in

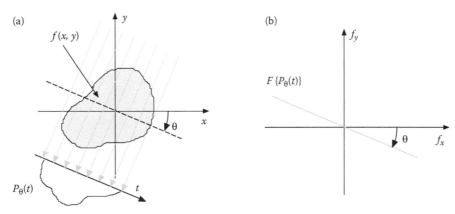

FIGURE 1.7 (a) Parallel projection of an object. (b) Fourier domain segment filled by the Fourier transform of $P_\theta(t)$.

$$f(x,y) = \int_{0}^{2\pi} \int_{0}^{\infty} F(w,\theta) e^{2\pi i w(x\cos\theta + y\sin\theta)} w \, dw \, d\theta \qquad (1.7)$$

and writing $t = x\cos\theta + y\sin\theta$ gives

$$f(x,y) = \int_{0}^{2\pi} \int_{0}^{\infty} F(w,\theta) e^{2\pi i w t} w \, dw \, d\theta. \qquad (1.8)$$

Without loss of generality, the integration limits can be changed to $\theta = 0 \cdots \pi$, $w = -\infty \cdots \infty$, since both sets of limits result in integration over the complete 2D surface. However, positivity of the area element $w \, dw \, d\theta$ needs to be ensured, thus

$$f(x,y) = \int_{0}^{\pi} \int_{-\infty}^{\infty} F(w,\theta) e^{2\pi i w t} \mid w \mid dw \, d\theta. \qquad (1.9)$$

Using the Fourier slice theorem, the Fourier transform of the object along θ, $F(w,\theta)$, is given by the Fourier transform of the line integral $S_\theta(w)$. Thus one obtains the filtered back-projection reconstruction formula

$$f(x,y) = \int_{0}^{\pi} \left[\int_{-\infty}^{\infty} S_\theta(w) e^{2\pi i w t} \mid w \mid dw \right] d\theta. \qquad (1.10)$$

In this equation, the inner integral over dw is a convolution of the line integral $P_\theta(t)$ with the Fourier transform of a ramp filter. The outer integration over $d\theta$ is a back-projection operation, where the filtered projection is smeared across the 2D plane for each angle θ. These two steps give rise to the name "filtered back-projection" (FBP). This algorithm is widely used for image reconstruction in computed tomography. Note that Equation 1.10 is exact in its analytic form. However, in order to actually reconstruct a tomographic image from projection data, Equation 1.10 needs to be solved numerically, which can cause image artifacts. Further, projection data are typically discrete and view sampling (i.e., the number of acquired projection views) is finite. Examples for reconstructed images of a uniform sphere with the BP or FBP algorithms are shown in Figure 1.8a–h for different imaging parameters such as view sampling and scan angle (also see Table 1.1).

In addition, projection data are discrete and corrupted with noise and physical factors in the imaging chain, which degrade the reconstructed image. Noise in the projection data tends to occur at high spatial frequencies, which is amplified by the ramp filter. Thus, when reconstructing from noisy projection data, in addition to filtering the projections with a ramp filter, an apodization filter is applied to suppress high spatial frequencies. Often, this is a Hann window with a frequency response of

$$H(f_x) = \frac{1}{2}\left(1 + \cos\left(\frac{Lf_x\pi}{2}\right)\right). \qquad (1.11)$$

To illustrate these concepts, image reconstruction was performed using different view sampling schemes (number of views, scan angle), and using different reconstruction filters. Projection data were generated by computing line integrals of a uniform sphere ($\mu = 1.4$). Noisy data were produced by adding 20% Gaussian noise to these data.

The BP algorithm (Figure 1.8a) results in a blurry, nonuniform sphere. This algorithm is equivalent to the shift-and-add algorithm from Section 1.1. In Figure 1.8b and c, an FBP reconstruction was performed according to Equation 1.10, from 360 or 18 views. While the higher view sampling reproduces the original sphere, reconstruction from few-view data results in severe artifacts even in the noise-free case. Noise amplification occurs when noisy data are reconstructed using a ramp filter only (Figure 1.8e), and is significantly used when an apodization filter is used (Figure 1.8f and g). Also shown is a reconstruction of the uniform sphere from a limited angle scan (60°), representative of a tomosynthesis acquisition. The object cannot be recovered from this limited scan. Reducing the number of views results in further artifacts.

The next section describes how to obtain the line integral $P_\theta(t)$ from measured projection data.

1.2.1.3 Preprocessing

Digital flat-panel detectors measure x-ray fluence as

$$\Phi(x_{\text{det}}, y_{\text{det}}) = \Phi_0(x_{\text{det}}, y_{\text{det}}) e^{-\int_{L(x_{\text{det}}, y_{\text{det}})} \mu(r) dr}, \qquad (1.12)$$

where Φ_0 is x-ray fluence emitted from the x-ray source toward the detector bin at $(x_{\text{det}}, y_{\text{det}})$, and $\mu(r)$ is the spatial distribution of attenuation coefficients (i.e., the object). Note that Equation 1.12 assumes a mono-energetic x-ray beam and neglects physical factors that may degrade the measurement, such as focal spot blur, x-ray scatter, detector blur, and noise.

Image reconstruction is generally performed based on a relationship between line integrals and image estimate such as that presented in Equation 1.10, rather than in terms of x-ray fluence. Rearranging Equation 1.12, the detector output can be converted to line integrals $P(x_{\text{det}}, y_{\text{det}})$ according to

$$P(x_{\text{det}}, y_{\text{det}}) = -\log(\Phi(x_{\text{det}}, y_{\text{det}}) / \Phi_0(x_{\text{det}}, y_{\text{det}})) \qquad (1.13)$$

$$= \int_{L(x_{\text{det}}, y_{\text{det}})} \mu(r) \, dr. \qquad (1.14)$$

1.2.2 FBP-Based Image Reconstruction for Tomosynthesis

An adaptation of filtered back-projection for tomosynthesis has been proposed by Lauritsch and Härer et al. (Lauritsch and

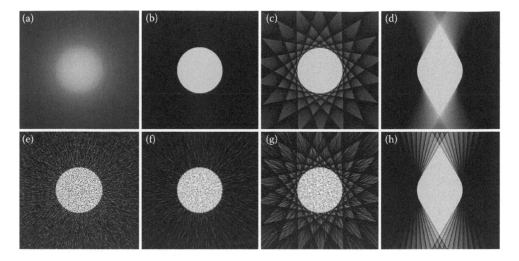

FIGURE 1.8 Reconstruction of a uniform sphere with different algorithms, from projections with different scan parameters and noise levels (see table below). The display window was [0,2] for all images.

TABLE 1.1 Reconstruction Algorithms, Scan Parameters and Noise Levels Used in Each Panel of Figure 1.8

Figure Panel	(a)	(b)	(c)	(d)
Reconstruction algorithm	BP	FBP	FBP	FBP + Hann
Scan length[°]	360	360	360	60
Number of views	360	360	18	61
% Noise	0	0	0	0
Figure Panel	(e)	(f)	(g)	(h)
Reconstruction algorithm	FBP	FBP + Hann	FBP + Hann	FBP + Hann
Scan length[°]	360	360	360	60
Number of views	360	360	18	61
% Noise	20	20	$20/\sqrt{(n_{\text{views}})}$	0

Härer, 1998; Mertelmeier et al., 2006). The image reconstruction is similar to FBP reconstruction (Section 1.2.1.2) in that a series of filters $H(\vec{f})$ is applied to the projection data. In a Fourier-domain representation, the reconstructed image $F(\vec{f})$ is obtained from the data $G(\vec{f})$ by

$$F(\vec{f}) = H(\vec{f})G(\vec{f}) \qquad (1.15)$$

with

$$H(\vec{f}) = H_r(f_x, f_z)H_a(f_x)H_{st}(f_x, f_z). \qquad (1.16)$$

Here, $H_r(f_x, f_z)$ is a ramp filter that operates in the plane spanned by the x-ray source motion,

$$H_r(f_x, f_z) = \begin{cases} 2\alpha\sqrt{f_x^2 + f_z^2} & \text{for}\ |f_z| < \tan(\alpha)\ |f_x| \\ 0 & \text{elsewhere,} \end{cases} \qquad (1.17)$$

where α is the tomosynthesis scan angle and $H_a(f_x)$ is an apodization filter to reduce the image noise, usually a Hann window given by Equation 1.11. So far the algorithm is identical to the original FBP reconstruction (Section 1.2.1.2). To account for the noise properties of the tomosynthesis volume, in particular, to suppress noise aliasing from the nonisotropic sampling (Figure 1.6), a *slice-thickness filter H_{st}* is introduced, which is given by

$$H_{st}(f_z) = \begin{cases} 0.5\left(1 + \cos\left(\dfrac{\pi f_z}{B}\right)\right) & \text{for}\ |f_z| < B\ \text{and}\ |f_z| < \tan(\alpha)\ |f_x| \\ 0 & \text{elsewhere.} \end{cases} \qquad (1.18)$$

The parameter B controls the reconstructed slice thickness and will depend on the scan range. Hu et al. have optimized this parameter for $\alpha = 50°$ (Hu and Zhao, 2011).

Note that use of FBP reconstruction assumes parallel-beam projection data. When the projection data are acquired using a cone-beam geometry, the FDK algorithm (Feldkamp and Davis,

1984) is more appropriate because it accounts for the cone-beam geometry. Other Fourier-domain reconstruction algorithms include matrix-inversion tomography (MITS) (Godfrey et al., 2001), which has been primarily applied to chest tomosynthesis.

1.2.3 Iterative Image Reconstruction for Tomosynthesis

A major difference between FBP-based image reconstruction and iterative reconstruction algorithms is that FBP reconstruction requires one single back-projection operation only, whereas many cycles of successive forward- and back-projections are performed in iterative reconstruction. Iterative reconstruction is therefore significantly more computationally expensive than FBP. On the other hand, iterative reconstruction is better suited to the tomosynthesis problem, an ill-defined inverse problem. The reconstruction problem can be converted into an optimization problem, with the objective of estimating an image that is consistent with the measurement, that is, the projection views. There are no assumptions about the existence or requirements on the sampling density of the measurements. This is in stark difference with the FBP reconstruction algorithm, which is based on the analytic inverse of the Radon transform, with the assumption that the data space has been completely sampled. There is no provision for unknown measurements; within the algorithm, missing data are assumed to be zero.

A widely used algorithm for iterative image reconstruction is the expectation-maximization algorithm (EM), which has been applied for tomosynthesis reconstruction. The EM algorithm minimizes the KL-distance $D_{KL}(\mathbf{g},\mathbf{g_0})$ (Barrett and Myers, 2004)

$$D_{KL}(\mathbf{g},\mathbf{g_0}) = \sum_{m=1}^{M} \left\{ g_{0m} - g_m + g_m \ln\left[\frac{g_m}{g_{0m}}\right] \right\}, \quad (1.19)$$

where, in the reconstruction context, $\mathbf{g}$ is the measurement (i.e., the log-projection data) and $\mathbf{g_0}$ is the data estimate, for $m = 1\cdots M$ measurements (i.e., detector pixels). For positive g_m, the KL-distance is always positive.

Given the tomosynthesis log-projection data $g(u,v,\theta)$, where (u,v) are detector coordinates and θ is the x-ray source angle, a tomosynthesis image $\hat{f}(x,y,z)$ is estimated iteratively using

$$\hat{f}^{k+1}(x,y,z) = \hat{f}^k(x,y,z)\frac{1}{s}\mathbf{H}^+\left\{\frac{g(u,v,\theta)}{\mathbf{H}\hat{f}^k(x,y,z)}\right\}, \quad (1.20)$$

where k is the iteration number and (x,y,z) are the tomosynthesis image coordinates. The operators $\mathbf{H}$ and $\mathbf{H}^+$ perform the forward-projection and back-projection, respectively, and $s = \Sigma\mathbf{H}$. At $k = 0$, the algorithm is initialized by setting $\hat{f}^0 = 1 \ \forall \ (x,y,z)$.

The EM algorithm ensures that the estimate $\mathbf{g_0}$ is always positive, under the condition that the data $\mathbf{g}$ and the operators $\mathbf{H}$, $\mathbf{H}^+$ are positive. This is a useful property in image reconstruction.

The forward-projection operator projects the discrete image array onto the detector, $f(x,y,z)\overset{\mathbf{H}}{\rightarrow}g(u,v,\theta)$; the reverse is true for the back-projection operator. These operations are computationally expensive and a number of implementations exist (Siddon, 1985; Man and Basu, 2004; Sidky et al., 2009).

Other iterative reconstruction algorithms include transmission EM (Lange and Fessler, 1995; Wu et al., 2003), where the projected image estimate is compared to the projection data in the measured x-ray fluence domain, as well as simultaneous algebraic reconstruction (SART) (Zhang et al., 2006).

More advanced iterative reconstruction algorithms include regularization of the image estimate $\hat{f}$, which imposes limits on the image noise (Sidky et al., 2009; Das et al., 2011). Minimization of the total variation (TV) in the image has produced images with superior visibility of microcalcifications, while maintaining the mass conspicuity (Sidky et al., 2009).

Advanced iterative image reconstruction increases the computational load substantially (Das et al., 2011), which has been alleviated by using graphics processing units (GPU) for computation (Sidky et al., 2009).

A comparison of the early reconstruction algorithms for tomosynthesis has been performed (Wu et al., 2004; Zheng et al., 2006). Figures 1.9 and 1.10 show a breast tomosynthesis slice reconstructed with the FDK algorithm, FDK with slice-thickness filter (SLT), EM, and TV.

1.3 Tomosynthesis Image Quality

1.3.1 Effect of Acquisition Parameters

The tomosynthesis image is greatly affected by the choice of acquisition parameters, already apparent in the ball phantom images shown in Figure 1.11, where the tomosynthesis scan angle α was varied. The scan angle determines the depth resolution of the system. The following rule of thumb has been proposed by Maidment et al. (Maidment et al., 2005):

$$\Delta z = \Delta x / \tan\left(\frac{\alpha}{2}\right), \quad (1.21)$$

where Δx is the extent of the object in the x–y plane, while Δz is the depth spread of the object in the tomosynthesis image. Thus, depth resolution is dependent on the tomosynthesis scan angle *and* the detail size of the structure-of-interest (Figure 1.11).

Figure 1.12 shows in-plane slices through tomosynthesis reconstructions of uniform volumes containing spherically symmetric signals of different radii ($r = 0.4$ cm, $r = 0.1$ cm, $r = 0.025$ cm). Slices are shown at two depths; $z = 2.6$ cm is centered on the signals, while $z = 4.1$ cm is located 1.5 cm higher than the signal center. For the 60° scan angle, no signal residual can be perceived at $z = 4.1$ cm. For the smaller scan angles, signal residual can be observed in planes that do not actually

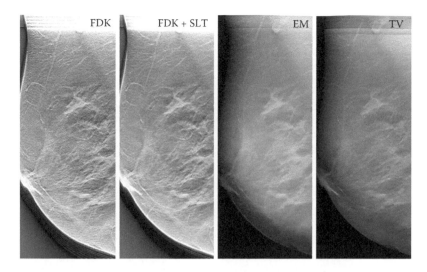

FIGURE 1.9 A breast tomosynthesis slice reconstructed by use of different reconstruction algorithms.

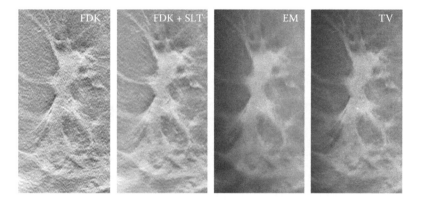

FIGURE 1.10 Appearance of a region-of-interest showing a mass that contains microcalcifications for different reconstruction algorithms.

contain any signal. In addition, signal contrast increases as the scan angle increases.

Figure 1.13 illustrates the effect of the number of projection views on tomosynthesis image quality. In the slice through the signal centers, there are no apparent differences as the view sampling changes. The signal in the slice away from the signal center has been multiplied by a factor of 5 for visual enhancement. The effect of view sampling becomes apparent in the signal residual, with increasing differences as the signal size decreases.

In addition to scan angle, there are a large number of system parameters that need to be considered in tomosynthesis. Probably the second most important parameter is the number of projection views. Ideally one might like to acquire projections with a small angular step, resulting in a large number of projection views with low x-ray fluence each. The potential trade-off however is electronic detector noise, which is incurred from large-area flat-panel integrating detectors (Section 1.4.3). Photon-counting detectors are generally not affected by this problem, since they are free of electronic noise, and a low count rate prevents pulse pile up. Further, the x-ray exposure could be distributed nonuniformly among the individual projection views.

There are several other design considerations, which are discussed in Section 1.4. The focus in this section is on the effect of scan angle, view sampling, and dose distribution on tomosynthesis image quality, all of which have been investigated by several research groups (Chawla et al., 2009; Gang et al., 2010; Park et al., 2010; Reiser and Nishikawa, 2010; Richard and Samei, 2010). So far, these studies indicate that a larger scan angle results in improved object contrast for larger scale objects of about 1 mm or greater. For structures at this scale, the view sampling did not affect image quality much. However, for small-scale structures of the order of several 100 μm, a larger number of views did improve image quality; with the amount of improvement depending on the amount of quantum noise—the effect was most pronounced in the noise-free case.

1.3.2 Image Artifacts

Due to the limited scan angle and number of projections, tomosynthesis images exhibit artifacts. It is important to recognize artifacts as such. While the actual appearance of the artifact will depend on the acquisition parameters and reconstruction

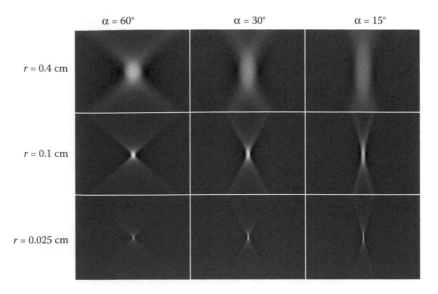

FIGURE 1.11 In-depth cuts through tomosynthesis images of spherically symmetric signals with different radii, reconstructed from projections acquired over $\alpha = 60°$, $\alpha = 30°$, $\alpha = 15°$. Projections were acquired every 1.5°. Reconstruction was performed using the EM algorithm (eight iterations).

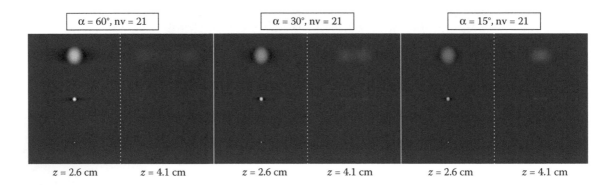

FIGURE 1.12 In-plane cuts through spherically symmetric signals reconstructed with the EM algorithm from projections taken across different angular spans. Projection views were taken every 1.5°. The signal centers are located at $z = 2.6$ cm.

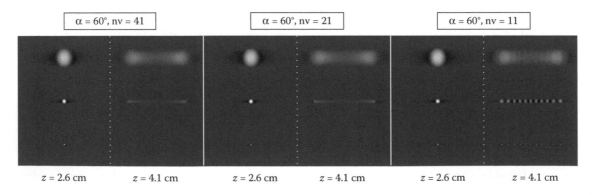

FIGURE 1.13 In-plane cuts through tomosynthesis images of spherically symmetric signals reconstructed with the EM algorithm from projections taken across a scan angle of 60° and 41, 21, or 11 views. The signal centers are located at $z = 2.6$ cm.

algorithm, the overall characteristics of two commonly observed artifacts are presented here.

1.3.2.1 High-Contrast Objects

When a high-contrast object is present in the object, its shadow can repeat through multiple or all slices of the reconstructed image. Each projection causes a shadow so that the appearance of this artifact is similar to a zipper and is therefore sometimes called a *zipper-artifact*. Figure 1.14 shows an example of this artifact, caused by a metallic surgical clip. It is clearly visible when in-focus, and also in all adjacent slices. From the number of low-contrast replications of the object, one can infer that this tomosynthesis image was reconstructed from 11 projection views. Large calcifications, or calcified vessels, produce a similar effect. In general, this artifact becomes less prominent as view sampling increases. While artifact reduction schemes have been proposed (Wu et al., 2006), the drawback of such schemes is that they can cause ambiguity when the artifact is not completely removed.

1.3.2.2 Truncation Artifacts

Truncation artifacts can be observed when the projection of the object falls off the detector at one or several angles. This is shown schematically in Figure 1.15.

This results in stepping artifacts in the reconstructed slice, as shown at the top of the tomosynthesis breast slice in Figure 1.15. This artifact is more pronounced when the scan angle is large and the object is thick. Ad hoc corrections to the reconstructed volume can alleviate the conspicuity of this artifact (Zhang et al., 2007). Simply excluding the affected regions from the displayed volume is not a solution because abnormalities may be located in these regions.

1.4 Design Considerations

As has become apparent in the discussion so far, there is no "tomosynthesis" system per se. "Tomosynthesis" encompasses a wide variety of x-ray imaging geometries that perform a limited

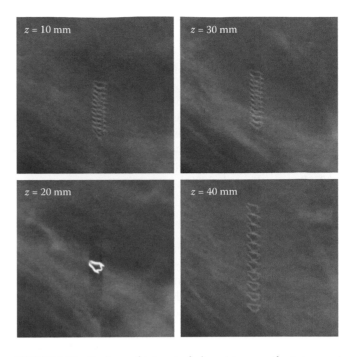

FIGURE 1.14 Regions-of-interest of a breast tomosynthesis image at different depths (region size: 2 × 2 cm). A metallic surgical clip is clearly visible in all slices.

angle scan. This section discusses common imaging geometries and hardware design considerations.

1.4.1 System Geometries

Tomosynthesis systems can be categorized by the relative motions of the x-ray source, x-ray detector, and object. There are three broad categories (Dobbins, III and Godfrey, 2003), consisting of (a) a linear scan, where the x-ray source and detector move along a linear trajectory (Maidment et al., 2006); (b) an isocentric motion where the x-ray source and detector rotate about the object in a CT-like fashion (Ren et al., 2005); and (c) a partial

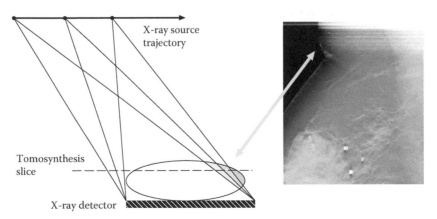

FIGURE 1.15 When the projection of the object exceeds the detector surface, the corresponding regions in the object receive less x-ray photons (illustrated by a lighter shade of gray in the line drawing). This produces stepping artifacts in the reconstruction, shown here at the top of a breast tomosynthesis slice.

isocentric motion where the x-ray source rotates and the detector remains stationary in relation to the object.

There is no clear advantage of any of these geometries. For a linear scan, the magnification of each image slice is constant. For both linear and isocentric scans, the x-ray source does not move relative to the detector, which provides mechanical stability. On the other hand, a rotating detector results in a bulky detector housing. In breast imaging, this may render patient positioning difficult. Therefore, several tomosynthesis prototypes for breast imaging employ a partial-isocentric scan geometry (Wu et al., 2003; Bissonnette et al., 2005a).

1.4.2 X-Ray Sources for Tomosynthesis

In planar radiography, the image is acquired while the x-ray source is stationary. In x-ray CT, the projection data are acquired while the x-ray source spins about the object. In tomosynthesis, both schemes have been implemented and are known as step-and-shoot mode and continuous-scan mode, respectively. The advantage of the step-and-shoot mode is the absence of focal-spot motion blur. However, step-and-shoot is mechanically more demanding as vibrations can occur, and can result in increased scan time. Continuous x-ray source motion avoids these problems, but does introduce focal-spot motion blur, which can be minimized by using a pulsed x-ray beam, where the goal is to shorten the path travelled by the x-ray source while an image is being acquired. While this could, in principle, be achieved by slowing the x-ray source, an increase in overall scan time increases patient motion, which is detrimental to the reconstructed tomosynthesis image as object correlations are suppressed by the motion (Acciavatti and Maidment, 2012).

For breast imaging, a harder x-ray beam tends to be used for tomosynthesis imaging than for projection mammography. Presumably the loss in contrast resolution due to the higher x-ray energy is offset by the increased contrast resolution of tomosynthesis imaging (Feng and Sechopoulos, 2012). In addition, novel x-ray source designs have been employed for tomosynthesis using stationary field-emission x-ray tubes (so-called nanotubes) (Sprenger et al., 2011). Further, researchers are exploring the use of view-dependent exposure levels to improve the visualization of microcalcifications in breast imaging (Nishikawa et al., 2007; Das et al., 2009; Hu and Zhao, 2011).

1.4.3 X-Ray Detectors for Tomosynthesis

A number of flat-panel, large-area x-ray detectors have been developed and are being used for full-field digital mammography (Zhao et al., 2003; Ghetti et al., 2008) and planar radiography. For mammography detectors in particular, the resolution needs to be high (typical pixel sizes are 100 μm or less) and the quantum efficiency needs to be large, in order to ensure adequate depiction of microcalcifications. The requirements for tomosynthesis x-ray detectors are similar to those of the corresponding planar radiography application. However, since multiple projection views are acquired, the exposure to the detector per view is low, requiring low electronic noise levels to ensure quantum-limited detection. Furthermore, electronic noise from each projection is accumulated in the reconstructed image.

To reduce the scan time, fast readout time is required. In breast tomosynthesis, scan times of the order of 10 s or less are being sought, with typically 10–25 projection views per scan. Image lag can also be problematic (Zhao and Zhao, 2005). In current clinical systems and prototypes, both indirect and direct conversion detectors have been employed. In addition, photon-counting detectors are well-suited for tomosynthesis, since the x-ray exposure to the detector at each projection view is low. Photon-counting detectors exhibit practically no electronic noise. Since they tend to be constructed as slot-scan detectors, x-ray scatter is intrinsically suppressed (Aslund et al., 2006; Maidment et al., 2006).

1.4.4 Effect of Oblique Incidence on Physical Factors

The keV-X-ray imaging is based on the attenuation of x-rays in matter, which varies with the chemical composition and density of the materials being imaged. However, in an actual imaging system, x-ray interactions occur that do not convey information about the object but do add quantum noise to the image, such as x-ray scatter. Furthermore, the actual construction of the imaging device results in finite structures that degrade image resolution, such as the finite focal spot size, or the finite size of detector elements. These effects are classified as *physical factors*. In the design of an imaging system, a balance must be struck between technical feasibility, radiation dose, and acceptable image quality. The impact of physical factors on image quality in planar radiography has been studied extensively (Beutel et al., 2000). In tomosynthesis, large source angles can result in oblique x-ray paths through the object, and subsequent oblique incidence on the x-ray detector. This affects both x-ray scatter and detector blur, as will be described later.

1.4.4.1 X-Ray Scatter

Since x-ray scatter depends primarily on the length of the x-ray path through the object (Boone and Cooper, 2000), the amount of scatter incurred depends on the x-ray source angle, and, for cone-beam projections, the scatter across the detector surface is nonuniform.

Figure 1.16 shows scatter point-spread functions (SPSF) for two x-ray pencil beams traversing a 5-cm-thick breast (assuming 50% glandularity and a 26 kVp Mo/Mo beam), for normal and oblique (30°) incidence. For oblique incidence, the SPSF is not isotropic, and the scatter magnitude differs from that at 0°. Comprehensive scatter measurements for tomosynthesis have been performed by Sechopoulos et al. using Monte Carlo simulation methods (Sechopoulos et al., 2007).

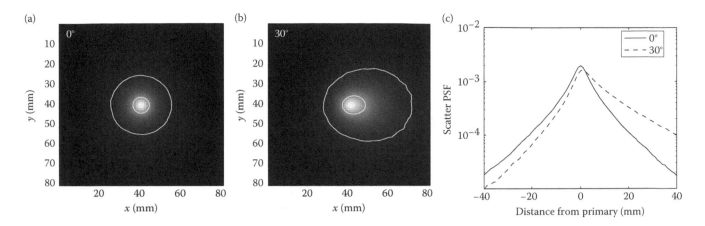

FIGURE 1.16 Scatter point-spread functions for an x-ray pencil beam incident at (a) 0° and (b) 30°. Contour lines of the SPSF are drawn at 90%, 50%, and 10% of the maximum SPSF value. (c) Line profiles through the SPSFs.

1.4.4.2 Detector Blur

A similar effect can be observed for the x-ray detector point-response functions (PRF) in large-area, flat-panel digital detectors. In both direct and indirect conversion detectors, x-rays are absorbed and converted into optical photons (indirect detector) or into electron–hole pairs (direct detector). These interactions occur across the depth of the conversion medium, with the number of interacting x-rays determined by the Lambert–Beer law.

For x-rays incident on the detector along the detector normal, all interactions occur at the same (x, y)-locations but at different depths. When the x-ray incidence is oblique, however, the (x, y)-location of interactions becomes depth dependent. The magnitude of this effect depends on several factors including thickness of the conversion medium, detector pixel size, x-ray absorption coefficient of the conversion medium, and the properties of the secondary quanta (optical photons for indirect detectors or charge transport in direct detectors). These effects have been measured and simulated for a number of detectors (Badano et al., 2006; Mainprize et al., 2006; Reiser et al., 2009; Acciavatti and Maidment, 2011).

1.5 Clinical Applications

The current dominant clinical applications of tomosynthesis are breast and chest imaging (Dobbins, III, 2009), which are discussed in the following sections. In addition, tomosynthesis has been used in image-guided radiation therapy (Godfrey et al., 2006; Wu et al., 2011), skeletal imaging (Gazaille et al., 2011; Geijer et al., 2011), as well as dental imaging (Kolehmainen et al., 2003; Ogawa et al., 2010). A review of tomosynthesis breast imaging can be found in Baker and Lo (2011).

1.5.1 Breast Imaging

Breast cancer is the second leading cause of death from cancer in women, causing an estimated 40,000 deaths in 2012 (American Cancer Society, 2012). Since 1990, the mortality rate from breast

cancer has decreased by about 2.2% annually, due to earlier detection through mammography screening and better treatments (American Cancer Society, 2012). Approximately, half of this decrease has been attributed to screening by mammography (Berry et al., 2005). Mammography, the current gold-standard for breast cancer screening, produces an x-ray projection image of the breast.

A major limitation of mammography is tissue superimposition due to the projection of the complex 3D breast anatomy onto a plane. Tissue overlap can hide cancerous lesions, producing a false-negative mammogram and thereby reducing sensitivity of mammography screening. On the other hand, overlaying normal fibroglandular tissues can mimic the appearance of a cancer, prompting a false-positive callback to perform further diagnostic imaging, which reduces the specificity of mammography screening.

An attempt to address this shortcoming was to image the breast under different compressions. A typical mammography screening examination consists of two x-ray projections of each breast, compressed in the cranio-caudal and medio-lateral oblique directions.

A novel tool to overcome this limitation is tomosynthesis breast imaging, which partially resolves overlapping tissue structures. Early work on breast tomosynthesis was carried out by D. Kopans and colleagues (Niklason et al., 1997; Wu et al., 2003), which has led to the FDA approval of a breast tomosynthesis system for breast cancer screening and diagnosis in early 2011.* Approval for units from other manufacturers is expected to follow.

The different tomosynthesis breast-imaging products and prototypes employ widely varying acquisition parameters (Table 1.2). Simulation studies assuming an ideal imaging system (i.e., no image degradation due to physical factors in the imaging chain) show a strong dependence of image quality on acquisition parameters (Gang et al., 2010; Reiser and Nishikawa, 2010).

Several studies have been performed to evaluate the clinical performance of tomosynthesis. Early studies compared

---

* Hologic Selenia Dimensions.

TABLE 1.2 Acquisition Parameters Used in a Selection of Breast Tomosynthesis Prototypes

α	n_{view}	Detector Type	Reference
13	48	Photon counting	Maidment et al. (2005)
15	11	a-Se direct conversion	Smith et al. (2006)
50	49–25	a-Se direct conversion	Bissonnette et al. (2005b)
50	11	CsI indirect conversion	Wu et al. (2004)

tomosynthesis to mammography based on subjective image-quality measures such as lesion conspicuity for a cohort of patients that were recalled due to an abnormal screening mammogram (Poplack et al., 2007). They found that mass lesion conspicuity was higher in tomosynthesis than in mammography, but microcalcification conspicuity was better in mammography than in tomosynthesis. They also estimated the effect of tomosynthesis on recall rate: Whenever a lesion was absent or clearly benign on tomosynthesis, the recall was deemed unnecessary. Already in this early study, the potential reduction in recall rate when using tomosynthesis in addition to mammography was estimated to be about 30%. This result was supported by a retrospective study from Gur and colleagues who found that a combination of FFDM and tomosynthesis did result in a 30% reduction in recall rate compared to FFDM alone (Gur et al., 2009). This study included eight radiologists reading 125 two-view FFDM and two-view tomosynthesis breast images of which 35 contained a cancer.

Sensitivity of tomosynthesis was found to be comparable to that of mammography, based on a study involving 513 women with an abnormal screening mammogram (Teertstra et al., 2010). Note that it is more difficult to find statistically significant effects for the cancer detection rate than it is for the recall rate: For 1000 women screened, the number of cancers detected will be between 3 and 4, while of the same number of women, 100 will be recalled (based on a recall rate of 10%) (Rosenberg et al., 2006). Thus, when assessing the impact of tomosynthesis on breast cancer screening, a larger cohort of women is required to show statistical significance for changes in sensitivity than for changes in specificity.

At the time of this writing, participant recruitment has begun for two clinical trials that aim to evaluate the recall rate in FFDM screening compared to combined FFDM and tomosynthesis screening. One is the Oslo Tomosynthesis Screening Trial, spearheaded by P. Skaane. This prospective trial will evaluate FFDM and tomosynthesis based on 7600 women and eight radiologists. The second trial is sponsored by the American College of Radiology Imaging Network (ACRIN) and led by E. Conant, with the goal of enrolling 550 women.

While much progress has been made in tomosynthesis breast imaging during the past decade, open questions remain. From Table 1.1, it is clear that there is no consensus on optimum tomosynthesis acquisition parameters.

An important aspect of tomosynthesis imaging is interpretation time, in particular in a screening setting, where a radiologist will encounter only three to four cancers in every 1000 screening exams. Early studies indicate that reading time might roughly double relative to mammography, but radiologists are probably still on a learning curve reading tomosynthesis since the modality is so new. Another important clinical question is the display and comparison of prior views. Clearly some kind of image registration will be needed to support the radiologist in this task.

Further, the imaging protocol has not been optimized. The Hologic Selenia Dimensions tomosynthesis unit acquires a tomosynthesis scan and a projection image while the breast is under compression, with the dose for the screening exam evenly divided between the tomosynthesis scan and the mammogram. Studies indicate that radiologists perform better reading the combined views, rather than tomosynthesis alone (Andersson et al., 2008; Gur et al., 2009). Will radiologists learn to interpret tomosynthesis without relying on the mammogram, or does the mammogram provide information that is not present in the tomosynthesis image? Is it necessary to perform FFDM + TOMO for both CC and MLO views, or is one TOMO view sufficient? Future studies will clarify the answers to these questions.

If tomosynthesis screening is to become commonplace, how will the radiologist compare with tomosynthesis images from previous years? Clearly some kind of image registration will be needed to support the radiologist in this task.

Mammography screening has a high sensitivity in predominantly fatty breasts, but a low sensitivity in heterogeneously dense and extremely dense breasts (Carney et al., 2003). Tomosynthesis imaging, with improved contrast resolution, may increase sensitivity for these women.

Lastly, there is a controversy about whether microcalcifications can be imaged with tomosynthesis as well as they can with mammography. Microcalcifications are tiny specs of calcium with a diameter from about 150 to 400 μm, which are an important early indicator of a potential cancer. Patient motion during the tomosynthesis scan can produce a misregistration of microcalcifications, such that they may become obscured in the reconstructed image, while larger-scale structures may still be visualized. Clearly, the goal is to minimize the scan time or the illumination time for a given region of the breast.

In conclusion, the introduction of tomosynthesis for breast cancer screening may cause a drastic improvement in screening sensitivity and specificity. While this discussion has focused on breast cancer screening, most of it also applies to diagnostic imaging. Future studies will show whether tomosynthesis will hold up to its promise.

1.5.2 Chest Imaging

The motivation for chest tomosynthesis is similar to that for breast tomosynthesis: While chest radiography is widely used for diagnosis of pulmonary diseases, preoperative examinations, and follow-up, normal anatomy can obscure pathologic findings, thereby lowering the sensitivity.

The current alternative to chest radiography is chest CT, which comes at a substantially increased dose and cost; the organ dose from an abdominal CT scan is almost two orders of magnitude

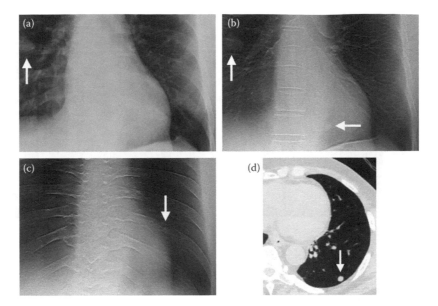

FIGURE 1.17 Chest nodule visualization in a PA radiograph (a), or in tomosynthesis. The nodule indicated by the arrow in the PA radiograph is seen in the tomosynthesis slice in (b), but two additional nodules are seen in tomosynthesis (horizontal arrow in (b), arrow in (c)) that are not seen in the PA radiograph. The additional lesions were confirmed with chest CT (d). (Reprinted with permission from Dobbins et al., 2008. *Medical Physics*, 35(6):2554–2557.)

larger than that from a lateral chest radiograph (Brenner and Hall, 2007). Chest tomosynthesis may be a viable alternative to chest CT. Two clinical studies have been performed to compare detection of pulmonary nodules in chest radiography and tomosynthesis (Dobbins, III et al., 2008; Vikgren et al., 2008). Both studies found that detection performance in tomosynthesis was better than that in radiography. The tomosynthesis systems in both studies came from the same manufacturer* but were operated using different acquisition parameters: In one study, the tomosynthesis image was reconstructed from 60 projection views acquired over a 35° scan, while the other used 71 views acquired over a 20° scan.

Figure 1.17 shows a posterior-anterior (PA)-radiograph, tomosynthesis slices, and a CT image. Nodule visibility in tomosynthesis slices is clearly improved compared to the radiograph. The nodules were confirmed with chest CT.

1.6 Tomosynthesis versus CT

A comparison of tomosynthesis and CT is difficult because it does depend on the clinical application. Diagnostic whole-body imaging with CT is a mature, well-established technology with a wide range of clinical applications. However, it comes at a high cost and at a relatively high patient dose (Brenner and Hall, 2007). For these reasons, planar radiography is still widely used despite its limitations. Tomosynthesis may overcome some of these limitations, while maintaining low patient dose, and presumably providing diagnostic information at a lower cost than

CT. In-plane resolution in tomosynthesis is greater than CT voxel dimensions, allowing for greater detail if the structure-of-interest is within the tomosynthesis planes. Tomosynthesis may therefore become an adjunct to planar radiography, potentially providing sufficient anatomic detail to render an additional CT scan unnecessary.

Dedicated breast CT prototypes have been developed recently (Lindfors et al., 2008; O'Connell et al., 2010; Prionas et al., 2010) that can produce breast CT images at high resolution (sub-200 μm voxel size) at a dose comparable to that of two screen-film mammograms (Boone et al., 2005). A visual comparison of full-field digital mammography (FFDM), tomosynthesis, and dedicated breast CT is shown in Figures 1.18 through 1.20. These figures show an infiltrating ductal carcinoma in a patient with scattered fibroglandular densities. This lesion was discovered in screening mammography (Figure 1.18, top row). Upon recalling the patient, spot-compression views were obtained (Figure 1.18, bottom row), which is the current clinical standard-of-care. In addition, tomosynthesis was performed in both the CC and MLO views (Figure 1.19). Lesion conspicuity is greatly improved. Figure 1.20 shows the precontrast as well as the contrast-enhanced dedicated breast CT image of the same lesion. The lesion morphology is visualized with great detail. The patient dose in one CT scan was estimated to be similar to that incurred from the combined tomosynthesis and FFDM images.

Lesion detection in breast imaging with mammography, tomosynthesis, and CT has been compared quantitatively by Gong et al. in a simulation study (Gong et al., 2006). They found that tomosynthesis significantly improved detectability compared to mammography, but that no significant difference for

---

* GE Healthcare.

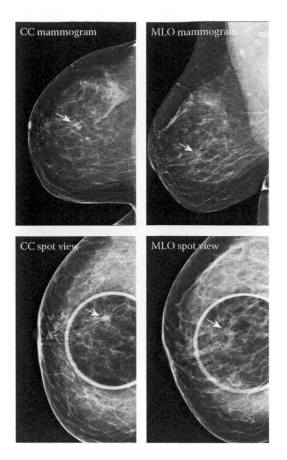

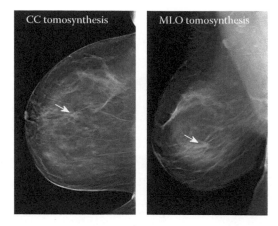

FIGURE 1.19 Tomosynthesis slices of an intraductal carcinoma (Hologic Dimensions). The corresponding projection mammograms are shown in Figure 1.18. (Images courtesy of Dr. K. K. Lindfors, The University of California at Davis.)

FIGURE 1.18 Full-field digital mammograms and spot-compression views of an intraductal carcinoma. (Images courtesy of Dr. K. K. Lindfors, The University of California at Davis.)

diagnosis, staging, and estimation of lesion extent in preoperative imaging.

While it is impossible to predict the future, it is quite certain that tomosynthesis will become a valuable imaging tool that will likely become more widespread as time progresses. Tomosynthesis is still in the early stages of development but with promising preliminary clinical benefits. Time will show whether it will hold up to this promise.

Acknowledgments

The author would like to thank R. M. Nishikawa for many helpful discussions on the topics covered in this chapter; D. B. Kopans for providing her with breast tomosynthesis image data, some of which are shown in this chapter; B. A. Lau for the angle-dependent scatter point response functions; and E. Y. Sidky for providing tomosynthesis images reconstructed by TV minimization.

detectability existed between tomosynthesis and CT imaging. Future investigations will more clearly define the clinical utility of both modalities. Dedicated breast CT may find its place as a low-cost alternative to magnetic resonance imaging (MRI), which is currently used for high-risk screening, breast-cancer

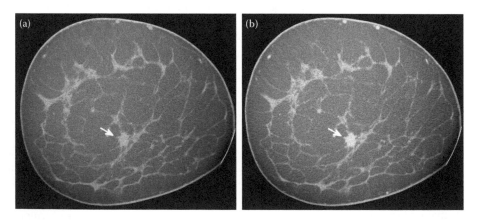

FIGURE 1.20 Dedicated breast CT images of an intraductal carcinoma. (a) Precontrast scan, (b) postcontrast scan. Mammograms, spot views, and tomosynthesis images of this case are shown in Figures 1.18 and 1.19, respectively. (Images courtesy of Dr. K. K. Lindfors, The University of California at Davis.)

References

Acciavatti, R. J. and Maidment, A. D. A. 2011. Optimization of phosphor-based detector design for oblique x-ray incidence in digital breast tomosynthesis. *Medical Physics*, 38(11):6188–6202.

Acciavatti, R. J. and Maidment, A. D. A. 2012. Optimization of continuous tube motion and step-and-shoot motion in digital breast tomosynthesis systems with patient motion. In *Proceedings of SPIE 8313*, 8313, pp. 831306.

American Cancer Society 2012. *Breast Cancer Facts & Figures 2011-2012*. Atlanta: American Cancer Society, Inc.

Andersson, I., Ikeda, D., Zackrisson, S., and Ruschin, M., 2008. Tomosynthesis and digital mammography: A comparison of breast cancer visibility and BIRADS classification in a population of cancers with subtle mammographic findings. *European Radiology*, 18:2817–2825.

Aslund, M., Cederstrom, B., Lundqvist, M., and Danielsson, M. 2006. Scatter rejection in multislit digital mammography. *Medical Physics*, 33(4):933.

Badano, A., Kyprianou, I. S., and Sempau, J. 2006. Anisotropic imaging performance in indirect x-ray imaging detectors. *Medical Physics*, 33(8):2698.

Baker, J. A. and Lo, J. Y. 2011. Breast tomosynthesis: State-of-the-art and review of the literature. *Academic Radiology*, 18(10):1298–1310.

Barrett, H. H. and Myers, K. J. 2004. *Foundations of Image Science*. Hoboken, NJ: Wiley-Interscience.

Berry, D. A., Cronin, K. A., Plevritis, S. K., Fryback, D. G., Clarke, L., Zelen, M., Mandelblatt, J. S., Yakovlev, A. Y., Habbema, J. D. F., and Feuer, E. J. 2005. Effect of screening and adjuvant therapy on mortality from breast cancer. *The New England Journal of Medicine*, 353(17):1784–1792.

Beutel, J., Kundel, H. L., and Van Metter, R. L., editors 2000. *Handbook of Medical Imaging: Physics and Psychophysics*, Volume 1. Bellingham, WA: SPIE Press.

Bissonnette, M., Hansroul, M., Masson, E., Savard, S., Cadieux, S., Warmoes, P., Gravel, D. et al. 2005a. Digital tomosynthesis using an amorphous selenium flat panel detector. In *Proceedings of SPIE*, 5745, pp. 529.

Bissonnette, M., Hansroul, M., Masson, E., Savard, S., Cadieux, S., Warmoes, P., Gravel, D. et al. 2005b. Digital breast tomosynthesis using an amorphous selenium flat panel detector. In Flynn, M. J., editor, *Proceedings of SPIE*, 5745, pp. 529–540.

Boone, J. M. and Cooper, V. N. 2000. Scatter/primary in mammography: Monte Carlo validation. *Medical Physics*, 27(8):1818–1831.

Boone, J. M., Kwan, A. L. C., Seibert, J. A., Shah, N., Lindfors, K. K., and Nelson, T. R. 2005. Technique factors and their relationship to radiation dose in pendant geometry breast CT. *Medical Physics*, 32(12):3767.

Brenner, D. J. and Hall, E. J. 2007. Computed tomography—An increasing source of radiation exposure. *The New England Journal of Medicine*, 357(22):2277–2284.

Carney, P. A., Miglioretti, D. L., Yankaskas, B. C., Kerlikowske, K., Rosenberg, R., Rutter, C. M., Geller, B. M. et al. 2003. Individual and combined effects of age, breast density, and hormone replacement therapy use on the accuracy of screening mammography. *Annals of Internal Medicine*, 138:168–175.

Chawla, A. S., Lo, J. Y., Baker, J. A., and Samei, E. 2009. Optimized image acquisition for breast tomosynthesis in projection and reconstruction space. *Medical Physics*, 36(11):4859.

Das, M., Gifford, H. C., O'Connor, J. M., and Glick, S. J. 2009. Evaluation of a variable dose acquisition technique for microcalcification and mass detection in digital breast tomosynthesis. *Medical Physics*, 36(6):1976–1984.

Das, M., Gifford, H. C., O'Connor, J. M., and Glick, S. J. 2011. Penalized maximum likelihood reconstruction for improved microcalcification detection in breast tomosynthesis. *IEEE Transactions on Medical Imaging*, 30(4):904–14.

des Plantes, Z. 1932. Eine neue Methode zur Differenzierung in der Roentgenographie (Planigraphie). *Acta Radiologica*, 13:182–192.

Dobbins, III, J. T. 2009. Tomosynthesis imaging: At a translational crossroads. *Medical Physics*, 36(6):1956.

Dobbins, III, J. T. and Godfrey, D. J. 2003. Digital x-ray tomosynthesis: Current state of the art and clinical potential. *Physics in Medicine and Biology*, 48(19):R65–106.

Dobbins, III, J. T., McAdams, H. P., Song, J.-W., Li, C. M., Godfrey, D. J., DeLong, D. M., Paik, S.-H., and Martinez-Jimenez, S. 2008. Digital tomosynthesis of the chest for lung nodule detection: Interim sensitivity results from an ongoing NIH-sponsored trial. *Medical Physics*, 35(6):2554–2557.

Feldkamp, L. and Davis, L. 1984. Practical cone-beam algorithm. *JOSA A*, 1(6):612–619.

Feng, S. S. J. and Sechopoulos, I. 2012. Clinical digital breast tomosynthesis system: Dosimetric characterization. *Radiology*, 263(1):35–42.

Gang, G. J., Tward, D. J., Lee, J., and Siewerdsen, J. H. 2010. Anatomical background and generalized detectability in tomosynthesis and cone-beam CT. *Medical Physics*, 37(5):1948–1965.

Gazaille, R. E., Flynn, M. J., Page, W., Finley, S., and van Holsbeeck, M. 2011. Technical innovation: Digital tomosynthesis of the hip following intra-articular administration of contrast. *Skeletal Radiology*, 40(11):1467–1471.

Geijer, M., Börjesson, A. M., and Göthlin, J. H. 2011. Clinical utility of tomosynthesis in suspected scaphoid fracture. A pilot study. *Skeletal Radiology*, 40(7):863–867.

Ghetti, C., Borrini, A., Ortenzia, O., Rossi, R., and Ordonez, P. L. 2008. Physical characteristics of GE senographe essential and DS digital mammography detectors. *Medical Physics*, 35(2):456.

Godfrey, D., Warp, R., and Dobbins III, J. 2001. Optimization of matrix inverse tomosynthesis. In *Proceedings of SPIE*, 4320, pp. 696.

Godfrey, D. J., Yin, F.-F., Oldham, M., Yoo, S., and Willett, C. 2006. Digital tomosynthesis with an on-board kilovoltage imaging device. *International Journal of Radiation Oncology, Biology, Physics*, 65(1):8–15.

Gong, X., Glick, S. J., Liu, B., Vedula, A. A., and Thacker, S. 2006. A computer simulation study comparing lesion detection accuracy with digital mammography, breast tomosynthesis, and cone-beam CT breast imaging. *Medical Physics*, 33(4):1041.

Gur, D., Abrams, G. S., Chough, D. M., Ganott, M. A., Hakim, C. M., Perrin, R. L., Rathfon, G. Y., Sumkin, J. H., Zuley, M. L., and Bandos, A. I. 2009. Digital breast tomosynthesis: Observer performance study. *AJR*, 193(2):586–591.

Hu, Y.-H. and Zhao, W. 2011. The effect of angular dose distribution on the detection of microcalcifications in digital breast tomosynthesis. *Medical Physics*, 38(5):2455.

Kak, A. C. and Slaney, M. 1988. *Principles of Computerized Tomographic Imaging*. New York, NY: IEEE Press.

Kolehmainen, V., Siltanen, S., Järvenpää, S., Kaipio, J. P., Koistinen, P., Lassas, M., Pirttilä, J., and Somersalo, E. 2003. Statistical inversion for medical x-ray tomography with few radiographs: II. Application to dental radiology. *Physics in Medicine and Biology*, 48(10):1465–1490.

Lange, K. and Fessler, J. A. 1995. Globally convergent algorithms for maximum a posteriori transmission tomography. *IEEE Transactions on Image Processing*, 4(10):1430–1438.

Lauritsch, G. and Härer, W. H. 1998. A theoretical framework for filtered backprojection in tomosynthesis. In *Proceedings of SPIE 3338*, February, pp. 1121.

Lindfors, K. K., Boone, J. M., Nelson, T. R., Yang, K., Kwan, A. L. C., and Miller, D. F. 2008. Dedicated breast CT: Initial clinical experience. *Radiology*, 246(3):725–733.

Maidment, A. D. A., Albert, M., Thunberg, S., Adelow, L., Blom, O., Egerstrom, J., Eklund, M. et al. 2005. Evaluation of a photon-counting breast tomosynthesis imaging system. In *Proceedings of SPIE*, 5745, pp. 572.

Maidment, A. D. A., Ullberg, C., Lindman, K., Adelöw, L., Egerström, J., Eklund, M., Francke, T. et al. 2006. Evaluation of a photon-counting breast tomosynthesis imaging system. In *Proceedings of SPIE*, 6142, pp. 61420B.

Mainprize, J. G., Bloomquist, A. K., Kempston, M. P., and Yaffe, M. J. 2006. Resolution at oblique incidence angles of a flat panel imager for breast tomosynthesis. *Medical Physics*, 33(9):3159.

Man, B. D. and Basu, S. 2004. Distance-driven projection and backprojection in three dimensions. *Physics in Medicine and Biology*, 49(11):2463–2475.

Mertelmeier, T., Orman, J., Haerer, W., and Dudam, M. K. 2006. Optimizing filtered backprojection reconstruction for a breast tomosynthesis prototype device. In *Proceedings of SPIE*, 6142, pp. 61420F.

Niklason, T., Christian, B. T., Niklason, L. E., Kopans, B., Opsahl-ong, B. H., Landberg, E., Giardino, A. et al. 1997. Digital tomosynthesis in breast imaging. *Radiology*, 205(2):399–406.

Nishikawa, R. M., Reiser, I., and Seifi, P. 2007. A new approach to digital breast tomosynthesis for breast cancer screening. *Proceedings of SPIE 6510*, pp. 65103C.

O'Connell, A., Conover, D. L., Zhang, Y., Seifert, P., Logan-Young, W., Lin, C.-F. L., Sahler, L., and Ning, R. 2010. Cone-beam CT for breast imaging: Radiation dose, breast coverage, and image quality. *AJR*, 195(2):496–509.

Ogawa, K., Langlais, R. P., McDavid, W. D., Noujeim, M., Seki, K., Okano, T., Yamakawa, T., and Sue, T. 2010. Development of a new dental panoramic radiographic system based on a tomosynthesis method. *Dento Maxillo Facial Radiology*, 39(1):47–53.

Park, S., Jennings, R., Liu, H., Badano, A., and Myers, K. 2010. A statistical, task-based evaluation method for three-dimensional x-ray breast imaging systems using variable-background phantoms. *Medical Physics*, 37(12):6253.

Poplack, S. P., Tosteson, T. D., Kogel, C. A., and Nagy, H. M. 2007. Digital breast tomosynthesis: Initial experience in 98 women with abnormal digital screening mammography. *AJR*, 189(3):616–23.

Prionas, N. D., Lindfors, K. K., Ray, S., Huang, S.-Y., Beckett, L. A., Monsky, W. L., and Boone, J. M. 2010. Contrast-enhanced dedicated breast CT: Initial clinical experience. *Radiology*, 256(3):714–723.

Reiser, I. and Nishikawa, R. M. 2010. Task-based assessment of breast tomosynthesis: Effect of acquisition parameters and quantum noise. *Medical Physics*, 37(4):1591–1600.

Reiser, I., Nishikawa, R. M., and Lau, B. A. 2009. Effect of non-isotropic detector blur on microcalcification detectability in tomosynthesis. In *Proceedings of SPIE*, 7258, pp. 72585Z.

Ren, B., Ruth, C., Stein, J., Smith, A., Shaw, I., and Jing, Z. 2005. Design and performance of the prototype full field breast tomosynthesis system with selenium-based flat panel detector. In *Proceedings of SPIE*, 5745, pp. 550–561.

Richard, S. and Samei, E. 2010. Quantitative imaging in breast tomosynthesis and CT: Comparison of detection and estimation task performance. *Medical Physics*, 37(6):2627–2637.

Rosenberg, R., Yankaskas, B., Abraham, L., Sickles, E., Lehman, C., Geller, B., Carney, P., Kerlikowske, K., Buist, D., Weaver, D., and Others 2006. Performance benchmarks for screening mammography. *Radiology*, 241(1):55–66.

Sechopoulos, I., Suryanarayanan, S., Vedantham, S., DOrsi, C. J., and Karellas, A. 2007. Scatter radiation in digital tomosynthesis of the breast. *Medical Physics*, 34(2):564.

Siddon, R. 1985. Fast calculation of the exact radiological path for a three-dimensional CT array. *Medical Physics*, 12:252.

Sidky, E. Y., Pan, X., Reiser, I. S., Nishikawa, R. M., Moore, R. H., and Kopans, D. B. 2009. Enhanced imaging of microcalcifications in digital breast tomosynthesis through improved image-reconstruction algorithms. *Medical Physics*, 36(11):4920.

Smith, A., Niklason, L., Ren, B. R., Wu, T., Ruth, C., and Jing, Z. X. 2006. Lesion Visibility in Low Dose Tomosynthesis, volume 4046 of *Lecture Notes in Computer Science*, pp. 160–166. Springer.

Sprenger, F., Calderon, X., Gidcumb, E., Lu, J., Qian, X., Spronk, D., Tucker, A., Yang, G., and Zhou, O. 2011. Stationary digital breast tomosynthesis with distributed field emission x-ray tube. In *Proceedings of SPIE 7961*, 7961, pp. 79615I.

Teertstra, H. J., Loo, C. E., van Den Bosch, M. A. A. J., van Tinteren, H., Rutgers, E. J. T., Muller, S. H., and Gilhuijs, K.

G. A. 2010. Breast tomosynthesis in clinical practice: Initial results. *European Radiology*, 20(1):16–24.

Vikgren, J., Zachrisson, S., and Svalkvist, A. 2008. Comparison of chest tomosynthesis and chest radiography for detection of pulmonary nodules: Human observer study of clinical cases. *Radiology*, 249(3):1034–1041.

Wu, Q. J., Meyer, J., Fuller, J., Godfrey, D., Wang, Z., Zhang, J., and Yin, F.-F. 2011. Digital tomosynthesis for respiratory gated liver treatment: Clinical feasibility for daily image guidance. *International Journal of Radiation Oncology, Biology, Physics*, 79(1):289–296.

Wu, T., Moore, R. H., and Kopans, D. B. 2006. Voting strategy for artifact reduction in digital breast tomosynthesis. *Medical Physics*, 33(7):2461.

Wu, T., Moore, R. H., Rafferty, E. A., and Kopans, D. B. 2004. A comparison of reconstruction algorithms for breast tomosynthesis. *Medical Physics*, 31(9):2636.

Wu, T., Stewart, A., Stanton, M., McCauley, T., Phillips, W., Kopans, D. B., Moore, R. H. et al. 2003. Tomographic mammography using a limited number of low-dose cone-beam projection images. *Medical Physics*, 30(3):365.

Zhang, Y., Chan, H.-P., Sahiner, B., Wei, J., Goodsitt, M. M., Hadjiiski, L. M., Ge, J., and Zhou, C. 2006. A comparative study of limited-angle cone-beam reconstruction methods for breast tomosynthesis. *Medical Physics*, 33(10):3781.

Zhang, Y., Chan, H.-P., Sahiner, B., Wu, Y.-T., Zhou, C., Ge, J., Wei, J., and Hadjiiski, L. M. 2007. Application of boundary detection information in breast tomosynthesis reconstruction. *Medical Physics*, 34(9):3603.

Zhao, W., Ji, W. G., Debrie, A., and Rowlands, J. A. 2003. Imaging performance of amorphous selenium based flat-panel detectors for digital mammography: Characterization of a small area prototype detector. *Medical Physics*, 30(2):254.

Zhao, B. and Zhao, W. 2005. Temporal performance of amorphous selenium mammography detectors. *Medical Physics*, 32(1):128.

Zheng, B., Lu, A., Hardesty, L. A., Sumkin, J. H., Hakim, C. M., Ganott, M. A., and Gur, D. 2006. A method to improve visual similarity of breast masses for an interactive computer-aided diagnosis environment. *Medical Physics*, 33(1):111.

2

Spectral/Photon-Counting Computed Tomography

Jochen Cammin
Johns Hopkins University

Jan S. Iwanczyk
DxRay, Inc.

Katsuyuki Taguchi
Johns Hopkins University

2.1 Introduction

For over 100 years, x-rays have played a crucial role in imaging the internal structure of objects nondestructively. The usefulness of x-rays for medical purposes was recognized as soon as Wilhelm C. Röntgen took the famous transmission image of his wife's hand in 1895 (Kevles 1996). For the next 7 decades, the same simple two-dimensional projection method, still known from dental x-rays, for example, remained the state of the art for medical imaging. It was not until the 1970s that x-ray imaging fundamentally changed with the invention of computed tomography (CT) (Ambrose 1973; Hounsfield 1973) and the necessary reconstruction algorithms (Cormack 1963, 1964). Hounsfield's CT scanner allowed visualization of two-dimensional slices of the human body rather than projection images where all the organs are superimposed and overlapped. The technology was refined quickly over the next 20 years, decreasing scan times and radiation dose and increasing the spatial resolution and the field-of-view size. Medical imaging experienced another boost in the 1990s with the advent of multislice CT scanners. They allowed acquisition of true three-dimensional data sets and reconstruction of volumetric images in a single scan. Even whole-body scans became possible with clinically acceptable scan times and radiation dose levels. Multislice CT (or MDCT for multidetector CT) is now the state-of-the-art technology for fast, high-resolution, and cost-effective imaging of the human body with a wide range of clinical applications. With gantry rotation times of <0.3 s and detectors with up to 320 rows, it is possible to scan entire organs like the heart in <1 s.

All current x-ray imaging systems are still based on the original idea of recording changes in the x-ray intensity due to attenuation in the scanned object or patient. However, the transmitted x-rays carry more information than just intensity changes. X-ray tubes produce x-rays spanning a wide spectrum of energies. Since the attenuation of x-ray photons depends not only on the traversed material but also on the energy of the photons, the transmitted energy spectrum contains valuable information about the structure and materials in the object. This energy information is currently not used as the detectors record only intensities. The advantage of using energy information to better distinguish different tissue types has been known for a long time, though, and was even mentioned in Hounsfield's original paper (Hounsfield 1973). But available technology limited the practical usefulness of energy-resolved imaging for many years. Due to a lack of energy-sensitive detectors that could handle the high x-ray flux in clinical CT scanners, the solution at the time was to perform two separate scans at different tube voltage (kVp) settings. This approach increases the scan time and radiation dose and can suffer from motion inconsistencies between the two scans. More recent developments allow recording of projection data sets at two different energies in a single scan and thus enable tissue-specific imaging. The approaches vary significantly between manufacturers. Scanners for dual-energy CT (DECT) comprise systems with two x-ray tubes and two detector arrays, mounted perpendicular to each other (Flohr et al. 2005, 2008), systems that rapidly switch the tube voltage between projections (Zou and Silver 2008), and systems with sandwich detectors where two layers detect different parts of the x-ray spectrum. The problem with all systems is that the energy separation is far from perfect as there is large spectral overlap between the data acquired at low and high energies. In addition, only two different energy data sets are practically feasible, which limits the number of simultaneously identifiable materials.

It is therefore desirable to image objects in more than two energy ranges with little or no spectral overlap. This goal can be achieved with a new generation of detectors that have the inherent ability to measure both the intensity and the energy of x-ray photons. As opposed to the current detector technology, which integrates the energy information of many photons, these new detectors measure individual photons. They are therefore called photon-counting detectors (PCDs). The technology has many potential benefits. Among them are improved noise, dose reduction, and material-specific imaging. A major obstacle so far has been the limited count-rate capability, which does not yet meet the requirements created by the high x-ray fluxes in clinical CT. PCDs were first developed and used in high-energy physics experiments and for nuclear medicine imaging where high count rates are not a concern. However, the research community has made steady progress over the last decade in improving the detector technology, making the readout electronics faster, and developing correction and compensation algorithms to handle these limitations.

Spectral CT, the method to produce energy-selective CT images, has the potential to improve existing imaging methods and enable completely new clinical applications. It provides a new dimension of information, similar to that provided by color photography or color TV when compared to gray-scale images. Spectral CT is thus sometimes also referred to as *color CT*.

Section 2.2 of this chapter discusses the technology of PCDs and compares it to the current detector technology. Section 2.3 explains imaging techniques for spectral CT and compensation methods for PCDs. Section 2.4 highlights some of the potential clinical applications that will become possible with spectral CT. Section 2.5 summarizes this chapter.

2.2 Detector Technologies

This section discusses two different technologies for detecting x-rays in CT imaging. The first are the energy-integrating detectors (EIDs) that are the state-of-the-art technology in all current commercial CT scanners. The second are the PCDs that are the subject of increasing interest in both academic and commercial research. PCDs have many potential advantages over EIDs but also require new approaches to correct or compensate for data distortions and to image reconstruction. We only briefly review the current detector technologies in Section 2.1 and then focus on PCDs as an emerging technology in Section 2.2.

2.2.1 Current Detector Technology: Energy-Integrating Detectors

Most of the detectors that currently operate in commercial x-ray detectors are so called *indirect-conversion* detectors. The mechanism is indirect because incident x-ray photons are first converted

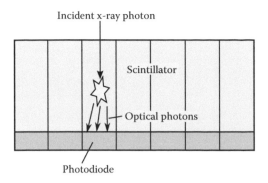

FIGURE 2.1 Mechanism of indirect signal conversion in conventional x-ray detectors.

into optical (visible) or ultraviolet (UV) photons in a scintillator material, and the optical/UV photons are then converted into an electrical signal. This process is shown in Figure 2.1. Typical scintillator materials are $CdWO_4$ and Gd_2O_2S (Hsieh 2009). The optical photons created in the scintillator are converted into an electrical current using a photosensor (usually photodiodes). The produced current is proportional to the x-ray photon intensity. The electric signal is also proportional to the energy of the incident photons, giving more weight to high-energy photons than to low-energy photons. The readout electronics do not process the signals from individual photons but they integrate signals over a certain time period and then amplify and convert them into a digital signal. Energy information is lost in this process, and it is impossible to distinguish, for example, whether a given signal came from a single 120 keV photon or two 60 keV photons. Hence, the detectors are called *energy-integrating detectors*. EIDs integrate not only the signal but also noise (electronic noise and Swank noise (Swank 1973)). The mechanism discussed above limits the ability of clinical scanners to resolve soft-tissue contrast. For further reading on EIDs and general principles of radiation detection, see Buzug (2010), Knoll (2010), Bushberg et al. (2011), Grupen and Buvat (2011).

2.2.2 Photon-Counting Detectors

The problem that valuable energy information is not recorded by EIDs can be overcome with PCDs. As the name suggests, these types of x-ray detectors measure signals from individual photons. They are capable of measuring both the number of photons in a given time interval as well as the signal height which is a monotonic function of the photon energy. The additional energy information is useful to improve tissue differentiation. PCDs discussed here are *direct-conversion* detectors and use a semiconductor material to convert the x-ray photon energy into an electrical signal without the extra step of producing optical/UV photons first. We do not discuss PCDs based on silicon photomultipliers (SiMP) as they cannot yet compete with the energy resolution of semiconductor-based PCDs. Whereas EIDs have been used for decades to detect x-ray photons, PCDs for medical imaging have been limited to applications in PET/SPECT imaging and some specialty x-ray

TABLE 2.1 Comparison of Prototype PCDs

#	Name/ASIC	MOCR Mcps/pixel	Pixel Size (μm^2)	MOCR (Mcps/mm²)	# Energy Windows	Tileup Capability
1	Medipix	0.1	55×55	33	2	No
2	GMI CA3	1–2	400×1000	2.5–5	6	No
3	Hamamatsu	1–2	1000×1000	1–2	5	No
4	CIX	3.3	250×500	26	1	n/a
5	ChromAIX	13.5	300×300	150	n/a	n/a
6	Siemens 2010	n/a	225×225	n/a	n/a	Yes
7	DXMCT-1	5.5	1000×1000	5.5	2	Yes
8	DXMCT-2	5.5	500×500	22	4	Yes

Note: Detectors #1 and #5 have not yet been measured with the crystals attached. "MOCR" stands for maximum output count rate and n/a indicates that the information was not available.

imaging modalities (e.g., breast tomosynthesis). The reason is that CT imaging produces rather high x-ray fluxes, and the resulting count-rate requirements are very challenging for PCDs with the technology currently available. Steady advancements in building faster PCDs and the expected benefits for dose reduction, image quality, and quantitative imaging, however, have made PCDs for x-ray CT an active field of research and development. There are at least eight PCD prototypes available as listed in Table 2.1.

2.2.2.1 Detection Mechanism

The direct conversion mechanism used in PCDs is illustrated in Figure 2.2. An incident x-ray photon creates electron–hole pairs through various physics processes (Knoll 2010). The created charges drift in a high-voltage bias field toward electrodes at the back side of the detector where they are collected. Electronics directly attached to the sensor amplify the signal pulse, shape it, and count the signals that are above a certain threshold. The efficiency of the detection mechanism is higher by about an order of magnitude compared to indirect detectors, mainly due to two reasons: First, the mean energy necessary to create an electron–hole pair in a semiconductor is almost an order of magnitude smaller than the energy necessary to create an optical/UV photon in a scintillator. Second, the strong bias field minimizes charge losses in the lateral direction. In contrast, optical/UV photons in an indirect-conversion scintillator are emitted in all directions, and the collection efficiency is significantly lower despite reflective shields around the scintillator.

A number of semiconductor materials exist that can be used in x-ray PCDs. Among them are high-purity germanium (HPGe), cadmium zinc telluride (CdZnTe), and cadmium telluride (CdTe). These materials can be finely pixelized and bump-bonded to the readout electronic circuitry. A disadvantage of HPGe is that

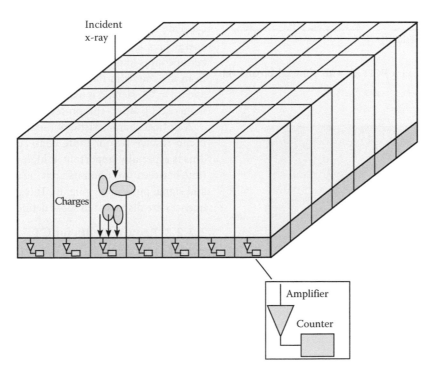

FIGURE 2.2 Mechanism of direct signal conversion in photon-counting detectors.

it needs cryogenic cooling to keep the thermal noise at acceptable levels. CdZnTe and CdTe, however, can be operated at room temperature. They are the material of choice for PCDs for x-ray CT, as they have high stopping power for x-rays, making them efficient, and low leakage currents at room temperature, making them low-noise detectors. The low leakage current is a result of the high bulk resistivity of the two materials ($10^{11}\ \Omega \cdot cm$ for CdZnTe and $10^9\ \Omega \cdot cm$ for CdTe). Due to the low noise, the achievable energy resolution for Cd(Zn)Te PCDs is suitable for many medical imaging applications. However, the best energy resolution also requires collecting as many as possible of the electric charge carriers produced in the sensor material. The number of collected charge carriers depends on the shaping time of the pulse-shaping electronics. But a longer shaping time also means that the detector is in an active state longer, which has an adverse effect on the maximum count rate and on signal distortions due to signal pileup if another photon arrives before the previous signal has vanished, as shown in Figure 2.3. In practice, a compromise needs to be found between sufficient energy resolution and acceptable count rate. Other materials have been considered for direct-conversion PCDs as well, such as HgI_2, $TlBr$, and PbI_2. But taking into account noise, effective charge collection, and also reliability in the production process, CdTe is currently the material of choice. Figure 2.4 shows a 3 mm thick CdTe crystal with a 256 pixel anode with 1 mm pitch (pixel size 1×1 mm). Pixels at the perimeter are slightly smaller

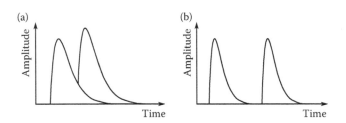

FIGURE 2.3 Signal pulses in a PCD (a) with pulse-pileup and (b) without pulse-pileup.

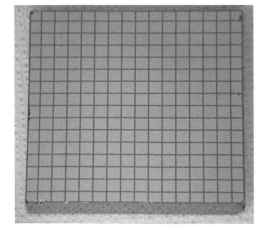

FIGURE 2.4 A 256 pixel CdTe detector with pixel dimensions of 1×1 mm. (Courtesy of DxRay, Inc., Northridge, CA.)

so that a 1 mm pitch can be maintained when several crystals are tiled together to create a larger detector area.

2.2.2.2 Readout Electronics

Besides the sensor material that converts x-ray photons into an electrical charge, the readout electronics play an equally important role in the functioning of a PCD. Application-specific integrated circuits (ASICs) that are connected to each sensor pixel through wire or bump-bonding perform the signal processing. Typical steps consist of (1) signal amplification, (2) signal shaping, (3) signal discrimination, (4) digital-to-analog conversion (DAC), and (5) signal counting. The basic architecture for a single channel is shown in Figure 2.5.

After amplification, the pulse-like signal of the collected charges is transformed by the shaper into a broader shape, which helps to reduce the noise. The maximum of the shape, or the pulse height, is a surrogate measure for the photon energy. The relation between pulse height and photon energy needs to be established in calibration measurements using photons with known energies. The PCD electronics contain a number M of discriminators chains, $M \geq 2$, which measure the signal pulse height (Figure 2.6). Each discriminator is followed by a digital counter that is increased by one if the signal pulse height is larger than the discriminator threshold. The counter measures the number of photons for a fixed small time frame, and is read out and then reset. The discriminator threshold can be controlled externally, allowing different energy thresholds to be set for each counter. In addition to these global threshold settings, each discriminator can also be fine adjusted individually in order to account for small pixel-to-pixel variations. The M counters then contain the number of photons, N_i, $i = 1 \cdots M$, with energy larger than their threshold values, $E_i \geq E_i^{th}$. From the number of photons in these one-sided, overlapping energy bins, one can calculate the number n_i of photons in two-sided, nonoverlapping energy bins $E_i^{th,\ low} \leq E < E_i^{th,\ high}$ by simply subtracting the number of counts in the one-sided bin with the next higher threshold from the counts in the current bin: $n_i = N_i - N_{i+1}$, $i = 1 \cdots M - 1$, and $n_M = N_M$. Here, it is assumed that the energy thresholds E_i are in increasing order. This procedure is also shown in Figure 2.6.

A critical characteristic of the readout electronics is whether it can resolve two or more consecutive signal pulses in time. This is especially important at high photon flux rates, when the time between two photons arriving at the detector is very short and signal pulses may pile up. This pileup effects and its consequences are discussed in more detail in the following sections.

2.2.2.3 Requirements for CT

The intended application of PCDs for routine clinical CT puts some stringent requirements on the detector design, specification, and performance. Photon count rates in clinical CT can reach up to 1 billion photons per square mm per second (10^9 photons/($mm^2 \cdot s$)) measured at 1 m away from the x-ray source. The current PCDs (Table 2.1) can operate at maximum count rates that are at least one order of magnitude smaller than this figure; data distortions in both count rates and energy measurement are thus unavoidable. One way to increase count rates is to reduce

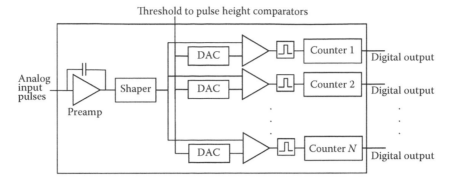

FIGURE 2.5 Readout electronics: typical components of the ASIC channel architecture.

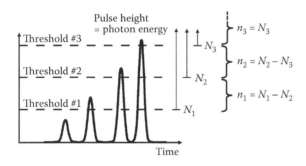

FIGURE 2.6 Concept of a pulse-height threshold counter. A photon is counted if its signal pulse exceeds a certain threshold. Counts of nonoverlapping bins are obtained by subtracting counts above adjacent thresholds.

the detection area per channel by making the pixel size smaller. For example, reducing the pixel pitch by half reduces the area and count rate per channel by 75%. However, a smaller pixel size has undesirable side effects as the cross talk between pixels increases. This also leads to distorted energy measurements and counts in neighboring pixels. Even with the rapid improvements in detector and electronics technologies, it will be challenging to meet the count-rate requirements through improved hardware alone in the near future. Therefore, other components of CT scanners need to be improved in parallel to enable clinical CT with PCDs. In total, four key areas for improvements can be identified: (1) detector technology, (2) beam-shapers (stationary and dynamic bowtie filters), (3) compensation for imperfect data (measurement-based and model-based corrections and compensations), and (4) image reconstruction methods.

Detector technology: Apart from increasing the speed of the readout electronics for handling higher count rates, the uniformity and stability of the detectors also needs to improve. Current prototypes exhibit pixel-to-pixel variations in the detector response and detection efficiency that lead to ring artifacts in the reconstructed images. Temporal stability of the detector behavior over the duration of the scan is another issue.

Beam-shapers: Specially shaped attenuating filters reduce the x-ray flux at the periphery of the scanned object or patient where the x-rays traverse only a small amount

of material. This improves the uniformity of the x-rays incident on the detector, which reduces both the noise inhomogeneity and the required dynamic range. Most importantly, they help to reduce the patient dose. For use with PCDs, these beam-shapers will need to be redesigned to avoid excessive count rates in air and near the periphery. It may be necessary to develop dynamic filters that change their position or shape during the gantry rotation to account for shape changes of the patient.

Calibration and compensation methods: The distortions from imperfect count rate and energy measurements need to be quantified and calibrated. If the PCD properties like count-rate losses and energy response are known, they can be incorporated into compensation schemes during the image-formation process.

Image reconstruction: Current image reconstruction algorithm will need to be adjusted and extended to take into account the data distortions that cannot be compensated for by other means. For example, despite beam shapers, the x-ray flux at the patient periphery may exceed the maximum detector count rate. The effect is similar to that of missing or inaccurate projection data. Recent research in this area provides options on how to overcome this problem (Taguchi et al. 2011). The additional information from several energy bins also requires more research on how to optimally combine the resulting projection data or reconstructed images (Giersch et al. 2004; Schmidt 2009).

Some general specifications for x-ray and CT are given in Table 2.2. Further requirements for the successful application of PCDs for clinical CT are: spatial resolution (pixel size) better than 1 mm, readout frame rate better than 3000 frames per second, small cross talk between neighboring pixels and consecutive photon signals, photon detection efficiency greater than 95%, geometrical efficiency better than 70%, low noise, large dynamic range (better than 20 bits or 10^6), linearity, uniformity smaller than 3% for the periphery and better than 0.1% for the object center, and the ability to tile-up individual detector modules to build a detector that covers a field-of-view of 50 cm in diameter and 3–16 cm in height, corresponding to a detector size of 700–1000 by 64–320 pixels.

TABLE 2.2   Specifications for X-Ray and Computed Tomography

	Mammography	General X-Ray Radiography	Computed Tomography
Count rate (photons per mm² · s)	5×10^7	10^6 to 5×10^8	10^9
Pixel size (μm)	Typically 85	Typically 150	55–1000
Energy range (keV)	28–40	70–120	80–140

Source: Adapted from Overdick, M. et al., 2008. Towards direct conversion detectors for medical imaging with x-rays. In *IEEE Nuclear Science Symposium Conference Record, 2008. NSS '08*, Dresden, Germany. IEEE, pp. 1527–1535.

2.2.2.4 Distortions in Photon-Counting Detectors

This section discusses the main physics effects of measuring x-ray photons with PCDs and how they lead to various distortions in both count rates and energy measurement:

1. Pulse-pileup
2. Charge sharing
3. K-escape x-ray
4. Compton scattering

The magnitude of the first effect (pulse-pileup) depends on the x-ray intensity. The higher the flux of the incident photons, the higher is the probability for pileup. The other effects degrade mainly the energy resolution and are independent of the flux.

2.2.2.4.1 Pulse-Pileup

If an x-ray photon hits a PCD, the detector and readout electronics need a small amount of time to collect the charges and process the signal. This time is called the *deadtime* of the detector. If another photon reaches the detector during this time, then the two signals cannot be separated and both photons will be counted as one. Since the energy is measured from the height of the electric pulse, the energy measurement may also be compromised as the result of the overlapping signal pulses (Figure 2.7). The effect of having n coincident photons within the deadtime

window is called n-th order pileup and the probability for each order decreases rapidly with n. We discuss first the effect of the lost counts due to pulse-pileup and then the effect of mismeasured energy.

The behavior of a PCD can be simplified by assuming a fixed deadtime τ. Two detector types can be distinguished: (1) nonparalyzable: additional photons hitting the detector during the deadtime will not extend the deadtime, (2) paralyzable: each time a new photon hits the detector during the deadtime, the deadtime will be extended by τ. In both cases, the counts from the additional photons are lost. If the photon rate increases and more and more photons hit the detector during the deadtime, then the nonparalyzable detector count rate saturates against a rate given by $r = 1/\tau$. In the paralyzable case, each new photon extends the deadtime further and the count rate goes to zero for very high incident rates. Mathematically, the recorded count rates as a function of the input count rates r_0 are given by (Knoll 2010)

$$r_{\text{nonparalyzable}} = \frac{r_0}{1 + r_0 \cdot \tau},$$
$$r_{\text{paralyzable}} = r_0 \cdot e^{-r_0 \cdot \tau}. \qquad (2.1)$$

The count rates for the two models are shown in Figure 2.8. If the detector follows either the nonparalyzable or the paralyzable model and the deadtime is known from calibration measurements, the true count rates can be recovered approximately from the recorded rates by solving Equation 2.1 for r_0. Some care must be taken though for the paralyzable detector because the function is nonmonotonic and there are two solutions for r_0. In reality, however, most detectors have a more complicated behavior and cannot be described by either the nonparalyzable or the paralyzable model. Corrections then cannot be derived from simple mathematical models and need to be based on calibration measurements or an accurate model of the pulse shapes and pulse-pileup.

The second distortion stems from the mismeasured energies due to pulse-pileup. In a very simplified model, the energies of the photons that arrive at the detector within the deadtime τ are added and a single photon is recorded with the sum energy. For example, two coincident photons with $E_1 = 40$ keV

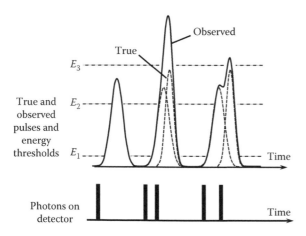

FIGURE 2.7   Concept of pulse-pileup. (Copyright AAPM (2010). Used with permission from Taguchi, K. et al., 2010. *Medical Physics*, 37(8), 3957–3969.)

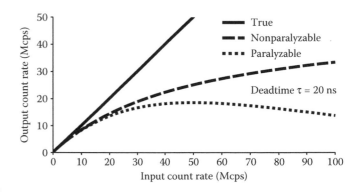

FIGURE 2.8   Count rates for a paralyzable and nonparalyzable detector. The unit Mcps stands for million counts per second.

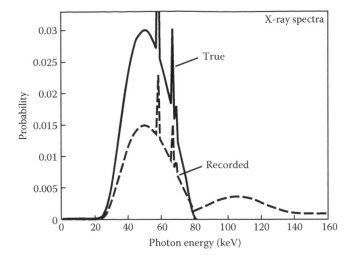

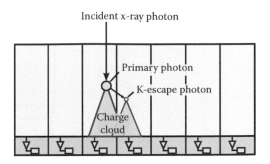

FIGURE 2.10 The charge sharing effect.

FIGURE 2.9 A simulated 80 kVp x-ray spectrum with and without distortions due to pulse-pileup.

and $E_2 = 60$ keV would be measured as a single photon with $E = 100$ keV. Three coincident photons, each with energy of 40 keV, would be measured as a single 120 keV photon and so on for higher-order pileup. To illustrate the effect, Figure 2.9 shows both a realistic 80 kVp x-ray spectrum and the recorded spectrum including up to third-order pulse-pileup. The recorded spectrum is severely distorted and extends to energies far beyond the maximum energy of the original spectrum. Since the PCD measures the spectrum in several energy bins, this means that the count rates in a given bin are mismeasured not only because photon counts are lost due to the detector deadtime, but also because the photon may be counted in a higher energy bin. However, the spectrum distortion in real detectors is more complex and depends on the specific pulse shape and the time interval between the photons. For some pulse shapes and certain pileup conditions, it is even possible that a photon is measured at a lower energy than its true energy (Taguchi et al. 2010).

The following sections describe distortions of the measured energy for a single photon due to physical effects in the detector sensor material. For a more in-depth discussion of the physical processes involved in x-ray detection, see, for example, Knoll (2010).

2.2.2.4.2 *Charge Sharing*

When an x-ray photon is absorbed in the detector, a charge cloud is created in the sensor. The charges drift to the pixel electrodes due to the applied high-voltage bias field. Diffusion effects cause the charges to spread out laterally. If the charges are created near a pixel boundary, the charges can spread into a neighboring pixel (Figure 2.10). This fraction of the charge, and hence the photon energy, is lost in the pixel and adds to the energy measurement in the neighbor pixel. The size of this effect depends on the sensor material (mobility of the charge carriers), pixel dimensions, and the applied high voltage. The charge sharing effect increases with smaller pixel area and larger pixel thickness.

2.2.2.4.3 *K-Escape X-Rays*

When the incident x-ray photon interacts with the detector through the photoelectric effect, an electron in the K-shell of the sensor material atoms can be ejected and leave the atom in an excited state. The atom returns to its ground energy state by emitting a *characteristic x-ray photon*. The energy of this photon is specific to the detector material. For a CdTe PCD, the K-shell energies are 26.7 and 31.8 keV. The characteristic x-rays are emitted in a random direction and often leave the detector pixel undetected, especially if they are created near the edge of the pixel. The measured energy in the pixel is then reduced by the K-shell energy. The escaped photon may also be absorbed in a neighbor pixel, increasing the cross talk between pixels (Figure 2.10).

2.2.2.4.4 *Compton Scatter*

The incident photon can also be scattered in the detector material, where it changes direction and loses some of its energy. This is the *Compton effect* or *Compton scatter*. The photon may leave the pixel and only the fraction of the energy deposited in the material is detected. Contrary to the discrete energy loss due to K-escape photons, the energy loss from Compton scatter is continuous.

2.2.2.4.5 *Summary of Energy Response*

A perfect detector would measure the energy of the incident x-ray photon exactly. The effect of charge sharing, K-escape x-rays, and Compton scatter, however, often leads to a measured energy that is significantly lower. In addition, the stochastic process of the signal generation in the sensor and noise in both sensor and readout-electronics degrade the precision of the energy measurement further. The result is a probability distribution that describes how likely it is to measure an incident photon with energy E_0 as an arbitrary energy E. This distribution is also called the *energy response* or the *spectral response function (SRF)* of the detector. The SRF is not constant but depends on the incident energy E_0. An example of an SRF with an explanation of the physics effects that cause the shape is shown in Figure 2.11. The steep increase toward very low energies is due to noise. Its effect can be avoided by setting the lowest energy threshold well above the noise level. The long continuum tail, however, is undesirable but unavoidable. The tail causes photons to be counted

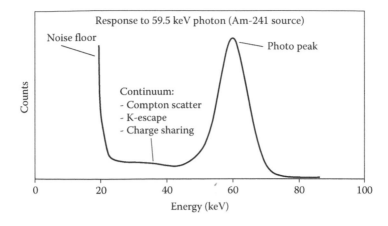

FIGURE 2.11   Example of a spectral response function measured with a radioisotope.

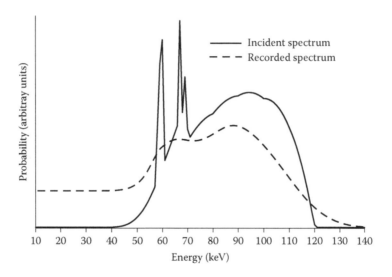

FIGURE 2.12   Comparison of an incident 120 kVp x-ray spectrum onto a semiconductor PCD and the measured spectrum. The incident spectrum was filtered with 18.5 mm aluminum.

in the wrong energy bin, which results in image artifacts and negatively impacts the accuracy of tissue identification. To demonstrate how severe the distortions due to the SRF can be, Figure 2.12 shows a 120 kVp x-ray spectrum incident on a PCD and the measured spectrum using a realistic SRF. The SRF smears out the spectrum mostly toward lower energies. Effectively, this creates bin overlap by blurring the sharp boundary between energy bins, diminishing one of the benefits of PCDs.

Noise and continuum tail effects can be influenced by the design of the sensor and the readout electronics within some limits. Calibration and algorithmic compensation schemes are needed to minimize the effects from the SRF and they are the subjects of active research.

2.3 Imaging Techniques

This section discusses the basics concepts of spectral CT imaging and two specialized imaging techniques.

2.3.1 Concepts of Spectral/Multienergy Imaging

All imaging techniques using x-rays make use of the fact that different materials attenuate x-rays in different ways. Attenuation of x-rays in the diagnostic energy range (30–140 keV) occurs predominantly through two physical phenomena, the photoelectric effect and Compton scatter (Knoll 2010). Photoelectric absorption dominates at lower energies and decreases rapidly with energy, approximately as $1/E^3$. Compton scatter depends only weakly on the energy and dominates at higher energies over the photoelectric effect. The strength of the two absorption mechanisms also depends significantly on the attenuating material. The photoelectric effect scales roughly with Z^3, where Z is the atomic number of the material. The linear attenuation coefficients for various tissue materials as a function of the x-ray energy are shown in Figure 2.13a. The graphs suggest that for a fixed energy, the difference in attenuation between materials is larger at low energies and smaller at high energies. Hence, one

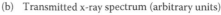

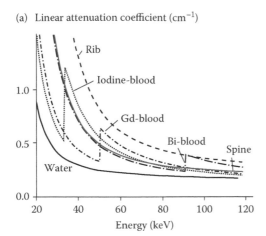

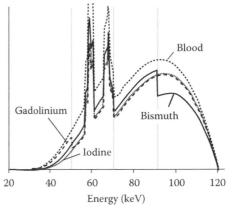

FIGURE 2.13 (a) Linear attenuation as a function of energy for various materials. (b) Transmitted x-ray spectra for various materials. Note that the K-edge is not visible for iodine. The vertical gray lines indicate energy bin boundaries suitable for K-edge imaging.

expects to see better contrast differences at low energies. Figure 2.13a also shows that in a certain energy range (between 70 and 90 keV) some of the materials have very similar attenuation values, resulting in the same pixel values in the reconstructed image. This energy range coincides with the typical effective energy of spectra used in current CT scanners and is the reason for insufficient contrast resolution. Although the attenuation curves have very different shapes, the current EIDs cannot resolve the materials. Spectral CT, however, measures the attenuation at two or more energies and allows for discrimination of different tissues based on the energy behavior of their attenuation. Figure 2.13b shows transmitted spectra for blood and iodine, gadolinium, and bismuth contrast agents. These contrast agents can potentially be identified using the technique described in Section 3.2. From the spectrum, however, one can see that the iodine K-edge has nearly vanished due to photon starvation and only the gadolinium and bismuth K-edges remain visible for clinical CT. In contrast, the iodine K-edge will be clearly visible for small animal CT because the attenuation of the x-ray spectrum is very small.

2.3.2 K-Edge Imaging

The contrast for certain materials can be enhanced using dual-energy or spectral-imaging techniques. Materials with a K-edge in the diagnostic energy range show a discontinuity with a sudden increase in the attenuation at the K-edge energy as seen in Figure 2.13a. The K-edge energies for some materials used in diagnostic CT imaging are 33 keV for iodine, 50 keV for gadolinium, and 90 keV for bismuth. If two images are acquired, one with an effective energy slightly below the K-edge and one with effective energy slightly above the K-edge, then the attenuation of the K-edge material will be significantly different. In contrast, the attenuation for other materials such as water or bone will change only slightly between the two energies. Therefore, subtraction of the two images reduces the presence of the other materials and enhances the K-edge material (Riederer 1977; Ruzsics et al. 2008).

K-edge imaging with conventional CT scanners requires two scans with different tube voltages or different beam filtration to achieve different effective energies. The disadvantages of this approach include increased dose and difficulties in properly aligning the two independent images, which will deteriorate the resulting contrast-enhanced image. The problem can somewhat be mitigated by using the dual-energy approaches discussed in Section 2.1, but they still suffer from large spectral overlap. Therefore, using PCDs is a promising way to improve the image quality for certain contrast agents and has been explored in the literature (Roessl and Proksa 2007; Feuerlein et al. 2008; Roessl et al. 2011). More accurate K-edge imaging with PCDs is possible using three energy bins and is discussed in the next section on material decomposition.

2.3.3 Material Decomposition

Material decomposition is a powerful way to extract quantitative information about the materials that are contained in a scanned object. The technique was explored soon after the advent of x-ray CT (Alvarez and Macovski 1976) and is based on the idea that any material can be modeled as the weighted sum of two other materials (or three materials if a K-edge material is involved). We start the discussion from the underlying physics involved in x-ray attenuation.

The photoelectric effect and Compton scatter are the two fundamental processes responsible for attenuation of diagnostic x-ray photons in matter. Hence, the energy-dependent attenuation curve for any material can be expressed as the linear combination of these two effects:

$$\mu(E) = c_1 \cdot \mu_{PE}(E) + c_2 \cdot \mu_C(E), \quad (2.2)$$

where c_1 and c_2 are weight factors, and μ, μ_{PE}, and μ_C are the total attenuation coefficient of the material, the attenuation due to the photoelectric effect, and the attenuation due to the Compton effect, respectively. This relation can be presented graphically

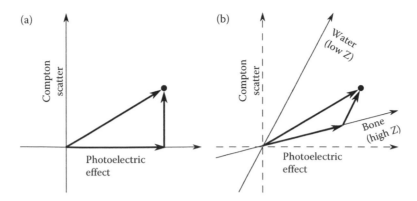

FIGURE 2.14   Concept of material decomposition with two basis materials.

as shown in Figure 2.14a, where the photoelectric effect and the Compton effect form a vector basis and the attenuation of a material at a specific energy is represented by simple vector addition. There are, however, an infinite number of other vector bases that can achieve the same value (Figure 2.14b). This graphical result can be interpreted such that the attenuation of an arbitrary material can be represented as a linear combination of two *basis materials:*

$$\mu(E) = \gamma_1 \cdot \mu_1(E) + \gamma_2 \cdot \mu_2(E). \tag{2.3}$$

For a reconstructed CT image, $\mu(E)$ represents the local linear attenuation coefficient. Hence, it depends not only on the energy, but also has spatial components: $\mu(E) \rightarrow \mu(x,y,z;E)$. It is convenient to separate the contribution of each basis material in Equation 2.3 into an energy-only and a location-only part:

$$\mu(x,y,z;E) = \rho_1(x,y,z) \cdot \mu_{M,1}(E) + \rho_2(x,y,z) \cdot \mu_{M,2}(E), \tag{2.4}$$

where ρ is the local density and μ_M is the mass attenuation coefficient of the basis material. It follows that if the μ_M are known, a CT image can be represented by the density images of the two basis materials. The task of material decomposition is to find the coefficients γ in Equation 2.3 or equivalently, the local densities ρ in Equation 2.4. Since there are two unknowns it takes at least two independent measurements of line integrals to find the densities. For monochromatic x-rays, this task is straightforward as the equations simplify to

$$\mu(E_A) = \rho_1 \cdot \mu_{M,1}(E_A) + \rho_2 \cdot \mu_{M,2}(E_A),$$
$$\mu(E_B) = \rho_1 \cdot \mu_{M,1}(E_B) + \rho_2 \cdot \mu_{M,2}(E_B). \tag{2.5}$$

Equation 2.5 is a linear system of equations which can be solved algebraically for ρ_1 and ρ_2. For polychromatic spectra, however, one needs to integrate the line integrals over the energy spectra to obtain a nonlinear system of equations for the photon counts:

$$N_A = \int N_0(E_A) e^{-\left(\int \left(\rho_1(t)\cdot\mu_{M,1}(E_A)+\rho_2(t)\cdot\mu_{M,2}(E_A)\right)dt\right)} dE_A,$$
$$N_B = \int N_0(E_B) e^{-\left(\int \left(\rho_1(t)\cdot\mu_{M,1}(E_B)+\rho_2(t)\cdot\mu_{M,2}(E_B)\right)dt\right)} dE_B, \tag{2.6}$$

where t is the path of a projection ray. In general, this system of equations can only be solved numerically. In practice, material decomposition works best if the two basis materials have largely different mass attenuation coefficients, for example, a high Z and a low Z material, and if the two measurements are taken at maximally separate energies.

The statements made so far apply if there is no material involved with a K-edge in the relevant energy range. Because of the discontinuity in the attenuation, a third basis material is required to represent objects containing a K-edge material. Consequently, a third energy measurement is needed to determine the basis material density images.

Typical choices for basis materials in clinical applications are water as a low-Z material and bone or aluminum as a high-Z material. If K-edge materials are involved, they are a natural choice for the third basis material. A simple example for material decomposition is shown in Figure 2.15. A cylindrical water phantom contains inserts of various materials, including bone and an iodine–blood mixture. After separation into the three basis materials (bone, water, and iodine), one finds that the spine and rib inserts show up predominantly in the bone basis image, water in the water basis image, and the iodine–blood mixture in the iodine basis image. Other materials are synthesized as combinations of the three basis materials. An example for an application of material decomposition is the imaging of the coronary vessels of the heart after injection of iodine contrast. The vessel structure can be isolated from other tissues by including iodine as a basis material. The separation is better compared to conventional CT where tissues with comparable attenuation, for example bone, appear with similar intensity. In contrast, bone is strongly suppressed in the material-decomposed iodine basis image.

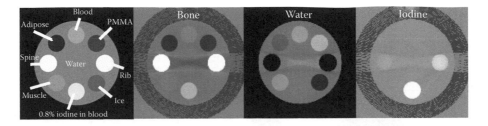

FIGURE 2.15 Demonstration of material decomposition. The simulated computer phantom on the left is decomposed into bone, water, and iodine basis images.

2.3.4 Spectral CT with PCDs

PCDs can improve the image quality for CT substantially and may lead to many new applications. But as discussed before (Sections 2.2.2.3 and 2.2.2.4), PCDs are not free from imperfections and many factors degrade the accuracy of the data. If they are not accounted for, many of the potential benefits of PCDs may not apply. This section discusses possible approaches to correct and compensate for distortions so that CT with PCDs becomes feasible. Degradation factors for PCDs are (1) the pulse-pileup effect (PPE) or deadtime losses, (2) charge sharing, (3) K-escape x-ray, (4) re-absorption of K-escape x-ray, (5) Compton scattering inside the PCD, (6) nonlinear photon energy-pulse height response, and (7) shift-variant finite energy resolution. Some of these effects can be corrected using calibration measurements and some effects need to be compensated for using accurate models of the underlying phenomena.

2.3.4.1 Compensation for Detector Degradation Factors

The spectrum distortions caused by the PPE are the most difficult to model because they are a complex phenomenon. They depend on the incident count rate, the transmitted x-ray spectrum (and thus on the scanned object), and on the specific detector pulse shape and readout electronics. It is imperative to model the pulse-pileup properly and realistically to achieve the accuracy and efficiency required to handle the large number of coincident photons expected for clinical CT. Models are available that take into account the pulse shape, the probability distribution of time intervals between random photon events, and the transmitted spectrum (Taguchi et al. 2010).

Parameterizations for the count-rate independent distortions (2–7) can be obtained by either calibration measurements or models. Although models have been developed for some of these distortions, either based on analytical approximations or Monte Carlo simulations, often they cannot accurately describe the true phenomena. Therefore, it seems advantageous to obtain the energy response function of the PCD directly from calibration measurements. These measurements must be performed using low-intensity x-ray beams to avoid distortions due to pulse-pileup. One option is to use radioisotopes with transitions in the relevant energy range such as Ba-133 (31.6 keV), Am-241 (59.5 keV), Cd-109 (88 keV), and Tc-99m (140 keV). Another option is to measure the response function at a synchrotron

(Schlomka et al. 2008). The response functions obtained at discrete energies then need to be parameterized and interpolated so that they are available at arbitrary energies.

Once models of the PPE and the energy response (or spectral response function) are available, compensation of the PCD degradation factors can be achieved by incorporating the models into the image reconstruction chain as a part of the forward imaging process. Either the imaged object or the projection data (line integrals) can then be estimated iteratively, for example, using a maximum likelihood approach based on the Poisson log-likelihood (Eggermont and LaRiccia 2010). The imaging chain for this compensation approach is illustrated in Figure 2.16.

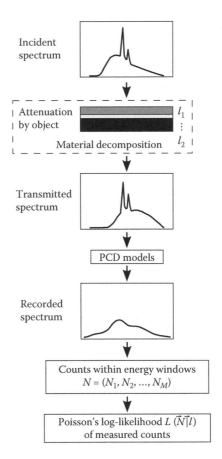

FIGURE 2.16 PCD compensation scheme.

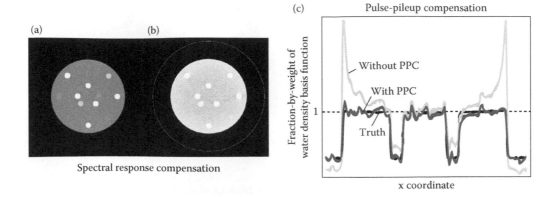

FIGURE 2.17 Computer phantom reconstructed (a) with SRF compensation and (b) without SRF compensation. (c) Profile through the phantom comparing the effect of the PPE compensation. (Copyright SPIE (2012). Used with permission from Srivastava, S. et al., 2012. Spectral response compensation for photon-counting clinical x-ray CT using sinogram restoration. In *Proceedings of SPIE. Medical Imaging 2012: Physics of Medical Imaging.* San Diego, California, USA.)

First, the x-ray spectrum exiting from the bowtie filter and incident on the patient and the parameters needed for the PCD models are obtained during prescan calibration. Then, the patient is scanned with multiple energy windows, obtaining counts $\vec{N} = (N_1, N_2, \ldots, N_M)$. The recorded data are distorted due to the degradation factors discussed above. Material decomposition is applied and the attenuation inside the scanned object is modeled by the thickness of multiple basis functions, $\vec{l} = (l_1, l_2, \ldots, l_L)$, where $L \leq M$ is the number of basis functions. Knowing the basis material thicknesses, the transmitted spectrum and counts can be calculated. Finally, the Poisson log-likelihood of the measured data, $L(\vec{N}|\vec{l})$, is calculated. The unknown values are the basis function thicknesses $\vec{l}$, which are determined iteratively by maximizing $L(\vec{N}|\vec{l})$. This is a standard optimization problem. An accurate estimate of $\vec{l}$ is all that is needed to represent the imaged object, eliminate beam-hardening effects, and quantify the material content.

In this approach, the accuracy of the PCD models is crucial to obtain accurate and unbiased estimators of the basis materials in the decomposition process. Computer simulations are a good tool to illustrate these effects. Figure 2.17a and b shows a simulated phantom with inserts of gadolinium and iodine contrast materials without and with compensation for the spectral response function. Without compensation, the image exhibits streaking artifacts between the inserts that shows that the beam-hardening effect has not been corrected, cupping artifacts, and biased pixel values. Figure 2.17c shows a profile through a similar phantom for three cases: (1) the true pixel values, (2) the reconstructed pixel values without compensating for the PPE, and (3) with compensation for the PPE. Without compensation, there are strong cupping artifacts whereas the full compensation of the PPE is able to reconstruct the pixel values accurately (Srivastava and Taguchi 2010). The accuracy of quantifying contrast agent densities is also affected by the compensation as shown in Figure 2.18 for the example of compensating for the SRF. Without compensation, the gadolinium insert

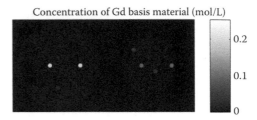

FIGURE 2.18 Basis material images with SRF compensation (left) and without SRF compensation (right). (Copyright SPIE (2012). Used with permission from Srivastava, S. et al., 2012. Spectral response compensation for photon-counting clinical x-ray CT using sinogram restoration. In *Proceedings of SPIE. Medical Imaging 2012: Physics of Medical Imaging.* San Diego, California, USA.)

concentration is underestimated and iodine inserts show up in the gadolinium basis image (Srivastava et al. 2012).

Computer simulations are also useful to study the effect of different detector properties on the reconstructed image quality. Figure 2.19 shows the contrast-to-noise ratio (CNR) measured in reconstructed images of the same simulated phantom for the iodine and gadolinium inserts. A conventional EID is compared to three different PCDs. The simulated PCDs vary in the ratio of the area of the "tail" part of the SRF to the total area, which is an indicator for how precise the PCD can measure the photon energy. A perfect detector has no tail. It is clear from the figure that PCDs outperform EIDs in terms of CNR and the simulation helps to quantify the difference between different levels of degradation from the SRF. This in turn helps to optimize the design of PCDs as the amount of tail is directly influenced by the choice of the sensor material and the pixel geometry.

2.3.4.2 Interior Reconstruction Methods

Despite stationary or even dynamic filters that are designed to reduce and equalize count rates, the input count rates near the periphery of an object and outside an object can be high and in excess of the maximum count rate of the PCD. This situation

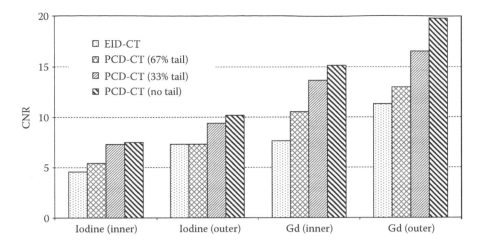

FIGURE 2.19 CNR for EID and PCD with different levels of tailing of the spectral response due to continuous energy loss (Compton scatter, K-escape, charge sharing). "Inner" and "outer" refer to different locations of the inserts within the phantom.

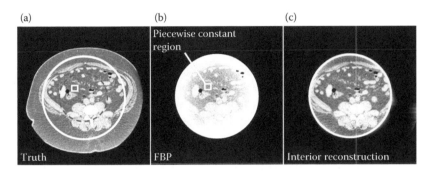

FIGURE 2.20 Reconstructed images (a) without truncation, (b) reconstructed with filtered-backprojection from truncated data, and (c) reconstructed with the interior ROI reconstruction method from Taguchi et al. (2011). Data inside the full circle was acquired from a moderately truncated detector. Data inside the dashed circle was acquired from a severely truncated detector. The rectangle indicates the *a priori* region. (Copyright AAPM (2011). Used with permission from Taguchi, K. et al., 2011. *Medical Physics*, 38(3), 1307–1312.)

may occur if the patient is positioned slightly off center or if the filter is not optimally adjusted to the varying patient size. The measured counts may then not be accurate even with sophisticated compensation schemes to account for pulse-pileup and other degradation effects. Reconstructed images from such inaccurate data can result in severe artifacts. The problem of high count rates outside the object boundaries can be seen as a particular case of the *interior reconstruction problem*. The projection data are not physically truncated but suffer from a shortage of sufficiently high-quality data. Recently, two important algorithms were developed for the interior problem. First, when a small region located inside the region of interest (ROI) is known *a priori*, then the ROI image can be reconstructed exactly using a differentiated backprojection framework (DBP) (Courdurier et al. 2008; Kudo et al. 2008; Yu et al. 2008; Wang et al. 2009). Second, if the ROI is piece-wise constant, an exact image can be reconstructed using a *total variation minimization* (TV-min) approach without a prior (Yu and Wang 2009; Yu et al. 2009). Unfortunately, clinical data satisfies neither of these two requirements. However, it has been shown that a

quasi-exact ROI image can be reconstructed even from noisy clinical CT projections by sequentially using filtered-backprojection (FBP), TV-min, and DBP (Taguchi et al. 2011). Pixel values from a very small flat region obtained by the TV-min image are used as *a priori* knowledge during the DBP reconstruction. Figure 2.20 shows the results of this algorithm. For comparison, a clinical image is reconstructed without data truncation outside the ROI (a). If the data is truncated, standard FBP results in severe cupping artifacts (b). The interior ROI reconstruction method results in an ROI image that is comparable to the untruncated image.

2.3.4.3 Optimal Energy Binning and Weighting

Spectral CT acquires image data in several energy bins. Whereas EIDs intrinsically weight each photon by its energy, thus giving more weight to the high-energy region which carries less contrast, PCDs intrinsically weight each energy bin equally. Since the largest contrast is contained in the lower-energy bins, one can conclude that neither EIDs nor PCDs are detectors with optimal energy weighting. In the case of PCDs,

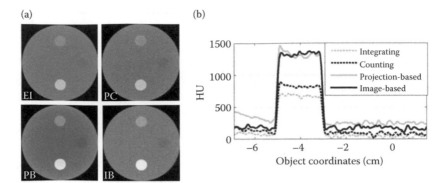

FIGURE 2.21 (a) Breast phantom images reconstructed with energy-integrating (EI), photon counting (PC), projection-based optimal weighting (PB), and image-based optimal weighting (IB). The optimal projection-based weights were optimized for the task of depicting $CaCO_3$ in breast tissue. Level (HU): EI = 105; PC = 180; PB = 450; IB = 185; width (HU): 2500. (b) Central vertical profile through the $CaCO_3$ element. (Copyright AAPM (2009). Used with permission from Schmidt, T.G., 2009. *Medical Physics*, 36(7), 3018–3027.)

however, a weight can easily be applied to each bin before reconstruction of the final image. This approach is effective if cross talk across the energy bins due to the PPE and the energy response or spectral response function is minimal. The question then arises about how to choose the energy thresholds and whether any weights should be assigned to each bin. There is no general answer to this question as the details depend on the specific imaging task, the relevant image quality index, the scanned materials, and many other factors. For example, in contrast imaging with gadolinium, a threshold should be set at the K-edge of the material to maximize the contrast difference. For other applications, it may be beneficial to choose the bins such that each bin contains about the same number of photons. Wang and Pelc (Wang and Pelc 2009) introduced a framework that allows calculation of both the optimal bin thresholds and the optimal bin weights if the detector properties (number of energy thresholds) and the materials are provided. An interesting finding is that it may be beneficial to leave gaps between the energy bins.

Other authors have found optimal energy weighting for specific tasks and assumptions about the detector for either weighting in the projection domain or the image domain (Giersch et al. 2004; Shikhaliev 2005, 2006, 2008a, Schmidt 2009). An illuminating example is given in Schmidt (2009), where a simulated breast phantom is reconstructed using EIDs and PCDs without weighting and with optimal weighting in the projection and image domain (Figure 2.21). It is obvious how the contrast is improved for the optimal weighting technique.

2.4 Clinical Benefits and Applications of Spectral CT with PCDs

In this section, we discuss some of the benefits and applications of spectral CT with PCDs for clinical imaging. There are five areas of interest: (1) CNR, contrast and dose of the reconstructed images, (2) quantitative CT imaging, (3) beam-hardening effects

with contrast, (4) K-edge imaging and simultaneous multiagent imaging, and (5) molecular CT with nanoparticle contrast agents.

2.4.1 Image Quality (CNR, Contrast, Dose)

PCDs introduce less noise and enhance the contrast between tissue types. As a result, PCDs lead to an improved CNR at the same radiation dose. Conversely, the radiation dose can be lowered while maintaining the CNR at the level of current systems, although the achievable dose savings depend strongly on the particular circumstances. Simulations of a specific case have shown that a dose reduction of up to 40% can be obtained using PCDs and optimal energy weighting (Cahn et al. 1999; Lundqvist et al. 2001).

2.4.2 Quantitative CT

Spectral CT will make CT imaging more quantitative. In CT, the projection data is first reconstructed as distributions of linear attenuation coefficients. For practical purposes, these values are then converted to Hounsfield units (HU) according to

$$HU = \frac{\mu_x - \mu_{water}}{\mu_{water}} \times 1000, \quad (2.7)$$

where μ_x is the linear attenuation coefficient of a particular image pixel and μ_{water} is the linear attenuation coefficient of water. However, the linear attenuation coefficient is energy dependent and the energy is not well defined. The effective energy of the spectrum incident on the patient is altered as the spectrum changes during transmission due to the energy- and material-dependent attenuation. Even for a homogeneous slab of material, the spectrum and effective energy would be different for each pixel, because each projection ray traverses different thicknesses of the (bowtie) filter material. It is therefore difficult to establish the absolute physical meaning of a given pixel value or to relate the value to the underlying physics of attenuation effects like the photoelectric effect and Compton scatter. Using the concepts of material decomposition, as discussed in Section 2.3.3, it is possible to extract quantitative information about the scanned object, such as the material composition as represented by the densities of a limited number of basis materials, the chemical composition by calculating both the atomic number and the electron density, or by representing the image at a chosen monochromatic energy (Alvarez and Macovski 1976; Heismann et al. 2003; Heismann and Balda 2009). An example of a clinical application that requires quantitative imaging is cardiac perfusion CT, where the concentration of the contrast agent needs to be determined accurately.

2.4.3 Beam-Hardening Effect with Contrast Agents

Although beam-hardening corrections for water and bone have reached a mature level (Hsieh et al. 2000), artifacts caused by contrast agents remain problematic for cardiac imaging (So et al. 2009; Stenner et al. 2010). Distortions on the order of 35 HU have been observed, which is larger than the typical enhancement of the myocardium due to contrast agents (So et al. 2009). The authors of So et al. (2009) point out that the attenuation time curve may be underestimated by 20–40 HU leading to an underestimation of myocardial perfusion. Spectral CT can address these beam-hardening problems and reduce artifacts for soft plaque, calcium and bone, and contrast-enhanced blood vessels.

2.4.4 K-Edge Imaging and Simultaneous Multiagent Imaging

K-edge imaging can be used to identify and quantify specific materials with a K-edge in the diagnostic energy range, for example, contrast agents, according to the mechanism described in Section 2.3.2. Due to the discontinuity in the linear attenuation, a third basis function is needed for accurate material decomposition and quantification of the K-edge material. Thus, spectral CT using a PCD that has at least three energy thresholds can provide contrast-enhanced images in a single scan. This eliminates the need of the traditional perfusion CT method to subtract a baseline image without contrast agent and also eliminates the related errors and noise from mapping the nonenhanced image to the contrast-enhanced image for subtraction. PCDs with more than three energy bins would allow imaging several K-edge contrast agents simultaneously (Schlomka et al. 2008). This way the assessment of multiple functional properties in a single scan becomes possible using a single imaging modality.

2.4.5 Molecular CT with Nanoparticle Contrast Agents

So far, x-ray-based CT has not been considered a modality for molecular or functional imaging due to a lack of targeted contrast agents. Instead, molecular imaging has been the realm of nuclear medicine imaging and magnetic resonance imaging (MRI). Recently, new types of contrast agents have been developed that specifically target certain biological processes and provide adequate attenuation to be visible in x-ray CT. Combined with the improved K-edge imaging capabilities of PCDs, these novel nanoparticles can make the growing field of molecular imaging accessible to CT. The nanoparticles are created by "coating" a high-Z material with a target-specific biomarker. Elements with high atomic numbers like gold or bismuth are preferable because they attenuate more photons and their K-edge is in an energy range where typical x-ray spectra used in clinical CT have a maximum of photons. Potential applications for nanoparticle spectral CT include tumor imaging (Popovtzer et al. 2008) and imaging of soft plaques in coronary arteries that may lead to a heart attack.

2.4.6 Imaging of Coronary Plaques

We elaborate on the previously mentioned application of using nanoparticle contrast agents to image coronary plaques with CT. With cardiac diseases being the leading cause of death in the Western world, the identification and differentiation of plaques has the potential to be a key application for spectral CT with PCDs. Whereas calcified plaques are stable and pose a smaller risk of causing ischemia or cardiac arrest, soft (atherosclerotic) plaque is prone to rupture unpredictably, leading to the formation of a thrombus with the risk of complete vessel occlusion and sudden cardiac death. Conventional CT has a high negative predictive value to exclude coronary diseases; however, if plaque is already present, CT has insufficient sensitivity to detect unstable plaques among calcium deposits and thrombi caused by ruptured unstable plaques. The challenge is therefore to selectively image atherosclerotic plaque and detect small thrombi that did not lead to any symptoms. Bismuth nanoparticles were developed recently (Pan et al. 2010) that target intravascular thrombi. The nanoparticles will overcome the inherent problem of low contrast between soft tissues and enhance the detectability of soft plaque using K-edge imaging with PCDs. It is challenging to find a good compromise between bio-compatibility of the biomarker and good imaging qualities. High-Z materials are preferred to push the K-edge into the diagnostic energy range and large concentrations are needed so that the material is detectable. Some heavy metals are potential options but they are toxic. Therefore, they need to be encapsulated in biocompatible materials that can be adjusted for specific targets. Examples are nanoparticles based on bismuth that attach to specific molecules of micro thrombi. Spectral imaging with material decomposition allows for separation of the bismuth component from tissues that otherwise have similar attenuation. Figure 2.22 shows a calcified clot target with bismuth nanoparticles (Pan et al. 2010). The bismuth image is overlaid in color and makes it easy to distinguish the soft plaque from the calcium deposit.

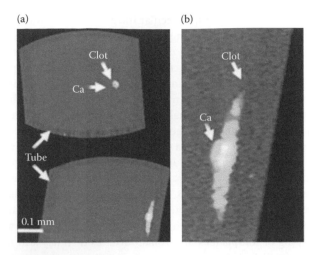

FIGURE 2.22 **(See color insert.)** Bismuth nanoparticle contrast agent highlighting soft plaque in a calcified clot. (Pan, D. et al. *Angewandte Chemie (International Ed. in English)* 9635–9639. 2010. Copyright Wiley-VCH Verlag GmbH & Co. KGaA. Reproduced with permission.)

2.4.7 Other Applications

To conclude, we provide an (incomplete) list of applications of potential interest for spectral CT that are discussed in the literature:

Identification of tissues and chemical composition of materials
Bone removal
Virtual contrast exams
Bone mineral analysis
Kidney stone characterization
Measurement of the fat content of organs
Coronary plaque detection and differentiation
Attenuation correction for nuclear medicine
Cardiac perfusion CT
Assessment of therapy success using biomarkers in oncology
Functional imaging using novel nanoparticle contrast agents

2.5 Conclusions

The underlying principles of spectral CT have been known for a long time and were developed at about the same time as the first CT scanner. Restrictions in detector technology, however, limited the clinical use of spectral CT and only the well-known intensity-based CT became clinical routine. It was not until recent years that dual-energy systems based on conventional detector technology reached a wider market. Despite this development, fully exploiting the energy information will only be possible with a new generation of x-ray detectors. So far, only PCDs are a viable option. Spectral CT with photon counting detectors is a promising technology for a wide range of improvements in clinical CT. They include dose reduction, enhanced image quality, quantitative, material-specific, and functional imaging and many other novel imaging applications for x-ray CT.

References

Alvarez, R.E. and Macovski, A., 1976. Energy-selective reconstructions in x-ray computerised tomography. *Physics in Medicine and Biology*, 21(5), 733–744.

Ambrose, J., 1973. Computerized transverse axial scanning (tomography): Part 2. Clinical application. *British Journal of Radiology*, 46(552), 1023–1047.

Bushberg, J.T. et al., 2011. *The Essential Physics of Medical Imaging*, 3rd Edition, North American edition, Lippincott Williams & Wilkins, Philadelphia, PA.

Buzug, T.M., 2010. *Computed Tomography: From Photon Statistics to Modern Cone-Beam CT.* Softcover reprint of hardcover 1st ed. 2008, Springer.

Cahn, R.N. et al., 1999. Detective quantum efficiency dependence on x-ray energy weighting in mammography. *Medical Physics*, 26(12), 2680.

Cormack, A.M., 1963. Representation of a function by its line integrals, with some radiological applications. *Journal of Applied Physics*, 34(9), 2722–2727.

Cormack, A.M., 1964. Representation of a function by its line integrals, with some radiological applications. II. *Journal of Applied Physics*, 35(10), 2908–2913.

Courdurier, M. et al., 2008. Solving the interior problem of computed tomography using a priori knowledge. *Inverse Problems*, 24(6), 065001.

Eggermont, P.P.B. and LaRiccia, V.N., 2010. *Maximum Penalized Likelihood Estimation: Volume I: Density Estimation.* Softcover reprint of hardcover 1st ed. 2001, Springer.

Feuerlein, S. et al., 2008. Multienergy photon-counting K-edge imaging: Potential for improved luminal depiction in vascular imaging. *Radiology*, 249(3), 1010–1016.

Flohr, T. G. et al., 2008. Image reconstruction and image quality evaluation for a dual source CT scanner. *Medical Physics*, 35(12), 5882–5897.

Flohr, T.G. et al., 2005. First performance evaluation of a dual-source CT (DSCT) system. *European Radiology*, 16(2), 256–268.

Giersch, J., Niederlöhner, D., and Anton, G., 2004. The influence of energy weighting on x-ray imaging quality. *Nuclear Instruments and Methods in Physics Research Section A: Accelerators, Spectrometers, Detectors and Associated Equipment*, 531(1–2), 68–74.

Grupen, C. and Buvat, I. eds., 2011. *Handbook of Particle Detection and Imaging.* 2012th edition, Springer, Berlin, Heidelberg, Germany.

Heismann, B. and Balda, M., 2009. Quantitative image-based spectral reconstruction for computed tomography. *Medical Physics*, 36(10), 4471.

Heismann, B.J., Leppert, J., and Stierstorfer, K., 2003. Density and atomic number measurements with spectral x-ray attenuation method. *Journal of Applied Physics*, 94(3), 2073.

Hounsfield, G.N., 1973. Computerized transverse axial scanning (tomography): Part 1. Description of system. *British Journal of Radiology*, 46(552), 1016–1022.

Hsieh, J. et al., 2000. An iterative approach to the beam hardening correction in cone beam CT. *Medical Physics*, 27(1), 23–29.

Hsieh, J., 2009. *Computed Tomography: Principles, Design, Artifacts, and Recent Advances*, 2nd Edition, 2nd revised ed., SPIE Publications, Bellingham, WA.

Kevles, B.H., 1996. *Naked to the Bone: Medical Imaging in the Twentieth Century*, Rutgers University Press, New Brunswick, NJ.

Knoll, G.F., 2010. *Radiation Detection and Measurement*, 4th edition, Wiley, Hoboken, NJ.

Kudo, H. et al., 2008. Tiny a priori knowledge solves the interior problem in computed tomography. *Physics in Medicine and Biology*, 53(9), 2207–2231.

Lundqvist, M. et al., 2001. Evaluation of a photon-counting x-ray imaging system. *IEEE Transactions on Nuclear Science*, 48(4), 1530–1536.

Overdick, M. et al., 2008. Towards direct conversion detectors for medical imaging with x-rays. In *IEEE Nuclear Science Symposium Conference Record, 2008. NSS '08*, Dresden, Germany. IEEE, pp. 1527–1535.

Pan, D. et al., 2010. Computed tomography in color: NanoK-enhanced spectral CT molecular imaging. *Angewandte Chemie (International Ed. in English)*, 49(50), 9635–9639.

Popovtzer, R. et al., 2008. Targeted gold nanoparticles enable molecular CT imaging of cancer. *Nano Letters*, 8(12), 4593–4596.

Riederer, S.J., 1977. Selective iodine imaging using K-edge energies in computerized x-ray tomography. *Medical Physics*, 4(6), 474.

Roessl, E. et al., 2011. Sensitivity of photon-counting based K-Edge imaging in x-ray computed tomography. *Medical Imaging, IEEE Transactions on*, PP(99), 1.

Roessl, E. and Proksa, R., 2007. K-edge imaging in x-ray computed tomography using multi-bin photon counting detectors. *Physics in Medicine and Biology*, 52(15), 4679–4696.

Ruzsics, B. et al., 2008. Dual-energy CT of the heart for diagnosing coronary artery stenosis and myocardial ischemia-initial experience. *European Radiology*, 18(11), 2414–2424.

Schlomka, J.P. et al., 2008. Experimental feasibility of multi-energy photon-counting K-edge imaging in pre-clinical computed tomography. *Physics in Medicine and Biology*, 53(15), 4031–4047.

Schmidt, T.G., 2009. Optimal "image-based" weighting for energy-resolved CT. *Medical Physics*, 36(7), 3018–3027.

Shikhaliev, P.M., 2005. Beam hardening artefacts in computed tomography with photon counting, charge integrating and energy weighting detectors: A simulation study. *Physics in Medicine and Biology*, 50(24), 5813–5827.

Shikhaliev, P.M., 2008a. Computed tomography with energy-resolved detection: A feasibility study. *Physics in Medicine and Biology*, 53(5), 1475–1495.

Shikhaliev, P.M., 2008b. Energy-resolved computed tomography: First experimental results. *Physics in Medicine and Biology*, 53(20), 5595–5613.

Shikhaliev, P.M., 2006. Tilted angle CZT detector for photon counting/energy weighting x-ray and CT imaging. *Physics in Medicine and Biology*, 51(17), 4267–4287.

So, A. et al., 2009. Beam hardening correction in CT myocardial perfusion measurement. *Physics in Medicine and Biology*, 54(10), 3031–3050.

Srivastava, S. et al., 2012. Spectral response compensation for photon-counting clinical x-ray CT using sinogram restoration. In *Proceedings of SPIE. Medical Imaging 2012: Physics of Medical Imaging*. San Diego, California, USA.

Srivastava, S. and Taguchi, K., 2010. Sinogram restoration algorithm for photon counting clinical x-ray CT with pulse pileup compensation. *Proceedings of the First International Conference on Image Formation in X-Ray Computed Tomography*. Salt Lake City, UT. Available at: http://www.ucair.med.utah.edu/CTmeeting/.

Stenner, P. et al., 2010. Dynamic iterative beam hardening correction (DIBHC) in myocardial perfusion imaging using contrast-enhanced computed tomography. *Investigative Radiology*, 45(6), 314–323.

Swank, R.K., 1973. Absorption and noise in x-ray phosphors. *Journal of Applied Physics*, 44(9), 4199–4203.

Taguchi, K. et al., 2010. An analytical model of the effects of pulse pileup on the energy spectrum recorded by energy resolved photon counting x-ray detectors. *Medical Physics*, 37(8), 3957–3969.

Taguchi, K. et al., 2011. Interior region-of-interest reconstruction using a small, nearly piecewise constant subregion. *Medical Physics*, 38(3), 1307–1312.

Wang, A.S. and Pelc, N.J., 2009. Optimal energy thresholds and weights for separating materials using photon counting x-ray detectors with energy discriminating capabilities. *Proceedings of SPIE*, 7258(1), 725821–725821–12.

Wang, G., Yu, H., and Ye, Y., 2009. A scheme for multisource interior tomography. *Medical Physics*, 36(8), 3575–3581.

Yu, H. et al., 2009. Supplemental analysis on compressed sensing based interior tomography. *Physics in Medicine and Biology*, 54(18), N425–N432.

Yu, H. and Wang, G., 2009. Compressed sensing based interior tomography. *Physics in Medicine and Biology*, 54(9), 2791–2805.

Yu, H., Ye, Y., and Wang, G., 2008. Interior reconstruction using the Truncated Hilbert transform via singular value decomposition. *Journal of X-Ray Science and Technology*, 16(4), 243–251.

Zou, Y. and Silver, M.D., 2008. Analysis of fast kV-switching in dual energy CT using a pre-reconstruction decomposition technique. *Proceedings of SPIE*, 6913(1), 691313–691313–12.

II

X-Ray Phase-Contrast Imaging

3

Grating-Based X-Ray Phase-Contrast Imaging

Peter Modregger
Paul Scherrer Institute
University of Lausanne

Bernd Pinzer
Paul Scherrer Institute

Zhentian Wang
Paul Scherrer Institute

Marco Stampanoni
Paul Scherrer Institute
ETH and University of Zürich

3.1 Introduction

Since its first experimental demonstration in 2002/2003 (David et al., 2002; Momose et al., 2003), the phase-sensitive, hard x-ray imaging technique called *grating interferometry* (GI) has attracted increasing interest from researchers in a wide variety of fields. This interest is due to the unique combination of advantageous imaging characteristics offered by GI. The following three characteristics may be identified as the major contributing factors. First, GI provides a particularly high sensitivity to electron-density variations, which allows for imaging subtle differences within a sample. It has been experimentally demonstrated that electron-density variations down to 0.18 e nm^{-3} are observable (Pfeiffer et al., 2007a). This renders GI especially suitable for soft-tissue specimens from the field of biomedical research. Destruction-free x-ray imaging with GI has been successfully applied to biological samples and tissues such as insects (Pfeiffer et al., 2007c; Weitkamp et al., 2005), rat brains (McDonald et al., 2009; Müller et al., 2008), rat liver (Momose et al., 2006), myocardial structures (Bech et al., 2009), and even human tissues such as breast tissue (Chen et al., 2010b; Stampanoni et al., 2011) and the cerebellum (Schulz et al., 2010).

Second, GI simultaneously provides three complementary types of information about the sample. The absorption contrast relates to the "classic" x-ray image, which is based on variations of the linear attenuation coefficient. The (differential) phase contrast (DPC) is proportional to the (first derivative of the) accumulated phase shift of the x-ray beam after transmission through the sample. The DPC signal is mainly responsible for the high sensitivity of GI (McDonald et al., 2009). The so-called scatter contrast

(also dark-field or visibility-reduction contrast) is a measure of the scatter strength of specific sample details (Pfeiffer et al., 2008), and its origin as well as its interpretation is the subject of ongoing research (Lynch et al., 2011; Yashiro et al., 2010). These complementary contrasts can be taken advantage of in order to increase the ability to detect and discriminate between minute sample variations (Bech et al., 2010; Stampanoni et al., 2011).

Third and most important for the potential broad application of the technique, GI features a comparably low requirement for the spectral width of the utilized x-ray beam, which allows GI to be nearly achromatic (Engelhardt et al., 2008; Weitkamp et al., 2005). This particular advantage yields the possibility of using an x-ray tube as the source (Engelhardt et al., 2007; Pfeiffer et al., 2006), which is accompanied by the amenities of low cost and ready accessibility.

It is the combination of high sensitivity, the availability of complementary contrasts, and the possibility of implementing the GI in a laboratory setup that strongly suggests its future widespread application in materials sciences (e.g., material testing), biomedical research (e.g., monitoring drug effects), or even clinical diagnostics (e.g., mammography).

The basic principle of any phase-sensitive imaging technique lies with the description of the contrast formation process in terms of the complex wave. A complex wave is characterized by the combination of intensity and phase, but in experimental practice, only the intensity is directly accessible. Thus, the experimental setup must offer the capability to transform phase contrast into a detectable intensity contrast in order to provide phase sensitivity. For GI, this is realized in the following way.

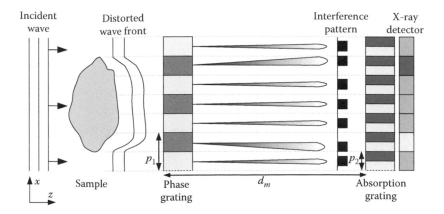

FIGURE 3.1   Sketch of an x-ray grating interferometer.

The GI consists of two line gratings, which are positioned downstream of the sample (see Figure 3.1). The first line grating acts as a beam splitter, dividing the incident beam into several diffraction orders. Due to the coherence of the beam, these diffraction orders interfere constructively at certain distances d downstream of the beam splitter. This provides a rectangular-shaped interference pattern. Placing a sample in front of the first line grating introduces refraction into the beam, which leads to a lateral offset of the interference pattern. Since the period of the interference pattern is too small to be resolved by standard x-ray detectors, a second line grating of purely absorbing structures is used to analyze the interference pattern. The pitch of the absorption grating matches the period of the interference pattern and a lateral offset of the interference pattern is translated into a change of intensity at the detector. Since the lateral offset is proportional to the refraction angle and the refraction angle is, in turn, proportional to the first derivative of the phase of the wave after transmission through the sample, DPC is observed at the detector.

The goal of this presentation is to introduce the reader to the basic concepts of x-ray imaging with GI, thus providing sufficient understanding in order to appreciate the exciting potential benefits of this imaging technique. In Section 3.2, we present the fundamentals of x-ray imaging with GI. We will focus on the theoretical description of the image formation process, on postdetection analysis algorithms and the interpretation of the complementary contrasts. In Section 3.3, the most common variants of experimental implementation of GI will be introduced. Finally, Section 3.4 will present prominent examples of the successful application of GI in the field of biomedical research and the potential of GI for mammography will be discussed.

3.2 Elements of the Image Formation Process

As described in the introduction, the principle of GI is based on the occurrence of a rectangular-shaped interference pattern whose formation process is presented in more detail in the following. Additionally, this section describes the scan and data analysis procedures that are utilized in order to retrieve the three complementary contrasts.

3.2.1 Fresnel Diffraction at Periodic Structures

First discovered by Talbot (1836) and later explained by Rayleigh (1881), the so-called *Talbot effect* relates to coherent diffraction at periodic structures. At certain distances downstream of the diffracting object, self-images of the periodic structure can be observed. In the case of GI, the beam splitter provides a periodic interference pattern and it can be realized by either an absorption grating, which is described by the integer Talbot effect, or more commonly by a phase grating, which is described by the fractional Talbot effect.

For reasons of simplicity, the discussion will be limited to the one-dimensional case and since only relative intensities are of interest, global phase terms as well as constant factors will be omitted. Further, in this section, a monochromatic and collimated incident beam (i.e., the case of perfect coherence) is assumed.

3.2.1.1 Integer Talbot Effect

The interaction of x-rays with matter is described by the complex refractive index $n = 1 - \delta + i\beta$, with δ the refractive index decrement, which corresponds to phase contrast, and β the extinction coefficient, which quantifies the absorption contrast. The complex wave amplitude directly after transmission through the sample, denoted by $D_{in}(x)$, is given by the integration of the refraction index over the beam path (see Figure 3.1):

$$D_{in}(x) = \exp\left(\int ik n(x,z)\, dz\right) \tag{3.1}$$

with $k = 2\pi/\lambda$ being the wavenumber and λ the wavelength. The Fresnel diffraction integral (Born and Wolf, 1999) relates $D_{in}(x)$ to the complex wave amplitude $D_{out}(x)$, which is observable after free-space propagation over the distance z:

$$D_{out}(x) = \int dq\, \hat{D}_{in}(q)\exp\left(-i\frac{q^2}{2k}z\right)\exp(iqx). \tag{3.2}$$

$\hat{D}_{in}(q)$ denotes the Fourier transform of $D_{in}(x)$. If $D_{in}(x)$ describes a periodic diffracting structure with period p_1 (e.g., the first line grating), then its Fourier transform is discrete:

$$\hat{D}_{in}(q) = \sum_m \hat{G}(q)\delta(q - q_m) \qquad (3.3)$$

with $\hat{G}(q)$ the Fourier transform of the structure within one period. $\hat{D}_{in}(q)$ is only nonzero for the spatial frequencies:

$$q_m = m\frac{2\pi}{p_1} \quad \text{with } m \in Z. \qquad (3.4)$$

Inserting Equation 3.3 in Equation 3.2 demonstrates that the propagated wave amplitude $D_{out}(x)$ differs from the input wave field $\hat{D}_{in}(q)$ only by the factor $\exp(-izq_m^2/2k)$. For multiples $z_n = n\tau$ ($n \in Z$) of the Talbot distance τ

$$\tau = \frac{p_1^2}{2\lambda}, \qquad (3.5)$$

this factor equals one and, thus, the original wave field D_{in} is again observable. If the diffracting structure comprises a line grating with purely absorbing structures, the corresponding self-image forms the desired rectangular-shaped interference pattern. Figure 3.2a shows the interference pattern of an absorption grating at different propagation distances (given in terms of n), which is—for obvious reasons—called *the Talbot carpet*. The visibility of the interference pattern is optimal for all integer multiples of the Talbot distance and, thus, these positions would be used in experimental practice.

3.2.1.2 Fractional Talbot Effect

In terms of detection efficiency, the utilization of a beam splitter with absorbing line structures is undesirable. Half of the photons that are transmitted through the sample would be absorbed in the beam-splitter grating and could not contribute to the signal at the detector. But fortunately, the beam splitter can also be realized by a phase-shifting grating, which avoids the unfavorable absorption.

This possibility is based on the *fractional Talbot effect*, which describes Fresnel diffraction at propagation distances that are given as fractions of the Talbot distance τ, that is

$$z = \frac{n_1}{n_2}\tau \quad \text{with } n_1, n_2 \in N. \qquad (3.6)$$

Inserting Equations 3.3 through 3.6 into Equation 3.2 and using Gauss' sum of number theory (i.e., $\exp(-i2\pi\frac{n_1}{n_2}m^2) = \sum_{s=0}^l c_s \exp(-i2\pi m\frac{s}{l})$ (Banaszek et al., 1997)) leads to

$$D_{out}(x) = \sum_m \hat{G}(q_m)e^{iq_m x}\sum_{s=0}^l c_s e^{-i2\pi m\frac{s}{l}} \qquad (3.7)$$

with $l \le n_2$. Equation 3.7 can be understood as a superposition of l copies of the original amplitude, each weighted with the factor

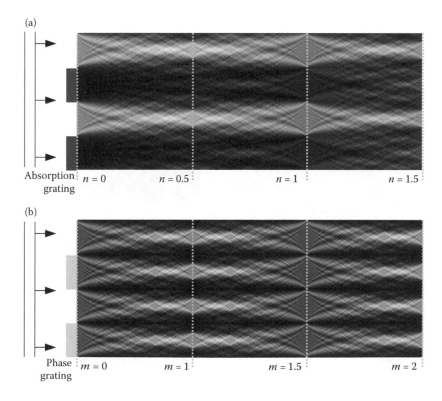

(a)

Absorption grating $n = 0$ $n = 0.5$ $n = 1$ $n = 1.5$

(b)

Phase grating $m = 0$ $m = 1$ $m = 1.5$ $m = 2$

FIGURE 3.2 Talbot carpets for (a) an absorption grating and (b) a phase grating.

$cs = c_s(n_1, n_2)$ and separated by the fraction s/l of the period. As an example, let us consider $l = 4$, which corresponds to $z = \tau/4$, and a beam-splitter grating that shifts the phase of the incident wave by π. In this case, $c_0 = 1$, $c_1 = 0$, $c_2 = 1$, and $c_3 = 0$ holds true and the observable wave amplitude is given by two copies of the original amplitude shifted by half a period. Due to the interference of the two copies, the result will be an observable rectangular-shaped interference pattern with half of the period. An overview of experimental conditions for phase gratings utilizing different phase shifts can be found in Suleski (1997).

In case of a π-shifting phase grating, the rectangular-shaped interference pattern is located at the so-called Lohmann distances d_m

$$d_m = \left(m - \frac{1}{2} \right) \frac{p_1^2}{4\lambda} \quad \text{with } m = 1, 2, \ldots \quad (3.8)$$

with m the Lohmann diffraction order. The corresponding Talbot carpet is shown in Figure 3.2b. Please note the varying definitions of the diffraction order in literature (see, e.g., Weitkamp et al., (2005)).

3.2.2 Data Analysis Procedures and Complementary Contrasts

GI realizes phase sensitivity in the following way. The sample introduces refraction into the incident wave and, consequently, shifts the interference pattern laterally. This lateral shift Δx is proportional to the refraction angle α (i.e., $\Delta x \approx d\alpha$). In turn, the refraction angle is proportional to the first derivative of the phase of the complex wave amplitude after transmission through the sample Φ (see Born and Wolf (1999)):

$$\alpha = -\frac{1}{k} \frac{\partial \Phi}{\partial x}. \quad (3.9)$$

Therefore, (differential) phase sensitivity is established if the lateral shift of the interference pattern can be detected. However, in order to provide high sensitivity, the period of the rectangular-shaped interference pattern is typically around a few microns, which is too small to be resolved by standard x-ray detectors. For this reason, an analyzer grating of purely absorbing line structures is utilized (see Figure 3.1), which essentially scales up the information that is encoded in the interference pattern. In order to maximize the visibility, the pitch p_2 of the analyzer grating should match the period of the interference pattern, that is, $p_2 = p_1/2$ for a π-shifting phase grating and a parallel-beam geometry. Refraction and, thus, phase information becomes visible as brightening and darkening of the corresponding detector pixels (Figure 3.1).

The so-called *phase-stepping technique* constitutes the most widely used scan and data analysis scheme for GI. The scan is performed by laterally scanning one of the gratings in steps that are a fraction of the grating's pitch and acquiring an image at each step. In doing so, at each pixel, one measures a phase-stepping curve (PSC) that shows an oscillatory behavior (Figure 3.3a). This can be understood by realizing that the lateral scan of a rectangular function (e.g., the interference pattern) over another rectangular function (e.g., the transmission function of the analyzer grating) is equivalent to the convolution of the two functions. Thus, the result is an oscillatory triangular function. However, due to the effects of finite spatial coherence, the

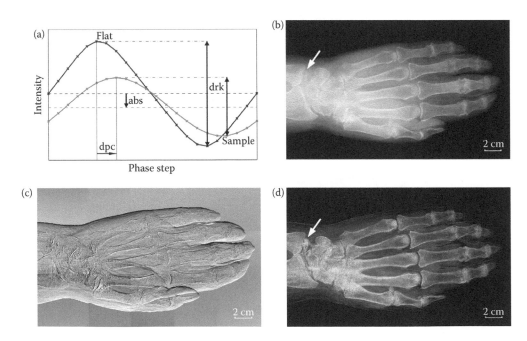

FIGURE 3.3 (a) Phase-stepping curves and extraction of contrasts. Radiographs for three different contrasts of a human hand: (b) absorption, (c) differential phase, and (d) scatter contrast.

interference pattern will not be perfectly rectangular, and thus the convolved pattern will be more sinusoidal than triangular, as depicted in Figure 3.3a.

The resulting PSC can be approximated by a first-order Fourier expansion (Momose, 2004):

$$\text{PSC}(\phi) \approx a + b \sin(\phi - p), \qquad (3.10)$$

where ϕ is the lateral offset of the phase step in terms of radians. The other parameters correspond to the complementary contrasts (Figure 3.3) that are provided by GI. a is the mean of the PSC and relates to the absorption contrast. b is connected to the scatter contrast and denotes the amplitude of the PSC. p constitutes the offset or phase of the PSC and relates to the (differential) phase contrast.

Usually, a flat-field PSC, $f(\phi)$, without the sample, and a sample PSC, $s(\phi)$, are acquired. The parameters a, b, and p are determined by a Fourier analysis approach, which is mathematically equivalent to a least-squares fit. The so-called Fourier-component analysis starts with a discrete Fourier transform of the PSCs with respect to the phase steps ϕ yielding $\hat{f}(q)$ and $\hat{s}(q)$. In the following, the subsequent steps to retrieve the individual contrasts will be discussed separately.

3.2.2.1 Absorption Contrast

The absorption contrast is calculated by

$$A = \frac{\hat{s}(q_0)}{\hat{f}(q_0)} \qquad (3.11)$$

with q_0 the zeroth harmonic Fourier component. By taking the negative logarithm of each projection, A is related to the attenuation coefficient β via the line integral

$$-\log A(x) = k \int \beta(x, z) \, dz, \qquad (3.12)$$

where the beam direction is along z. Thus, the Fourier slice theorem is applicable (Kak and Slaney, 1988), and the 2D distribution of the attenuation coefficient β can be tomographically reconstructed. In Qi et al. (2010), it was presented that β may be expressed as

$$\beta = \left(\bar{m} Z^{\bar{n}} + \bar{b}\right) \rho_e \qquad (3.13)$$

with ρ_e, the electron density, Z, the effective atomic number, and $\bar{m}, \bar{n}, \bar{b}$ parameters that depend only on the details of experimental implementation and not on sample properties.

3.2.2.2 Differential Phase Contrast

The DPC is determined by

$$P = \arg(\hat{s}(q_n)) - \arg(\hat{f}(q_n)) \qquad (3.14)$$

with q_n the n-th harmonic Fourier component and n equal to the number of scanned PSC periods. The obtained DPC value is related to the refraction angle by

$$P = 2\pi \frac{d_m \alpha}{p_2}. \qquad (3.15)$$

However, if the lateral offset Δx of the interference pattern becomes larger than the pitch of the analyzer grating, phase wrapping occurs (Modregger et al., 2011a). This means that P-values outside of the interval $[-\pi, \pi]$ are falsely translated into this interval. An example of phase wrapping is visible in Figure 3.3c as the strong dark/bright contrast at the edges of the hand. Naturally, phase wrapping leads to severe artifacts and, thus, should be avoided if possible. Recently, an approximate correction algorithm has been proposed that utilizes the absorption contrast as a predictor for the phase contrast (Jerjen et al., 2011).

Equation 3.15 also reveals the importance of realizing an interference pattern with the smallest possible period p_2. Since the maximum of P is π and d_m is typically several decimeters, the interference pattern must have a pitch of few micrometers in order to observe nano-radiant refraction angles. By fulfilling this experimental condition, it was demonstrated that minute refraction angles of only 17 nrad are detectable by GI (Pfeiffer et al., 2007a).

The DPC value P is related to the first derivative of the line integral over the refractive index decrement δ, that is

$$P(x) = \frac{\partial}{\partial x} k \int \delta(x, z) \, dz. \qquad (3.16)$$

It has been shown that a modified version of the Fourier slice theorem can account for the differential nature of the phase contrast (Pfeiffer et al., 2007b). Thus, the 2D distribution of the refractive index decrement δ can be reconstructed. For a given photon energy E, δ is given by

$$\delta = \frac{r_0 \hbar^2 c^2}{2\pi E^2} \rho_e \qquad (3.17)$$

with r_0 the classical electron radius, $\hbar$ the reduced Planck's constant, and c the speed of light. Comparing Equation 3.13 with Equation 3.17 reveals the complementary nature of the absorption and the phase contrast. While the absorption contrast is proportional to the product $\rho_e Z^n$, the phase contrast depends only on the electron density ρ_e.

3.2.2.3 Scatter Contrast

Finally, the scatter contrast (also called dark-field (Pfeiffer et al., 2008) or visibility reduction contrast (Yashiro et al., 2010)) is retrieved by

$$B = \frac{\hat{s}(q_n) \hat{f}(q_0)}{\hat{s}(q_0) \hat{f}(q_n)}. \qquad (3.18)$$

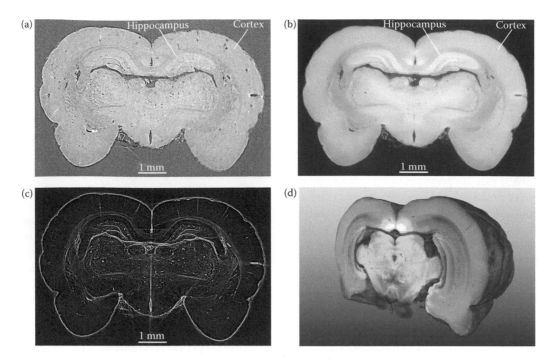

FIGURE 3.4 Tomographic reconstructions of the complementary contrasts provided by grating interferometry of a paraffin-embedded rat brain: (a) absorption contrast, (b) phase contrast, (c) scatter contrast, and (d) a 3D visualization of the phase-contrast data.

Based on the assumption that the scattering of the x-rays is isotropic and describable by a Gaussian distribution, it was demonstrated (Khelashvili et al., 2006) that the scattering width σ fulfills the line integral

$$\sigma(x)^2 = \int S(x,z)\,\mathrm{d}z \qquad (3.19)$$

with S a generalized scattering parameter that quantifies the local scatter strength. Since the scattering width σ is related to the scatter contrast by (Wang et al., 2009)

$$\sigma^2 = -\frac{1}{2\pi^2}\left(\frac{p_2}{d_m}\right)^2 \ln B, \qquad (3.20)$$

the 2D distribution of S may be tomographically reconstructed in a way that closely resembles the procedure for the absorption contrast. However, if the assumption of isotropic Gaussian scattering is violated, artifacts appear in the reconstructions (Modregger et al., 2011b).

The physical interpretation of the scatter contrast is still being investigated. So far, it is clear that the scatter contrast corresponds to the width of the underlying but unresolved scatter distribution within a detector pixel (Modregger et al., 2012a; Yashiro et al., 2010). Recent investigations also suggest that the scatter contrast provides particle size sensitivity (Lynch et al., 2011).

Figure 3.3 shows the radiographs for the complementary contrasts of a human hand, which were retrieved by GI. Soft tissue is enhanced in the DPC image (Figure 3.3c), while the hand

bones are clearly visible in the absorption as well as in the scatter image. While the soft tissue provides a still well-defined signal in the absorption contrast, the scattering of the tissue is obviously too weak to produce a significant contribution to the scatter contrast. So, generally, although bones are already well represented in absorption contrast, scatter imaging can potentially yield a complementary and even enhanced contrast, as shown by the soft-tissue calcification in the region of the triangular fibrocartilage (indicated by the arrows).

Figure 3.4 shows tomographic reconstructions of the complementary contrasts for a paraffin-embedded rat brain. The sample consists mainly of low-Z material and in such a case the superiority of the phase contrast over the absorption contrast is apparent by the visible noise in the images. Further, the contrast between morphological regions of the brain (e.g., cortex and hippocampus) is much better defined in the phase image, which significantly simplifies quantitative data analysis. The scatter contrast is dominated by the edges of the samples. This is understandable since the scattering distribution within a pixel is broad at the edges of the sample, which leads to a significant scatter contrast in the projections and consequently in the reconstruction.

3.2.2.4 Alternative Scan Procedures

In addition to the phase-stepping method, several scan and data analysis procedures have been suggested. These approaches aim at reducing the necessary number of acquisitions for phase retrieval. They avoid the need for scanning one grating, which potentially simplifies the experimental setup and image acquisition.

The reverse-projection method (RPM) takes advantage of the fact that the refraction direction of a certain sample feature is inverted if the sample is rotated by 180° (Zhu et al., 2010). Utilizing this fact in combination with a linear approximation of the PSC yielded the basis for a postdetection algorithm that allows for the retrieval of the absorption and the phase signal (but not the scatter image) from only two projection images.

The Moiré fringe method (MFM) is based on the occurrence of a Moiré pattern in the experimental raw images, when the lines of the two gratings are not parallel to each other. Information about all three contrasts is encoded in the Moiré pattern, which can be retrieved by two-dimensional Fourier analysis of a single image (Momose et al., 2009b). But at the same time, the MFM unavoidably leads to a loss of spatial resolution in the direction perpendicular to the Moiré fringes.

However, the reduced number of acquisitions required for phase retrieval does not necessarily imply a reduction of delivered dose. In fact, the defining quantity for dose efficiency (i.e., sensitivity per delivered dose) is not the number of acquired images, but the number of absorbed photons. It stands to reason that the dose efficiency for the alternative scan procedures is comparable to the phase-stepping technique.

3.3 Experimental Implementation

In this section, details about the experimental implementation of GI are introduced. After presenting the influence of finite coherence, the experimental conditions at a synchrotron beamline (i.e., parallel-beam geometry) as well as with a laboratory source (i.e., cone-beam geometry) are discussed. Further, the manufacturing of high-quality gratings, which are essential for the performance of GI, will be briefly introduced.

3.3.1 Influence of Finite Coherence

A successful experimental implementation of GI requires accounting for the effects of finite coherence and a divergent beam. The implications are briefly discussed in the following.

3.3.1.1 Requirements for Temporal Coherence

A major advantage of the grating interferometer is the low requirements on temporal coherence, which opens up the possibility of phase-sensitive x-ray imaging with x-ray tubes with reasonable acquisition times (Engelhardt et al., 2007; Pfeiffer et al., 2006). It has been shown that the spectral width of the incident beam ζ can be as large as (Weitkamp et al., 2005)

$$\zeta = \frac{1}{2m - 1},\tag{3.21}$$

which is 1 for the first Lohmann order and still as large as 6×10^{-2} for the 5th Lohmann order.

3.3.1.2 Requirements for Spatial Coherence

The requirement for the spatial coherence is more demanding. Spatial coherence relates to the lateral blurring of interference patterns, which is due to an extended source size (Born and Wolf, 1999). Simple geometrical considerations show that lateral blurring can be quantified by the projected source size s at the position of the detector (Weitkamp et al., 2006). The projected source size is given by

$$s = dS/l_s\tag{3.22}$$

with d, the beam splitter to detector distance, S, the source size, and l_s, the source to beam splitter distance. In the case of GI, the projected source size s must not be larger than the period of the interference pattern p_2 because otherwise the interference pattern would be smeared out and no contrast would be observable. For synchrotron radiation-based setups with typical lengths of several 10 m, this restriction does not play a crucial role. However, the situation is different for laboratory-based setups.

3.3.1.3 Corrections for Divergent Beams

Up to now, all equations assumed parallel-beam geometry. However, for a cone-beam geometry, the effect of a curved wavefront must be accounted for by adjusting the pitches of the gratings and the intergrating distance. It has been shown (Engelhardt et al., 2008) that the Talbot distance rescales to

$$d_m^* = \frac{l_s d_m}{l_s - d_m}\tag{3.23}$$

and the pitch of the gratings are related by

$$p_2^* = \frac{l_s}{l_s - d_m} \frac{p_1}{2}.\tag{3.24}$$

3.3.2 Instruments Utilizing Synchrotron Radiation

Synchrotron radiation-based setups utilize highly brilliant x-ray sources that offer the advantage of a small beam divergence and high photon flux, which leads to small scan times. In recent years, GI was implemented at several beamlines around the world: at ID19 of the European Synchrotron Radiation facility in France (Weitkamp et al., 2010), at 20XU of Spring-8 in Japan (Momose et al., 2006), at the beamlines 32-ID and 2-BM of the Advanced Photon Source in the United States (Lynch et al., 2011), and at TOMCAT of the Swiss Light Source in Switzerland (McDonald et al., 2009). As an example, the latter will be described in more detail.

Figure 3.5 shows the experimental implementation of GI at the TOMCAT beamline. The x-ray source (not shown) is a 2.9 T bending magnet 25 m upstream from the phase grating. The photon energy (e.g., 25 keV) is selected by a double multilayer monochromator. Both the phase grating (pitch: $p_1 = 3.981$ μm) and the absorption grating (pitch: $p_2 = 2$ μm) were manufactured by the Laboratory for Micro- and Nanotechnology (David et al., 2002). Typically, a Lohmann diffraction order of $m = 2$ is chosen,

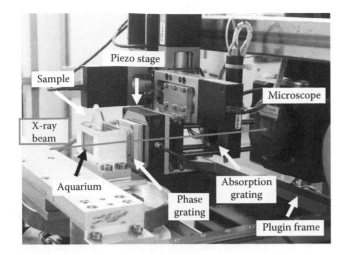

FIGURE 3.5 Grating interferometry at the TOMCAT beamline (Swiss Light Source).

which corresponds to an intergrating distance of $d_2 = 119$ mm. The sample is embedded in 2% agarose gel for mechanical fixation and located in an Eppendorf cylinder. The sample holder is submerged in water for refractive index matching in order to avoid phase wrapping at the edges of the sample. After transmission through the sample and the gratings, the x-ray beam is converted to visible light by a scintillator and subsequently imaged to a charged-coupled device (CCD). A pco.2000 (PCO AG, Kelheim, Germany) CCD camera is used for the detection of the optical photons. It offers an effective dynamic range of approximately 12 bits and a pixel size of 7.4 μm. The field of view is 12×3.7 mm, which results in a total scanned volume of $12 \times 12 \times 3.7$ mm with an isotropic voxel size of 7.4 μm. The images shown in Figures 3.4, 3.10, and 3.11 were acquired with the described setup.

3.3.3 Instruments Based on X-Ray Tubes

In comparison to synchrotron radiation-based setups, instruments utilizing an x-ray tube as the source provide the advantage of low cost and ready accessibility. Thus, for widespread use of any kind of x-ray imaging technique, its feasibility using x-ray tubes is crucial. In this section, two variants of laboratory-based GI setups will be presented.

3.3.3.1 Microfocus X-Ray Tube Setup

As discussed above, the projected source size s at the position of the analyzer grating (Equation 3.22) must be significantly smaller than the period of the interference pattern p_2. For instance, a total setup length $L = l_s + d$ of 2 m, an intergrating distance of $d = 0.1$ m and a period of $p_2 = 2$ μm restrict the source size to values smaller than 45 μm. Thus, microfocus x-ray tubes with typical source sizes in the order of a few microns are the best choice for setups that do not use additional optical elements.

The cone-beam configuration of a microfocus setup (Figure 3.6) opens up the possibility to utilize the geometrical magnification. The closer the sample is placed to the source, the larger it will appear on the detector, with the magnification factor M given by $M = L/l_s$. Since magnification factors of ten and more are feasible, it is possible to achieve micrometer resolution with x-ray detectors providing pixel sizes on the order of several 10 μ. However, increasing the magnification inevitably decreases the sensitivity to refraction angles. This effect is due to the fact that the lateral shift of the interference pattern decreases with a decreasing source-to-sample distance (Engelhardt et al., 2007). Thus, in experiments, a compromise between the spatial resolution and sensitivity must be found. Figure 3.7 shows example radiographs of a wasp that were obtained with a microfocus setup (Thüring et al., 2011b).

A significant issue of GI with a microfocus tube is the limited field of view (FoV) in the direction perpendicular to the grating lines (i.e., the direction of phase sensitivity). Due to the inclined illumination direction at the borders of the flat analyzer grating, the curved wavefront leads to changes in its transmission function. Consequently, the visibility drops significantly at the border, thus limiting the useable FoV. Therefore, experimental setups were proposed that utilize bent gratings in order to account for the wavefront curvature.

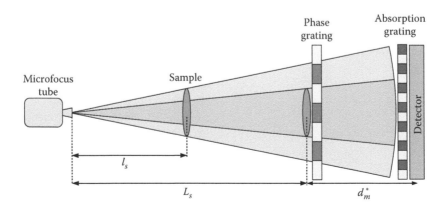

FIGURE 3.6 Grating interferometry with a microfocus x-ray tube.

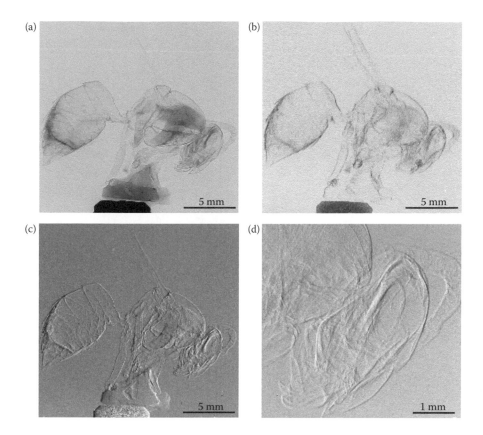

FIGURE 3.7 X-ray radiographs of a wasp acquired with a compact microfocus setup: (a) absorption contrast, (b) scatter contrast, (c) DPC, and (d) DPC with fourfold magnification. (Image courtesy of T. Thüring; from Thüring, T. et al. 2011b. *Proc. SPIE: Phys. Med. Imaging* 7691:79611G.)

A doubling of the usable FoV in the direction of phase sensitivity was already demonstrated (Revol et al., 2010; Thüring et al., 2011a).

3.3.3.2 Talbot–Lau Interferometer

The Talbot–Lau interferometer constitutes an experimental implementation of GI in a laboratory-based setup that allows for larger sources (Chen et al., 2010a; Momose et al., 2009a; Pfeiffer et al., 2006). By including an additional absorption grating close to the source, the limitation of source size, which is implied by Equation 3.22, can be circumvented (Figure 3.8). The source grating creates an array of individually coherent line sources. The line sources are mutually incoherent but if the pitch of the source grating p_0 fulfills the condition

$$p_0 = p_2 \frac{l}{d_m}, \tag{3.25}$$

neighboring line sources provide interference patterns that are shifted by exactly one period. Therefore, all line sources produce the same interference pattern, thus preserving spatial coherence. Simultaneously, the photon flux is increased by a factor that is equal to the number of illuminated periods of the source grating. For instance, a horizontal source size of 0.8 mm and source

grating pitch of 127 µm (Pfeiffer et al., 2006) lead to an increase of photon flux by a factor of 6.

3.3.4 Grating Manufacturing

The availability of high-quality gratings is crucial for the performance of a GI setup in terms of sensitivity, resolution, and artifact-free imaging. A major challenge in manufacturing the gratings is the required high aspect ratio of micrometer line structures. For example, a phase grating that is made of Si and is designed for a photon energy of 40 keV implies a structure height $h = 57$ µm in order to realize a phase shift of π. Thus, a typical pitch of $p_1 = 4$ µm leads to an aspect ratio of $A_r = 2h/p_1 \approx 28$. Since the pitch of the analyzer grating is even smaller, aspect ratios of 50 and higher are required for an excellent performance.

In recent years, the production process of the gratings was optimized with respect to the specific combinations of utilized materials (Au for the source and absorption gratings, Si for the phase gratings) and desired aspect ratios. Using a multistep procedure, which consists of photo lithography, anisotropic wet edging, and electroplating (David et al., 2007), high-quality line gratings are now available to fulfill the requirements of GI. Figure 3.9 shows the scanning electron micrographs for two different types of line gratings.

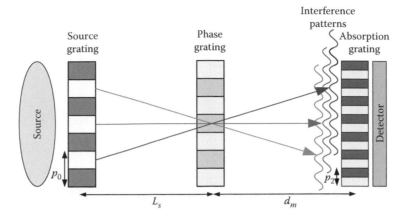

FIGURE 3.8 The Talbot–Lau interferometer.

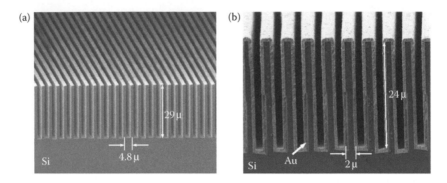

FIGURE 3.9 Scanning electron micrographs of different line gratings: (a) phase grating and (b) analyzer gratings. (Image courtesy of C. David and S. Rutishauser.)

3.4 Biomedical Research and Potential Clinical Application

Due to the unprecedented sensitivity to electron-density variations, GI has already demonstrated its potential for biomedical research. For hard x-rays (i.e., photon energy between 10 and 120 keV) and soft tissues, δ is typically three orders of magnitude larger than β, which directly translates to a corresponding increase in contrast and tissue detectability. Simultaneously, GI offers access to three complementary contrasts (see Section 3.2.2), which opens up the possibility to identify otherwise undetectable sample features. Obviously, this significantly increases the applicability of GI to a wider range of scientific questions. In order to demonstrate these beneficial characteristics of GI, three examples of utilizing GI in the context of biomedical research and potential clinical applications are presented in this section.

3.4.1 Imaging Aortic Heart Valve Calcifications

While the aortic heart valve normally has three leaflets, about 1–2% of the human population have a genetic defect that results in the formation of only two leaflets (termed bicuspid aortic valve (Tzemos et al., 2008)). This is a serious condition that may

severely affect quality of life; heart surgery is often necessary in order to replace the defective heart valve with an artificial valve. Bicuspid aortic valves commonly show calcifications, which constitute the root cause of performance reduction.

Up to now, immunohistochemistry has been the imaging technique of choice for *ex vivo* microscale tissue analysis of extracted heart valves. Although immunohistochemistry offers the capability to readily distinguish between different tissues types, it constitutes a destructive method and provides only limited access to three-dimensional information. In order to overcome these limitations, GI was recently utilized for imaging nodules derived from bicuspid aortic heart valves (Modregger et al., 2012b).

The major challenge of this work was to demonstrate that GI provides a sufficient distinction between tissue types of interest. This was done by taking advantage of the complementary contrasts. More specifically, the availability of the absorption contrast ($\beta \propto Z^n \rho_e$, Equation 3.13) and the phase contrast ($\delta \propto \rho_e$, Equation 3.17) opens up the possibility to retrieve the effective atomic number map $\tilde{Z}$ with

$$\tilde{Z} = \frac{\beta}{\delta} = f(Z),$$

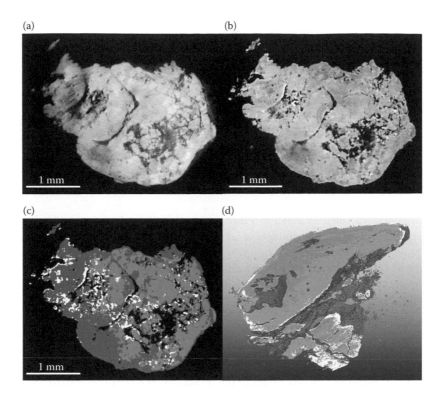

FIGURE 3.10 **(See color insert.)** Microstructural analysis of a calcified nodule derived from a bicuspid aortic heart valve: (a) electron density map, (b) effective atomic number map, (c) segmented tissue of the same slice, and (d) 3D visualization of the entire segmented sample. Blue: myxoid tissue, red: dense tissue, and white: regions of elevated calcium concentration.

which is only a function of the material property Z (i.e., the atomic number). By combining the information of the accessible contrasts (i.e., β, δ, and $\tilde{Z}$), it was possible to segment otherwise indistinguishable tissue types.

Figure 3.10 shows the approach and the result of the segmentation process. It is obvious that the electron density map (a) and the effective atomic number map (b) provide complementary information about the inner morphology of a calcified nodule. On the one hand, dense and myxoid (i.e., loose) connective tissue appear as bright and dark areas in (a), while being almost indistinguishable in (b). On the other hand, regions of elevated calcium concentration appear as bright spots in (b), while being absent in (a). Figure 3.10c and d presents the result of combining these complementary contrasts into a single segmentation. This example demonstrates that GI provides the possibility of destruction-free, three-dimensional microstructural analysis with sufficient sensitivity to distinguish among tissue types. Therefore, GI could play a significant role in future medical research of cardiovascular diseases with the potential to be used complementary or even as an alternative to histology.

3.4.2 Imaging Alzheimer's Plaque Deposits in a Mouse Model

The unique pathological characteristic of Alzheimer's disease is the accumulation of a specific peptide from the amyloid precursor protein into spherical agglomerates—so-called amyloid plaques—in the extracellular space of the brain (Reitz et al., 2011). These plaques are believed to play an important role in the biochemical chain that finally leads to cognitive impairment and dementia. The problem with plaque detection and visualization is their small size: in humans, amyloid plaques are typically on the order of 25 μm. Imaging techniques like positron emission tomography provide an overview of the plaque load in a living patient, but imaging single plaques remains a big challenge.

Mouse models with the same pathology have been developed to study various aspects of plaques, including possible treatments to dissolve them. But in order to reveal the precise amount and locations of plaques, the brain needs to be sliced and treated with dyes. In certain mouse models with larger plaques (~75 μm and larger), magnetic resonance imaging has been shown to be able to detect some plaques, but this is at the limit of resolution.

With the sensitivity and resolution provided by GI, the plaque population of a mouse brain can be measured and automatically segmented *ex vivo*, revealing complete structural information. Figure 3.11a shows a close up of the phase-data reconstruction, which was acquired by GI with eight phase steps at 25 keV. The coronal cross section of the entire brain fit into the field of view, which permitted acquisition of a complete overview of the plaque distribution. The plaque pathology is clearly visible (i.e., small, dark spots). Image processing for normalization of the images and automated segmentation revealed the set of amyloid plaques, which is shown within a region of interest—the cerebral cortex—in panel (b). The measurements as well

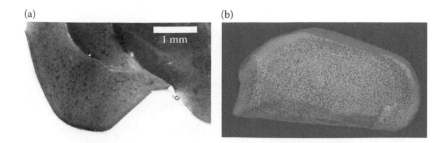

FIGURE 3.11 (**See color insert.**) Amyloid plaque deposit in the cerebral cortex of a mouse model as revealed by GI: (a) zoom into a phase reconstruction slice and (b) a 3D rendering of the segmented mouse brain with the cerebral cortex in gray and amyloid plaques in red.

as the image-processing steps are described in detail in Piner et al. (2012). This example again elucidates the advantages of GI for biomedical imaging: the sensitivity to distinguish different types of soft tissue, coupled with high resolution and a large field of view.

3.4.3 Potential for Mammography

So far, in clinical applications like radiography, mammography, or computed tomography, only x-ray attenuation has been used for image generation. As demonstrated in this chapter, GI offers a significantly higher sensitivity to density variations in soft tissues at clinically relevant photon energies. Thus, GI can potentially revolutionize the radiological approach to medical imaging. Further, the availability of the scatter contrast can be used to increase the diagnostic utility of minute sample features (see Figure 3.3). In the following, the exciting potential of GI for future mammography is suggested with an example.

Breast cancer is the most common cancer in women and the second leading cause of cancer deaths (Kamangar et al., 2006). International standards for diagnostics and treatment are not stringently followed and it is estimated that 35% of all breast cancer deaths in Europe could be avoided if optimal diagnostic and therapeutic procedures were always applied. In the United States, the age-adjusted incidence rates of breast cancer rose rapidly between 1980 and 1987, and they are also rising rapidly in several Asian countries (e.g., Japan), which currently have the lowest incidence rates. This may indicate that, besides genetic predisposition, environmental factors, for example, estrogen-replacement therapy, oral contraceptives, dietary factors, or alcohol consumption may play an additional role (Kelsey, 1993). Improving the sensitivity and specificity of the mammographic examination without increasing the dose applied to the patient is therefore of crucial importance. Recently, an exploratory study showed the world's first phase-contrast mammography images of native (not fixed) human breast using a conventional x-ray source (Stampanoni et al., 2011).

Figure 3.12 clearly shows how the high-frequency components in a mammographic image can be enhanced when the differential phase-contrast signal is merged into the conventional absorption signal, to yield a much sharper image. The novel approach yields complementary and otherwise inaccessible information on the electron density distribution and the small-angle scattering power of the sample at the microscopic

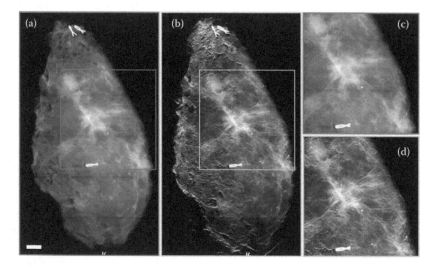

FIGURE 3.12 Freshly dissected tissue (mastectomy specimens) imaged with conventional (a) and phase-contrast-enhanced mammography (b). Panels (c) and (d) show the corresponding magnified regions marked in (a) and (b), respectively. Images (a) and (b) have been obtained with the same dose. Scale bar in (a) is 2 cm. Data has been acquired with a conventional x-ray source with a Talbot–Lau interferometer operated at 28 keV.

scale. Potentially, this could answer clinically relevant, yet unresolved, questions such as the capability to unequivocally discern between malignant, premalignant changes, and postoperative scars or to identify cancer-invaded regions within healthy tissue.

Acknowledgments

The authors like to acknowledge Thomas Thüring, Simon Rutishauser, and Christian David of the Paul Scherrer Institut (Villigen, Switzerland) for their contributions in the form of raw data and images.

References

Banaszek, K., K. Wodkiewicz, and W. Schleich. 1997. Fractional Talbot effect in phase space: A compact summation formula. *America* 117:4.

Bech, M., O. Bunk, T. Donath et al. 2010. Quantitative x-ray dark-field computed tomography. *Phys. Med. Biol.* 55:5529.

Bech, M., T.H. Jensen, R. Feidenhans'l et al. 2009. Soft-tissue phase-contrast tomography with an x-ray tube source. *Phys. Med. Biol.* 54:2747.

Born, M., and E. Wolf. 1999. *Principles of Optics.* Cambridge University Press, Oxford, England.

Chen, G.-H., N. Bevins, J. Zambelli, and Z. Qi. 2010a. Small-angle scattering computed tomography (SAS-CT) using a Talbot-Lau interferometer and a rotating anode x-ray tube: Theory and experiments. *Opt. Express* 18:12960.

Chen, R.C., R. Longo, L. Rigon et al. 2010b. Measurement of the linear attenuation coefficients of breast tissues by synchrotron radiation computed tomography. *Phys. Med. Biol.* 55:4993.

David, C., J. Bruder, T. Rohbeck et al. 2007. Fabrication of diffraction gratings for hard x-ray phase contrast imaging. *Microelectron. Eng.* 84(5–8):1172.

David, C., B. Nöhammer, H.H. Solak, and E. Ziegler. 2002. Differential x-ray phase contrast imaging using a shearing interferometer. *Appl. Phys. Lett.* 81:3287.

Engelhardt, M., J. Baumann, M. Schuster et al. 2007. High-resolution differential phase contrast imaging using a magnifying projection geometry with a microfocus x-ray source. *Appl. Phys. Lett.* 90:224101.

Engelhardt, M., C. Kottler, O. Bunk et al. 2008. The fractional Talbot effect in differential x-ray phase-contrast imaging for extended and polychromatic x-ray sources. *J. Microsc.* 232:145.

Jerjen, I., V. Revol, P. Schuetz et al. 2011. Reduction of phase artifacts in differential phase contrast computed tomography. *Opt. Express* 19(14):13604.

Kak, A.C., and M. Slaney. 1988. *Principles of Computerized Tomographic Imaging.* IEEE Press, New York, USA.

Kamangar, F., G.M. Dores, and W.F. Anderson. 2006. Patterns of cancer incidence, mortality, and prevalence across five continents: Defining priorities to reduce cancer disparities in different geographic regions of the world. *J. Clin. Oncol.* 24(14):2137.

Kelsey, J.L. 1993. Breast-cancer epidemiology—Summary and future-directions. *Epidemiol. Rev.* 15(1):256.

Khelashvili, G., J.G. Brankov, D. Chapman et al. 2006. A physical model of multiple-image radiography. *Phys. Med. Biol.* 51:221.

Lynch, S.K., V. Pai, J. Auxier et al. 2011. Interpretation of dark-field contrast and particle-size selectivity in grating interferometers. *Appl. Optics* 50:4310.

McDonald, S.A., F. Marone, C. Hintermüller et al. 2009. Advanced phase-contrast imaging using a grating interferometer. *J. Synchrotron Radiat.* 16(Pt 4):562.

Modregger, P., B.R. Pinzer, T. Thüring et al. 2011a. Sensitivity of x-ray grating interferometry. *Opt. Express* 19:18324.

Modregger, P., F. Scattarella, B.R. Pinzer et al. 2012a. Imaging the ultrasmall-angle x-ray scattering distribution with grating interferometry. *Phys. Rev. Lett.* 108(4):048101.

Modregger, P., Z. Wang, T. Thüring, B. Pinzer, and M. Stampanoni. 2011b. Artifacts in x-ray dark-field tomography. *AIP Conf. Proc.* 1365:269.

Modregger, P., B. Weber, B.R. Pinzer et al. 2012b. Micro-structual analysis of human bicuspid aortic valve derived nodules using x-ray grating interferometry. *Phys. Med. Biol.* submitted.

Momose, A. 2004. Phase tomography using an x-ray Talbot interferometer. *Proc. SPIE.* 5535:352.

Momose, A., S. Kawamoto, I. Koyama et al. 2003. Demonstration of x-ray Talbot interferometry. *Jpn. J. Appl. Phys.* 42:L866.

Momose, A., W. Yashiro, H. Kuwabara, and K. Kawabata. 2009a. Grating-based x-ray phase imaging using multiline x-ray source. *Jpn. J. Appl. Phys.* 48:076512.

Momose, A., W. Yashiro, H. Maikusa, and Y. Takeda. 2009b. High-speed x-ray phase imaging and x-ray phase tomography with Talbot interferometer and white synchrotron radiation. *Opt. Express* 17:12540.

Momose, A., W. Yashiro, Y. Takeda, Y. Suzuki, and T. Hattori. 2006. Phase tomography by x-ray Talbot interferometry for biological imaging. *Jpn. J. Appl. Phys.* 45:5254.

Müller, B., S. Lang, M. Dominietto et al. 2008. High-resolution tomographic imaging of microvessels. *Proc. SPIE.* 7078:70780B.

Pfeiffer, F., M. Bech, O. Bunk, P. Kraft, and E.F. Eikenberry. 2008. Hard-x-ray dark-field imaging using a grating interferometer. *Nat. Mat.* 7(2):134.

Pfeiffer, F., O. Bunk, C. David et al. 2007a. High-resolution brain tumor visualization using three-dimensional x-ray phase contrast tomography. *Phys. Med. Biol.* 52:6923.

Pfeiffer, F., O. Bunk, C. Kottler, and C. David. 2007b. Tomographic reconstruction of three-dimensional objects from hard x-ray differential phase contrast projection images. *Nucl. Instrum. Meth. A* 580:925.

Pfeiffer, F., C. Kottler, O. Bunk, and C. David. 2007c. Hard x-ray phase tomography with low-brilliance sources. *Phys. Rev. Lett.* 98:1.

Pfeiffer, F., T. Weitkamp, O. Bunk, and C. David. 2006. Phase retrieval and differential phase-contrast imaging with low-brilliance x-ray sources. *Nature Phys.* 2:258.

Pinzer, B.R., M. Cacquevel, P. Modregger, S.A. McDonald, J.C. Bensadoun, T. Thuering, P. Aebischer, M. Stampanoni. July 2012. Imaging brain amyloid deposition using grating-based differential phase contrast tomography. *NeuroImage* 61(4):1336–1346.

Qi, Z., J. Zambelli, N. Bevins, and G.-H. Chen. 2010. Quantitative imaging of electron density and effective atomic number using phase contrast CT. *Phys. Med. Biol.* 55:2669.

Rayleigh, L. 1881. Experiments with and theory of diffraction gratings. *Phil. Mag.* 11:196.

Reitz, C., C. Brayne, and R. Mayeux. 2011. Epidemiology of Alzheimer disease. *Nat. Rev. Neurol.* 7(3):137.

Revol, V., C. Kottler, R. Kaufmann et al. 2010. X-ray interferometer with bent gratings: Towards larger fields of view. *Nucl. Instrum. Meth. A* 648:302.

Schulz, G., T. Weitkamp, I. Zanette et al. 2010. High-resolution tomographic imaging of a human cerebellum: Comparison of absorption and grating-based phase contrast. *J. R. Soc. Interface* 7:1665.

Stampanoni, M., Z. Wang, T. Thüring et al. 2011. The first analysis and clinical evaluation of native breast tissue using differential phase-contrast mammography. *Invest. Radiol.* 46(12):801.

Suleski, T.J. 1997. Generation of Lohmann images from binary-phase Talbot array illuminators. *Appl. Optics* 36:4686.

Talbot, H.F. 1836. LXXVI. Facts relating to optical science. No. IV. *Phil. Mag. Series 3* 9:401.

Thüring, T., P. Modregger, T. Grund et al. 2011a. High resolution, large field of view x-ray differential phase contrast imaging on a compact setup. *Appl. Phys. Lett.* 99:041111.

Thüring, T., P. Modregger, B.R. Pinzer et al. 2011b. Towards x-ray differential phase contrast imaging on a compact setup. *Proc. SPIE: Phys. Med. Imaging* 7691: 79611G.

Tzemos, N., J. Therrien, J. Yip et al. 2008. Outcomes in adults with Bicuspid Aortic Valves. *Jama-J. J. Am. Med. Assoc.* 300(11): 1317.

Wang, Z.-T., K.-J. Kang, Z.-F. Huang, and Z.-Q. Chen. 2009. Quantitative grating-based x-ray dark-field computed tomography. *Appl. Phys. Lett.* 95:094105.

Weitkamp, T., C. David, C. Kottler, O. Bunk, and F. Pfeiffer. 2006. Tomography with grating interferometers at low-brilliance sources. *Proc. SPIE.* 63180S.

Weitkamp, T., A. Diaz, C. David et al. 2005. X-ray phase imaging with a grating interferometer. *Opt. Express* 13:6296.

Weitkamp, T., I. Zanette, C. David et al. 2010. Recent developments in x-ray Talbot interferometry at ESRF-ID19. *Proc. SPIE.* 7804:780406.

Yashiro, W., Y. Terui, K. Kawabata, and A. Momose. 2010. On the origin of visibility contrast in x-ray Talbot interferometry. *Opt. Express* 18(16):16890.

Zhu, P., K. Zhang, Z. Wang et al. 2010. Low-dose, simple, and fast grating-based x-ray phase-contrast imaging. *P. Natl. Acad. Sci. U.S.A.*

Analyzer-Based X-Ray Phase-Contrast Imaging

Jovan G. Brankov
Illinois Institute of Technology

Adam M. Zysk
Illinois Institute of Technology

4.1 Introduction

In conventional radiography, the x-ray absorption due to an object is measured by projecting a shadow of the object onto a detector. This method, which has traditionally been used to image highly absorbing tissue structures like bones, yields relatively limited contrast from soft-tissue structures, which are not highly absorbing. Phase-contrast x-ray methods comprise a class of emerging medical-imaging techniques that aims to employ additional contrast mechanisms to more effectively visualize the soft-tissue structures.

In conventional radiography, x-ray photons are assumed to travel in straight lines except when scattered by interaction with an object. X-rays are, in fact, a form of short-wavelength electromagnetic radiation (wavelength range 0.1–10 nm) that can exhibit classical optical phenomena such as refraction. When an x-ray beam passes through an object, it is attenuated and deflected (i.e., refracted) by a small but measurable angle (typically <10 μrad in a biological tissue) that can be detected using phase-contrast imaging modalities, including analyzer-based imaging (ABI), propagation-based imaging, and grating-based imaging (see Chapters 3 and 5). ABI systems and grating-based systems can also be used to measure a phenomenon called ultra-small-angle x-ray scattering (USAXS), which is

the result of x-ray refraction by many tiny structures within the object. These structures are small when compared to the imaging system resolution (e.g., digital detector pixel size, typically 10–400 μm) but much bigger than the x-ray wavelength (e.g., scattering structures of 1–10 μm in size for a 50 μm detector pixel). The potential significance of ABI lies in its ability to visualize soft tissue in great detail; preliminary studies have shown great promise for medical and biological applications including cancer and joint imaging.

ABI is a computational medical imaging that uses measurements to calculate parametric images representing the object absorption, scattering, and USAXS. These images are shown in Figure 4.1, which shows images of an infiltrating lobular carcinoma breast tumor and demonstrates clear visualization of the branching fibril structures emanating from the tumor, a feature that is relatively indistinct in the conventional radiograph. For reference, a conventional radiograph is also shown. The attenuation image is similar to a conventional radiograph, but exhibits much greater contrast because the ABI technique removes wide angle scattering caused by subnanometer-sized structures. The refraction image shows the effect of very small beam deflections caused by refractive index variations in the object that are slowly varying compared to the system resolution. Finally, the USAXS image shows the presence of textural

Conventional
mammogram

Analyzer-based images

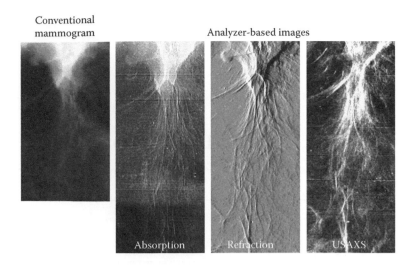

Absorption   Refraction   USAXS

FIGURE 4.1 Conventional radiograph and ABI images of a breast tumor. From left to right, the conventional radiograph, ABI absorption, ABI refraction, and ABI USAXS images are shown. (Reprinted with permission from IEEE. Brankov, J. G., and L. C. Cobo. 2006b. *Med Phys* 33:278–289; Marquet, B., J. G. Brankov, and M. N. Wernick. 2006. Noise and sampling analysis for multiple-image radiography. In *3rd IEEE International Symposium on Biomedical Imaging: Nano to Macro*, 2006, Arlington, Virginia, USA. 1232–1235; Cobo, L. C. and J. G. Brankov. 2007. Evaluation of model based parametric image estimation in MIR. In *4th IEEE International Symposium on Biomedical Imaging: From Nano to Macro*, 2007, Arlington, Virginia, USA. 452–455.)

structure within the object at a scale smaller than the system resolution (e.g., digital detector pixel size) and much larger than the x-ray wavelength.

Unlike propagation- or grating-based x-ray phase-contrast imaging methods (see Chapters 3 and 5), which rely on the measurement of wavefield interference effects due to the object, ABI systems angularly filter the x-ray beam to simultaneously measure and then calculate multiple object properties. ABI of absorption, refraction, and USAXS effects is enabled by the use of a semiconductor crystal, similar to those used to fabricate integrated circuits, that permits the measurement of the angular direction of x-rays after they pass through an object (see the system diagram in Figure 4.2). The x-ray intensity behind this crystal, called the analyzer crystal, is dependent on the angular position of the crystal. The measured angularly dependent intensity is called an angular intensity profile (AIP). The effects of absorption, refraction, and USAXS respectively manifest in the measured angularly

dependent data as an attenuation, shift, and broadening of the characteristic response.

Early ABI research was conducted at synchrotron facilities, where the x-ray source is very bright and the use of x-ray crystal optics is routine. Significant recent progress has been made to translate ABI for benchtop use to enable broader application for biological and medical imaging. In this chapter, the fundamental principles of ABI are presented, including experimental system configurations, a model of beam–object interactions, and methods for object parameter estimation. Biological and medical applications are also reviewed and an analysis of issues related to the development of benchtop systems is presented.

4.2 Imaging Principles

The first demonstration of ABI was the Bonse–Hart camera Bonse and Hart 1965, 1966, also referred to as a double-crystal diffractometer, which was developed in the 1960s as a method for USAXS measurement (Long et al. 1991, Brumberger 1995). The system consists of two channel-cut perfect crystals (e.g., Si or Ge) positioned so that the direction of beam propagation is aligned with the crystal Bragg angle (Born and Wolf). The Bragg angle, the angle at which constructive coherent scattering (diffraction) from a crystal lattice occurs, is predicted by Bragg's law (see Appendix A), which was formulated by Sir William Henry Bragg and his son William Lawrence Bragg, winners of the 1915 Nobel Prize in physics. Their work was inspired by the earlier work of Max von Laue, who was awarded the 1914 Nobel Prize in physics. These principles form the basis for modern ABI techniques (Förster et al. 1980, Davis et al. 1995), which were first

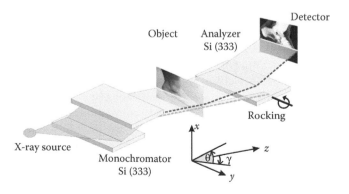

FIGURE 4.2 Schematic of an ABI system in the Bragg configuration.

applied to biological and medical investigations by Chapman and colleagues (Chapman et al. 1997). The biomedical applications of ABI are described in detail in Chapman et al. (1997) and Wernick et al. (2003).

Like conventional x-ray imaging systems, ABI systems direct an x-ray beam through the object to form an image. ABI systems, however, make use of a highly collimated (directional) and quasi-monochromatic (containing x-rays with a narrow range of energies) input beam that can be created by the use of a crystal monochromator. As the collimated x-ray beam passes through the object (e.g., a patient's body), the beam will be deflected due to refraction, attenuated due to absorption processes, and directed into a broadened range of angular directions due to USAXS. After passing through the object, the beam strikes the analyzer crystal, which reflects primarily the portion of the beam traveling in a specific narrow direction according to Bragg's law (see Appendix A). By rotating the analyzer and collecting images at multiple angles, the AIP of the transmitted beam is obtained at each detector pixel. The beam deflection and broadening are small enough to eliminate significant crosstalk between adjacent pixels (Wernick et al. 2003). Consequently, the measured AIP describes the object interaction with the beam as a function of crystal angle and thus contains the information needed to calculate the absorption, refraction, and USAXS properties of the object at each detector pixel. The absorption coefficient is estimated from the integral of the measured AIP (i.e., the total transmitted intensity), the refraction is estimated from the measured AIP centroid shift, and the angular divergence due to USAXS is estimated from the second central moment of the measured AIP. Measurements of these three parameters yield three distinct parametric images, which convey different physical characteristics of the object.

The Bragg configuration (see Figure 4.2), in which a double-crystal monochromator and a single analyzer crystal are positioned at the Bragg angle with respect to the straight-beam path, is the most commonly used ABI configuration (Chapman et al. 1997). The Laue configuration (Hasnah et al. 2002a, Kitchen et al. 2011) and other configurations shown in Hasnah et al. (2002a) and Modregger et al. (2007) have also been investigated on a more limited basis.

As an example, consider a simple phantom consisting of a stack of papers and a plastic rod, as shown in Figure 4.3. Here, the ABI absorption, refraction, and USAXS are shown along with a conventional radiograph for reference. In the absorption image, which is similar to the conventional radiograph, the attenuation due to the rod is clearly visible, as is some attenuation due to the paper. The refraction image clearly shows the edge contrast due to the refractive index of the rod, which causes a beam deflection, but relatively little refraction contrast is evident from the paper. Finally, the USAXS image derives the majority of its contrast from the paper, which has fibers that are smaller than the system resolution.

In Figure 4.4, the AIP is shown for several regions of the object. In a region without an object (see "Background" region), the AIP is simply due to the response of the system of crystals. That response is reduced in amplitude due to absorption in the object. In addition, regions with refraction contrast will result in beam deflection and, thus, a shift in the AIP (see "Rod, off center" region). Finally, USAXS results in a broadening of the AIP (see "Thick paper" region). Finally, these effects manifest simultaneously in the AIP of each detector pixel region (see "Rod and paper" region).

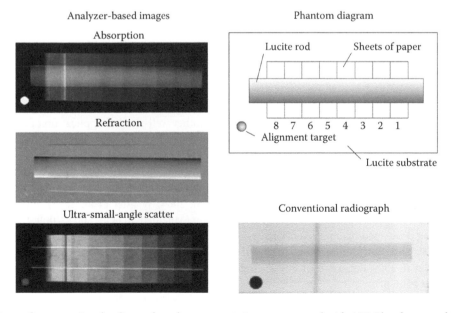

FIGURE 4.3 Comparison of a conventional radiograph to the parametric images generated with ABI. The phantom object shown here consists of a glass rod and a stack of paper containing an increasing number of sheets from right to left. (Reprinted with permission from the Institute of Physics Publishers. Wernick, M. N. et al. 2003. *Phys Med Biol* 48:3875–3895.)

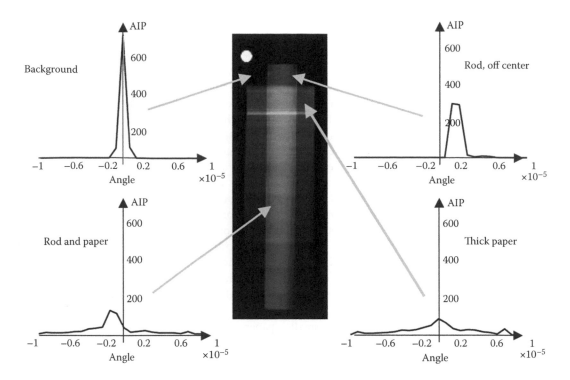

FIGURE 4.4 The relationship between object parameters and the AIP. (Reprinted with permission from the Institute of Physics Publishers. Wernick, M. N. et al. 2003. *Phys Med Biol* 48:3875–3895.)

4.3 Biological and Medical Applications

ABI has been investigated for a variety of medical and biological applications. Tumor imaging and skeletal imaging, for example, are particularly well-studied applications, demonstrating the improved visualization of tissue features and the additional tissue property information offered by ABI techniques when compared to conventional radiography.

4.3.1 Tumor Imaging

One extensively investigated application is the imaging of breast tissue, which has been studied over the last two decades (Chapman et al. 1997, 1998, Arfelli et al. 1998, Ingal et al. 1998, Moeckli et al. 2000, Pisano et al. 2000, Hasnah et al. 2002b, Wernick et al. 2003, 2006, Fiedler et al. 2004, Kiss et al. 2004, Briedis et al. 2005, Fernández et al. 2005, Keyriläinen et al. 2005). These studies, which were carried out at synchrotron facilities in order to make use of the very bright x-ray sources, have been limited to *ex vivo* tissue studies in a highly controlled experimental environment. To enable future clinical studies, benchtop systems are being developed using conventional x-ray tubes (Faulconer et al. 2009, 2010, Muehleman et al. 2009b, 2010, Nesch et al. 2009, Parham et al. 2009, Fogarty et al. 2011), a topic that will be addressed in Section 4.7.

An example of parametric ABI images (obtained with multiple-image radiography (MIR) (Wernick et al. 2003)) of the breast is shown in Figure 4.1 with a conventional radiograph for comparison. These images show features of an infiltrating lobular carcinoma that are difficult to visualize in the conventional radiograph. It is also important to note that ABI imaging can achieve these images while depositing a reduced x-ray dose, as shown by Chapman et al. (1997), Pisano et al. (2000), and Kiss et al. (2004).

A number of other ABI investigations have been carried out to visualize tumor tissues, including studies of lung lesion models in mice and rats (Kitchen et al. 2005, 2010, Li et al. 2009a, Connor et al. 2011a), a study of intraocular tumors (Tan et al. 2010), and a study of the microstructure of uterine mouse and rat leiomyomas (Liu et al. 2005).

4.3.2 Skeletal Imaging

Another well-studied application of ABI is for the evaluation of bones and joints. Extensive ABI investigations of human joints (see, e.g., Figure 4.5) have shown features of soft tendon injuries in the foot, ankle cartilage degradation (Mollenhauer et al. 2002, Muehleman et al. 2002, 2006b, Li et al. 2004, 2005, Majumdar et al. 2004, Wagner et al. 2005), muscle tissue adjacent to the joint (Muehleman et al. 2004, 2009b), and lesions in the Achilles tendon (Muehleman et al. 2009a).

ABI studies include investigations of bone damage and implant gaps (Connor et al. 2005, 2006), spinal fusion in animal models (Kelly et al. 2006, 2008), synovial joints and calcifications (Li et al. 2009b), trabecular bone degradation (Connor et al. 2009a), human patella cartilage (Coan et al. 2010), equine hooves (Fogarty et al. 2011), and galline bone degradation *in vivo* (Olkowski et al. 2010).

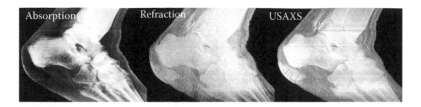

FIGURE 4.5 MIR images of a human ankle. (Reprinted with permission from the Institute of Physics Publishers. Wernick, M. N. et al. 2003. *Phys Med Biol* 48:3875–3895; Khelashvili, G. et al. 2006. *Phys Med Biol* 51:221–236.)

4.3.3 Other Applications

In addition to studies of general soft tissue contrast (Lewis et al. 2003, Muehleman et al. 2006a, Rao et al. 2010), applications as varied as eye imaging, which includes studies of canine (Antunes et al. 2005, 2006), porcine (Kelly et al. 2007), and leporine eye structures (Yin et al. 2010), inner ear imaging (Hu et al. 2008), liver imaging (Zhang et al. 2009, Li et al. 2009a), plant and seed development (Kao et al. 2007, Young et al. 2007), tissue engineering scaffolds (Brey et al. 2010), and amyloid plaques (Connor et al. 2009b) have been investigated. ABI studies have even extended beyond medicine and biology to the study of geologic artifacts (Krug et al. 2008).

4.4 Image Formation Model

To understand the relationship between the object properties and the measured AIP data, a mathematical model of the interaction between the x-ray beam and the object being imaged must be employed. A number of image formation models have been developed, but few (Rigon et al. 2003, Khelashvili et al. 2006, Nesterets 2008) account for angular divergence of the beam caused by USAXS. Image formation models include geometrical optics (GO) models (Rigon et al. 2003, Kitchen et al. 2007), various wave propagation models (Pavlov et al. 2000, 2001, 2004, 2005, Paganin et al. 2004, Paganin and Gureyev 2008, Nesterets et al. 2005, 2006a, Bravin et al. 2007b, Nesterets 2008), a Fourier optics model (Guigay et al. 2007), and a radiative transport theory model (Khelashvili et al. 2006).

The image formation model used in this chapter employs a stratified medium object model containing scattering structures, the GO approximation, and the radiative transport theory (Khelashvili et al. 2006) to elucidate the origins of the AIP, parameter estimation techniques, noise properties, and angular sampling strategies. The stratified medium model treats the object as a stack of homogeneous (or slowly changing) planar layers (Born and Wolf 1999) containing discrete scattering centers. In ABI, beam interactions with the object are dominated by structures that are large compared with the x-ray wavelength, a regime in which the GO approximation is valid. Finally, beam propagation is modeled using the radiative transport theory (Ishimaru 1997), an approach that accounts for average net energy flows within a medium. Note that, unlike most other models, this model does not neglect the angular divergence of the beam caused by USAXS and gives an explicit explanation of its origin.

The object model is described in detail in Appendix B. The main results are the following:

1. The absorption, refraction, and USAXS parametric images are line integrals of various object properties through the object.
2. The measured AIP can be approximated by a function that is obtained as a convolution of a Gaussian function with the system intrinsic AIP.

4.5 Methods for Estimation of Parametric Images

Several methods to compute ABI parametric images from measurements of the AIP have been proposed, including diffraction-enhanced imaging (DEI) (Chapman et al. 1997), multiple image radiography (MIR) (Wernick et al. 2003), scatter DEI (S–DEI) (Rigon et al. 2003), extended DEI (E–DEI) (Chou et al. 2007, Rigon et al. 2007), and model-based parameter estimation (Brankov and Cobo 2006b, Cobo and Brankov 2007, Majidi et al. 2008a). Here, the most common parameter estimation methods are reviewed. Details of alternative methods (Oltulu et al. 2003, Paganin et al. 2004, Nesterets et al. 2006b, Maksimenko 2007, Hu et al. 2008, Chen et al. 2009) and experimental comparisons of these methods (Diemoz et al. 2010a) are available in the literature.

In addition to methods for estimation of absorption, refraction, and USAXS object parameters, there are several alternative approaches that estimate wavefield phase delay (Nesterets et al. 2004), object density (Wernick et al. 2006), and refractive index (Maksimenko et al. 2005, Sunaguchi et al. 2010).

4.5.1 Multiple Image Radiography

The MIR method, which in 2003 was independently proposed by multiple investigators (Pagot et al. 2003, Wernick et al. 2003), approximates absorption, refraction, and USAXS from knowledge of a collection of intensity data $I_m[\vec{\rho},\theta_l]$ at pixel location $\vec{\rho}$ with analyzer crystal angles θ_l, where $l = 1,\ldots,L$. In the ultrasmall-angle regime, attenuation is caused by absorption and scattering into angles outside the measured angular range. These attenuation processes are expressed collectively by the estimated absorption

$$\tilde{A}[\vec{\rho}] = -\ln\left(\frac{\sum_{l=1}^{L} I_m[\vec{\rho},\theta_l]}{\tilde{I}_0}\right), \tag{4.1}$$

where $\tilde{I}_0$ is the estimated intensity measured in the absence of an object and $l = 1,2,\ldots,L$ are indices of the analyzer crystal angular positions. Refraction, which induces an overall deflection of the beam that produces an angular shift of the beam centroid, is expressed by the estimated refraction

$$\Delta\tilde{\theta}[\vec{\rho}] = \sum_{l=1}^{L} \theta_l \cdot P[\vec{\rho},\theta_l] - \Delta\tilde{R}, \tag{4.2}$$

where $\Delta\tilde{R}$ is the estimated AIP shift of the imaging system and $P[\vec{\rho},\theta_l]$ is the normalized AIP defined as

$$P[\vec{\rho},l] = \frac{I_m[\vec{\rho},\theta_l]}{\sum_{l=1}^{L} I_m[\vec{\rho},\theta_l]}.$$

Finally, USAXS causes angular broadening of the AIP about the angle $\Delta\tilde{\theta}[\vec{\rho}]$ and is expressed by the second central moment of the normalized AIP. The estimated USAXS is

$$\tilde{w}[\vec{\rho}] = \sum_{l=1}^{L} \left(\theta_l - \Delta\tilde{\theta}[\vec{\rho}]\right)^2 \cdot P[\vec{\rho},\theta_l] - \tilde{W}_R, \tag{4.3}$$

where $\tilde{W}_R$ is the estimated second central moment of the imaging system.

4.5.2 Diffraction-Enhanced Imaging

DEI was the first planar ABI technique to produce separate images of absorption and refraction (Chapman et al. 1997). The DEI method is derived using a first-order Taylor expansion of the system intrinsic AIP (i.e., the rocking curve) $R(\theta)$, which describes the incoming beam AIP and analyzer crystal AIP. Two measurements, $I_m[\vec{\rho},\theta_L]$ and $I_m[\vec{\rho},\theta_H]$ (measured at the full-width-half-maximum angular position $R(\theta)$), and knowledge of the system intrinsic AIP are used to estimate the attenuated intensity

$$\tilde{I}_{DEI}[\vec{\rho}] = \frac{I_m[\vec{\rho},\theta_L]\frac{dR}{d\theta}(\theta_H) - I_m[\vec{\rho},\theta_H]\frac{dR}{d\theta}(\theta_L)}{R(\theta_L)\frac{dR}{d\theta}(\theta_H) - R(\theta_H)\frac{dR}{d\theta}(\theta_L)} \tag{4.4}$$

and the refraction angle

$$\Delta\tilde{\theta}_{DEI}[\vec{\rho}] = \frac{I_m[\vec{\rho},\theta_H]R(\theta_L) - I_m[\vec{\rho},\theta_L]R(\theta_H)}{I_m[\vec{\rho},\theta_L]\frac{dR}{d\theta}(\theta_H) - I_m[\vec{\rho},\theta_H]\frac{dR}{d\theta}(\theta_L)}. \tag{4.5}$$

Note that these results assume that USAXS effects are not present in the object. This failure to incorporate USAXS effects leads to DEI artifacts, which are most noticeable when the data are used for computed tomography (CT) (Brankov et al. 2006a, Diemoz et al. 2010b). Note also that the DEI formulation is the maximum-likelihood solution of the image estimation problem for the case of Poisson's noise (Brankov et al. 2004).

4.5.3 Extended Diffraction-Enhanced Imaging

The E–DEI method (Chou et al. 2007, Rigon et al. 2008) incorporates higher-order Taylor expansions of the rocking curve to estimate object parameters. In this section, the formulation from Chou et al. (2007) is also used to describe DEI and S–DEI. First, a second-order Taylor expansion is used to estimate the rocking curve

$$R(\theta_0 - \theta) \cong R(\theta_0) - \frac{dR(\theta)}{d\theta}\bigg|_{\theta_0} \theta + \frac{1}{2}\frac{d^2R(\theta)}{d\theta^2}\bigg|_{\theta_0} \theta^2. \tag{4.6}$$

Applying this expansion to Equation 4.20 yields

$$I_m(\vec{\rho},\theta_i) = I_r(Z,\vec{\rho})\bigg(R(\theta_i) - \frac{dR}{d\theta}(\theta_i)\Delta\theta(\vec{\rho})$$
$$+ \frac{1}{2}\frac{d^2R(\theta_i)}{d\theta^2}\left(\Delta\theta(\vec{\rho})^2 + w(\vec{\rho}) + W_R\right)\bigg). \tag{4.7}$$

Note that if $(d^2R(\theta_i)/d\theta^2) \approx 0$ or $\Delta\theta(\vec{\rho})^2 \approx 0$, the previously shown DEI equations can be obtained with the use of two measurements, $I_m[\vec{\rho},\theta_L]$ and $I_m[\vec{\rho},\theta_H]$.

The S–DEI method uses two measurements, $I_m[\vec{\rho},\theta_{TOP}]$ (measured at the angular position of maximum $R(\theta)$) and $I_m[\vec{\rho},\theta_{TOE}]$ (measured at the angular position where the second derivative of $R(\theta)$ is maximized), knowledge of the system intrinsic AIP $R(\theta)$, and the assumption that $\Delta\theta(\vec{\rho}) = 0$ to obtain

$$\tilde{I}_{S-DEI}[\vec{\rho}] = \frac{I_m[\vec{\rho},\theta_{TOE}]\frac{d^2R(\theta_{TOP})}{d\theta^2} - I_m[\vec{\rho},\theta_{TOP}]\frac{d^2R(\theta_{TOE})}{d\theta^2}}{R(\theta_{TOE})\frac{d^2R(\theta_{TOP})}{d\theta^2} - R(\theta_{TOP})\frac{d^2R(\theta_{TOE})}{d\theta^2}} \tag{4.8}$$

and

$$\tilde{w}_{S-DEI}[\vec{\rho}] = 2\frac{I_m[\vec{\rho},\theta_{TOP}]R(\theta_{TOE}) - I_m[\vec{\rho},\theta_{TOE}]R(\theta_{TOP})}{I_m[\vec{\rho},\theta_{TOE}]\frac{d^2R(\theta_{TOP})}{d\theta^2} - I_m[\vec{\rho},\theta_{TOP}]\frac{d^2R(\theta_{TOE})}{d\theta^2}}. \tag{4.9}$$

Further investigation of reconstruction methods with higher-order Taylor expansion of $R(\theta)$ using Equations 4.6 and 4.7 and additional measurements has been performed in Chou et al. (2007), showing improved E–DEI estimation accuracy of absorption, refraction, and USAXS with increased Taylor expansion order.

4.5.4 Model-Based Parameter Estimation

Model-based methods estimate object parameters $\mathbf{v}[\vec{\rho}] = \left[\tilde{A}[\vec{\rho}], \Delta\tilde{\theta}[\vec{\rho}], \tilde{w}[\vec{\rho}]\right]^T$ by cost function minimization. One can construct a general cost functional

$$J[\vec{\rho}] = \min_{v[\vec{\rho}]} \sum_{l=1}^{L} \left[I_m[\vec{\rho}, l] - I_a\left(\vec{\rho}, \theta_1; v[\vec{\rho}]\right)\right]^2, \qquad (4.10)$$

subject to $v[\vec{\rho}]$ constraints

where $I_m[\vec{\rho}, l]$ is a measured intensity and $I_a(\vec{\rho}, \theta_1; v[\vec{\rho}])$ is a proposed model.

A solution to this nonlinear minimization problem can be found by employing the modified conjugate gradient method suggested by Powell (Powell 1986) and using the approximate model in Equation 4.28 (Brankov and Cobo 2006b, Cobo and Brankov 2007, Majidi et al. 2008a). Other alternative-model methods are available in the literature (Oltulu et al. 2003, Maksimenko 2007, Hu et al. 2008, Chen et al. 2009).

4.6 Computed Tomography Methods

ABI parametric images represent line integrals through the object (see Appendix B) and are, thus, two-dimensional (2D) projections of three-dimensional (3D) objects. CT methods have been investigated to extend ABI for 3D volumetric imaging. In addition to revealing the object's 3D x-ray attenuation coefficient distribution, as in conventional x-ray CT, ABI–CT produces volumetric images of the refraction and USAXS properties, as well. The complementary 3D information contained in these ABI–CT images yields dramatically improved diagnostic capabilities over conventional x-ray CT methods.

In the classical CT mode, the object is rotated about the *x*-axis (see Figure 4.6a). At each tomographic view angle, the 2D parametric images produced by the ABI method represent the raw projection data from which the 3D object property images are reconstructed. An alternative imaging geometry, in which the object is rotated about the *y*-axis (see Figure 4.6b), has also been implemented (Maksimenko et al. 2005, Huang et al. 2006, 2007, Wang et al. 2006, Sunaguchi et al. 2010), yielding volumetric images of the refractive index, as opposed to the refractive index gradient.

DEI–CT (Dilmanian et al. 2000, Fiedler et al. 2004) and CT–MIR (Brankov et al. 2006a) are ABI methods operating in the classical CT mode that are extensions of the DEI and MIR estimation methods, respectively. Experimental validation of CT–MIR in Brankov et al. (2006a) has demonstrated the predicted linearity of USAXS (see Appendix B) and the reduction of USAXS artifacts in the absorption and refractive index gradient images produced by DEI–CT. A comprehensive evaluation and comparison of ABI–CT methods is given in Diemoz et al. (2010b) and clinical comparisons between conventional x-ray CT methods and ABI–CT are given in Bravin et al. (2007a), Keyriläinen et al. (2008), Connor et al. (2009a), and Kao et al. (2009). Additional work has also incorporated wave-propagation image formation models (Pavlov et al. 2000) instead of the radiative transport models used in the methods described above.

4.7 Benchtop Systems

Experimental ABI investigations originated in synchrotron facilities due to the need for very bright synchrotron x-ray radiation sources. In order to apply these methods for clinical investigations, efforts are underway to implement ABI with a benchtop system employing a conventional x-ray tube source. A summary of available systems and their performance is given in Table 4.1.

A number of barriers to benchtop ABI implementation remain. For instance, conventional laboratory x-ray tube sources are limited by thermal loading constraints of the anode material; alternative tube configurations are under investigation (Kim et al. 2006). In addition, system alignment stability problems have been reported due to vibration and thermal variations (Vine et al. 2007, Nesch et al. 2009, Connor et al. 2011b).

In the development of a benchtop ABI system, a number of outstanding technical issues exist. The trade-off between source anode spot size and beam intensity, for example, must be optimized and its effect on imaging system resolution investigated. The potential use of x-ray focusing optics to increase beam intensity from a laboratory source is also relevant to these issues, but carries trade-offs that must be investigated. It is also important to consider the thermal and spatial stability of system components, including crystal stages and the anode source spot itself,

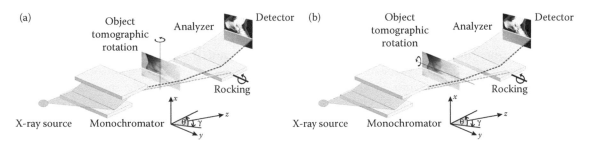

FIGURE 4.6 Schematic of an ABI system in CT mode. A classical CT configuration (a) and an alternative CT configuration (b) are shown.

TABLE 4.1   Summary of Available Benchtop ABI Systems and Their Performance

Publication	Anode Material	Tube Power	X-Ray Energy	Analyzer Crystal	Imaging Area	Imaging Time	Image Resolution
Vine et al. (2007)	Cu	1.2 kW	30 keV	Ge 220	0.2 mm × 1.5 mm	4 h	NA
Parham et al. (2009) Faulconer et al. (2009)	Stationary anode W	1 kW	~60 keV	Ge 111 and 333	12 cm × 12 cm	23 + h	50 μm
Muehleman et al. (2009)	Stationary anode W	1 kW	~60 keV	Ge 111 and 333	NA	NA	50 μm
Nesch et al. (2009) Fogarty et al. (2011)	Ag	2.2 kW	22 keV		1.2 cm × 6 cm	240 s	~155 μm
Connor et al. (2011)	Rotating anode W	10 kW	~60 keV	Ge 111	12 cm × 12 cm	3 h (480 s of tube on)	50 μm

especially when employing rotating-anode sources. Finally, the angular sampling pattern and its impact on parametric image estimation methods must be considered and optimized.

These technical issues are not well studied in the literature and are under consideration by a number of research groups. In the following sections, ABI system noise limitations and the trade-offs governing the selection of angular sampling locations are presented.

4.7.1 Noise Properties

As the flux from conventional x-ray tubes is a known performance constraint in benchtop ABI systems, the acquired data are photon-limited and, thus, are dominated by a Poisson's process, which is multiplicative and object-dependent (Wernick et al. 2003). In this section, an analysis of noise in ABI is presented (Brankov et al. 2004, 2005, Marquet et al. 2006). For an alternative analysis using a wave-based image formation model, a weak-object approximation, and the ideal observer signal-to-noise ratio image quality metric, see Anastasio et al. (2010).

Under Poisson's noise model, the likelihood function of the measured intensity $I_m[\vec{\rho}, \theta_l]$, object properties $\mathbf{v}(\vec{\rho}) = [A(\vec{\rho}), \Delta\theta(\vec{\rho}), w(\vec{\rho})]^T$, and true intensity $I_m(\vec{\rho}, \theta_l)$ is

$$p(I_m[\vec{\rho}, \theta_l]; \mathbf{v}(\vec{\rho})) = \frac{I_m(\vec{\rho}, \theta_l)^{I_m[\vec{\rho}, \theta_l]} e^{-I_m(\vec{\rho}, \theta_l)}}{I_m[\vec{\rho}, \theta_l]!}, \quad (4.11)$$

where $I_m(\vec{\rho}, \theta_l) = E[I_m[\vec{\rho}, \theta_l]]$. For notational simplicity, the quantities $i_l \triangleq I_m(\vec{\rho}, \theta_l)$, $i[l] \triangleq I_m[\vec{\rho}, \theta_l]$, and $i_l = E[i[l]]$ are used here and the position $\vec{\rho}$ is omitted in $\mathbf{v}(\vec{\rho})$.

To investigate the noise properties of each measurement, the theoretical noise variance limit in the estimated object parameters is considered for one angular sample and a simultaneous estimate of the three object parameters. By the Cramer–Rao lower bound (CRLB) theorem for vector estimation (Kay 1993), the variance of any unbiased estimator $\hat{v}_i$, i.e., the ith element of the vector $\tilde{v}$, is bounded such that $\mathrm{Var}(\hat{v}_i) \geq \left\{\mathbf{I}^{-1}(\mathbf{v}; \theta_l)\right\}_{ii}$, where $\mathbf{I}(\mathbf{v})$ is the Fisher information matrix. Applying the Gaussian object model approximation (see Appendix B) yields

$$\mathbf{I}(\mathbf{v}; \theta_l) =$$

$$
\begin{bmatrix}
\dfrac{i_l}{A^2} & \dfrac{h_l - \Delta\theta \cdot i_l}{wA} & \dfrac{q_l - w \cdot i_l}{2w^2 A} \\[2ex]
\dfrac{h_l - \Delta\theta \cdot i_l}{wA} & \dfrac{(h_l - \Delta\theta \cdot i_l)^2}{w^2 i_l} & \dfrac{(h_l - \Delta\theta \cdot i_l)(q_l - w \cdot i_l)}{2w^2 i_l} \\[2ex]
\dfrac{q_l - w \cdot i_l}{2w^2 A} & \dfrac{(h_l - \Delta\theta \cdot i_l)(q_l - w \cdot i_l)}{2w^3 i_l} & \dfrac{(q_l - w \cdot i_l)^2}{4w^4 i_l}
\end{bmatrix},
$$

where

$$
\begin{aligned}
i_l &\triangleq I_r(Z, \vec{\rho}) \big[R(\theta) * P(Z, \vec{\rho}, \theta) \big] \Big|_{\theta = \theta_l} \\
h_l &\triangleq I_r(Z, \vec{\rho}) \big[R(\theta) * \theta P(Z, \vec{\rho}, \theta) \big] \Big|_{\theta = \theta_l} \quad (4.12) \\
q_l &\triangleq I_r(Z, \vec{\rho}) \big[R(\theta) * (\theta - \Delta\theta)^2 P(Z, \vec{\rho}, \theta) \big] \Big|_{\theta = \theta_l}.
\end{aligned}
$$

4.7.2 Angular Sampling

The scheme used to angularly sample measurements of the AIP has a significant impact on the development of practical ABI systems, as the noise properties, dose, and total imaging time are highly dependent on this configuration. When the rocking curve is known and the total dose exposure to the object is fixed, the dominant factor affecting the noise is the angular sampling pattern, i.e., the angular positions θ_l of the analyzer crystal. In this section, the effects of analyzer crystal position selection, or sampling strategy, on the CRLB is analyzed.

To gain an intuition about the effect of sampling strategy, the true bounds $\left\{\mathbf{I}^{-1}(\mathbf{v}; \theta_l)\right\}_{ii}$ are approximated with the inverse of the diagonal elements of the Fisher information matrix, $\left\{\mathbf{I}(\mathbf{v}; \theta_l)\right\}_{ii}^{-1}$, i.e., the bounds that would have existed if each parameter were estimated individually while the other parameters were known. The functions $\left\{\mathbf{I}(\mathbf{v}; \theta_l)\right\}_{ii}$ can be used as information functions that show the value of information available from i_l at θ_l analyzer positions, as they are inversely proportional to the estimator variance. Examples for three different object parameter vectors are given in Figure 4.7. These object parameters, which represent the mean attenuation parameter, the mean refraction parameter, and the mean, minimum, and maximum USAXS parameter measured from MIR images of the human thumb shown in Figure 4.8 are given in Table 4.2.

Each point in Figure 4.7 shows the relative information at one analyzer crystal position. To minimize the CRLB in an estimate of v_i, one should sample at the location where the information function $\{I(v;\theta_l)\}_{ii}$ has a maximum value.

A CRLB analysis of two examples is shown here for optimization of the sample quantity and sampling pattern for MIR. The behavior of the CRLB in the MIR estimation method varies as a function of the number of analyzer angular positions L (Marquet et al. 2006, Majidi et al. 2008b). Here, a set of analyzer angular positions, $\Theta = [\theta_1, \theta_2, \ldots, \theta_L]$ uniformly distributed in the range of $[-5:5]$ μrad is considered. In this analysis, total exposure dose to the object is fixed, that is, if more samples are acquired each angular sample receives fewer photons and the total number of photons is constant. Angular samples are independent, so the total CRLB for a Θ sampling set and a given $\hat{v}_i$ can be estimated by

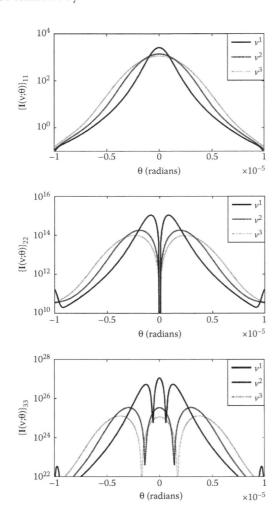

FIGURE 4.7 Examples of information functions for sample parameter vectors plotted on a logarithmic scale. Each function indicates the relative value of obtaining a data sample at a given angular position θ_l for estimating a single v_i parameter.

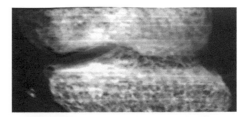

Attenuation

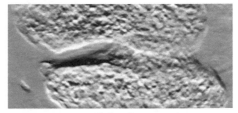

Refraction

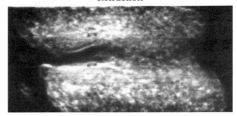

USAXS

FIGURE 4.8 MIR images of a human thumb acquired at 40 keV with 25 analyzer positions over $[-4.8$ to $4.8]$ μrad with 0.4 μrad increments. (Reprinted with permission from IEEE. Majidi, K., J. G. Brankov, and M. N. Wernick. 2008b. Sampling strategies in multiple-image radiography. In *5th IEEE International Symposium on Biomedical Imaging: From Nano to Macro*, 2008, Paris, France. 688–691.)

$$\text{CRLB}_i(\Theta) = E\left[\sum_{l=1}^{L}\{I(v;\theta_l)\}_{ii}^{-1}\right]_{w.r.t\ v^c}, \qquad (4.13)$$

where $E[\cdot]_{w.r.t\ v}$ is the expectation over the probability density function (PDF) of MIR parameters v^c (i.e., over all possible object parameters values). Direct evaluation of the expectation in Equation 4.13 is intractable, but it can be approximated by the arithmetic average $\text{CRLB}_i(\Theta) \simeq (1/C)\sum_{c=1}^{C}\sum_{l=1}^{L}\{I(v;\theta_l)\}_{ii}^{-1}$, where v^c, $c = 1, \ldots, C$ represents the values of the measured MIR parameter vector v at the cth pixel.

The results shown in Figure 4.9 demonstrate the improvement in the CRLB as the number of analyzer samples is increased. When the number of samples is low, the bound for even sample

TABLE 4.2 Parameters Obtained from the Human Thumb Images Shown in Figure 4.8

	A	$\Delta\theta$ (μrad)	w (μrad²)
v^1	0.6	0	0.1
v^2	0.6	0	1.4
v^3	0.6	0	2.4

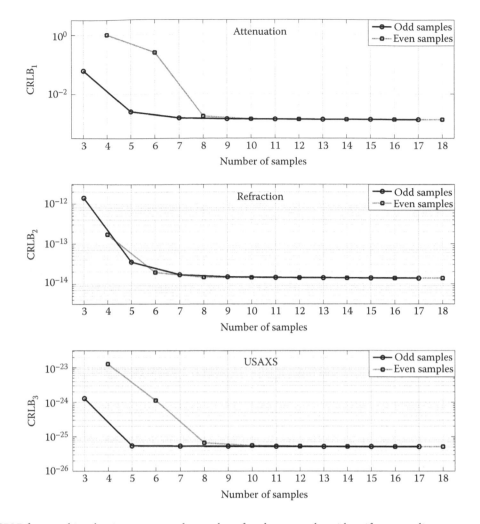

FIGURE 4.9 The CRLB for an unbiased estimator versus the number of analyzer samples with uniform sampling.

quantities is higher than that for odd sample quantities. This is due to the fact that the maximum amount of information for $\{I(v;\theta_1)\}_{11}$ and $\{I(v;\theta_1)\}_{33}$, as shown in Figure 4.7, lies at the center of the curves, a point that is not sampled when an even number of samples is used. The center sample does not contain any information from $\{I(v;\theta_1)\}_{22}$, however, and, therefore, the CRLB for the refraction image lies on the same curve for odd and even samples.

After a certain number of samples (11 for this data set), there is very little improvement in the bound. In the case of a minimum variance unbiased estimator, there is no benefit in acquiring more than 11 samples from the analyzer if the total exposure dose is fixed. It is important to note that this analysis does not define the best estimator for v, but serves as a basis for the comparison of sampling procedure and estimator performance.

The possibility of lowering the CRLB and hence achieving better estimation performance by using nonuniform sampling patterns has also been investigated (Majidi et al. 2008b). In this case, the minimization problem can be posed as $\Theta^* = \arg\min_{\Theta} J(\Theta)$, where $J(\Theta) = E[\sum_{i=1}^{3}(\mathrm{CRLB}_i(v,\Theta)/\mathrm{CRLB}_i(v,\Theta_u))]_{w.r.t\ v}$ and the expectation is calculated as described before. As the magnitude

scales, the cost function is normalized by dividing each CRLB value from a given sampling pattern, Θ, by the CRLB value obtained from the uniform sampling pattern, Θ_u, with the same number of samples. For Θ_u, a uniform sampling pattern in the range of $[-5{:}5]$ μrad is used.

To optimize the objective function, an exhaustive search method can be used. For each sampling quantity, the object function is evaluated at every possible combination of sampling points (with a minimum distance, e.g., 0.2 μrad) to find the pattern with the smallest $J(\Theta)$. To decrease the computation time, the number of possible patterns is constrained such that $\theta_1 = 0$ is always sampled and the sampling positions are symmetrical around zero. These constraints are reasonable because the object function is always symmetrical around the refraction parameter and the measured mean value for the refraction parameter is zero in the data set shown in Figure 4.8. The CRLB for a uniform sampling pattern and the CRLB for an optimal nonuniform sampling pattern are shown for different sampling quantities in Figure 4.10. Attenuation and refraction estimates demonstrate clear improvement with nonuniform sampling patterns, while

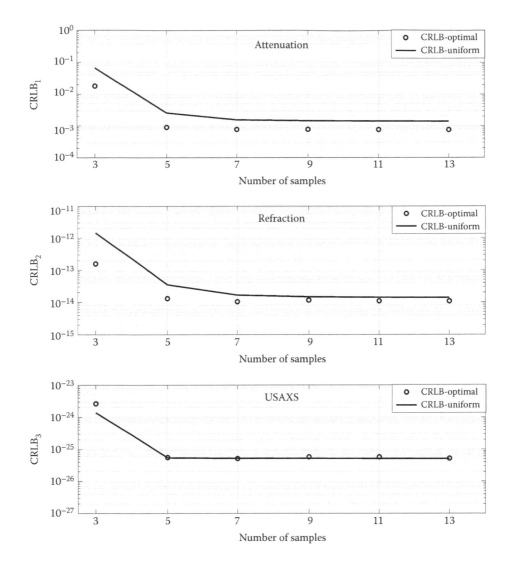

FIGURE 4.10 The CRLB for an unbiased estimator versus the number of analyzer samples with uniform and nonuniform sampling patterns.

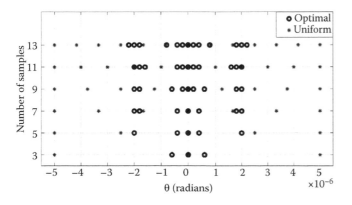

FIGURE 4.11 Uniform and nonuniform sampling patterns for different sample quantities.

the USAXS estimate does not. This result shows that it is possible to reduce the estimation variance by employing nonuniform sampling patterns.

The optimal sampling patterns shown in Figure 4.11 demonstrate that samples tend to converge to the three dominant angular positions that correspond to the locations in Figure 4.7 with the maximum information.

4.8 Future Challenges

The work reviewed in this chapter has demonstrated the biological and medical applications of ABI. The future of ABI clinical application is still limited by the general reliance on synchrotron x-ray sources. This reliance is a clear hurdle since synchrotron facilities are generally remote to the majority of the patient population and, with some exceptions (Wysokinski et al. 2007, Dreossi et al. 2008), remain inaccessible for medical use.

The majority of future applications are expected to be performed with benchtop systems that, as described herein, are currently under development. A number of engineering challenges exist that must be solved to provide a robust system for clinical application. The most significant hurdle is the low flux offered by conventional tube sources. Tube flux is limited by the melting point of the metal anode, but the improved flux offered by using a rotating anode source can induce a drift in the source anode position. A further concern is that ABI measurements rely on highly stable source and crystals' positions, which may be shifted by temperature fluctuations and system vibration (Nesch et al. 2009). As these technical barriers to benchtop implementation are addressed, a promising future awaits for ABI application in a number of clinical and biological fields.

Appendix A: Bragg's Law

Consider a crystal composed of discrete parallel atomic lattice planes separated by a constant distance d. Incident x-ray radiation produces a Bragg peak when reflections off the planes interfere constructively. By Bragg's law, this occurs when the phase shift is a multiple of 2π (see Figure 4.12) as expressed by $n\lambda = 2d\sin(\theta)$, where n is an integer, λ is the wavelength of the incident x-rays, and θ is the angle between the incident x-rays and the crystal atomic lattice planes. This constructive interference varies with angle θ, giving rise to a unique angular intensity profile $\phi(\theta)$ for each set of crystal planes (see Figure 4.13).

Appendix B: Image Formation Model

The object is characterized by a linear absorption coefficient $\mu(\vec{r}) = \mu(z)$ that is slowly varying in z (the beam propagation direction) and by a refractive index $n(\vec{r}) = n_0 + n_x(z)x$, where $\vec{r} = (x,y,z)^T$ is the spatial coordinate. The object contains a collection of identical discrete scatterers having a number density $\rho_n(\vec{r}) = \rho_n(z)$ that is slowly varying in z. Each scatterer is characterized by an extinction cross section $\sigma_{ext} = \sigma_s + \sigma_a$, where σ_s and σ_a are due to scattering and absorption, and a phase function $p(\vec{r},\hat{\vec{s}},\hat{\vec{s}}')$, which is the incremental fraction of radiation scattered from solid angle $d\omega$ in the incoming direction $\hat{\vec{s}}$ into solid angle $d\omega'$ in the outgoing direction $\hat{\vec{s}}'$ (see Figure 4.14). Assuming isotropic scattering

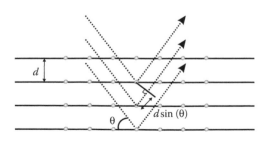

FIGURE 4.12 Diagram of the interaction between incident x-rays and the crystal lattice planes that gives rise to Bragg's law.

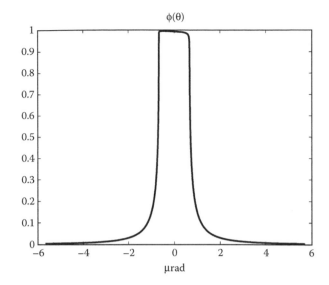

FIGURE 4.13 Example of a single crystal AIP for a Si crystal and (333) crystal planes at an x-ray energy of 40 keV.

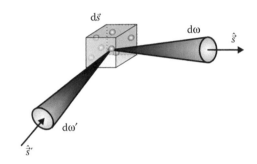

FIGURE 4.14 Diagram of the incoming and outgoing scattering directions and corresponding incremental solid angles.

$$p(\vec{r},\hat{\vec{s}},\hat{\vec{s}}') = p(\hat{\vec{s}} - \hat{\vec{s}}')$$
$$= 4\pi W_0 \frac{1}{2\pi|\alpha|^{1/2}}\exp\left(-\frac{1}{2}(\hat{\vec{s}} - \hat{\vec{s}}')^T\alpha^{-1}(\hat{\vec{s}} - \hat{\vec{s}}')\right), \quad (4.14)$$

where $\alpha = 1/2\,\mathrm{diag}\,(1/\alpha_p,1/\alpha_p)$ and $\mathrm{diag}(\cdot)$ denotes a diagonal matrix.

The incoming x-ray beam is represented by the specific intensity $I(\vec{r},\hat{\vec{s}})$, which describes the radiation density at position $\vec{r}$ in direction $\hat{\vec{s}} = (\sin\theta\cos\phi,\sin\theta\sin\phi,\cos\theta)$ (Figure 4.14). Propagation through the object is described by the radiative transport equation (RTE)

$$\hat{\vec{s}}\cdot\nabla_{\vec{r}}I(\vec{r},\hat{\vec{s}}) = -\mu(\vec{r})I(\vec{r},\hat{\vec{s}}) - \rho_n(\vec{r})\sigma_a I(\vec{r},\hat{\vec{s}}) - \rho_n(\vec{r})\sigma_s I(\vec{r},\hat{\vec{s}})$$
$$+ \vec{b}(\vec{r})\cdot\nabla_{\hat{\vec{s}}}I(\vec{r},\hat{\vec{s}}) + \frac{\rho_n(\vec{r})\sigma_{ext}}{4\pi}\int_{4\pi} p(z,\hat{\vec{s}} - \hat{\vec{s}}')I(\vec{r},\hat{\vec{s}}')\,d\omega',$$

$$(4.15)$$

where $\vec{b}(\vec{r}) = \nabla_{\vec{r}} \ln n(\vec{r}) \approx (n_x(z)/(n_0 + n_x(z)x)\vec{i} \approx (n_x(z)/n_0)\vec{i} \triangleq b_x(z)\vec{i}$ is a constant vector and $\vec{i}$ is the x-directed unit vector (see Figure 4.15). The first three terms on the RTE right-hand side respectively account for radiation loss due to absorption by the medium surrounding the scatterers, absorption by the scatterers, and scattering from the forward direction into other directions. The last two terms account for refraction into the forward direction from other directions and scattering into the forward direction from other directions. A similar RTE equation appears in standard treatments of transport theory (Ishimaru 1997), but the fourth term (Khelashvili et al. 2006) is generally omitted.

A general solution of Equation 4.15 is obtained by the Fourier transform method (Ishimaru 1997). A Gaussian form is assumed for the incident beam $I_0(\vec{\rho},\vec{s}) = I_0(1/2\pi|\beta|^{1/2})\exp(-(1/2)\vec{s}^T\beta^{-1}\vec{s})$, where $|\beta|$ is the determinant of the matrix $\beta = 1/2\,\mathrm{diag}(1/\beta_x,1/\beta_y)$. The final solution of the RTE (Khelashvili et al. 2006) is given by

$$I(z,\vec{\rho},\vec{s}) = I_0 e^{-\tau'(z)+\tau(z)W_0} \sum_{k=0}^{\infty} \frac{(\tau(z)W_0)^k}{k!} e^{-\tau(z)W_0}$$

$$\frac{1}{2\pi|\mathbf{C}(k)|^{1/2}} \exp\left(-\frac{1}{2}(\vec{s}-\Delta\vec{s}(z))^T \mathbf{C}(k)^{-1}(\vec{s}-\Delta\vec{s}(z))\right) \quad (4.16)$$

and

$$\mathbf{C}(k) \triangleq [\beta + \alpha k] = \frac{1}{2}\,\mathrm{diag}\left(\frac{1}{\beta_x}+\frac{k}{\alpha_p},\frac{1}{\beta_y}+\frac{k}{\alpha_p}\right), \quad (4.17)$$

where $\tau(z) = \int_0^z \rho_n(z')\sigma_{\mathrm{ext}}\,dz'$ is an optical distance, $\tau'(z) = \int_0^z(\rho_n(z')\sigma_{\mathrm{ext}} + \mu(z'))\,dz'$ is the extinction length, $\Delta s(z) = \int_0^z(n_x(z')/n_0)\,dz'$ is the total beam deflection, and the albedo is defined as $W_0 \triangleq (1/4\pi)\int_{4\pi} p(\hat{s}-\hat{s}')\,d\omega' = \sigma_s/\sigma_{\mathrm{ext}} = \sigma_s/(\sigma_a + \sigma_s)$. Note that this solution assumes an infinite medium and is, therefore, not strictly valid at object edges. Equation 4.16 can be decomposed as

$$I(z,\vec{\rho},\vec{s}) = I_r(z,\vec{\rho})P(z,\vec{\rho},\vec{s}) * I_0(\vec{\rho},\vec{s}), \quad (4.18)$$

where $*$ denotes convolution with respect to $\vec{s}$, $I_r(z,\vec{\rho}) = e^{-\tau'(z)+\tau(z)W_0}$ describes apparent absorption, and $P(z,\vec{\rho},\vec{s})$ is a PDF governing the angular distribution of photons participating in USAXS or refraction.

ABI measurements of beam intensity

$$I_m(\vec{\rho},s_x) = \int_{-\infty}^{\infty} ds_x' \int_{-\infty}^{\infty} ds_y' I(Z,\vec{\rho},\vec{s}'')\phi(s_x - s_x') \quad (4.19)$$

are made at one or more analyzer crystal angular position s_x, where $\phi(s_x)$ is the analyzer crystal rocking curve, or AIP, and Z is the object thickness (Figure 4.13). The approximation $s_x \cong \theta$ can be made in the ultra-small-angle regime. From Equations 4.18 and 4.19, the measured intensity can be written as a convolution

$$I_m(\vec{\rho},s_x) = I_r(Z,\vec{\rho}) \int_{-\infty}^{\infty} ds_x'' P_{s_x}(Z,\vec{\rho},s_x)R(\vec{\rho},s_x - s_x''), \quad (4.20)$$

where $P_{s_x}(Z,\vec{\rho},s_x) = \int_{-\infty}^{\infty} ds_y'' P(Z,\vec{\rho},\vec{s}'')$ and the system rocking curve (Figure 4.16) describing the intrinsic AIP due to the incoming beam and analyzer crystal AIP is $R(\vec{\rho},s_x) = \int_{-\infty}^{\infty} ds_x' \int_{-\infty}^{\infty} ds_y' \phi(s_x - s_x') I_0(\vec{\rho},\vec{s}' - \vec{s}'')$. The system rocking curve is simplified $R(s_x) = R(\vec{\rho},s_x)$ as it usually is not a function of position $\vec{\rho}$. The moments of the system intrinsic AIP can be defined by

$$I_0 \triangleq \int_{-\infty}^{\infty} R(s_x)\,ds_x, \quad \Delta R \triangleq \frac{1}{I_0}\int_{-\infty}^{\infty} s_x R(s_x)\,ds_x, \quad \text{and}$$

$$W_R \triangleq \frac{1}{I_0}\int_{-\infty}^{\infty} (s_x - \Delta R)^2 R(s_x)\,ds_x, \quad (4.21)$$

yielding a model of the measured AIP at the detector

$$I_m[\vec{\rho},l] = I_m(\vec{\rho},\theta_l) + n, \quad (4.22)$$

where $l = 1, 2, \ldots, L$ represents the index of the analyzer crystal angular position θ_l and n is the measurement noise.

From the predicted AIP, the absorption, refraction, and USAXS images can be derived. By manipulation of Beer's law (the Beer–Lambert–Bouguer law), the absorption image is

$$A(\vec{\rho}) \triangleq -\ln\left(\frac{I_T(\vec{\rho})}{I_0}\right) = \int_0^Z \bar{\mu}(z')\,dz', \quad (4.23)$$

where $I_T(\vec{\rho})$ is the total measured intensity of the beam behind the object at all angular analyzer positions and $\bar{\mu}(z) \triangleq \rho_n(z)\sigma_a + \mu(z)$ is the net linear absorption of the object. Thus, the absorption image is a line integral of the net linear absorption coefficient of the object.

The refraction-angle image is found by estimating the angular shift of the AIP centroid. From Equation 4.21, the centroid is at $s_x = 0$ (i.e., $\Delta R = 0$) and the shift is the measured intensity centroid when the object is present. The refraction-angle image is

$$\Delta s(\vec{\rho}) \triangleq \frac{1}{I_T(\vec{\rho})}\int_{-\infty}^{\infty} s_x I_m(\vec{\rho},s_x)\,ds_x - \Delta R \quad (4.24)$$

and applying Equation 4.20 yields

$$\Delta s(\vec{\rho}) = \int_0^Z \frac{n_x(z')}{n_0}\,dz'. \quad (4.25)$$

Thus, the refraction-angle image is equal to the line integral of the scaled refractive index gradient $n_x(z)/n_0$. Note that in the

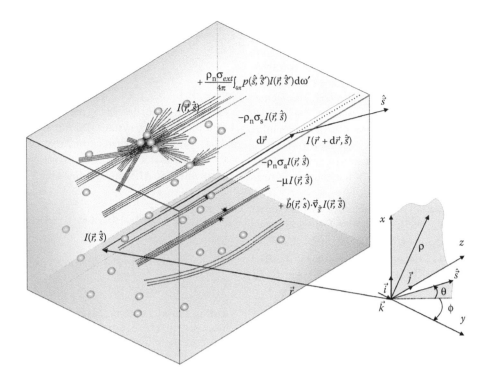

FIGURE 4.15   Illustration of terms in the RTE.

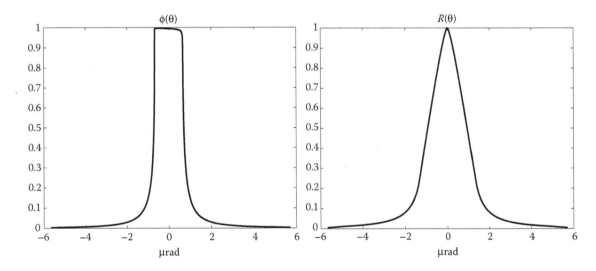

FIGURE 4.16   Example of a single crystal and system rocking curve for a Si (333) crystal at 40 keV. Note that in the ultra-small-angle regime $S_x \cong \theta$.

ultra-small-angle regime $\Delta\theta(\vec{\rho}) = \Delta s(\vec{\rho})$ and that the second term in Equation 4.24 accounts for the possibility that the rocking curve angular shift is not zero. The USAXS image is found from the second central moment of the beam AIP, which represents the beam divergence induced by scatterers. The USAXS image (Brankov et al. 2006a) is

$$w(\vec{\rho}) \triangleq \frac{1}{I_T(\vec{\rho})} \int_{-\infty}^{\infty} \left(s_x - \Delta s(\vec{\rho})\right)^2 I(\vec{\rho}, s_x) \, ds_x - W_R, \quad (4.26)$$

where $W_R = (1/\beta_x) + w_\phi$ is the second central moment of the analyzer crystal intrinsic rocking curve (i.e., the response with the object absent). Applying Equation 4.19 (Khelashvili et al. 2006) yields

$$w(\vec{\rho}) = \int_0^Z \left(\frac{\rho_n(z')\sigma_s}{2\alpha_p}\right) dz'. \quad (4.27)$$

Thus, the USAXS parameter is also a line integral through the object.

Approximation of Measured Angular Intensity Profile

Using Equation 4.16 and the small angle regime approximation, Equation 4.20 can be rewritten as

$$I_{\mathrm{m}}(\vec{\rho},\theta) =$$

$$e^{-\int_0^z \bar{\mu}(\vec{r})\,\mathrm{d}z} \sum_{k=0}^{\infty} \left[\frac{(\tau(Z)W_0)^k}{k!} e^{-\tau(Z)W_0} \sqrt{\frac{\alpha_{\mathrm{p}}}{k\pi}} e^{\left(-\frac{\alpha_{\mathrm{p}}}{k}(\theta-\Delta\theta)^2\right)} \right] * R(\theta). \quad (4.28)$$

For large values of $\tau(Z)W_0$, the summation can be approximated by $k = \tau(Z)W_0(Z)$ yielding

$$I_{\mathrm{app}}(\vec{\rho},\theta) = I_T(\vec{\rho})\sqrt{\frac{1}{2\pi\sigma^2(\vec{\rho})}} e^{\left(-\frac{(\theta-\Delta\theta(\vec{\rho}))^2}{2\sigma^2(\vec{\rho})}\right)} * R(\theta), \quad (4.29)$$

where

$$I_T(\vec{\rho}) = e^{-\int_0^Z \bar{\mu}(\vec{r})\,\mathrm{d}z},$$

$$\Delta\theta(\vec{\rho}) = \int_0^Z b(\vec{s})\,\mathrm{d}z' = \int_0^Z \frac{n_{\mathrm{x}}(z')}{n_0}\,\mathrm{d}z', \quad \text{and} \quad (4.30)$$

$$\sigma^2(\vec{\rho}) = \frac{\tau(Z)W_0}{2\alpha_{\mathrm{p}}} = \int_0^Z \frac{\rho_{\mathrm{n}}(z)\sigma_{\mathrm{s}}}{2\alpha_{\mathrm{p}}}\,\mathrm{d}z.$$

Therefore, by comparing Equations 4.23, 4.25, 4.27, and 4.30, one observes that the calculated AIP properties are identical. Therefore, the measured AIP after the object can be accurately approximated by a Gaussian PDF convolved with the system rocking curve $R(\theta)$.

References

Anastasio, M. A., C. Y. Chou, A. M. Zysk, and J. G. Brankov. 2010. Analysis of ideal observer signal detectability in phase-contrast imaging employing linear shift-invariant optical systems. *J Opt Soc Am A—Opt Image Sci Vision* 27:2648–2659.

Antunes, A., M. G. Hönnicke, C. Cusatis, and S. L. Morelhão. 2005. High contrast radiography of normal and cataractous canine lenses. *J Phys D Appl Phys* 38:A85–A88.

Antunes, A., A. M. V. Safatle, P. S. M. Barros, and S. L. Morelhão. 2006. X-ray imaging in advanced studies of ophthalmic diseases. *Med Phys* 33:2338–2343.

Arfelli, F., M. Assante, V. Bonvicini et al. 1998. Low-dose phase contrast x-ray medical imaging. *Phys Med Biol* 43:2845–2852.

Bonse, U. and M. Hart. 1965. Tailless x-ray single-crystal reflection curves obtained by multiple reflection. *Appl Phys Lett* 7:238.

Bonse, U. and M. Hart. 1966. Small angle x-ray scattering by spherical particles of polystyrene and polyvinyltoluene. *Zeitschrift für Physik* 189:151–162.

Born, M. and E. Wolf. 1999. *Principles of Optics : Electromagnetic Theory of Propagation, Interference and Diffraction of Light*. Cambridge University Press, Cambridge.

Brankov, J. G., A. Sáiz-Herranz, and M. N. Wernick. 2004. Noise analysis for diffraction enhanced imaging. In *IEEE International Symposium on Biomedical Imaging: Nano to Macro*, 2004, Arlington, Virginia, USA. 1428–1431 Vol. 1422.

Brankov, J. G., A. Sáiz-Herranz, and M. N. Wernick. 2005. Task-based evaluation of diffraction-enhanced imaging. In *Nuclear Science Symposium Conference Record*, 2005, Fajardo, Puerto Rico, IEEE. 3 pp.

Brankov, J. G., M. N. Wernick, Y. Y. Yang et al. 2006a. Computed tomography implementation of multiple-image radiography. *Med Phys* 33:278–289.

Brankov, J. G., and L. C. Cobo. 2006b. Iterative method for parametric image estimation in MIR. In *IEEE Nuclear Science Symposium Conference Record*, 2006, San Diego, California, USA. 1684–1686.

Bravin, A., J. Keyriläinen, M. Fernández et al. 2007a. High-resolution CT by diffraction-enhanced x-ray imaging: Mapping of breast tissue samples and comparison with their histo-pathology. *Phys Med Biol* 52:2197–2211.

Bravin, A., V. Mocella, P. Coan, A. Astolfo, and C. Ferrero. 2007b. A numerical wave-optical approach for the simulation of analyzer-based x-ray imaging. *Opt Express* 15:5641–5648.

Brey, E. M., A. Appel, Y. C. Chiu et al. 2010. X-Ray imaging of poly(ethylene glycol) hydrogels without contrast agents. *Tissue Eng Part C Methods* 16:1597–1600.

Briedis, D., K. K. W. Siu, D. M. Paganin, K. M. Pavlov, and R. A. Lewis. 2005. Analyser-based mammography using single-image reconstruction. *Phys Med Biol* 50:3599–3611.

Brumberger, H. 1995. *Modern Aspects of Small-Angle Scattering*. Kluwer Academic Publishers, Dordrecht.

Chapman, D., W. Thomlinson, R. E. Johnston et al. 1997. Diffraction enhanced x-ray imaging. *Phys Med Biol* 42:2015–2025.

Chapman, D., E. Pisano, W. Thomlinson et al. 1998. Medical applications of diffraction enhanced imaging. *Breast Disease* 10:197–207.

Chen, Z. -Q., F. Ding, Z. -F. Huang et al. 2009. Polynomial curve fitting method for refraction-angle extraction in diffraction enhanced imaging. *Chinese Phys C* 33:969–974.

Chou, C. Y., M. A. Anastasio, J. G. Brankov et al. 2007. An extended diffraction-enhanced imaging method for implementing multiple-image radiography. *Phys Med Biol* 52:1923–1945.

Coan, P., F. Bamberg, P. C. Diemoz et al. 2010. Characterization of osteoarthritic and normal human patella cartilage by computed tomography x-ray phase-contrast imaging a feasibility study. *Invest Radiol* 45:437–444.

Cobo, L. C. and J. G. Brankov. 2007. Evaluation of model based parametric image estimation in MIR. In *4th IEEE International Symposium on Biomedical Imaging: From Nano to Macro*, 2007, Arlington, Virginia, USA. ISBI 2007. 452–455.

Connor, D. M., H. Benveniste, F. A. Dilmanian et al. 2009b. Computed tomography of amyloid plaques in a mouse model of Alzheimer's disease using diffraction enhanced imaging. *Neuroimage* 46:908–914.

Connor, D. M., H. D. Hallen, D. S. Lalush, D. R. Sumner, and Z. Zhong. 2009a. Comparison of diffraction-enhanced computed tomography and monochromatic synchrotron radiation computed tomography of human trabecular bone. *Phys Med Biol* 54:6123–6133.

Connor, D. M., E. Pisano, Z. Zhong et al. 2011b. *Second Generation Diffraction Enhanced Imaging Prototype: Progress towards a Clinical Imaging System.* In Radiologocal Society of North America, Chicago.

Connor, D. M., D. Sayers, D. R. Sumner, and Z. Zhong. 2005. Identification of fatigue damage in cortical bone by diffraction enhanced imaging. *Nucl Instrum Meth A* 548:234–239.

Conor, D. M., D. Sayers, D. R. Sumner, and Z. Zhong. 2006. Diffraction enhanced imaging of controlled defects within bone, including bone-metal gaps. *Phys Med Biol* 51:3283–3300.

Connor, D. M., Z. Zhong, H. D. Foda et al. 2011a. Diffraction enhanced imaging of a rat model of gastric acid aspiration pneumonitis. *Acad Radiol* 18:1515–1521.

Davis, T. J., D. Gao, T. E. Gureyev, A. W. Stevenson, and S. W. Wilkins. 1995. Phase-contrast imaging of weakly absorbing materials using hard x-rays. *Nature* 373:595–598.

Diemoz, P. C., P. Coan, C. Glaser, and A. Bravin. 2010a. Absorption, refraction and scattering in analyzer-based imaging: Comparison of different algorithms. *Opt Express* 18:3494–3509.

Diemoz, P. C., A. Bravin, C. Glaser, and P. Coan. 2010b. Comparison of analyzer-based imaging computed tomography extraction algorithms and application to bone-cartilage imaging. *Phys Med Biol* 55:7663–7679.

Dilmanian, F. A., Z. Zhong, B. Ren et al. 2000. Computed tomography of x-ray index of refraction using the diffraction enhanced imaging method. *Phys Med Biol* 45:933–946.

Dreossi, D., A. Abrami, F. Arfelli et al. 2008. The mammography project at the SYRMEP beamline. *Eur J Radiol* 68:S58–S62.

Faulconer, L., C. Parham, D. M. Connor et al. 2009. Radiologist evaluation of an x-ray tube-based diffraction-enhanced imaging prototype using full-thickness breast specimens. *Acad Radiol* 16:1329–1337.

Faulconer, L. S., C. A. Parham, D. M. Connor et al. 2010. Effect of breast compression on lesion characteristic visibility with diffraction-enhanced imaging. *Acad Radiol* 17:433–440.

Fernández, M., J. Keyriläinen, R. Serimaa et al. 2005. Human breast cancer in vitro: Matching histo-pathology with small-angle x-ray scattering and diffraction enhanced x-ray imaging. *Phys Med Biol* 50:2991–3006.

Fiedler, S., A. Bravin, J. Keyriläinen et al. 2004. Imaging lobular breast carcinoma: Comparison of synchrotron radiation DEI-CT technique with clinical CT, mammography and histology. *Phys Med Biol* 49:175–188.

Fogarty, D. P., B. Reinhart, T. Tzvetkov, I. Nesch, and C. Williams. 2011. In-laboratory diffraction-enhanced x-ray imaging of an equine hoof. *J Equine Vet Sci* 31:365–369.

Förster, E., K. Goetz, and P. Zaumseil. 1980. Double crystal diffractometry for the characterization of targets for laser fusion experiments. *Kristall und Technik* 15:937–945.

Guigay, J. P., E. Pagot, and P. Cloetens. 2007. Fourier optics approach to x-ray analyser-based imaging. *Opt Commun* 270:180–188.

Hasnah, M., O. Oltulu, Z. Zhong, and D. Chapman. 2002a. Single-exposure simultaneous diffraction-enhanced imaging. *Nucl Instrum Meth A* 492:236–240.

Hasnah, M. O., Z. Zhong, O. Oltulu et al. 2002b. Diffraction enhanced imaging contrast mechanisms in breast cancer specimens. *Med Phys* 29:2216–2221.

Hu, C. H., L. Zhang, H. Li, and S. Q. Luo. 2008. Comparison of refraction information extraction methods in diffraction enhanced imaging. *Opt Express* 16:16704–16710.

Huang, Z. F., K. J. Kang, Z. Li et al. 2006. Direct computed tomographic reconstruction for directional-derivative projections of computed tomography of diffraction enhanced imaging. *Appl Phys Lett* 89:041124.

Huang, Z. F., K. J. Kang, P. P. Zhu et al. 2007. Strategy of extraction methods and reconstruction algorithms in computed tomography of diffraction enhanced imaging. *Phys Med Biol* 52:1–12.

Ingal, V. N., E. A. Beliaevskaya, A. P. Brianskaya, and R. D. Merkurieva. 1998. Phase mammography—A new technique for breast investigation. *Phys Med Biol* 43:2555–2567.

Ishimaru, A. 1997. *Wave Propagation and Scattering in Random Media.* IEEE Press and Oxford University Press, New York.

Kao, T., C. J. Liu, X. H. Yu et al. 2007. Characterization of diffraction enhanced imaging contrast in plants. *Nucl Instrum Meth A* 582:208–211.

Kao, T., D. Connor, F. A. Dilmanian et al. 2009. Characterization of diffraction-enhanced imaging contrast in breast cancer. *Phys Med Biol* 54:3247–3256.

Kay, S. M. 1993. *Fundamentals of Statistical Signal Processing.* Prentice-Hall PTR, Englewood Cliffs, NJ.

Kelly, M. E., R. C. Beavis, D. R. Fourney et al. 2006. Diffraction-enhanced imaging of the rat spine. *Can Assoc Radiol J* 57:204–210.

Kelly, M. E., D. J. Coupal, R. C. Beavis et al. 2007. Diffraction-enhanced imaging of a porcine eye. *Can J Ophth* 42: 731–733.

Kelly, M. E., R. C. Beavis, D. Fiorella et al. 2008. Analyzer-based imaging of spinal fusion in an animal model. *Phys Med Biol* 53:2607–2616.

Keyriläinen, J., M. Fernández, S. Fiedler et al. 2005. Visualisation of calcifications and thin collagen strands in human breast tumour specimens by the diffraction-enhanced imaging technique: A comparison with conventional mammography and histology. *Eur J Radiol* 53:226–237.

Keyriläinen, J., M. Fernández, M. L. Karjalainen-Lindsberg et al. 2008. Toward high-contrast breast CT at low radiation dose. *Radiology* 249:321–327.

Khelashvili, G., J. G. Brankov, D. Chapman et al. 2006. A physical model of multiple-image radiography. *Phys Med Biol* 51:221–236.

Kim, C. H., M. A. Bourham, and J. M. Doster. 2006. A wide-beam x-ray source suitable for diffraction enhanced imaging applications. *Nucl Instrum Meth A* 566:713–721.

Kiss, M. Z., D. E. Sayers, Z. Zhong, C. Parham, and E. D. Pisano. 2004. Improved image contrast of calcifications in breast tissue specimens using diffraction enhanced imaging. *Phys Med Biol* 49:3427–3439.

Kitchen, M. J., R. A. Lewis, N. Yagi et al. 2005. Phase contrast x-ray imaging of mice and rabbit lungs: A comparative study. *Brit J Radiol* 78:1018–1027.

Kitchen, M. J., K. M. Pavlov, K. K. W. Siu et al. 2007. Analyser-based phase contrast image reconstruction using geometrical optics. *Phys Med Biol* 52:4171–4187.

Kitchen, M. J., D. M. Paganin, K. Uesugi et al. 2010. X-ray phase, absorption and scatter retrieval using two or more phase contrast images. *Opt Express* 18:19994–20012.

Kitchen, M. J., D. M. Paganin, K. Uesugi et al. 2011. Phase contrast image segmentation using a Laue analyser crystal. *Phys Med Biol* 56:515–534.

Krug, K., L. Porra, P. Coan et al. 2008. Relics in medieval altarpieces? Combining x-ray tomographic, laminographic and phase-contrast imaging to visualize thin organic objects in paintings. *J Synchrotron Radiat* 15:55–61.

Lewis, R. A., C. J. Hall, A. P. Hufton et al. 2003. X-ray refraction effects: Application to the imaging of biological tissues. *Brit J Radiol* 76:301–308.

Li, H., L. Zhang, X. Y. Wang et al. 2009a. Investigation of hepatic fibrosis in rats with x-ray diffraction enhanced imaging. *Appl Phys Lett* 94:124101.

Li, J., Z. Zhong, R. Lidtke et al. 2004. Radiography of soft tissue of the foot and ankle with diffraction enhanced imaging. *J Am Podiat Med Assn* 94:315–322.

Li, J., J. M. Williams, Z. Zhong et al. 2005. Reliability of diffraction enhanced imaging for assessment of cartilage lesions, ex vivo. *Osteoarthr Cartilage* 13:187–197.

Li, J., Z. Zhong, D. Connor, J. Mollenhauer, and C. Muehleman. 2009b. Phase-sensitive x-ray imaging of synovial joints. *Osteoarthr Cartilage* 17:1193–1196.

Liu, C. L., Y. Zhang, X. Y. Zhang et al. 2005. X-ray diffraction-enhanced imaging of uterine leiomyomas. *Med Sci Monitor* 11:Mt33–Mt38.

Long, G. G., P. R. Jemian, J. R. Weertman et al. 1991. High-resolution small-angle x-ray-scattering camera for anomalous scattering. *J Appl Crystallogr* 24:30–37.

Majidi, K., J. G. Brankov, J. Li, C. Muehleman, and M. N. Wernick. 2008a. Parameter estimation in multiple-image radiography. In *IEEE Nuclear Science Symposium Conference Record*, 2008, Dresden, Germany. 4214–4217.

Majidi, K., J. G. Brankov, and M. N. Wernick. 2008b. Sampling strategies in multiple-image radiography. In *5th IEEE International Symposium on Biomedical Imaging: From Nano to Macro*, 2008, Paris, France. 688–691.

Majumdar, S., A. S. Issever, A. Burghardt et al. 2004. Diffraction enhanced imaging of articular cartilage and comparison with micro-computed tomography of the underlying bone structure. *Eur Radiol* 14:1440–1448.

Maksimenko, A., M. Ando, S. Hiroshi, and T. Yuasa. 2005. Computed tomographic reconstruction based on x-ray refraction contrast. *Appl Phys Lett* 86:124105.

Maksimenko, A. 2007. Nonlinear extension of the x-ray diffraction enhanced imaging. *Appl Phys Lett* 90:154106.

Marquet, B., J. G. Brankov, and M. N. Wernick. 2006. Noise and sampling analysis for multiple-image radiography. In *3rd IEEE International Symposium on Biomedical Imaging: Nano to Macro*, 2006, Arlington, Virginia, USA. 1232–1235.

Modregger, P., D. Lubbert, P. Schafer, and R. Kohler. 2007. Two dimensional diffraction enhanced imaging algorithm. *Appl Phys Lett* 90:193501.

Moeckli, R., F. R. Verdun, S. Fiedler et al. 2000. Objective comparison of image quality and dose between conventional and synchrotron radiation mammography. *Phys Med Biol* 45:3509–3523.

Mollenhauer, J., M. E. Aurich, Z. Zhong et al. 2002. Diffraction-enhanced x-ray imaging of articular cartilage. *Osteoarthr Cartilage* 10:163–171.

Muehleman, C., D. Fogarty, B. Reinhart et al. 2010. In-laboratory diffraction-enhanced x-ray imaging for articular cartilage. *Clin Anat* 23:530–538.

Muehleman, C., J. Li, K. E. Kuettner, and Z. Zhong. 2004. Diffraction enhanced x-ray imaging of musculoskeletal lesions. *Osteoarthr Cartilage* 12:S117.

Muehleman, C., J. Li, D. Connor et al. 2009b. Diffraction-enhanced imaging of musculoskeletal tissues using a conventional x-ray tube. *Acad Radiol* 16:918–923.

Muehleman, C., J. Li, A. Schiff, and Z. Zhong. 2009a. Diffraction-enhanced imaging for achilles tendon lesions a preliminary study. *J Am Podiat Med Assn* 99:95–99.

Muehleman, C., M. Whiteside, Z. Zhong et al. 2002. Diffraction enhanced imaging for articular cartilage. *Biophys J* 82:470A–470A.

Muehleman, C., J. Li, Z. Zhong, J. G. Brankov, and M. N. Wernick. 2006a. Multiple-image radiography for human soft tissue. *J Anat* 208:115–124.

Muehleman, C., J. Li, and Z. Zhong. 2006b. Preliminary study on diffraction enhanced radiographic imaging for a canine model of cartilage damage. *Osteoarthr Cartilage* 14:882–888.

Nesch, I., D. P. Fogarty, T. Tzvetkov et al. 2009. The design and application of an in-laboratory diffraction-enhanced x-ray imaging instrument. *Rev Sci Instrum* 80:093702.

Nesterets, Y. I. 2008. On the origins of decoherence and extinction contrast in phase-contrast imaging. *Opt Commun* 281:533–542.

Nesterets, Y. I., P. Coan, T. E. Gureyev et al. 2006b. On qualitative and quantitative analysis in analyser-based imaging. *Acta Crystallogr A* 62:296–308.

Nesterets, Y. I., T. E. Gureyev, D. Paganin, K. M. Pavlov, and S. W. Wilkins. 2004. Quantitative diffraction-enhanced x-ray imaging of weak objects. *J Phys D Appl Phys* 37:1262–1274.

Nesterets, Y. I., T. E. Gureev, K. M. Pavlov et al. 2005. Comparison of three theoretical approaches to quantitative analyzer-based imaging. *Bulletin—Russian Academy of Sciences Physics C/C* 69:223–227.

Nesterets, Y. I., T. E. Gureyev, K. M. Pavlov, D. M. Paganin, and S. W. Wilkins. 2006a. Combined analyser-based and propagation-based phase-contrast imaging of weak objects. *Opt Commun* 259:19–31.

Olkowski, A. A., D. Chapman, T. W. Wysokinski et al. 2010. Study of Bone Degeneration in a Chicken Model: Comprehensive Analysis Using Traditional Analytical Approach and Novel Synchrotron-Based Techniques. In 2010 Activity Report, Canadian Light Source Inc. 54–56.

Oltulu, O., Z. Zhong, M. Hasnah, M. N. Wernick, and D. Chapman. 2003. Extraction of extinction, refraction and absorption properties in diffraction enhanced imaging. *J Phys D Appl Phys* 36:2152–2156.

Paganin, D., T. E. Gureyev, K. M. Pavlov, R. A. Lewis, and M. Kitchen. 2004. Phase retrieval using coherent imaging systems with linear transfer functions. *Opt Commun* 234:87–105.

Paganin, D. M. and T. E. Gureyev. 2008. Phase contrast, phase retrieval and aberration balancing in shift-invariant linear imaging systems. *Opt Commun* 281:965–981.

Pagot, E., P. Cloetens, S. Fiedler et al. 2003. A method to extract quantitative information in analyzer-based x-ray phase contrast imaging. *Appl Phys Lett* 82:3421–3423.

Parham, C., Z. Zhong, D. M. Connor, L. D. Chapman, and E. D. Pisano. 2009. Design and implementation of a compact low-dose diffraction enhanced medical imaging system. *Acad Radiol* 16:911–917.

Pavlov, K. M., C. M. Kewish, J. R. Davis, and M. J. Morgan. 2000. A new theoretical approach to x-ray diffraction tomography. *J Phys D Appl Phys* 33:1596–1605.

Pavlov, K. M., C. M. Kewish, J. R. Davis, and M. J. Morgan. 2001. A variant on the geometrical optics approximation in diffraction enhanced tomography. *J Phys D Appl Phys* 34:A168–A172.

Pavlov, K. M., T. E. Gureyev, D. Paganin et al. 2004. Linear systems with slowly varying transfer functions and their application to x-ray phase-contrast imaging. *J Phys D Appl Phys* 37:2746–2750.

Pavlov, K. M., T. E. Gureyev, D. Paganin et al. 2005. Unification of analyser-based and propagation-based x-ray phase-contrast imaging. *Nucl Instrum Meth A* 548:163–168.

Pisano, E. D., R. E. Johnston, D. Chapman et al. 2000. Human breast cancer specimens: Diffraction-enhanced imaging with histologic correlation—Improved conspicuity of lesion detail compared with digital radiography. *Radiology* 214:895–901.

Powell, M. J. D. 1986. Convergence properties of algorithms for nonlinear optimization. *SIAM Rev* 28:487–500.

Rao, D. V., M. Swapna, R. Cesareo et al. 2010. Use of synchrotron-based diffraction-enhanced imaging for visualization of soft tissues in invertebrates. *Appl Radiat Isotopes* 68:1687–1693.

Rigon, L., F. Arfelli, and R. H. Menk. 2007. Three-image diffraction enhanced imaging algorithm to extract absorption, refraction, and ultrasmall-angle scattering. *Appl Phys Lett* 90:114102.

Rigon, L., A. Astolfo, F. Arfelli, and R. H. Menk. 2008. Generalized diffraction enhanced imaging: Application to tomography. *Eur J Radiol* 68:S3–S7.

Rigon, L., H. J. Besch, F. Arfelli et al. 2003. A new DEI algorithm capable of investigating sub-pixel structures. *J Phys D Appl Phys* 36:A107–A112.

Sunaguchi, N., T. Yuasa, Q. K. Huo, S. Ichihara, and M. Ando. 2010. X-ray refraction-contrast computed tomography images using dark-field imaging optics. *Appl Phys Lett* 97:153701.

Tan, G., H. Q. Wang, Y. Chen et al. 2010. X-ray diffraction enhanced imaging study of intraocular tumors in human beings. *Chinese Phys C* 34:237–243.

Vine, D. J., D. M. Paganin, K. M. Pavlov et al. 2007. Analyzer-based phase contrast imaging and phase retrieval using a rotating anode x-ray source. *Appl Phys Lett* 91:254110.

Wagner, A., M. Aurich, N. Sieber et al. 2005. Options and limitations of joint cartilage imaging: DEI in comparison to MRI and sonography. *Nucl Instrum Meth A* 548:47–53.

Wang, J. Y., P. P. Zhu, Q. X. Yuan et al. 2006. Design and construction of an x-ray phase contrast CT system at BSRF. *Radiat Phys Chem* 75:1986–1989.

Wernick, M. N., O. Wirjadi, D. Chapman et al. 2003. Multiple-image radiography. *Phys Med Biol* 48:3875–3895.

Wernick, M. N., Y. Yang, I. Mondal et al. 2006. Computation of mass-density images from x-ray refraction-angle images. *Phys Med Biol* 51:1769–1778.

Wysokinski, T. W., D. Chapman, G. Adams et al. 2007. Beamlines of the biomedical imaging and therapy facility at the Canadian light source—Part I. *Nucl Instrum Meth A* 582:73–76.

Yin, H. X., Z. F. Huang, Z. C. Wang et al. 2010. [Application research of DEI technique based on synchrotron x-ray source in imaging rabbit eyeball in vitro]. *Zhonghua yi xue za zhi* 90:777–781.

Young, L. W., C. Parham, Z. Zhong, D. Chapman, and M. J. T. Reaney. 2007. Non-destructive diffraction enhanced imaging of seeds. *J Exp Bot* 58:2513–2523.

Zhang, X., Q. X. Yuan, X. R. Yang et al. 2009. Medical application of diffraction enhanced imaging in mouse liver blood vessels. *Chinese Phys C* 33:986–990.

5

Propagation-Based Imaging

Glenn R. Myers
The Australian National University

Karen K.W. Siu
Monash University
Australian Synchrotron

Kaye S. Morgan
Monash University

5.1 Overview and Clinical Potential

In conventional x-ray radiography (e.g., a chest radiograph), a patient is illuminated by the penetrating x-rays. The patient attenuates the x-rays, and so casts a shadow on the detector. Stronger shadows correspond to strongly attenuating (usually dense) materials such as bone, while many soft-tissue features cause little to no attenuation and so do not cast meaningful shadows. Conventional x-ray radiography is thus an example of "attenuation contrast" imaging: our ability to resolve features in the radiograph is entirely dependent on the (harmful) absorption of x-rays by the patient.

Many features of interest (e.g., the air-filled lung) are weakly attenuating, and so cannot be reliably imaged using conventional, attenuation-contrast methods. Although these features do not attenuate x-rays, they will refract them (i.e., they bend/focus the x-rays in the same way a transparent lens bends light). If the detector is placed some distance (~1 m is typical for clinical applications) away from the sample, this refraction can lead to patterns of light and dark lines, or "fringes," in the radiographs. These fringes can be used to provide information about weakly attenuating features in the sample, expanding the potential range of applications for clinical x-ray radiography. This is an example of "refraction contrast" imaging [also referred to as phase-contrast imaging (PCI)]: our ability to resolve features in the radiograph is no longer entirely dependent on the existence of an absorbed dose, allowing for faster, safer imaging. Improvements in imaging speed can be leveraged to study the dynamic processes, such as tracking transport of airway contaminants.

The crucial step in the imaging method outlined above is the introduction of the gap between the sample and detector: it is the propagation of x-rays through this gap that renders the refraction of x-rays visible as light and dark lines in the radiograph.

For this reason, it is referred to as propagation-based phase-contrast imaging (PB–PCI).

The increased resolving power offered by PB–PCI is isometric (i.e., not dependent on feature orientation), but favors the high spatial frequencies (i.e., small/sharp features). Unlike many other PCI methods, PB–PCI does not require any expensive x-ray optics between the patient and detector; guidelines for calculating the ideal patent-detector separation may be found in Section 5.2.4. The penumbra effects associated with large patient-detector separation mean that PB–PCI must be implemented with either a microfocus x-ray source (source spot size on the order of 10 μm or less) or a very large source-patient separation (~100 m, as is found at a synchrotron). The resolving power of PB–PCI decays as x-ray wavelength is decreased, but does so more slowly than the absorbed dose.

Quantitative interpretation of attenuation-contrast images is fairly straightforward. In contrast, quantitative interpretation of PB–PCI images requires the use of reconstruction algorithms discussed in Section 5.3. Under certain circumstances, it is possible to perform quantitative PB–PCI with a clinical microfocus source. In fact, it has been demonstrated that the combination of PB–PCI and computed axial tomography (CAT) to achieve three dimensional (3D) PB–PCI is not only possible with a microfocus source; it is less vulnerable to noise than conventional attenuation-based CAT (in addition to the "standard" benefits of PB–PCI, listed above).

The enhanced contrast offered by PB–PCI, coupled with its relative ease of implementation, has meant that it has found widespread use in preclinical and clinical applications (see Section 5.4). Visualization of soft-tissue structures at high spatial resolution has proved useful for furthering understanding the pathologies of the musculoskeletal system, and improving the diagnostic capabilities of mammography. PB–PCI capabilities have grown in parallel on synchrotron and laboratory sources, with clinical trials of phase-contrast mammography being conducted on both types of sources in recent years. The impact and interest of this imaging modality is ably demonstrated by the fact that a commercial phase-contrast mammography system was released in 2005.

Preclinical studies of the dynamics of biological processes are similarly benefiting from the advantages of PB–PCI. Synchrotron-source-based imaging has permitted the short exposures required for imaging both cardiological and respiratory processes. The ability to noninvasively observe blood flow, air flow, the motion of tissue, and of liquid layers, in high resolution, means the direct effect of treatments may be assessed in real time. This has already led to results in the biomedical research on the first breaths of infants and lung conditions like cystic fibrosis. The translation of these dynamic imaging techniques into the clinic would be valuable in both early detection and monitoring of many biological processes.

Sections 5.2 and 5.3 comprise a detailed review of the physics underlying PB–PCI, with a focus on how this physics influences the implementation of the technique: ideal imaging setup, suitable samples, and so on. Section 5.4 reviews some of the current clinical and preclinical uses of PB–PCI, covering the orthopedic, mammography, cardiovascular, and respiratory applications.

5.2 Qualitative Imaging: Phase Contrast

5.2.1 The Formation of Propagation-Based Phase Contrast

In a typical x-ray imaging system, a sample of interest is illuminated with x-rays. The sample interacts with the illumination, and introduces structured perturbations into the x-ray beam. In the wave-optics formulation, the perturbed x-ray beam may be completely described by two scalar quantities: intensity and phase (Born and Wolf 1999). These correspond to the rate, and direction of energy flow respectively: the intensity at a given point is proportional to the energy flux through that point, and surfaces of constant phase are defined to be perpendicular to the direction of energy flow (see Figure 5.1) (Born and Wolf 1999). Identifying these "surfaces of constant phase" with wavefronts, we see that introducing a perturbation into the x-ray phase corresponds to the refraction of the x-rays. As intensity is proportional to the energy flux, introducing a perturbation into the x-ray intensity corresponds to the attenuation/emission of the x-rays.

Direct measurement of the x-ray phase is impractical, as it requires a time resolution on the order of approximately 10^{-18} s (Paganin 2006). Consequently, conventional biomedical x-ray imaging systems ignore both the x-ray phase and the corresponding refraction within the sample.* As a result, these imaging

---

* In clinical imaging, scattering/refraction is commonly removed by anti-scattering (or "bucky") grids immediately in front of the detector because it contributes to elevated background in the images (i.e., reduces SNR). These work on essentially geometric principles; their collimation is designed to match the point source at a fixed source-to-detector distance. The problem with this scatter rejection approach is that an elevated dose to the patient is required because less flux reaches the detector.

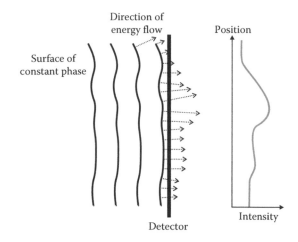

FIGURE 5.1 Energy flow is represented by the dotted lines, with length corresponding to magnitude. Surfaces of constant phase (solid lines) are perpendicular to the direction of energy flow.

systems rely on the attenuation of x-rays to resolve the features of interest. This becomes problematic if (i) we desire to minimize the absorbed radiation dose during imaging due to either time constraints (e.g., imaging dynamic systems), or a desire to avoid damaging the sample (e.g., any medical application); (ii) the sample of interest is transparent to x-rays, or is weakly attenuating (e.g., most soft-tissue samples); and/or (iii) our imaging setup is such that refractive effects are significant, and will lead to image artifacts if ignored (e.g., high-resolution imaging with a microfocus x-ray source).

Weakly attenuating samples may still significantly refract the x-ray beam (Fitzgerald 2000; Řeháček et al. 2006). In these circumstances, we may attempt indirect measurement of the x-ray phase. This may be done with any optical system that converts phase variations in the x-rays exiting the sample into measurable intensity variations at the detector. In general, the images produced by such an optical system will display both attenuation contrast and refraction contrast (a.k.a. phase contrast) (Paganin and Gureyev 2008). Any system capable of producing such images is referred to as a PCI system. Quantitative interpretation of attenuation-contrast x-ray images is relatively straightforward. In PCI, this is complicated by the mix of attenuation and phase contrast in the images. "Phase retrieval" refers to the inverse problem of calculating the x-ray phase from these phase-contrast measurements.

PCI is sensitive to both attenuation and refraction within the sample, and thus has many advantages over conventional techniques: an absorbed x-ray dose is no longer the prime determinant of image contrast, allowing for rapid imaging of dynamic systems and safe imaging of easily damaged samples. Furthermore, one can better distinguish features in weakly attenuating samples (e.g., soft tissues) (Paganin 2006).

The optical system used in PB–PCI is a gap between the sample and detector (see Figure 5.2). The wave-like nature of x-rays ensures that they undergo diffraction as they propagate across this gap, meaning that the detected intensity depends

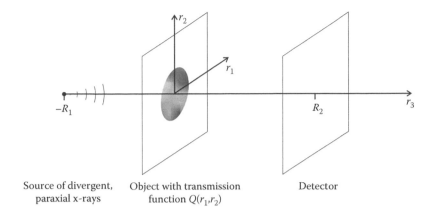

Source of divergent,   Object with transmission   Detector
paraxial x-rays   function $Q(r_1,r_2)$

FIGURE 5.2 A thin object is illuminated by x-rays from a point source. The perturbed x-rays travel through free space before being recorded by a detector.

on both the intensity and phase of the x-rays leaving the sample. Qualitatively speaking, these effects manifest as fringes that emphasize material boundaries, which become more pronounced as the sample–detector distance is increased. PB–PCI is relatively easy to implement, and requires no specialized optical equipment beyond the source and detector (Snigirev et al. 1995, 1996; Cloetens et al. 1996; Nugent et al. 1996; Wilkins 1996; Wilkins et al. 1996; Paganin 2006).

In this section, we form a quantitative model of PC–PCI, and explore the factors affecting phase-contrast formation. We begin by introducing the mathematical notation used to describe a generalized biomedical PB–PCI setup. We quantify the interactions between x-rays and the sample, in order to illustrate which broad categories of sample are most suitable for PCI. We then discuss the effects of free-space propagation on the intensity of the x-rays, and use this model to quantify the effects of source–sample distance, source spot size, and other physical factors on phase-contrast visibility.

5.2.2 Modeling Biomedical Imaging

In this subsection, we introduce a simplified biomedical imaging setup (see Figure 5.2). Defining $\mathbf{r} = (r_1, r_2, r_3)$ as a 3D Cartesian coordinate system, we are free to orient our coordinate system such that the x-ray beam is directed along the r_3 axis. The x-rays are emitted from a point source, located at $r_3 = -R_1$. We note that in both synchrotron (i.e., essentially nondiverging) and lab-based (i.e., diverging) x-ray sources, the x-rays travel approximately parallel to a single axis (Paganin 2006). This is known as the paraxial approximation or the "15 degree" approximation, as it begins to break down for diverging beams with a half-cone angle of more than 15°. The sample of thickness T is located between the planes $r_3 = -T$ and the "exit plane" (a.k.a. object plane) $r_3 = 0$. Finally, a detector is located some distance downstream in the plane $r_3 = R_2$. This model will be used throughout this chapter. To model nondiverging beams (i.e., synchrotron illumination), one needs only to explore the limiting case as $R_1 \to \infty$. Complications such as

finite source size and finite detector pixel size will be discussed in future subsections.

Assuming a linear, isotropic, nonmagnetic sample that does not vary on the length scale of the x-ray wavelength λ, we may describe x-rays as a superposition of complex, scalar, monochromatic (i.e., single-wavelength) waves $\psi(\mathbf{r},\lambda)$ (Born and Wolf 1999; Nieto-Vesperinas 2006). The phase $\phi(\mathbf{r},\lambda)$ of a given monochromatic component will be equal to the complex argument of $\psi(\mathbf{r},\lambda)$, and the time-averaged intensity $I(\mathbf{r})$ will be equal to the sum of the squared magnitudes:

$$I(\mathbf{r}) = \int_0^\infty |\psi(\mathbf{r},\lambda)|^2 \ d\lambda. \tag{5.1}$$

The intensity contribution of a single monochromatic component is referred to as the spectral density $S(\mathbf{r},\lambda) = |\psi(\mathbf{r},\lambda)|^2$ (Goodman 1985; Mandel and Wolf 1995; Gureyev 1999). For a strictly monochromatic x-ray beam, we find

$$\psi(\mathbf{r},\lambda) = I(\mathbf{r})^{1/2}\exp[i\phi(\mathbf{r},\lambda)]. \tag{5.2}$$

The presence of a sample introduces a perturbation $Q(r_1,r_2,\lambda)$, referred to as the transmission function, into the exit-plane x-ray wavefield

$$\psi(r_1,r_2,0,\lambda) = S_{\text{in}}(\lambda)^{1/2}Q(r_1,r_2,\lambda)\exp\left[\frac{ik}{2R_1}(r_1^2 + r_2^2)\right], \tag{5.3}$$

where $S_{\text{in}}(\lambda)$ is the spectral density of the incident illumination, and the exponent represents the curvature of a paraxial, diverging x-ray beam. The transmission function may be separated into (real) intensity $q(r_1,r_2,\lambda)$ and phase $\varphi(r_1,r_2,\lambda)$ disturbances

$$Q(r_1,r_2,\lambda) = q(r_1,r_2,\lambda)^{1/2}\exp\left[i\varphi(r_1,r_2,\lambda)\right]. \tag{5.4}$$

In the next subsection, we present an expression showing how the transmission function $Q(r_1, r_2, \lambda)$ depends on sample properties. This allows us to discuss what samples are most suitable for PB–PCI. We then move on to quantitative discussion of how the measured intensity in the plane $r_3 = R_2$ depends on the transmission function and the properties of the imaging system.

5.2.3 Matter/X-Ray Interactions

In biomedical applications, there is a preponderance of transparent and weakly attenuating samples, that is, samples which imprint little or no information on the intensity of the x-ray beam and are thus poorly resolved using conventional attenuation-based imaging techniques. For example, the majority of soft-tissue features cannot be resolved based on the x-ray attenuation alone at the x-ray energies used in clinical imaging (Fitzgerald 2000; Řeháček et al. 2006). Although they are almost transparent to x-rays, the light elements (hydrogen, carbon, oxygen, etc.) composing soft tissues may measurably refract the x-ray beam. In this subsection, we introduce a quantitative model of the interactions between the sample and the illuminating x-rays, in order to shed light on which samples are best suited for PCI. In CAT, this model is inverted to derive the quantitative sample properties from measurements of the scattered x-rays (Natterer 2001).

A linear, isotropic, nonmagnetic sample that does not vary significantly over length scales comparable to the x-ray wavelength λ may be characterized by the position- and wavelength-dependent complex refractive index (Born and Wolf 1999; Paganin 2006)

$$n(\mathbf{r}, \lambda) = [\varepsilon(\mathbf{r}, \lambda)\mu_0]^{1/2} c, \qquad (5.5)$$

defined in terms of the wavelength-dependent electrical permittivity $\varepsilon(\mathbf{r}, \lambda)$ of the sample, the magnetic permittivity of free space μ_0, and the speed of light c. Some typical values for $n(\mathbf{r}, \lambda)$ are found in Table 5.1. Just as x-rays will undergo diffraction

while moving through free space, they will also undergo diffraction while moving through a sample. In the vast majority of clinical applications (e.g., CAT scans), the sample is sufficiently thin that diffraction within the sample may be neglected.[*] This amounts to the geometric optics approximation: we treat x-rays as discrete rays that follow straight line paths through the sample, accumulating intensity and phase disturbances as they do so. The transmission function of the sample thus corresponds to the accumulated refractive index along the ray-path (Paganin 2006)

$$Q(r_1, r_2, \lambda) = \exp\left\{ -ik \int_{\overline{SR}} [1 - n(\mathbf{r}, \lambda)]\, ds \right\}, \qquad (5.6)$$

$$q(r_1, r_2, \lambda) = \exp\left[-2k \int_{\overline{SR}} \beta(\mathbf{r}, \lambda)\, ds \right], \quad \beta(\mathbf{r}, \lambda) \equiv \mathrm{Im}[n(\mathbf{r}, \lambda)], \quad (5.7)$$

$$\varphi(r_1, r_2, \lambda) = -k \int_{\overline{SR}} \Delta(\mathbf{r}, \lambda)\, ds, \quad \Delta(\mathbf{r}, \lambda) \equiv 1 - \mathrm{Re}[n(\mathbf{r}, \lambda)], \quad (5.8)$$

$$(\mathcal{P}\Delta)(r_1, r_2, \lambda) \equiv \int_{\overline{SR}} \Delta(\mathbf{r}, \lambda)\, ds, \qquad (5.9)$$

where $\overline{SR}$ is the ray-path connecting the source at $(0, 0, -R_1)$ and the point $(r_1, r_2, 0)$, $(\mathcal{P}n)$ is referred to as the projection operator, and s is the position along this path.

This result allows us to conclude our interpretation of the complex refractive index $n(\mathbf{r}, \lambda)$. We can see that

1. The 3D distribution of complex refractive index $n(\mathbf{r}, \lambda)$ completely characterizes the information imprinted by the sample on the illuminating x-rays. Consequently, the goal of any quantitative PB–PCI imaging system is to retrieve either $(\mathcal{P}n)(r_1, r_2, \lambda)$ (2D imaging), or $n(\mathbf{r}, \lambda)$ (3D imaging).
2. The real decrement to the refractive index $\Delta(\mathbf{r}, \lambda)$ alters the complex argument of the transmission function, and thus the phase of the x-rays. Recalling the definition of phase, we see that the real decrement to the refractive index describes the refractive properties of the sample.
3. Similarly, the imaginary component of the refractive index $\beta(\mathbf{r}, \lambda)$ alters the amplitude of the transmission function, and thus the intensity of the incoming radiation. Thus, $\beta(\mathbf{r}, \lambda)$ describes the attenuative properties of a sample.[†]

We can now identify what samples are best suited for PCI, and what samples are best suited for conventional (i.e., attenuation-based) methods. Referring again to Table 5.1, we see that cortical

TABLE 5.1 Refractive Index of Some Common Materials at 65 keV

Material	$n(\mathbf{r}) - 1.0$ ($\times 10^{-10}$)
Free space	0.0
Adipose	$-507 + 0.271i$
Air	$-0.591 + 0.333 \times 10^{-3}i$
Blood	$-573 + 0.322i$
Breast	$-553 + 0.303i$
Cortical bone	$-971 + 0.822i$
Lung	$-568 + 0.318i$
Muscle (skeletal)	$-567 + 0.317i$
Skin	$-587 + 0.327i$
Soft tissue	$-573 + 0.320i$
Water	$-545 + 0.304i$

Source: Values obtained from CSIRO Imaging Services at http://www.ts-imaging.net/Services/Default.aspx.

---

[*] For a detailed examination of the validity of this approximation, we refer the reader to Morgan et al. (2010).

[†] Identifying the imaginary refractive index with the linear attenuation coefficient $\mu(\mathbf{r}, \lambda) = 2k\beta(\mathbf{r}, \lambda)$, and considering a monochromatic beam incident upon a single material with linear attenuation coefficient μ and thickness T, equation 6.7 becomes Beer's law of absorption: $I(r_1, r_2, 0) = I_{\mathrm{in}}\exp[\mu T]$.

bone and skeletal muscle have very different imaginary refractive indices. Consequently, we expect to be able to distinguish these materials using a conventional x-ray imaging system. Indeed, hospital-based x-ray imaging systems are commonly used to image the skeletal system. Attenuation-based imaging is unable to reliably differentiate skeletal muscle from blood; these two materials have very similar imaginary refractive indices, and thus cannot be reliably differentiated based on attenuation alone. Note, however, that differences in the real decrement to the refractive index are several orders of magnitude greater than the differences in the imaginary part. We thus conclude that skeletal muscle and blood may be reliably differentiated based on their refractive properties. PCI is thus best-suited for imaging soft-tissue (or other weakly attenuating) features, such as damaged ligaments, cartilage tears, airway contaminants, and so on (Fitzgerald 2000; Řeháček et al. 2006).

It is also worth noting that there exists a special category of samples for which the "phase-attenuation duality" holds true, that is, the ratio

$$\epsilon(\lambda) = \beta(\mathbf{r},\lambda)/\Delta(\mathbf{r},\lambda) \tag{5.10}$$

is independent of position. For these samples, the refractive and attenuative properties of the sample are linked, significantly simplifying the imaging process (see, e.g., Section 5.3.2). This ratio is valid for x-ray-illuminated samples composed of a single material (Paganin et al. 2002; Mayo et al. 2003), as well as for objects composed of light elements ($Z < 10$) illuminated by high energy (60–500 keV) x-rays (Wu and Liu 2005).

Even if strong attenuation-contrast renders PCI unnecessary, PB–PCI may still occur as an unavoidable consequence of the imaging setup. In the next subsection, we introduce a quantitative model of x-ray propagation, in order to investigate the circumstances in which PB–PCI arises.

5.2.4 Contrast in the Near Field

We now seek to understand the propagation of x-rays through free space, in order to properly discuss the experimental factors affecting the visibility of propagation-based phase contrast. To do so, we again return to our model PB–PCI system. We note that there is a region of free space between the sample and detector. In this section, we are interested in how the propagated intensity $I(r_1,r_2,R_2)$ at the detector depends on the unpropagated x-ray wavefield $\psi(r_1,r_2,0,\lambda)$ in the exit plane.

A rigorous treatment of x-ray propagation will show that the behavior of the wave-field is characterized in broad terms by the Fresnel number (Born and Wolf 1999)

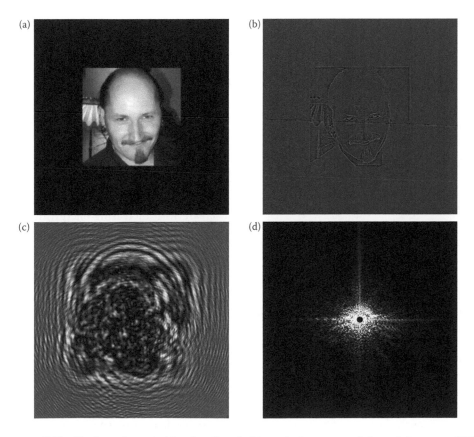

FIGURE 5.3 An x-ray wave-field with phase shown in (a) and undisturbed intensity is propagated through free space. Diffraction patterns typical of (b) the near field ($N_F \approx 10^4$); (c) the mid-field ($N_F \approx 10$); and (d) the far-field ($N_F \approx 10^{-4}$) regions are shown. In (d), a beam stop blocks out the strong, undiffracted central beam.

$$N_F = \frac{kb^2}{2\pi R_2}, \qquad (5.11)$$

where b is a characteristic length scale of the transmission function, and thus the features of interest in the sample. PB–PCI naturally separates into three regimes (see Figure 5.3): (i) the near-field or "Fresnel" region where $N_F \gg 1$, in which propagation-based phase contrast manifests as localized fringes and edge-enhancement; (ii) the far-field or "Fraunhoffer" region where $N_F \ll 1$, in which propagation-based phase contrast manifests as diffraction patterns typical of those found in protein crystallography; and (iii) the intermediate region $N_F = 1$, in which complex sets of wide fringes appear (Paganin 2006).

A back-of-the-envelope calculation shows that, for the length scales and wavelengths typical of clinical applications, near-field PB–PCI is the most practical. Assuming $R_2 \ll kb^2/2\pi$, we find that x-ray propagation in our simplified biomedical imaging setup may be described by the transport-of-intensity equation (TIE) (Teague 1983; Gureyev et al. 2006a)

$$M^2 S(Mr_1, Mr_2, R_2, \lambda)$$
$$= S(r_1, r_2, 0, \lambda) - \frac{R'}{k} \nabla_\perp \cdot [S(r_1, r_2, 0, \lambda) \nabla_\perp \varphi(r_1, r_2, 0, \lambda)], \quad (5.12)$$

where $M = (R_1 + R_2)/R_1$ is the geometric magnification, R' is the effective propagation distance R_2/M, and $\nabla_\perp = \vec{i}\,\partial_{r_1} + \vec{j}\,\partial_{r_2}$ is the gradient operator in the $r_1 - r_2$ plane. We can see that the edge-enhancement fringes typical of PB–PCI will arise from the second term in the TIE, and are modeled by the transverse gradient operator. Equation 5.12 assumes a point source of x-rays, and does not account for the finite size of the detector pixels. A more nuanced approach gives the following expressions for fringe signal-to-noise (as distinct from attenuation signal-to-noise) and resolution (Gureyev et al. 2008)

$$\text{SNR} \propto (R_1 + R_2) \frac{M-1}{M^{1/2}[M^2\sigma_{sample}^2 + (M-1)^2\sigma_{src}^2 + \sigma_{det}^2]^{3/4}}, \quad (5.13)$$

$$\text{Res} \propto [(M-1)^2 M^{-2}\sigma_{src}^2 + M^{-2}\sigma_{det}^2]^{1/2}, \qquad (5.14)$$

assuming a Gaussian source spot (the size of the region from which x-rays are emitted) with standard deviation σ_{src}, and a detector with a Gaussian point-spread function of standard deviation σ_{det}. σ_{sample} refers to the width of the sample edge producing the fringes; a gradual transition from one material to another produces less contrast than a sharp boundary.

A synchrotron source has an intrinsically small source size σ_{src}, and this is further demagnified at the detector position by the large magnitude of R_1. In contrast, any lab-based or clinical x-ray imaging system has a source spot size of finite width. This results in penumbra effects, causing a blurring in the final image. Blurring of this sort can completely obscure the phase-contrast fringes typical of the near field, making a microfocus x-ray source

(spot size < 10 μm) a requirement for lab-based or clinical PB–PCI (Wilkins et al. 1996; Gureyev et al. 2006a, 2008).

In transmission or reflection x-ray sources, there is a trade-off between source spot size and x-ray flux. X-rays are generated by focusing an electron beam on a target. The tighter the focus of the electron beam, the smaller the spot size and the greater the energy-per-area delivered to the target material.[*] To avoid damaging the target, one must limit the energy delivered per unit area. In attenuation-contrast imaging, contrast is determined by absorbed dose, so one typically opts for a high-flux, large spot-size source. This fundamentally limits the resolution of attenuation-contrast imaging: the detector must be placed as close as possible to the object plane in order to minimize penumbra effects, limiting the geometric magnification of the system. In PCI, our priorities are reversed as discussed in the previous paragraph: absorption is no longer the sole source of contrast, and a small spot size is necessary to resolve the phase-contrast fringes. The small source spot, in combination with the geometric magnification associated with free-space propagation, increases the potential resolution of PCI methods. Higher flux will improve PB–PCI signal-to-noise ratios (SNR); however since contrast is not dependent on the existence of an absorbed dose, PB–PCI places less emphasis on flux than attenuation-based methods (see Section 5.4 for examples). This makes it more suitable for imaging dynamic systems (where exposure time is limited), and fragile samples (where total x-ray dose is limited) (Fitzgerald 2000; Řeháček et al. 2006).

The most significant factor affecting fringe visibility (and hence phase-contrast visibility) is the effective propagation distance R', which is proportional to the sample–detector distance. Figure 5.4 shows fringe SNR and resolution for a few different systems; note that the penumbra effects associated with large sources necessitate a small sample–detector separation and do not produce the strong phase contrast.

In practice, a combination of absorption and refraction contributes to the contrast in any real image. It is important when seeking to minimize dose to also consider the energy dependence of the complex refractive index $n(\mathbf{r},\lambda)$: while the real decrement, $\Delta(\mathbf{r},\lambda)$, and the imaginary component, $\beta(\mathbf{r},\lambda)$, both decrease with increasing energy, the decrease in Δ is slower (Figure 5.5). Hence, while both phase and absorption contrast are increased at lower energies, at higher energies the relatively higher contribution of phase effects compared to absorption effects provides opportunities for improved image contrast and/or reduced dose (Lewis 2004; Gureyev et al. 2009). As with all x-ray imaging, dose, image contrast, spatial resolution, SNR, exposure times, and other parameters must be traded against one another. A fuller consideration of the optimization of the PB–PCI system is given in Gureyev et al. (2009).

At a synchrotron-based x-ray source, the high flux allows for essentially monochromatic (i.e., single-wavelength) illumination. At lab-based or clinical sources, a filter is used to remove extremely long wavelengths from the x-ray beam, as these

---

[*] Rotating anode sources can achieve high flux with a small spot size, but at the cost of an unstable source position.

(a)

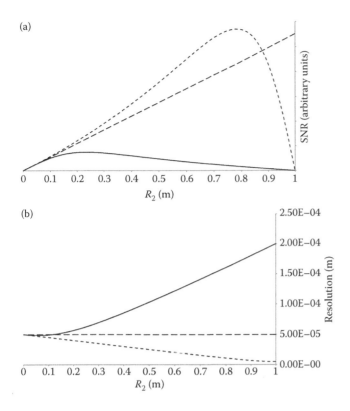

(b)

FIGURE 5.4 Curves showing the dependence of phase-contrast signal-to-noise (a) and resolution (b) on the imaging geometry. The object-edge width is 10 μm, and the detector pixel size 50 μm. Curves are for (i) a clinical system with a microfocus source (short dashes, σ_{src} = 5 μm, $R_1 + R_2$ = 1 m); (ii) a similar system with a larger source spot size (solid line, σ_{src} = 200 μm, $R_1 + R_2$ = 1 m); and (iii) a synchrotron source (long dashes, R_1 = 100 m).

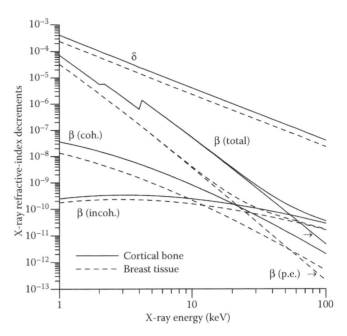

FIGURE 5.5 The complex refractive index decrements Δ and β for cortical bone and breast tissue, as a function of x-ray energy. (Reprinted with permission from Gureyev, T. et al. 2009. Refracting roentgens rays: Propagation-based x-ray phase contrast for biomedical imaging. *J. Appl. Phys. 105*, 102005. Copyright 2009, American Institute of Physics.)

wavelengths are typically completely absorbed by the sample (thus contributing to absorbed dose, but not image contrast). Removing too much of the initial x-ray spectrum will lower the x-ray flux below acceptable levels; consequently, the resulting beam is only quasi-monochromatic (Mayo et al. 2003; Pfeiffer et al. 2007a,b). Fortunately, under certain circumstances PB-PCI can be implemented with quasi-monochromatic x-ray sources (see Section 5.3.2) (Wilkins et al. 1996).

In summary, PB-PCI requires a small x-ray spot size (or equivalently a very large source-to-sample distance), and the largest possible sample-detector propagation distance that still satisfies Equation 5.12. An ideal magnification may be determined from Equation 5.13. Flux and spectrum are comparatively less important, and are largely dictated by sample composition and x-ray source type.

5.3 Quantitative Imaging: Phase Retrieval

Having established some guidelines for PB-PCI, it is natural to ask whether we may retrieve quantitative information from these phase-contrast images. This inverse problem of calculating the

x-ray phase from phase-contrast measurements is referred to as phase retrieval. In conventional attenuation-contrast x-ray imaging, interpretation of images is straightforward: the projected linear attenuation coefficient (i.e., the projected imaginary component of the complex refractive index) may be retrieved directly from the intensity measurements via Equation 5.7. In this section, we present several methods for determining the projected complex refractive index (a quantitative measure of both the attenuative and refractive properties of the sample) from the phase-contrast measurements.

As we are trying to calculate both the real and imaginary components of the refractive index, general phase-retrieval methods require at least two images. We begin with an example of a two-image phase-retrieval method; this method is general in the sense that it makes no assumptions about sample structure or composition, but is relatively difficult to implement experimentally (Teague 1983), requiring precise alignment of components and a monochromatic beam. Single-image phase-retrieval techniques make use of some *a priori* information to simplify the phase-retrieval process, are comparatively straightforward, and can be implemented in a lab-based/clinical setting (Paganin et al. 2002; Gureyev et al. 2006a). We discuss when single-image methods may be used, and explore the advantages they have over the two-image method. Finally we discuss the integration of phase-retrieval and computed tomography, to achieve 3D PB-PCI (Raven et al. 1996;

Bronnikov 1999; Cloetens et al. 1999; McMahon et al. 2003; Mayo et al. 2003).

5.3.1 A Two-Image Method for Near-Field Phase Retrieval

We again return to our simplified biomedical imaging setup (see Figure 5.2), this time with the intent of recovering the projected complex refractive index of the sample. After some manipulation, we may rewrite the TIE (Equation 5.12) and the projection approximation (Equation 5.7) as

$$M^2 S(Mr_1, Mr_2, R_2, \lambda) - S(r_1, r_2, 0, \lambda)$$
$$= R' \nabla_\perp \cdot [S(r_1, r_2, 0, \lambda) \nabla_\perp (\mathcal{P}\Delta)(r_1, r_2, \lambda)], \quad (5.15)$$

$$S(r_1, r_2, 0, \lambda) = S_{in}(\lambda) \exp[-2k(\mathcal{P}\beta)(r_1, r_2, \lambda)]. \quad (5.16)$$

Given knowledge of $S(Mr_1, Mr_2, R_2, \lambda)$ and $S(Mr_1, Mr_2, 0, \lambda)$, we can solve for the projected complex refractive index: the projected real decrement to the refractive index may be calculated using standard methods for solving the elliptical partial differential equations, and the projected imaginary refractive index may be calculated directly.

Provided the incident x-ray beam is sufficiently monochromatic (e.g., synchrotron radiation), the recorded intensity at the detector is proportional to the spectral density $S(\mathbf{r}, \lambda)$ (see Section 5.2.2). Thus, we can obtain the projected complex refractive index from: one attenuation-contrast image taken with the detector in the exit plane and one phase-contrast image with the detector at a distance R_2 from the sample. Guidelines for choosing R_2 may be found in the previous section.

In order to solve Equation 5.15, one must take the difference between the exit-plane and phase-contrast images; this difference map will contain only propagation-induced phase-contrast fringes. Consequently, the two images must be very precisely aligned. Any mismatch in the position of edges will produce "false" fringes in the difference image, contaminating the results. Figure 5.6 shows the effect of a small (~0.2 pixel)

misalignment on the retrieved phase. If different detectors are used to collect each image, they must be accurately calibrated against one another so that a meaningful difference can be taken. If the same detector is used to collect both images, then the effective resolution of the detector is reduced by a factor of M due to magnification effects: a sample that fills an N pixel wide detector at a distance of R_2 will be N/M pixels wide in the exit plane. The higher-resolution phase-contrast image must then be downscaled to match the lower-resolution exit-plane image.

5.3.2 Single-Image Techniques

As mentioned earlier in the chapter, there is a special category of samples for which the phase-attenuation duality holds true, that is, the ratio $\epsilon(\lambda) = \beta(\mathbf{r}, \lambda)/\Delta(\mathbf{r}, \lambda)$ is independent of the position. In these circumstances, the interdependence of the attenuative and refractive properties of the sample simplifies the phase retrieval process (Gureyev et al. 2006a):

$$(\mathcal{P}\Delta)(r_1, r_2, \lambda) = -\frac{1}{2k\epsilon(\lambda)} \ln$$
$$\left(\mathcal{F}_2^{-1} \left\{ \frac{\epsilon(\lambda) M^2}{\epsilon(\lambda)\cos(\pi\lambda R'\xi^2) + \sin(\pi\lambda R'\xi^2)} \mathcal{F}_2 \left[\frac{S(Mr_1, Mr_2, R_2, \lambda)}{S_{in}(\lambda)} \right] \right\} \right),$$
$$(5.17)$$

allowing the phase-retrieval from only a single phase-contrast image.

As phase-contrast images are collected at a single propagation distance and magnification, the full resolution of the detector may be used and precise alignment is now unnecessary. Furthermore, examination of the Fourier-space filter in Equation 5.17 shows that single-image phase retrieval is stable with respect to high-frequency noise in the measurements (Gureyev et al. 1995a,b): propagation-based phase-retrieval is a blurring operation, to reverse the sharpening effects of free-space propagation. If the ratio $\epsilon(\lambda)$ is nonzero (i.e., if there is any attenuation in the images), the phase-retrieval process is also stabilized at low frequencies. In contrast, two-image phase

(a)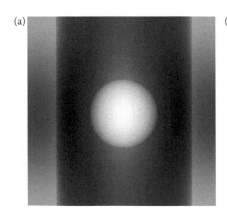

(b)

FIGURE 5.6 Simulated, noise-free two-image phase retrieval in the near field, with (a) perfectly aligned images and (b) for two images with a horizontal offset of 0.2 pixels. Images are 512 × 512 pixels.

retrieval is comparatively vulnerable to noise. Finally, we note that if attenuation in the sample is not too strong, and the ratio $\varepsilon(\lambda)$ does not depend strongly on wavelength (i.e., the spectrum $S_{in}(\lambda)$ of the incident x-ray beam avoids absorption-edges associated with the materials in the sample), the phase-retrieval process is essentially achromatic, that is, independent of wavelength (Myers et al. 2007b). This makes single-image phase retrieval ideal for polychromatic, clinical sources.

Strictly speaking, the phase-attenuation duality is valid for x-ray-illuminated samples composed of a single material (Paganin et al. 2002; Mayo et al. 2003), as well as for objects composed of light elements ($Z < 10$) illuminated by high energy (60–500 keV) x-rays (Wu and Liu 2005), such as soft tissues. In practice, single-image phase retrieval is quite tolerant to errors and variations in $\varepsilon(\lambda)$. Equation 5.17 assumes *a priori* knowledge of the constant $\varepsilon(\lambda)$, which can be difficult to obtain for complex samples. More sophisticated methods allow $\varepsilon(\lambda)$ to be determined from the phase-contrast image (Beltran et al. 2010; Chen et al. 2011; Eastwood et al. 2011).

5.3.3 Computed Tomography

In conventional attenuation-based CAT, attenuation-contrast radiographs are collected as per our simplified imaging model (Figure 5.2) with the detector in the exit plane. The sample is rotated about the r_2 axis, and a number of radiographs collected at different, evenly spaced viewing angles (Cormack 1963; Compton 1964; Hounsfield 1973; Kak and Slaney 2001; Natterer 2001). Ideally, $\pi N/2$ radiographs are collected, where N is the detector width in pixels (Crowther et al. 1970; Bracewell 1979). At each viewing angle (θ), these attenuation-contrast radiographs give us direct knowledge of the projected imaginary refractive index $(P_\theta\beta)(r_1, r_2, \theta, \lambda)$. Tomographic reconstruction is the process of determining the 3D distribution of imaginary refractive index $\beta(\mathbf{r},\lambda)$ from this collection of radiographs. There are several well-established methods for tomographic reconstruction (Bracewell and Riddle 1967; Ramachandran and Lakshminarayanan 1971; Shepp and Logan 1974; Feldkamp et al. 1984).

In phase-contrast tomography, the same well-established tomographic reconstruction methods are used to calculate a 3D map of the refractive properties of the sample (i.e., the 3D distribution of $\Delta(\mathbf{r},\lambda)$), given the projected decrement $(P_\theta \Delta)(r_1, r_2, \theta, \lambda)$. This projected decrement may be calculated as outlined in the previous sections. Phase-contrast tomography is thus a two-step process: (i) phase-contrast images are collected at each viewing angle, and phase-retrieval is used to calculate the projected decrement $(P_\theta \Delta)(r_1, r_2, \theta, \lambda)$; (ii) conventional CT reconstruction techniques are used to construct a 3D map of the attenuative properties of the object (Wolf 1969; Devaney 1986; Momose et al. 1995; Bonse and Busch 1996; Raven et al. 1996; Bronnikov 1999, 2007; Cloetens et al. 1999; Dilmanian et al. 2000; Natterer 2001; Pavlov et al. 2001; Gbur and Wolf 2002; Anastasio et al. 2003, 2004, 2005; Mayo et al. 2003; Anastasio and Shi 2004; Wu and Liu 2005; Groso et al. 2006; Gureyev et al. 2006b, Gureyev et al.

2007a,b; Miao et al. 2006; Momose et al. 2006; Paganin 2006; Engelhardt et al. 2007; Myers et al. 2007a,b; Pfeiffer et al. 2007a). The alignment problems facing two-image phase retrieval are compounded by the need to collect the radiographs at hundreds/thousands of viewing angles. For this reason, we shall focus on phase-and-amplitude computed tomography (PACT): the combination of CAT and single-material phase retrieval (Gureyev et al. 2006b; Myers et al. 2007b).

Many of the practical guidelines from previous sections still apply; spot size, ideal magnification, and sample–detector distance are all determined as laid out in Section 5.2.4, and the sample must satisfy the phase-attenuation duality described in Section 5.3.2. Phase-contrast tomography possesses the same advantages as 2D PCI: reduced dose, improved contrast in light materials, improved time resolution due to higher contrast, and so on. As discussed in Section 5.3.2, single-image phase retrieval is essentially achromatic, meaning that PACT is less vulnerable to beam-hardening artifacts[*] than attenuation-based CAT (Myers et al. 2007b). As a result, less filtering of the x-ray beam is required, improving flux, imaging time, and temporal resolution.

The CAT reconstruction process involves a sharpening operation and is consequently mildly unstable with respect to high-frequency noise in the radiographs. Thus, in attenuation-contrast CAT, there is a need for high SNR in each of the hundreds/thousands of radiographs, leading to a comparatively large delivered dose and long exposure times. In PACT, the formation of Fresnel fringes during free-space propagation sharpens the image "in hardware," removing the need for the sharpening operation in the reconstruction. As a result of this beneficial cancelation between tomographic reconstruction and single-image phase retrieval, PACT reconstruction is stable with respect to noise (Myers et al. 2007a). This permits faster, lower dose 3D imaging, even in cases where PCI is not strictly necessary.

5.4 Biomedical Applications of Propagation-Based Imaging

We have already seen in Section 5.2.3 that PCI methods are well suited to imaging the soft tissues where absorption contrast is relatively poor, and hence they have found keen interest in biomedical research. PB–PCI offers distinct advantages over alternative methods of PCI for clinical and preclinical applications. First, imaging of living systems demands a single-image method, since natural movements due to the cardiac and respiratory cycles are unavoidable, making coregistration in multiple image methods extremely difficult. As we have seen, the phase-attenuation duality holds true for the soft tissues and consequently single-image phase-retrieval

---

[*] Beam-hardening artifacts are caused by the differential attenuation of long and short wavelength x-rays. Longer ("softer") wavelengths are more readily absorbed, hardening the beam as it passes through the sample (Kak and Slaney 2001). This causes cupping and streaking artifacts in the 3D image.

for quantitative imaging is possible (5.3.2). Second, phase contrast is rendered in all directions in a single image, unlike in alternative methods of PCI such as analyzer-crystal-based imaging (Chapter 4). Third, exposure times are minimized by the absence of intermediate optical elements between the source and detector. In phase-contrast imaging methods requiring such elements, for example, crystal analyzers or gratings, to translate phase changes into intensity changes, there is an inevitable loss of incident intensity at either the detector or the sample, requiring not only longer exposures but also correspondingly more dose being delivered to the sample. Maintaining the mechanical stability of these optics over time is also critical. Exposure times as short as possible are highly desirable to eliminate or at least minimize the movement artifacts (image blurring). Fourth, and critical for more widespread uptake of PB–PCI, the method tolerates the use of polychromatic sources, provided the source size is sufficiently small (Wilkins et al. 1996; Pogany et al. 1997).

More generally in a biomedical or clinical context, PCI methods offer the attractive advantage that the photons generating the typical edge-enhancement fringes are not contributing to the absorbed dose. Any real image will display a mix of attenuation and phase contrast. As discussed in Section 5.2.4, decreasing the x-ray wavelength reduces phase contrast more slowly than it reduces absorbed dose (and the associated attenuation contrast). Proper selection of the x-ray wavelength thus allows for improved contrast and/or reduced dose.

In recent years, propagation-based phase-contrast x-ray imaging has progressed beyond imaging static samples, and has been used to generate movies of dynamic processes. Much of this work has been undertaken at synchrotrons, where the intrinsically small source size gives excellent phase contrast and the high flux enables very short exposures, even with monochromatic radiation. As cameras increase in sensitivity, ever shorter exposures and higher frame rates permit imaging of peak and average blood flows as a function of the cardiac cycle (Jamison et al. 2011). It is this ability to observe the processes deep within the body at high spatial resolution, and to track the physiological processes and the effect of treatments in real time, that makes PB–PCI so powerful for biomedical imaging.

In this section, we will examine how some of the key advantages of PB–PCI have been exploited in biomedical and clinical applications. These applications can be broadly separated into those which utilize the increased edge enhancement and hence contrast offered by PB–PCI to reveal new or improved information on anatomy and pathologies, generally in static systems, and those that use these advantages to investigate dynamic processes or function in the living body. By way of examples, we review PB–PCI as used in orthopedic imaging and mammography in the former category, and vascular and respiratory systems in the second category.

5.4.1 Orthopedic Imaging

While cortical bone is clearly visible in conventional absorption x-ray imaging, it tends to be difficult to visualize the

surrounding cartilage, as well as the fine structure of trabecular bone. The sensitivity of PB–PCI to boundaries makes it particularly useful for visualizing the cartilage–bone interface, the trabecular bone detail, and even small fractures below the spatial-resolution limit of conventional x-ray images (Mori et al. 2002, 1999; Ismail et al. 2010). In an excised rib sample, Mori et al. were able to distinguish trabecular bone from the surrounding bone marrow by the characteristic dark fringe around the periphery of the denser trabeculae using synchrotron PB–PCI (Mori et al. 2002). They also observed calcifications and a small fracture within the cartilage layer that was not visible on a corresponding absorption image. In a human sample, the extent of an osteosarcoma was inferred from the destruction of trabecular bone (Figure 5.7). The ability to visualize the cartilage structure at high spatial resolutions that are unachievable using the alternative modalities of magnetic resonance imaging and ultrasound is beginning to be exploited using laboratory-based microfocus sources, for investigation of degenerative joint diseases such as osteoarthritis (Ismail et al. 2010; Lee et al. 2010). Phase-contrast tomography has been trialed for identification of articular cartilage anomalies in a mouse model of collagen-induced arthritis, indicating promise as an assessment tool for preclinical studies.

A potentially rich area for exploration with PB–PBI is in the study of biomaterials and tissue engineering, particularly

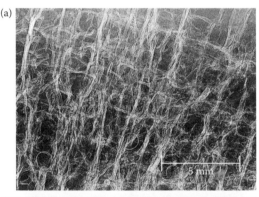

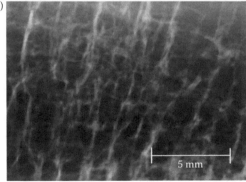

FIGURE 5.7 Synchrotron images of a specimen of human osteosarcoma taken using (a) $r_2 = 5$ m, and (b) $r_2 = 0$ m. (Images from Mori, K. et al. 2002. *J. Synchrotron Rad. 9*, 143–147. Used with permission from IUCr Journals, England.)

using 3D imaging. The bone–biomaterial interface for various bone substitutes or scaffolds has been studied using synchrotron phase-contrast tomography, using both *in vivo* (Weiss et al. 2003; Komlev et al. 2010; Yeom et al. 2010) and *in vitro* models (Langer et al. 2010) of bone cell growth. Knowledge of the microstructure and morphology of regenerated bone and its integration with the bone substitute are critical to assessing the quality of the implant. PB–PCI proved to be particularly useful for studying the fibrous nature of the immature bone matrix, which is typically non- or poorly calcified (Langer et al. 2010).

5.4.2 Mammography

Phase-contrast mammography for breast cancer detection has attracted significant interest, both at synchrotrons and using microfocus laboratory sources. The motivation to improve image contrast in mammograms without increasing the absorption dose is high; the breast is a radiosensitive organ but the similar absorption of breast tissue dictates that conventional mammography is typically performed at relatively low x-ray energies (for clinical practice) of approximately 30 kVp. Many developed countries have implemented breast screening programs in recognition of the fact that the benefits of early detection outweigh the risk of adverse consequences such as induced cancers (e.g., Smith et al. 2004), so any dose reduction via PCI could have a profound impact on current clinical practice by, for example, lowering the recommended age for screening the commencement.

Early investigations into synchrotron phase-contrast mammography used both propagation and analyzer-based methods on excised breast tissue or phantoms (Johnston et al. 1996; Arfelli et al. 1998, 2000). The value of PCI for mammography lies particularly in edge enhancement that serves to better distinguish the diagnostic markers of breast disease (e.g., Arfelli et al. 2000). Important diagnostic features in mammograms include the lesion density, lesion margins (smooth, irregular, or spiculated), and the presence and morphology of calcifications.

Although analyzer-based techniques have shown some additional enhancement of tissue components over PB–PCI, PB–PCI has been used for the first clinical trial for synchrotron phase-contrast mammography. This recently completed study at the SYnchrotron Radiation for MEdical Physics (SYRMEP) beamline of the ELETTRA synchrotron, Italy, examined 47 patients with questionable or suspicious breast abnormalities previously identified by conventional digital mammography or ultrasound (Abrami et al. 2005; Castelli et al. 2011), see Figure 5.8. For this cohort of patients for which further dose above the conventional clinical assessments could be justified, the sensitivity using synchrotron mammography was 81% and the specificity was 94%. This represents an improvement in the true-negative rate over conventional mammography, suggesting that the technique has value as a second-level examination to reduce the number of subsequent invasive procedures such as biopsy. Also under

investigation is the use of phase-contrast tomography for the breast (Pani et al. 2004).

In parallel with the earliest efforts at synchrotrons, investigations of conventional mammography x-ray sources revealed that phase-contrast enhancements were evident even with a modest distance between the sample and detector ($r_2 = 0.3$ m) and a moderate focal spot size (0.15 mm) (Kotre and Birch 1999; Kotre et al. 2002). The magnification that resulted from the use of a divergent source, as well as the scatter reduction provided by the sample–detector gap, served to also effectively increase the spatial resolution and contrast of the resulting images. Subsequently, there was a rapid move of phase-contrast mammography into clinical practice, with the release of the Konica Minolta REGIUS PureView system* in 2005 (Honda et al. 2002). The system is a modified conventional mammography unit, permitting a sample to detector distance 0.49 m, while maintaining a conventional source to sample distance of 0.65 m. Clinical trials to date of this system as well as excised tissue studies have suggested that, in common with synchrotron mammography, phase-contrast mammography offers improved true-negative and lower recall rates over conventional mammography (Tanaka et al. 2005; Morita et al. 2008; Williams 2008). Subsequent research studies have sought to optimize the imaging parameters phase contrast of systems suited to clinical application and quantify the image quality (Donnelly and Price 2002; Honda et al. 2006; Williams et al. 2008; Tanaka et al. 2009).

5.4.3 Cardiovascular System

Blood has a complex refractive index which is fairly similar to that of tissue, so in the past a highly absorbing contrast agent, like iodine, was required to provide contrast for absorption-based imaging, generally employed to locate the blood vessels. The injection of a contrast agent into the body is both invasive and not ideal when a researcher is seeking to isolate the effects of a particular treatment. PB–PCI work is moving toward using more natural contrast agents, or no agents at all, to visualize not only the structure of the blood vessels, but also the movement of blood inside them. For example, blood may be replaced with saline (Zhang et al. 2008), or pockets of air may be introduced into the blood, either in hollow beads (Kim et al. 2009) or in microbubbles (Jamison et al. 2011; Tang et al. 2011), to increase the visibility of blood vessels. When comparing blood and air, the substantial difference in the real part of the complex refractive index makes the air-pockets highly visible with PB–PCI. Visualizing the morphology of a blood vessel system in this way can be informative in studying the health of tissue, for example, locating cancerous tissue. "Seeding" blood with this kind of particle tracer also means that the motion of blood may be measured from consecutive high-resolution x-ray images. Given a sufficiently coherent source and sufficiently small detector pixels, the density variations within blood, due to red blood cells,

---

* http://www.konicaminolta.eu/healthcare/products/digital-mammography/regius-pureview-system/features.html

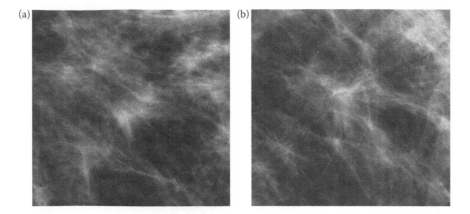

FIGURE 5.8 Images from a 62-year-old woman participating in the clinical trail conducted at the SYRMEP beamline at ELETTRA. A suspicious mass with spiculated margins was identified on the initial conventional digital mammogram (a), and later confirmed with superior depiction on (b), synchrotron phase-contrast image. (Images from Castelli, E. et al. 2011. *Radiology 259*, 684–694. Used with permission.)

can create a speckle intensity pattern, eliminating the need for seeding particles (Lee and Kim 2005; Fouras et al. 2007). Particle image velocimetry (PIV) techniques have been applied to such images in order to map the blood flow velocity across vessels and determine the pressure on vessel walls (Irvine et al. 2008, 2010; Jamison et al. 2011). 3D mapping of flow within a vessel can be achieved by combining this method with computed tomography (Dubsky et al. 2010). These same analysis techniques are now being applied to track the motion of the lungs.

5.4.4 Respiratory System

In vivo imaging of the airways, using synchrotron sources, has already advanced studies looking at a number of respiratory processes and conditions. While some studies have utilized inhaled xenon with K-edge subtraction to visualize the airways and lungs (Bayat et al. 2001; Porra et al. 2011), the contrast provided by PB–PCI alone enables a range of studies. Early work utilized PB–PCI to reveal the breathing cycle of insects (Westneat et al. 2003), but focus quickly moved to small mammals as models of

the human respiratory system (Yagi et al. 1999; Lewis et al. 2005; Sera et al. 2005). The ability of PB–PCI to reveal both the airway structure leading down to the lungs and the lungs themselves is seen in Figure 5.9. In the first panel, the liquid-filled lungs of a newborn rabbit pup are barely visible, but as the pup takes their first breaths and fills the airways with air, exquisite contrast is achieved (center panel). When the lungs are fully inflated (final panel), the lung outline is easily seen and the thickest sections exhibit a speckle pattern. This strong contrast has also proven valuable in visualizing the intricate 3D structure of the lungs (Kono et al. 2001; Schuster et al. 2004; Sera et al. 2005; Beltran et al. 2011) using the computed tomography.

Figure 5.9 was taken as part of studies looking at lung problems commonly associated with preterm birth (Lewis et al. 2005; Kitchen et al. 2008; Hooper et al. 2009). Hooper, Kitchen, Lewis et al. have imaged mouse and rabbit pups in the minutes after birth, assessing how patient orientation and a range of ventilation strategies and delivered treatments will affect the long-term health of the new lungs (Kitchen et al. 2007; Hooper et al. 2007, 2009; Siew et al. 2009). Phase retrieval can

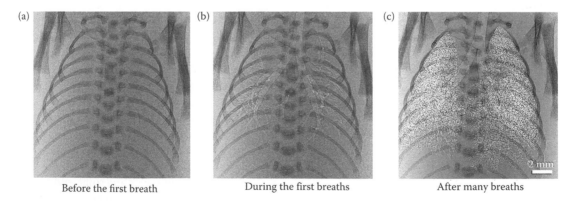

| Before the first breath | During the first breaths | After many breaths |

FIGURE 5.9 PB–PCI of the lungs of a rabbit pup visualizes the lungs when filled with air, seen here as a neonate rabbit (a) first has lungs filled with liquid, then (b) clears liquid from the airways until (c) the lungs fill with air. 24 keV x-rays are used with 3 m propagation, imaged with 16.2 μm pixels. (Images courtesy of Marcus Kitchen.)

be applied to these images to measure regional lung volume, achieved by partitioning the image into different lobes of the lungs (Kitchen et al. 2008). Results from these studies, which show the most effective methods of helping preterm infants to breath, have already transitioned into clinical practice (Hooper et al. 2011).

The speckle pattern seen in Figure 5.9c, produced by the superposition of the many alveoli (air sacs) in the lungs, has also been a topic of research interest (Kitchen et al. 2004). Measurements of the characteristic speckle size and visibility can be used to infer the health of the lungs (Kitchen et al. 2005; Liu et al. 2010). Even more subtle lung injuries or conditions can be detected by tracking the movement of this unique speckle pattern during a complete breath, to reveal the dynamic movement of each region of the lungs (Fouras et al. 2009). This analysis uses the same PIV techniques as mentioned when looking at blood flow. The measurement can be extended to a full 3D model by doing a CT scan, and taking a time series of images over a complete breath at each projection. This will result in a 3D visualization of the motion of the lungs at each time-point of a complete breath cycle (Dubsky et al. 2012). This detailed data is of particular use in early detection and/or better understanding of conditions such as asthma, emphysema, lung cancer, and cystic fibrosis.

Further research on cystic fibrosis has made use of high spatial resolution PB–PCI to look at the surface of the airways leading down to the lungs (Parsons et al. 2008; Siu et al. 2008). The depth of the liquid lining the airways and the resulting ability of those airways to clear inhaled debris are indicators of the airway health and may be used as a direct measure of the effectiveness of new treatments (Donnelley et al. 2009, 2010b). PB–PCI enables researchers to noninvasively observe these micron-scale processes in real time.

Across these studies, in order to achieve sharp images of the constantly moving airways, animals are usually anesthetized and intubated, with image capture synchronized to a particular point in the mechanically ventilated breath cycle (Donnelley et al. 2010a). This is particularly important in PIV–CT, where exactly the same point in the breath cycle must be captured from each projection angle around the lungs. The use of an immovable synchrotron source has led to specialized animal positioning mounts, as well as remote monitoring and treatment delivery instrumentation for when the animal is within the lead-shielded imaging "hutch." Radiation dose is also a consideration when capturing time-sequences of images, due to the large number of exposures taken over the period of treatment and beyond.

5.5 Summary

The ability to noninvasively image the living body, and furthermore to image biological processes in action, is not only of value as a future clinical tool, but is already advancing biomedical research.

Transitioning PB–PCI from biomedical research to clinical application requires not only a decrease in radiation dose, but also translation to less expensive sources (Wu and Liu 2003; Gureyev et al. 2009). However, these problems are not insurmountable, and may soon be addressed by increasingly sensitive detectors and more readily available microfocus x-ray sources.

References

Abrami, A., F. Arfelli, R. C. Barroso, A. Bergamaschi, F. Bille, P. Bregnant, F. Brizzi et al. 2005. Medical applications of synchrotron radiation at the syrmep beamline of elettra. *Nucl. Instrum. Meth. A 548*, 221–7.

Anastasio, M. A. and D. Shi. 2004. On the relationship between intensity diffraction tomography and phase-contrast tomography. *Proc. SPIE 5535*, 361–368.

Anastasio, M. A., D. Shi, D. D. Carlo, and X. Pan. 2003. Analytic image reconstruction in local phase-contrast tomography. *Phys. Med. Biol. 49*, 121–144.

Anastasio, M. A., Y. H. D. Shi, and G. Gbur. 2004. Phase-contrast tomography using incident spherical waves. *Proc. SPIE 5535*, 724–732.

Anastasio, M., D. Shi, Y. Huang, and G. Gbur. 2005. Image reconstruction in spherical-wave intensity diffraction tomography. *J. Opt. Soc. Am. A 22*, 2651–2661.

Arfelli, F., V. Bonvicini, A. Bravin, G. Cantatore, E. Castelli, L. D. Palma, M. D. Michiel et al. 2000. Mammography with synchrotron radiation: Phase-detection techniques. *Radiology 215*, 286–293.

Arfelli, F., M. A., V. Bonvicini, A. Bravin, G. Cantatore, E. Castelli, L. D. Palma, M. D. Michiel et al. 1998. Low-dose phase contrast x-ray medical imaging. *Phys. Med. Biol. 43*, 2845–2852.

Bayat, S., G. L. Duc, L. Porra, G. Berruyer, C. Nemoz, S. Monfraix, S. Fiedler et al. 2001. Quantitative functional lung imaging with synchrotron radiation using inhaled xenon as contrast agent. *Phys. Med. Biol. 46*, 3287–3299.

Beltran, M. A., D. M. Paganin, K. K. W. Siu, A. Fouras, S. B. Hooper, D. H. Reser, and M. J. Kitchen. 2011. Interface-specific x-ray phase retrieval tomography of complex biological organs. *Phys. Med. Biol. 56*, 7353.

Beltran, M., D. Paganin, K. Uesugi, and M. Kitchen. 2010. 2D and 3D x-ray phase retrieval of multi-material objects using a single defocus distance. *Opt. Exp. 18*, 6423–6436.

Bonse, U. and F. Busch. 1996. X-ray computed microtomography (µct) using synchrotron radiation (sr). *Prog. Biophys. Molec. Biol. 65*, 133–169.

Born, M. and E. Wolf. 1999. *Principles of Optics* (7 ed.). Cambridge University Press, Cambridge, UK.

Bracewell, R. N. 1979. Image reconstruction in radio astronomy. In G. T. Herman (Ed.), *Image Reconstruction from Projections*. Springer, Berlin.

Bracewell, R. N. and A. C. Riddle. 1967. Inversion of fan-beam scans in radio astronomy. *Astrophysical J. 150*, 427–434.

Bronnikov, A. V. 1999. Reconstruction formulas in phase-contrast tomography. *Opt. Commun. 171*, 239–242.

Bronnikov, A. 2007. Phase contrast ct: Fundamental theorem and fast image reconstruction algorithms. *Proc. of SPIE 6318*, 63180Q.

Castelli, E., M. Tonutti, R. Longo, E. Quaia, L. Rigon, D. Sanabor, F. Zanconati et al. 2011. Mammography with synchrotron radiation: First clinical experience with phase-detection technique. *Radiology 259*, 684–694.

Chen, R. C., H. L. Xie, L. Rigon, R. Longo, E. Castelli, and T. Q. Xiao. 2011. Phase retrieval in quantitative x-ray microtomography with a single sample-to-detector distance. *Opt. Lett. 36*, 1719–1721.

Cloetens, P., R. Barrett, J. Baruchel, J.-P. Guigay, and M. Schlenker. 1996. Phase objects in synchrotron radiation hard x-ray imaging. *J. Phys. D: Appl. Phys. 29*, 133–146.

Cloetens, P., W. Ludwig, J. Baruchel, D. V. Dyck, J. V. Landuyt, J. P. Guigay, and M. Schlenker. 1999. Holotomography: Quantitative phase tomography with micrometer resolution using hard synchrotron radiation x rays. *Appl. Phys. Lett. 75*, 2912–2914.

Compton, A. H. 1964. Representation of a function by its line integrals, with some radiological applications ii. *J. Appl. Phys. 35*, 195–207.

Cormack, A. M. 1963. Representation of a function by its line integrals, with some radiological applications. *J. Appl. Phys. 34*, 2722–2727.

Crowther, R. A., D. J. D. Rosier, and A. Klug. 1970. The reconstruction of a three-dimensional structure from projections and its applications to electron microscopy. *Proc. R. Soc. London Ser. A 317*, 319–340.

Devaney, A. J. 1986. Reconstructive tomography with diffracting wavefield. *Inv. Problems 2*, 161–183.

Dilmanian, F. A., Z. Zhong, B. Ren, X. Y. Wu, L. D. Chapman, I. Orion, and W. C. Thomlinson. 2000. Computed tomography of x-ray index of refraction using the diffraction enhanced imaging method. *Phys. Med. Biol. 45*, 933–946.

Donnelley, M., K. Morgan, A. Fouras, W. Skinner, K. Uesugi, N. Yagi, K. Siu, and D. Parsons. 2009. Real-time non-invasive detection of inhalable particulates delivered into live mouse airways. *J. Synchrotron Radiat. 16*, 553–561.

Donnelley, M., D. Parsons, K. Morgan, and K. Siu. 2010a. Animals in synchrotrons: Overcoming challenges for high-resolution, live, small-animal imaging. *AIP Proc. 1266*, 30–34.

Donnelly, E. F. and R. R. Price. 2002. Quantification of the effect of kvp on edge-enhancement index in phase-contrast radiography. *Med. Phys. 29*(6), 999–1002.

Donnelley, M., K. Siu, K. Morgan, W. Skinner, Y. Suzuki, A. Takeuchi, K. Uesugi, N. Yagi, and D. Parsons. 2010b. A new technique to examine individual pollutant particle and fibre deposition and transit behaviour in live mouse trachea. *J. Synch. Rad. 17*, 719–729.

Dubsky, S., R. A. Jamison, S. C. Irvine, K. K. W. Siu, K. Hourigan, and A. Fouras. 2010. Computed tomographic x-ray velocimetry. *Appl. Phys. Lett. 96*, 023702.

Dubsky, S., S. B. Hooper, K. K. W. Siu, and A. Fouras. 2012. Synchrotron-based dynamic computed tomography of tissue motion for regional lung function measurement. *J. R. Soc. Interface.* Published online April 4, doi: 10.1098/? rsif.2012.0116

Eastwood, S. A., A. C. Y. Liu, and D. M. Paganin. 2011. Automated phase retrieval of a homogeneous object using a single out-of-focus image. *Opt. Lett. 36*, 1878–1880.

Engelhardt, M., J. Baumann, M. Schuster, C. Kottler, F. Pfeiffer, O. Bunk, and C. David. 2007. High-resolution differential phase contrast imaging using a magnifying projection geometry with a microfocus x-ray source. *Appl. Phys. Lett. 90*, 224101.

Feldkamp, L. A., L. C. Davis, and J. W. Kress. 1984. Practical cone-beam algorithm. *J. Opt. Soc. Am. A 1*, 612–619.

Fitzgerald, R. 2000. Phase-sensitive x-ray imaging. *Phys. Today 53*, 23–26.

Fouras, A., J. Dusting, R. Lewis, and K. Hourigan. 2007. Three-dimensional synchrotron x-ray particle image velocimetry. *J. Appl. Phys. 102*, 064916.

Fouras, A., M. Kitchen, S. Dubsky, R. Lewis, S. Hooper, and K. Hourigan. 2009. The past, present, and future of x-ray technology for *in vivo* imaging of function and form. *J. Appl. Phys. 105*, 102009:1–14.

Gbur, G. and E. Wolf. 2002. Diffraction tomography without phase information. *Opt. Lett. 27*, 1890–1892.

Goodman, J. W. 1985. *Statistical Optics*. John Wiley and Sons, Inc, New York.

Groso, A., R. Abela, and M. Stampanoni. 2006. Implementation of a fast method for high resolution phase contrast tomography. *Opt. Exp. 14*, 8103–8110.

Gureyev, T. E. 1999. Transport of intensity equation for beams in an arbitrary state of temporal and spatial coherence. *Optik 110*, 263–266.

Gureyev, T., S. Mayo, D. Myers, Y. Nesterets, D. Paganin, A. Pogany, A. Stevenson, and S. Wilkins. 2009. Refracting roentgens rays: Propagation-based x-ray phase contrast for biomedical imaging. *J. Appl. Phys. 105*, 102005.

Gureyev, T. E., Y. I. Nesterets, D. M. Paganin, A. Pogany, and S. W. Wilkins. 2006a. Linear algorithms for phase retrieval in the fresnel region 2. Partially coherent illumination. *Opt. Commun. 259*, 569–580.

Gureyev, T. E., Y. I. Nesterets, K. M. Pavlov, and S. W. Wilkins. 2007. Computed tomography with linear shift-invariant optical systems. *J. Opt. Soc. Am. A 24*, 2230–2241.

Gureyev, T. E., Y. I. Nesterets, A. W. Stevenson, P. R. Miller, A. Pogany, and S. W. Wilkins. 2008. Some simple rules for contrast, signal-to-noise and resolution in in-line x-ray phase-contrast imaging. *Opt. Exp. 16*, 3223–3241.

Gureyev, T. E., D. M. Paganin, G. R. Myers, Y. I. Nesterets, and S. W. Wilkins. 2006b. Phase-and-amplitude computer tomography. *Appl. Phys. Lett. 89*, 034102.

Gureyev, T. E., A. Roberts, and K. A. Nugent. 1995a. Partially coherent fields, the transport-of-intensity equation and phase uniqueness. *J. Opt. Soc. Am. A 12*, 1942–1946.

Gureyev, T. E., A. Roberts, and K. A. Nugent. 1995b. Phase retrieval with the transport-of-intensity equation: Matrix solution with use of zernike polynomials. *J. Opt. Soc. Am. A 12*, 1932–1940.

Honda, C., H. Ohara, and T. Gido. 2006. Image qualities of phase-contrast mammography. In S. Astley, M. Brady, C. Rose, and

R. Zwiggelaar (Eds.), *Digital Mammography*, Volume 4046 of *Lecture Notes in Computer Science*, pp. 281–288. Springer, Berlin/Heidelberg.

Honda, C., H. Ohara, A. Ishisaka, F. Shimad, and T. Endo. 2002. X-ray phase imaging using a x-ray tube with a small focal spot—Improvement of image quality in mammography. *Igaku Butsuri 22*, 21–9.

Hooper, S. B., M. J. Kitchen, A. Fouras, M. J. Wallace, S. Dubsky, K. K. W. Siu, M. L. Siew, N. Yagi, K. Uesugi, and R. A. Lewis. 2011. Combined lung imaging and respiratory physiology research at spring-8. *Synchrotron Radiation News 24*, 19–23.

Hooper, S., M. Kitchen, M. Wallace, N. Yagi, K. Uesugi, M. Morgan, C. Hall et al. 2007. Imaging lung aeration and lung liquid clearance at birth. *FASEB J. 21*, 3329.

Hooper, S. B., M. J. Kitchen, M. L. Siew, R. A. Lewis, A. Fouras, A. B. te Pas, K. K. W. Siu, N. Yagi, K. Uesugi, and M. J. Wallace. 2009. Imaging lung aeration and lung liquid clearance at birth using phase contrast x-ray imaging. *Clin. Exp. Pharmacol. Physiol. 36*, 117–125.

Hounsfield, G. N. 1973. Computerized transverse axial scanning tomography: Part I, description of the system. *Br. J. Radiol. 46*, 1016–1022.

Irvine, S. C., D. M. Paganin, S. Dubsky, R. A. Lewis, and A. Fouras. 2008. Phase retrieval for improved three-dimensional velocimetry of dynamic x-ray blood speckle. *Appl. Phys. Lett. 93*, 153901.

Irvine, S., D. Paganin, R. Jamison, S. Dubsky, and A. Fouras. 2010. Vector tomographic x-ray phase contrast velocimetry utilizing dynamic blood speckle. *Opt. Exp. 18*, 2368–2379.

Ismail, E., W. Kaabar, D. Garrity, O. Gundogdu, O. Bunk, F. Pfeiffer, M. Farquharson, and D. A. Bradley. 2010. X-ray pahse contrast imaging of the bone-cartilage interface. *Appl. Radiat. Isotopes 68*, 767–771.

Jamison, R. A., S. Dubsky, K. K. W. Siu, K. Hourigan, and A. Fouras. 2011. X-ray velocimetry and haemodynamic forces within a stenosed femoral model at physiological flow rates. *Ann. Biomed. Eng. 39*, 1643–1653.

Johnston, R. E., D. Washburn, E. Pisano, C. Burns, W. C. Thomlinson, L. D. Chapman, F. Arfeli, N. F. Gmur, Z. Zhong, and D. Sayers. 1996. Mammographic phantom studies with synchrotron radiation. *Radiology 200*, 659–663.

Kak, A. C. and M. Slaney. 2001. *Principles of Computerized Tomographic Imaging*. SIAM, Philadelphia.

Kim, G. B., N. Y. Lim, and S. J. Lee. 2009. Hollow microcapsules for sensing micro-scale flow motion in x-ray imaging method. *Microfluid. Nanofluid. 6*, 419–424.

Kitchen, M., R. Lewis, S. Hooper, M. Wallace, K. K. W. Siu, I. Williams, S. Irvine et al. 2007. Dynamic studies of lung fluid clearance with phase contrast imaging. *AIP Conference Proceedings 879*, 1903–1907.

Kitchen, M. J., R. A. Lewis, M. J. Morgan, M. J. Wallace, M. L. Siew, K. K. W. Siu, A. Habib, A. Fouras, N. Yagi, K. Uesugi, and S. B. Hooper. 2008. Dynamic measures of regional lung air volume using phase contrast x-ray imaging. *Phys. Med. Biol. 53*, 6065d.

Kitchen, M., D. Paganin, R. Lewis, N. Yagi, and K. Uesugi. 2005. Analysis of speckle patterns in phase-contrast images of lung tissue. *Nucl. Instrum. Meth. A 548*, 240–246.

Kitchen, M., D. Paganin, R. Lewis, N. Yagi, K. Uesugi, and S. Mudie. 2004. On the origin of speckle in x-ray phase contrast images of lung tissue. *Phys. Med. Biol. 49*, 4335–4348.

Komlev, V., M. Mastrogiacomo, R.C.Pereira, F. Peyrin, F. Rustichelli, and R. Cancedda. 2010. Biodegration of porous calcium phosphate scaffolds in an ectopic bone formation model studied by x-ray computed microtomography. *Eur. Cells Mater. 19*, 136–146.

Kono, M., C. Ohbayashi, K. Yamasaki, Y. Ohno, S. Adachi, K. Sugimura, and Y. Suzuki. 2001. Refraction imaging and histologic correlation in excised tissue from a normal human lung: preliminary report. *Acad. Radiol. 8*, 898–902.

Kotre, C. and I. Birch. 1999. Phase contrast enhancement of x-ray mammography: A design study. *Phys. Med. Biol. 44* (11), 2853–2866.

Kotre, C., I. Birch, and K. Robson. 2002. Anomalous image quality phantom scores in magnification mammography: Evidence of phase contrast enhancement. *Br. J. Radiol. 75* (890), 170–173.

Langer, M., Y. Liu, F. Tortelli, P. Cloetens, R. Cancedda, and F. Peyrin. 2010. Regularized phase tomography enables study of mineralized and unmineralized tissue in porous bone scaffold. *J. Microsc. 238*, 230–239.

Lee, S.-J. and G. B. Kim. 2005. Synchrotron microimaging technique for measuring the velocity fields of real blood flows. *J. Appl. Phys. 97*, 064701.

Lee, Y. S., E.-A. Heo, H. Y. Jun, S. H. Kang, H. S. Kim, M. S. Lee, S.-J. Byun, S. H. Lee, S. H. Park, and K.-H. Yoon. 2010. Articular cartilage imaging by the use of phase-contrast tomography in a collagen-induced arthritis mouse model. *Acad. Radiol. 17*, 244–250.

Lewis, R. A. 2004. Medical phase contrast x-ray imaging: Current status and future prospects. *Phys. Med. Biol. 49*, 3573–3583.

Lewis, R. A., N. Yagi, M. J. Kitchen, M. J. Morgan, D. Paganin, K. K. W. Siu, K. Pavlov, I. Williams, K. Uesugi, M. J. Wallace, C. J. Hall, J. Whitley, and S. B. Hooper. 2005. Dynamic imaging of the lungs using x-ray phase contrast. *Phys. Med. Biol. 50*, 5031–5040.

Liu, X., J. Zhao, J. Sun, X. Gu, T. Xiao, P. Liu, and L. Xu. 2010. Lung cancer and angiogenesis imaging using synchrotron radiation. *Phys. Med. Biol. 55*, 2399–2409.

Mandel, L. and E. Wolf. 1995. *Optical Coherence and Quantum Optics*. Cambridge University Press, Cambridge, UK.

Mayo, S. C., T. Davis, T. E. Gureyev, P. Miller, D. M. Paganin, A. Pogany, A. Stevenson, and S. W. Wilkins. 2003. X-ray phase-contrast microscopy and microtomography. *Opt. Exp. 11*, 2289–2302.

McMahon, P. J., A. G. Peele, D. Paterson, J. J. A. Lin, T. H. K. Irving, I. McNulty, and K. A. Nugent. 2003. Quantitative x-ray phase tomography with sub-micron resolution. *Opt. Commun. 217*, 53–58.

Miao, J., C.-C. Chen, C. Song, Y. Nishino, Y. Kohmura, T. Ishikawa, D. Ramunno-Johnson, T.-K. Lee, and S. Risbud. 2006. Three-dimensional gan-ga$_2$o$_3$ core shell structure revealed by x-ray diffraction microscopy. *Phys. Rev. Lett. 97*, 215503.

Momose, A., T. Takeda, and Y. Itai. 1995. Phase-contrast x-ray computed tomography for observing biological specimens and organic materials. *Rev. Sci. Instrum. 66*, 1434–1436.

Momose, A., W. Yashiro, Y. Takeda, Y. Suzuki, and T. Hattori. 2006. Phase tomography by x-ray talbot interferometry for biological imaging. *Jpn. J. Appl. Phys. 45*, 5254–5262.

Morgan, K. S., K. K. W. Siu, and D. M. Paganin. 2010. The projection approximation and edge contrast for x-ray propagation-based phase contrast imaging of a cylindrical edge. *Opt. Exp. 18*, 9865–9878.

Mori, K., K. Hyodo, N. Shikano, and M. Ando. 1999. First observation of small fractures on a human dried proximal phalanx by synchrotron x-ray interference radiography. *Jpn. J. Appl. Phys. 238*, L1339–41.

Mori, K., N. Sekine, H. Sato, D. Shimao, H. Shiwaku, K. Hyodo, H. Sugiyama, M. Ando, K. Ohashi, M. Koyama, and Y. Nakajima. 2002. Application of synchrotron x-ray imaging to phase objects in orthopedics. *J. Synchrotron Rad. 9*, 143–147.

Morita, T., M. Yamada, A. Kano, S. Nagatsuka, C. Honda, and T. Endo. 2008. A comparison between film-screen mammography and full-field digital mammography utilizing phase contrast technology in breast cancer screening programs. In E. Krupinski (Ed.), *Digital Mammography*, Volume 5116 of *Lecture Notes in Computer Science*, pp. 48–54. Springer, Berlin/Heidelberg.

Myers, G. R., T. E. Gureyev, and D. M. Paganin. 2007a. Stability of phase-contrast tomography. *J. Opt. Soc. Am. A 24*, 2516–2526.

Myers, G. R., S. C. Mayo, T. E. Gureyev, D. M. Paganin, and S. W. Wilkins. 2007b. Polychromatic cone-beam phase-contrast tomography. *Phys. Rev. A 76*, 045804.

Natterer, F. 2001. *The Mathematics of Computerized Tomography*. Society for Industrial and Applied Mathematics, Philadelphia.

Nieto-Vesperinas, M. 2006. *Scattering and Diffraction in Physical Optics* (2 ed.). World Scientific Publishing Co, Singapore.

Nugent, K. A., T. E. Gureyev, D. F. Cookson, D. M. Paganin, and Z. Barnea. 1996. Quantitative phase imaging using hard x-rays. *Phys. Rev. Lett. 77*, 2961–2964.

Paganin, D. M. 2006. *Coherent X-Ray Optics*. Oxford University Press, New York.

Paganin, D. M. and T. E. Gureyev. 2008. Phase contrast, phase retrieval and aberration balancing in shift-invariant linear imaging systems. *Opt. Commun. 281*, 965–981.

Paganin, D. M., S. Mayo, T. E. Gureyev, P. R. Miller, and S. W. Wilkins. 2002. Simultaneous phase and amplitude extraction from a single defocused image of a homogeneous object. *J. Microsc. 206*, 33–40.

Pani, S., R. Longo, D. Dreossi, F. Montanari, A. Olivo, F. Arfelli, A. Bergamaschi, P. Poropat, L. Rigon, F. Zanconati, L. D. Palma, and E. Castelli. 2004. Breast tomography with synchrotron radiation: Preliminary results. *Phys. Med. Biol. 49*, 1739–1754.

Parsons, D., K. Morgan, M. Donnelley, A. Fouras, J. Crosbie, I. Williams, R. Boucher, K. Uesugi, N. Yagi, and K. Siu. 2008. High-resolution visualization of airspace structures in intact mice via synchrotron phase-contrast x-ray imaging (pcxi). *J. Anat. 213*, 217–227.

Pavlov, K. M., C. M. Kewish, J. R. Davis, and M. J. Morgan. 2001. A variant on the geometrical optics approximation in diffraction enhanced tomography. *J. Phys. D: Appl. Phys. 34*, A168–A172.

Pfeiffer, F., O. Bunk, C. Kottler, and C. David. 2007a. Tomographic reconstruction of three-dimensional objects from hard x-ray differential phase contrast projection images. *Nucl. Instrum. Meth. Phys. Res. A 580*, 925–928.

Pfeiffer, F., C. Kottler, O. Bunk, and C. David. 2007b. Hard x-ray phase tomography with low-brilliance sources. *Phys. Rev. Lett. 98*, 108105.

Pogany, A., D. Gao, and S. W. Wilkins. 1997. Contrast and resolution in imaging with a micro-focus x-ray source. *Rev. Sci. Instrum. 68*, 2774.

Porra, L., H. Suhonen, P. Suortti, A. Sovijrvi, and S. Bayat. 2011. Effect of positive end-expiratory pressure on regional ventilation distribution during bronchoconstriction in rabbit studied by synchrotron radiation imaging. *Crit. Care Med. 39*, 1731–1738.

Ramachandran, G. N. and A. V. Lakshminarayanan. 1971. Three-dimensional reconstruction from radiographs and electron micrographs: Application of convolution instead of fourier transforms. *Proc. Nat. Acad. Sci. US 68*, 2236–2240.

Raven, C., A. Snigirev, I. Snigireva, P. Spanne, A. Souvorov, and V. Kohn. 1996. Phase-contrast microtomography with coherent high-energy synchrotron x rays. *Appl. Phys. Lett. 69*, 1826–1828.

Řeháček, J., Z. Hradil, J. Peřina, S. Pascazio, P. Facchi, and M. Zawisky. 2006. Advanced neutron imaging and sensing. *Adv. Imag Electron. Phys. 142*, 53–157.

Schuster, D., A. Kovacs, J. Garbow, and D. Piwnica-Worms. 2004. Recent advances in imaging the lungs of intact small animals. *Am. J. Respir. Cell Mol. Biol. 30*, 129–38.

Sera, T., K. Uesugi, and N. Yagi. 2005. Refraction-enhanced tomography of mouse and rabbit lungs. *Med. Phys. 32*, 2787.

Shepp, L. A. and B. F. Logan. 1974. The fourier reconstruction of a head section. *IEEE Trans. Nucl. Sci. NS-21*, 21–43.

Siew, M. L., A. B. te Pas, M. J. Wallace, M. J. Kitchen, R. A. Lewis, A. Fouras, C. J. Morley et al. 2009. Positive end-expiratory pressure enhances development of a functional residual capacity in preterm rabbits ventilated from birth. *J. Appl. Physiol. 106*, 1487–1493.

Siu, K. K. W., K. S. Morgan, D. M. Paganin, R. Boucher, K. Uesugi, N. Yagi, and D. W. Parsons. 2008. Phase contrast x-ray imaging for the non-invasive detection of airway surfaces and lumen characteristics in mouse models of airway disease. *Eur. J. Radiol. 68*, S22–6.

Smith, R. A., S. W. Duffy, R. Gabe, L. Tabar, A. Yen, and T. H. Chen. 2004. The randomized trials of breast cancer screening: what have we learned? *Radiol. Clin. North Am.* *42*, 793–806.

Snigirev, A., I. Snigireva, V. G. Kohn, and S. M. Kuznetsov. 1996. On the requirements to the instrumentation for the new generation of the synchrotron radiation sources. beryllium windows. *Nucl. Instr. Meth. Phys. Res. A 370*, 634–640.

Snigirev, A., I. Snigireva, V. G. Kohn, S. M. Kuznetsov, and I. Schelekov. 1995. On the possibilities of x-ray phase contrast microimaging by coherent high-energy synchrotron radiation. *Rev. Sci. Instrum. 66*, 5486–5492.

Tanaka, T., C. Honda, S. Matsuo, K. Noma, H. Oohara, N. Hitta, S. Ota et al. 2005. The first trial of phase contrast imaging for digital full-field mammography using a practical molybdenum x-ray tube. *Invest. Radiol. 40*, 385–96.

Tanaka, T., N. Nitta, S. Ohta, T. Kobayashi, A. Kano, K. Tsuchiya, Y. Murakami et al. 2009. Evaluation of computer-aided detection of lesions in mammograms obtained with a digital phase-contrast mammography system. *Eur. Radiol. 19*, 2886–2895.

Tang, R., Y. Xi, W.-M. Chai, Y. Wang, Y. Guan, G.-Y. Yang, H. Xie, and K.-M. Chen. 2011. Microbubble-based synchrotron radiation phase contrast imaging: Basic study and angiography applications. *Phys. Med. Biol. 56*, 3503–3512.

Teague, M. R. 1983. Deterministic phase retrieval: A green's function solution. *J. Opt. Soc. Am. 73*, 1434–1441.

Weiss, P., L. Obadia, D. Magne, X. Bourges, C. Rau, T. Weitkamp, I. Khairoun et al. 2003. Synchrotron x-ray microtomography (on a micron scale provides three-dimensional imaging represenation of bone ingrowth in calcium phosphate biomaterials). *Biomaterials 24*, 4591–4601.

Westneat, M. W. and O. Betz, R. W. Blob, K. Fezzaa, W. J. Cooper, and W.-K. Lee. 2003. Trachael respiration in insects visualised with synchrotron x-ray imaging. *Science 24*, 555–580.

Wilkins, S. W. 1996. Simplified conditions and configurations for phase-contrast imaging with hard x-rays. *Australian Patent Application PN 2112/95 (1995); PCT Patent Application PCT/AU96/00178.*

Wilkins, S. W., T. E. Gureyev, D. Gao, A. Pogany, and A. W. Stevenson. 1996. Phase-contrast imaging using polychromatic hard x-rays. *Nature 384*, 335–338.

Williams, I. M. 2008. *Optmising phase contrast imaging of breast tissue with conventional sources.* PhD thesis, School of Physics, Monash University.

Williams, I. M., K. K. W. Siu, G. Runxuan, X. He, S. A. Hart, C. B. Styles, and R. A. Lewis. 2008. Towards the clinical application of x-ray phase contrast imaging. *Eur. J. Radiol. 68S*, S73–S77.

Wolf, E. 1969. Three-dimensional structure determination of semi-transparent objects from holographic data. *Opt. Commun. 1*, 153–156.

Wu, X. and H. Liu. 2003. Clinical implementation of x-ray phase-contrast imaging: Theoretical foundations and design considerations. *Med. Phys. 30*, 2169–2179.

Wu, X. and H. Liu. 2005. X-ray cone-beam phase tomography formulas based on phase-attenuation duality. *Opt. Exp. 13*, 6000–6014.

Yagi, N., Y. Suzuki, K. Umetani, Y. Kohmura, and K. Yamasaki. 1999. Refraction-enhanced x-ray imaging of mouse lung using synchrotron radiation source. *Med. Phys. 26*, 2190–2193.

Yeom, J., S. Chang, J. K. Park, J. H. Je, D. J. Yang, S. K. Choi, H. Shin, S. J. Lee, J. H. Shim, D. W. Cho, and S. K. Hahn. 2010. Synchrotron x-ray bioimaging of bone regeneration by artificial bone substitute of megagen synthetic bone and hyaluronate hydrogels. *Tissue Eng. Part C 16*, 1059–1068.

Zhang, X., X.-S. Liu, X.-R. Yang, S.-L. Chen, P.-P. Zhu, and Q.-X. Yuan. 2008. Mouse blood vessel imaging by in-line x-ray phase-contrast imaging. *Phys. Med. Biol. 53*, 5735–5743.

III

Photoacoustic Imaging and Tomography

6

Photoacoustic Mammography

Robert Kruger
OptoSonics, Inc.

Richard Lam
OptoSonics, Inc.

Daniel Reinecke
OptoSonics, Inc.

Stephen Del Rio
OptoSonics, Inc.

The field of photoacoustic mammography (PAM) is a nascent imaging modality for breast cancer screening and diagnosis. In this chapter, we discuss currently available breast imaging technologies and their shortcomings for specific populations. Then we present the theory behind photoacoustic mammography based on hemoglobin absorption of infrared light. Example photoacoustic imaging results on murine cancer models illustrate the basis for further exploration of PAM in human breast cancer screening. Finally, we show some exciting imaging results on human breast tissue using a modified photoacoustic tomography scanner.

6.1 Introduction

After skin cancers, breast cancer is the most commonly diagnosed malignancy in women in the United States, with a one-in-eight lifetime risk.[1] The American Cancer Society estimates that, in 2009, 192,370 women were diagnosed with invasive breast cancer and a further 62,280 with ductal carcinoma *in situ*. It is expected that 40,170 of these women will die of the disease.[2]

X-ray mammography remains today's imaging standard for breast cancer screening, demonstrating an overall sensitivity of ~80% and a specificity of ~90% for women over 50.[3] X-ray mammography depends on detecting often subtle differences in breast tissue density and qualitatively assessing certain morphologic features of suspect masses, such as irregular shapes or spiculated masses, which are reported to have a 32–81% positive predictive value for cancer.[4]

However, x-ray mammographic screening does not perform adequately in all patient populations. Breast fibroglandular tissue with higher densities, more prominent in younger women (<50 years), limits mammographic sensitivity for breast cancer to as low as 45%, compared with nearly 100% for predominantly fatty breasts.[5–7] Thus, alternative imaging modalities, such as MRI (with demonstrably higher sensitivity than mammography,

albeit with decreased specificity), have been sought to improve the screening performance in these and other populations.[8] Recognizing the limitations of screening mammography, in 2007, the American Cancer Society recommended annual screening with breast MR imaging for women with greater than 20% lifetime risk of breast cancer.[9]

Of the MR imaging techniques used for assessment of breast cancer, dynamic contrast-enhanced MR imaging (DCE-MRI) using gadolinium-based agents is the most established and widely used.[10] The rationale for employing these dynamic contrast-enhanced methodologies is to look for angiogenic biomarkers of cancer. Breast cancers that reach ~3 mm (10^6 cells) can no longer be adequately nourished by diffusion, and hypoxia ensues. Hypoxia induces expression of cytokines like VEGF, leading to creation of tumor vessels that are characterized by their disorganized structure and leaky capillaries.[11] Such malignant tumors are expected to display different *morphology* and *opacification and washout* patterns than do benign masses or normal glandular tissue, with MRI positive predictive values of up to 91%.[12] Unfortunately, MRI is an expensive technology, so effective and less costly imaging alternatives are desired.

6.2 Photoacoustic Alternatives

Hemoglobin is a strong optical absorber in the near infrared, and its presence in breast masses correlates strongly with angiogenesis and elevated microvessel density, both being consistent biomarkers of malignancy.[13] Average levels of hemoglobin are reportedly higher in malignant masses than in "normal" breast tissues.[14] Ntziachristos et al. found that malignant tumors had a 26-fold increase in hemoglobin concentration over normal background breast tissue and a twofold increase compared to fibroadenomas.[15] Investigators have noted that feeding arteries that flow directly into a tumor are often characterized by tortuosity and flexion.[16] Other investigators have found positive correlations between the number of feeding vessels and tumor size.[17] Still others have observed elevated vascularity in the ipsilateral breast when cancer is present.[18,19]

Several groups have hypothesized that breast cancer can be detected *without the need for exogenous contrast agents* using photoacoustic imaging of hemoglobin at near-infrared wavelengths. This hypothesis is based on the expectation that a malignant mass will exhibit an elevated concentration of hemoglobin,[20,21] whose absorption coefficient is over an order of magnitude higher than normal breast parenchyma.[22–24] Combined with an effective scattering coefficient in all breast tissues of $8–12$ cm^{-1}, the resulting optical attenuation coefficient for an average breast is therefore $1.0–1.3$ cm^{-1}, only slightly greater than that reported at mammographic x-ray energies ($0.5–0.8$ cm^{-1}).[25] Yet the differential contrast between hemoglobin-rich (tumors, blood vessels) and hemoglobin-poor (fat, glandular) tissue is expected to be *significantly greater* than the differential density between tumor and surrounding breast tissue at x-ray energies.

Technologies have been developed over the years to investigate the optical absorption and scattering properties of breast tissue via diffuse optical transmission and/or reflectance spectroscopy, usually incorporating multispectral optical sources.[26] Average levels of hemoglobin have been reported higher in malignant masses than in "normal" breast tissues,[14] but spatial resolution is very poor, and volume-averaging effects mask the heterogeneous distribution of hemoglobin within malignant tumors, reduce SNR, and render characterizing a tumor's morphology moot for all but large tumors.

A laser-based optoacoustic imaging system (LOIS) for detecting breast cancer was proposed over a decade ago by Oraevsky et al.[27] The original device has undergone several iterations since that time. A recent publication describes a 64-element annular array of rectangular transducer elements (20×3 mm) surrounding a hemi-cylindrical volume in which a breast is suspended.[28] An Alexandrite laser operating at 755 nm illuminates the breast normal to the imaging plane, and a single slice through the breast is captured at a time. Spatial resolution within the imaging plane is reported to be 0.5 mm, a significant improvement over diffuse optical imaging, but cross-plane spatial resolution is relatively poor. Imaging multiple planes requires rotating the device relative to the breast, thereby extending the imaging time for an entire breast to many minutes.

Another common approach to photoacoustic imaging incorporates a pulsed light source into a conventional medical ultrasound probe, for example, the OPUS (optoacoustic plus ultrasound) system developed by Haisch et al.[29] This device has the convenient property that the ultrasound and photoacoustic images are automatically coregistered spatially. Dynamic acquisition is also possible, albeit over a single imaging plane. But like the LOIS device, cross-plane spatial resolution is poor.

The Twente Photoacoustic Mammoscope, which produces 3D images over a 90 mm field of view, was introduced by Manohar et al. in 2004.[30] It consists of a planar array of 590 PVDF detectors. The side of the compressed breast opposite the detector plane is scanned slowly with a 16-mm-diameter laser beam as the array is read out, a single element at a time. Scan time is reported as 30–45 min and spatial resolution is 2.3–3.9 mm, which is poor compared to MRI and CT.

More recently, Kyoto University and Canon Inc. published results comparing a prototype photoacoustic mammography scanner with MRI.[31] They used mild breast compression with a patient lying prone with a pendant-suspended breast. Their scanner used a rectangular array of 345 detector elements with a 1 MHz center frequency and a 2×2 mm element size. Spatial resolution was reported as <3 mm. Tumors were detected photoacoustically in 8 out of 10 breast cancer patients, with two cases diagnosed as DCIS. The photoacoustic results compared favorably with both DCE-MRI and pathological findings.

6.3 Photoacoustic Tomography

In 1994, we first suggested that optical absorption could be imaged deep within turbid media, such as soft tissue, by stimulating photoacoustic interactions with a pulsed light source and recording the echoes with a piezoelectric transducer.[32] The following year we introduced a reconstruction algorithm for 2D PAT image reconstruction,[33] and in 1999, we introduced a fully 3D PAT breast imaging system.[34] In 2003, we introduced the first fully 3D PAT system for imaging small animals.[35]

More recently, we developed a second-generation PAT scanner that employs a sparse hemispherical array of transducers to capture 128 uniformly spaced radial "projections" at a time. The conceptual design for this scanner is illustrated in Figure 6.1a. The device consists of an array of 128 3-mm-diameter transducers (5 MHz center frequency) laid out in a spiral pattern on a hemispherical surface with a 100 mm radius of curvature. To first-order approximation, each element of the array samples a radial "projection" through k-space. Rotating the array to multiple angular positions allows us to increase the density of k-space sampling while maintaining uniform angular coverage (Figure 6.1b).

With current testing, the animal to be imaged is placed in a customized plastic holder suspended above the bowl array. The bowl is filled with fluid and illuminated from below through a 12 mm plano-convex lens by a tunable (680–950 nm) OPO laser operating at 10 Hz, 20 mJ/pulse, as illustrated in Figure 6.2. The physical width of the diverging light beam measures ~20 mm at the center-of-curvature of the bowl array. The fluid-filled bowl array is rotated continuously and radial projections are collected at a rate of 1280 per second, while the animal remains stationary. A full set of data over 360° of bowl rotation is acquired in 6, 12, or 24 s, corresponding to 7680, 15,360, or 30,720 radial projections.

Time-dependent signals, $p'_{ij}(t)$, are recorded for each transducer i and angle j ($i = 1, ..., 128$; $j = 1, ..., N$; $N = 60, 120$, or 240) following each laser pulse. These signals are captured in parallel at a rate of 20 MHz and digitized to 12-bit precision. These recorded signals, $p'_{ij}(t)$, are related to the actual photoacoustic pressure signals, $p_{ij}(t)$, according to $p'_{ij}(t) = p_{ij}(t) * h(t)$, where $h(t)$ is the impulse response of the transducers, assuming a point

(a) 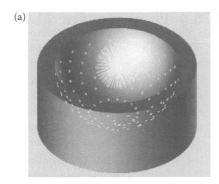 (b)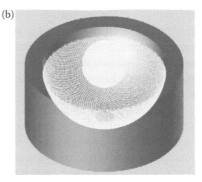

FIGURE 6.1 (a) Transducer array captures 128 projections through k-space (gray lines). (b) Increased k-space sampling after rotation of the array about its vertical axis.

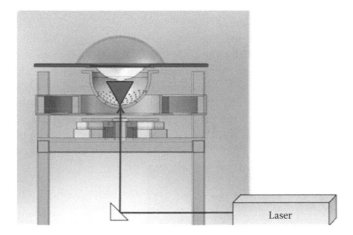

FIGURE 6.2 PAT scanner designed to capture radial projections, spaced uniformly in k-space.

$$\oiint_{|r_{ij}|=ct} A(r)\mathrm{d}S = \frac{4\pi C_p kt}{\beta}\,\mathrm{IFFT}\!\left(\frac{P'_{ij}(\omega)}{P'_0(\omega)}\right), \qquad (6.1)$$

where $\oiint_{|r_{ij}|=ct} A(\mathbf{r})\mathrm{d}S$ are the "projections" of the optical absorption distribution, $A(\mathbf{r})$, over spherical surfaces, and $P'_{ij}(\omega)$ and $P'_0(\omega)$ are the Fourier transforms of $p'_{ij}(t)$ and $p'_{ij}(t)$, respectively. Note that knowledge of the actual impulse response $h(t)$ of the system is not required. The parameters C_p, β, c, k are the specific heat, coefficient of thermal volume expansion, speed of sound, and a proportionality constant that depends on the illumination geometry and the absorption and scattering properties of the tissue. The optical-absorption distribution can then be reconstructed using a filtered back-projection algorithm, provided a sufficient number of projections have been acquired. We use a Fourier filter function of the form: $F(\omega) = K(\omega/\omega_C)^\alpha(1 + \cos(\pi(\omega/\omega_C)))$, where K is an arbitrary constant, ω_C is the cut-off frequency of the filter, and $1 \le \alpha \le 2$. The value of α controls the degree of high spatial frequency enhancement in the final image. Filtered projections are then back-projected over spherical surfaces.

Example PAT images of a murine breast cancer model are shown in Figure 6.3. Tumors initiated with MCF7 breast cancer cells transfected with the VEGF165 gene were imaged *in vivo* using our preclinical PAT scanner.[37] Figure 6.3 displays maximum intensity projections (MIPs) at two different projection angles, where image intensity is proportional to hemoglobin concentration. Within

detector, and "*" indicates convolution. Assume we do not know the impulse response of the transducer directly, but rather $p'_{ij}(t)$, the signal recorded due to the photoacoustic pressure, $p_0(t)$, from a photoacoustic point source. This can readily be recorded by illuminating a small point with a short light pulse. We can then write $p'_{ij}(t) = p_0(t) * h(t)$. Assuming we measure $p'_{ij}(t)$ and $p'_{ij}(t)$, Wang et al. have shown that the photoacoustic imaging equation takes the following form:[36]

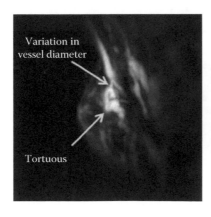 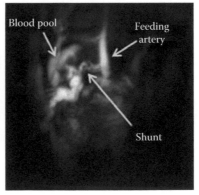

FIGURE 6.3 PAT images of a murine model of breast cancer (MCF7)—12.8 mm FOV.

these images, key mophologic features of the vasculature associated with this cancer can be identified, such as variations in vessel diameter, tortuous vascular structure, potential blood pools, feeding arteries, and vascular shunts. These are the first such PAT images to be made that depict angiogenic heterogeneity within a breast tumor model in 3D with such clarity.

To illustrate the applicability of DCE-PAT, a xenograft mouse model with an MDA-MB-231 breast tumor was injected with indocyanine green (ICG), a near-infrared fluorochrome.[38] The ICG binds to albumin in blood plasma and is filtered by the biliary system.[39] In effect, it acts as a blood pool agent during the initial wash-in phase through the vascularized tumor. DCE-PAT images were acquired prior to injection and every 12 s after injection. Figure 6.4 shows the resulting PCT images of the tumor over time, and Figure 6.5 plots the PCT intensity values versus time for three selected voxels. Note that voxel 2 represents a high-perfusion region due to its proximity to an artery feeding the tumor. Voxels 1 and 6 represent high- and low-perfusion regions. Based on these preliminary results, we hypothesize that dynamic perfusion studies of suspicious areas in human breast tissue, imaged through DCE-PAT with ICG contrast, will perform comparably to DCE-MRI using gadolinium for the detection of breast cancers.

6.4 Photoacoustic Mammography

We modified our PAT scanner in two ways to image human breasts. First, we installed a 60°, holographic diffuser (Edmunds, PN 53871) at the base of the imaging bowl array, just above the plano-convex lens, to increase the area of the illumination beam. Second, we installed a 5″-diameter, hemispherical "cup" molded from 0.020″ thick clear PETG above the imaging bowl. The cup was immersed in the degassed-water-filled imaging tank so that the apex of the cup lay 25 mm below the center of curvature of the imaging tank. The holographic diffuser spread the beam to a diameter of ~50 mm at the input to the cup. The mean optical fluence of the beam at the breast skin surface was ~1.0 mJ/cm² (800 nm), which is a factor of 32 less than the maximum permissible exposure (MPE) according to the *American National Standard for Safe Use of Lasers*.[40]

Prior to breast imaging, the water in the imaging bowl was heated to 30–32°C for the patient comfort. A small amount of water was placed in the cup prior to breast positioning to couple the breast to the bowl array acoustically. The patient volunteer stood next to the PAT scanner, bent over and placed one of her breasts into the plastic cup. The patient was instructed to remain still, but was allowed to breathe normally during the

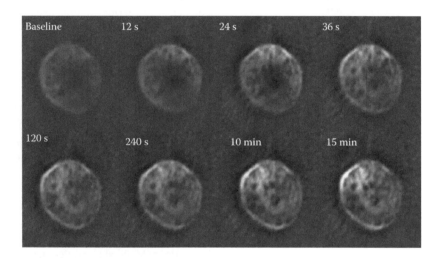

FIGURE 6.4 DCE-PCT images of ICG uptake in a tumor.

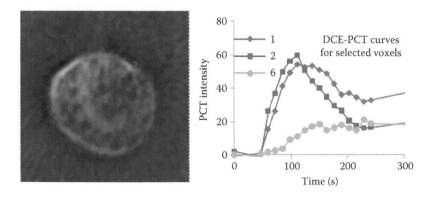

FIGURE 6.5 DCE-PCT curves for blood flow through different tumor regions versus time.

imaging procedure. The breast was then imaged at 755 nm using a 24-scan protocol (240 angles).

A two-component speed-of-sound algorithm was used to calculate the times-of-flight used in the breast reconstructions. We first calculated the velocity of sound within the water bath, based on a temperature calibration we performed in our laboratory. We then assumed that the velocity of sound within the breast was uniform, but different from that of water. Since the location of the breast–water boundary was determined by the geometry of the supporting hemispherical cup, we were able to calculate the times-of-flight based on the distances traversed through water *and* breast tissue along paths between each voxel and each detector location. We chose an "optimum" breast sound velocity by comparing vessel visibility qualitatively following the image reconstruction. Breast PAT images were reconstructed on a grid of $0.125 \times 0.125 \times 0.125$ mm^3 voxels. Two maximum intensity projections (MIPs) of the left breast of a 57-year-old patient volunteer are shown in Figure 6.6a and b.[41]

These initial 3D breast volumes demonstrate vascular detail that rivals the published DCE-MRI images, owing to the intrinsically high contrast of hemoglobin and the high spatial resolution of our technology. While we have yet to achieve a field of view sufficient for whole-breast imaging, we have been able to visualize the breast vasculature down to 0.35-mm diameter at a depth of 35–40 mm, *without contrast agents,* using a small fraction of the maximum permissible light exposure. These images represent a breakthrough for 3D PAT imaging performance in terms of depth of penetration and hemoglobin detail.

6.5 Toward Whole-Breast PAM

The next task is to modify the design of our scanner to image a range of breast sizes up to 1250 mL, which is a volume sufficiently large to image 90% of women in the United States.[33,34] We have obtained phantom data and performed Monte Carlo simulations to model the light distribution patterns that we have

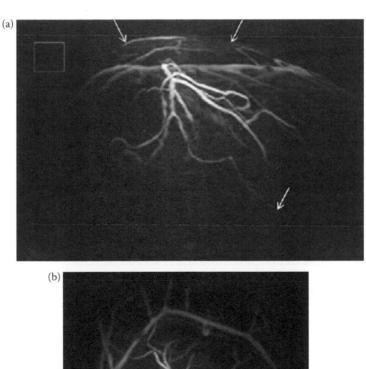

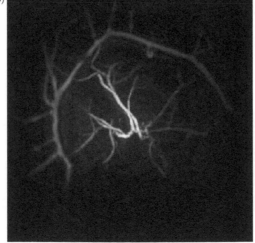

FIGURE 6.6 MIPs of the left breast of a 57-year-old patient volunteer in (a) lateral projection and (b) anterior-posterior projection. Top two arrows in (a) indicate front surface of breast, while the single arrow indicates a depth of 40 mm. Two MIP movies, depicting rotation of the 3D breast around two orthogonal axes, can be viewed at http://www.optosonics.com/NIH/grantfigures.html.

observed clinically with our current illumination geometry. These data suggest that the average effective attenuation coefficient of the breast corresponding to Figure 6.5 was 1.3 cm⁻¹, which is consistent with optical transport properties that others have reported.[25,26] The light distribution then is a function of both the optical transport properties of the breast and the incident illumination pattern. In our preliminary results, the optical diffuser produced a Gaussian beam profile with a full width half maximum of 48 mm and a total energy of 20 mJ per pulse. Our Monte Carlo simulations predict that we can increase this penetration depth to 5 cm over a breast area of 6.0×6.0 cm² by increasing the photon fluence to 400 mJ/pulse ($\lambda = 755$–797 nm), and illuminating the breast with a uniform, 32° light beam using a custom-engineered diffuser (RPC Photonics).

To achieve full-breast imaging, we envision a simple breast compression approach employing four interchangeable, flat-bottomed, clear-plastic breast "cups," ranging in volume from 250 to 1250 mL, as illustrated in Figure 6.7. These cups will be affixed to a patient-positioning table. By restraining the breast within these cups, the breast tissue will be spread out from a curvilinear shape, thereby reducing the maximum penetration depth

needed to image a 1250 mL volume to <50 mm. Natural breast compression is achieved by the weight of the patient pressing down from above the breast, and the compressed breast thickness that is achieved in this manner is anticipated to be *less* than that achieved via mammographic-style compression between parallel plates, as evidenced by ultrasonic breast imaging experience; as such, an exam is performed on a breast compressed against the chest wall in a supine position. The penetration depth required for mass detection in these exams is routinely <4.5 cm.[42,43] The immobilization provided by these cups will also provide some immunity from physiologic motion.

The imaging volume will be increased by translating the imaging bowl laterally, to multiple positions during data acquisition, and acquiring multiple PAT images, one subvolume at a time; the greater the volume of the breast, the greater the number of subvolumes that will be scanned. Each subvolume will be reconstructed individually and assembled into a larger composite image covering the entire breast, as demonstrated in Figure 6.8. The data acquisition time required for each subvolume will be 12 s and an entire exam will require <4 min (16 overlapping subvolumes) for the largest breast.

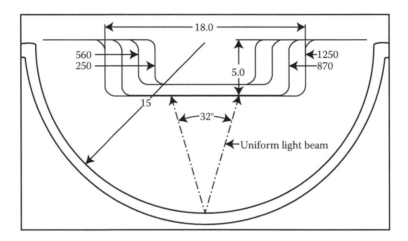

FIGURE 6.7   PAM scanner breast immobilization bowls with volumes ranging from 250 to 1250 mL. Maximum penetration depth needed to image 1250 mL breast = 5 cm.

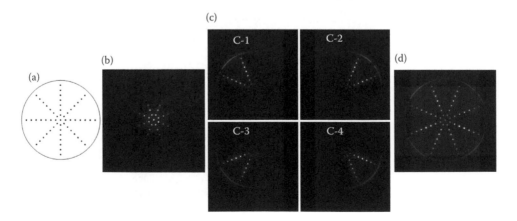

FIGURE 6.8   (a) Uniformity phantom (80 mm diameter). (b) PAT image of phantom. (c) Four PAT images of phantom off center from light beam. (d) Composite PAT image of uniformity phantom formed from four PAT component images (9C).

6.6 Summary

Photoacoustic mammography is a young technology, yet recent results suggest that photoacoustic imaging has the ability to visualize hemoglobin within the breast with submillimeter spatial resolution. While PAM has yet be developed to a point where the hemoglobin distribution can be visualized throughout the entire breast, there does not appear to be any physical reason why this cannot be accomplished with modified PAM data-acquisition protocols. We anticipate that PAM will soon be able to document angiogenic changes associated with breast cancer growth throughout the entire breast and eventually become a competitive imaging modality for human breast cancer screening and diagnosis.

References

1. Ries LAG, Melvert D, Krapcho M et al. SEER cancer statistics review, 1975–2005. National Cancer Institute, Bethesda, MD. Available at: http://seer.cancer.gov/scr/1975_2005/.

2. American Cancer Society. *Cancer Facts & Figures* 2009–2010. Atlanta (GA): American Cancer Society; 2010.

3. IARC, *Screening techniques. IARC Handbooks of Cancer Prevention: Breast Cancer Screening*, ed. B.F. Vainio H. Vol. 7. 2002, Lyon: IARC Press.

4. Nunes LW, Schnall MD, and Orel SG. Update of breast MR imaging architectural interpretation model. *Radiology* 2001; 219:484–494.

5. Carney PA, Miglioretti DL, Yankaskas BC et al. Individual and combined effects of age, breast density, and hormone replacement therapy use on the accuracy of screening mammography. *Ann. Intern. Med.* 2003; 138(3):168–175.

6. Rosenberg RD, Hunt WC, Williamson MR et al. Effects of age, breast density, ethnicity, and estrogen replacement therapy on screening mammographic sensitivity and cancer stage at diagnosis: Review of 183,134 screening mammograms in Albuquerque, New Mexico. *Radiology* 1998; 2009(2):511–518.

7. Kerlikowske K, Grady D, Barclay J et al. Likelihood ratios for modern screening mammography. Risk of breast cancer based on age and mammographic interpretation. *JAMA* 1996; 276(1):39–43.

8. Berg WA, Gutierrez L, NessAvier MS et al. Diagnostic accuracy of mammography clinical examination, US, and MR Imaging in pre-operative assessment of breast cancer. *Radiology* 2004; 233:830–849.

9. Saslow D, Boetes C, Burke W et al. American Cancer Society guidelines for breast screening with MRI as an adjunct to mammography. *CA Cancer J. Clin.* 2007; 57:75–89.

10. Moon M, Cornfeld D, and Weinreb J. Dynamic contrast-enhanced breast MR imaging. *Magn. Reson. Imaging Clin. N. Am.* 2009; 17:351–362.

11. Folkman J. Tumor angiogenesis: Therapeutic implications. *N. Engl. J. Med.* 1971; 285:1182–1186.

12. Lee CH. Problem solving MR imaging of the breast. *Radiol. Clin. N. Am.* 2004; 42:919–934.

13. Hoelen CGA, de Mul, FFM Pongers R, and Dekker A. Three-dimensional photoacoustic imaging of blood vessels in tissue. *Opt. Lett.* 1998; 23:648–650.

14. Tromberg BJ, Pogue BW, Paulsen KD et al. Assessing the future of diffuse optical imaging technologies for breast cancer management. *Med. Phys.* 2008; 35(6):2443–2451.

15. Ntziachristos V, Yodh AG, Schnall MD, and Chance B. MRI-guided diffuse optical spectroscopy of malignant and benign breast lesions. *Neoplasia* 2002; 4(4): 347–354.

16. Ueno E. Breast ultrasound. *Gan To Kagaku Ryoho* 1996 Mar; 23(Suppl 1):14–23.

17. Siewert C, Oellinger H, Sherif HK, Blohmer JU, Hadijuana J, and Felix R. Is there a correlation in breast carcinomas between tumor size and number of tumor vessels detected by gadolinium-enhanced magnetic resonance mammography? *MAGMA* 1997 Mar; 5(1):29–31.

18. Mahfouz AE, Sherif H, Saad A, Taupitz M, Filimonow S, Kivelitz D, and Hamm B. Gadolinium-enhanced MR angiography of the breast: is breast cancer associated with ipsilateral higher vascularity? *Eur. Radiol.* 2001; 11(6):965–969.

19. Sardanelli F, Fausto A, Menicagli L, and Esseridou A. Breast vascular mapping obtained with contrast-enhanced MR imaging: Implications for cancer diagnosis, treatment, and risk stratification. *Eur. Radiol.* 2007 Dec;17(Suppl 6):F48–F51.

20. G. Ku, X. Wang, X. Xie, G. Stoica, and L.-H. Wang. Imaging of tumor angiogenesis in rat brains *in vivo* by photoacoustic tomography. *Appl. Opt.* 2005; 44:770–775.

21. Oraevsky AA, Savateeva EV, Solomatin SV et al. Optoacoustic imaging of blood for visualization and diagnostics of breast cancer. *Biomed. Optoacoustics III* 2002; 4618:81–94.

22. Wray S, Cope M, Delpy DT et al. Characteristics of the near infrared absorption spectra of cytochrome aa3 and hemoglobin for the noninvasive monitoring of cerebral oxygenation. *Biochim. Biophys. Acta* 1988; 933:184–192.

23. Pifferi A, Swartling J, Chkiodze E. et al. Spectroscopic time-resolved diffuse reflectance and transmittance measurements of the female breast at different interfiber distances. *J. Biomed. Opt.* 2004; 9(6):1143–1151.

24. A. Cerussi, N. Shah, D. Hsiang, A. Durkin, J. Butler, and B. J. Tromberg. In vivo absorption, scattering, and physiologic properties of 58 malignant breast tumors determined by broadband diffuse optical spectroscopy. *J. Biomed. Opt.* 2006; 11(4):044005.

25. Heine JJ and Thomas JA. Effective x-ray attenuation coefficient measurements from two full field digital mammography systems for data calibration applications. *BioMed. Eng. OnLine* 2008; 7:13.

26. Volynskaya Z, Haka AS, Bechtel KL et al. Diagnosing breast cancer using diffuse reflectance spectroscopy and intrinsic fluorescence spectroscopy. *J. Biomed. Opt.* 2008;13(2):024012.

27. A. A. Oraevsky, R. O. Esenaliev, S. L. Jacques, F. K. Tittel, and D. Medina. Breast cancer diagnostics by laser optoacoustic tomography. *Trends in Optics and Photonics*, eds. R. R. Alfano and J. G. Fujimoto. 1996, pp. 316–321. Washington, DC: OSA Publishing House.

28. Ermilov SA, Khamapirad T, Conjusteau A et al. Laser opto-acoustic imaging system for detection of breast cancer. *J. Biomed. Opt.* 2009; 14(2):024007.

29. Haisch C, Eilert-Zell K, Vogel MM et al. Combined opto-acoustic/ultrasound system for tomographic absorption measurements: Possibilities and limitations. *Anal. Bioanal. Chem.* 2010; 397:1503–1510.

30. Manohar S, Kharine A, van Hespen JCG, Steenbergen W, and van Leeuwen TG. Photoacoustic mammography laboratory prototype: Imaging of breast tissue phantoms. *J. Biomed. Opt.* 2004; 9:1172–1181.

31. Kitai T, Asao Y, Sugie T et al. Photoacoustic mammography for the diagnosis of breast cancer. *Proc. 50th Annual Conf. Japanese Society for Medical and Biological Engineering*, 2011.

32. Kruger RA and Liu P-Y. Photoacoustic ultrasound: Pulse production and detection in 0.5% liposyn. *Med. Phys.* 1994; 21(7):1179–1184.

33. Kruger RA, Liu P-Y, and Fang Y. Photoacoustic ultrasound (PAUS)—Reconstruction tomography. *Med. Phys.* 1995; 22(10):1605–1609.

34. Kruger RA, Reinecke DR, and Kruger GA. Thermoacoustic computed tomography—Technical considerations. *Med. Phys.* 1999; 26(9):1832–1837.

35. Kruger RA, Kiser Jr WL, Reinecke DR, Kruger GA, and Miller KD. Thermoacoustic optical molecular imaging of small animals. *Mol. Imaging* 2003; 2(2):113–123.

36. Wang Y, Zeng Y, and Chen Q. Photoacoustic imaging with deconvolution algorithm. *Phys. Med. Biol.* 2004; 49:3117–3124.

37. Morgan TG, Kruger RA, Picot PA, Liu B, and Stantz KM. In Vivo molecular imaging applications of volume photoacoustic tomography for small animals. *World Medical Imaging Congress*, Montreal, September 23–26, 2009.

38. Stantz KM and Kruger RA. NIH grant 1R44CA102891-06: Technical Progress Report, 2011.

39. Cherrick GR, Stein SW, Leevy CM, and Davidson, CS. Indocyanine green: Observations on its physical properties, plasma decay, and hepatic extraction. *J. Clin. Invest.* 1960; 39(4):592–600. doi: 10.1172/JCI104072.

40. ANSI Z136.1–2007.

41. Kruger RA, Lam RB, Reinecke DR, DelRio SP, and Doyle RP. Photoacoustic mammography of the breast. *Med. Phys.* 2010; 37:6096–7000.

42. Madsen EL, Berg WA, Mendelson EB, and Fran GR. Anthropomorphic breast phantoms for qualification of investigators for ACRIN protocol 6666. *Radiology* 2006; 239(3):869–874.

43. Berg WA, Blume JD, Cormack JB, Mendelson EB, and Madsen EL. Lesion detection and characterization in a breast US phantom: Results of the ACRIN 6666 investigators. *Radiology* 2006; 239(3):693–702.

7

Multiparameter Photoacoustic Tomography

Song Hu
Washington University

Christopher Favazza
Washington University

Lihong V. Wang
Washington University

7.1 Introduction

Physiology is the study of mechanical, physical, and chemical functions of organelles, cells, tissues, and organs in living systems. Disorders in physiological functions can lead to disease and death. Thus, studying physiology and pathophysiology in a quantitative manner is important for both fundamental and translational biomedicine.

However, there are many outstanding limitations and challenges in quantitative physiological imaging. In vascular physiology, a variety of parameters are of interest, including vascular diameter, length, and tortuosity, the total concentration (HbT) and oxygenation (sO_2) of hemoglobin, and blood flow; however, none of the existing imaging technologies can provide a comprehensive assessment. Clinical modalities, including magnetic resonance imaging, x-ray computed tomography, positron emission tomography, and ultrasonography, lack either sufficient spatial resolution or satisfactory physiological contrast, or both. High-resolution optical microscopy, including confocal microscopy, multiphoton fluorescence microscopy, and optical coherence tomography, has difficulty in assessing the hemoglobin-associated physiological parameters (i.e., HbT and sO_2), even with the aid of exogenous contrast agents. More importantly, the lack of linkage between clinical modalities and optical microscopy poses a practical challenge in the translation of basic research findings to clinical practice.

These limitations and challenges are overcome by photoacoustic tomography (PAT). In PAT, biological tissues are excited by short-pulsed or intensity-modulated continuous-wave laser beams and emit ultrasound waves through thermoelastic expansion. Hemoglobin, the predominant endogenous optical absorber in the visible spectral region and the primary oxygen carrier in the circulatory system, allows PAT to assess the oxygen-related physiological parameters. Frequency-scalable ultrasound detection enables PAT to operate at various length scales, ranging in size from organelles to organs [1]. Further, the shared similarities between PAT and conventional ultrasonography promote its integration into clinical practice.

This chapter focuses on photoacoustic imaging and quantification of a variety of important physiological parameters *in vivo*. Detailed methodologies are discussed and promising applications of multiparameter PAT are reviewed. Although there is a particular emphasis on vascular physiology, some of the methods discussed in this chapter are applicable to quantitative imaging and characterization of other endogenous optical absorbers, including DNA, RNA, hemoglobin, melanin, lipid, and water.

7.2 Photoacoustic Quantification of Physiological Parameters

7.2.1 Vascular Anatomy

High-resolution PAT can be used to quantify the dimensions of optically absorbing tissue structures, such as blood-perfused vasculature, dye-perfused sentinel lymph nodes, melanin-rich skin melanomas, and nanoparticle-targeted tumors.

As an initial demonstration for diameter quantification, a 6-μm-thick carbon fiber immersed in water was imaged using

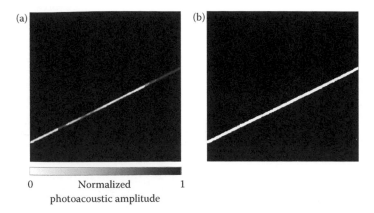

FIGURE 7.1 Optical-resolution photoacoustic microscopy of a 6-μm-diameter carbon fiber immersed in water. (a) Raw image. (b) Background-free binary image with an amplitude threshold.

optical-resolution photoacoustic microscopy (OR-PAM) [2,3]. After image acquisition, the photoacoustic amplitude was extracted via the Hilbert transform using the "hilbert" function in MATLAB® (MathWorks) to generate a maximum-amplitude-projection (MAP) image (Figure 7.1a). An empirically defined amplitude threshold that is 3 dB above the average signal in a non-fiber region was applied to remove the background. After Figure 7.1a was converted into a background-free binary image using the amplitude threshold (Figure 7.1b), the fiber cross section was identified by eight-connected labeling using the Image Processing Toolbox in MATLAB, and the average cross-sectional diameter was estimated to be 8.8 μm. Taking into account the 2.6 μm optical focal diameter, the imaged fiber diameter was estimated to be

5.9 μm by computing the convolution between the light fluence profile of the Gaussian optical focus and the optical absorption profile of the circular fiber cross section, which was in good agreement with the actual value.

With more sophisticated image registration and vessel segmentation techniques, longitudinal quantification of neovascularization in terms of chronic changes in vessel length, diameter, and tortuosity was recently demonstrated in a transgenic mouse model with controlled expression of hypoxia-inducible factor 1 (HIF-1) [4,5].

Serial images of the ear vasculature in a transgenic mouse (Figure 7.2a) were three-dimensionally coregistered and segmented with a semiautomatic algorithm developed from the

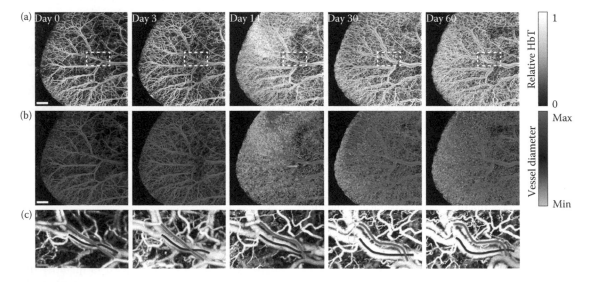

FIGURE 7.2 **(See color insert.)** Vessel-segmentation-enabled multiparameter OR-PAM of neovascularization in a transgenic mouse model. (a) Representative OR-PAM maximum-amplitude-projection images of the entire ear vasculature of a longitudinally imaged vascular endothelial growth factor floxed (VEGF$^{f/f}$) knock-in mouse treated with doxycycline for 60 days. (b) Differentiation between capillaries and trunk vessels via vessel segmentation. Capillaries are false-colored green and trunk vessels are red. (c) Quantification of trunk vessel remodeling (i.e., length, vasodilation, and tortuosity) in a selected pair of arterial-venous segments. Red and blue solid curves indicate the vessel axes of the arterial and venous segments, respectively. Scale bars in Panels (a)–(c) are 1 mm, 1 mm, and 200 μm, respectively.

MATLAB Image Processing Toolbox and Signal Processing Toolbox. First, the images acquired at different time points were coregistered to the baseline image acquired on day 0, using a three-dimensional (3-D) rigid registration (linear translation and/or rotation). Based on the similarity in vascular morphology between the baseline image and the image to be registered, 10–20 pairs of points were manually selected to compute the spatial translation and rotation matrix required for image registration. Second, each single vessel was identified and isolated from the vasculature network by using vessel segmentation, where each vessel was identified and labeled by tracking its axis in 3-D space. Third, each segmented vessel was identified as either a trunk vessel or a capillary, primarily based on the vessel diameter (Figure 7.2b). Since capillaries have an average diameter of 5–10 µm, 10 µm was selected to be the threshold. Additionally, depth information and signal intensity were also extracted to help identify densely packed capillary beds.

Based on vessel segmentation, the vessel length and tortuosity were further quantified by tracing the vessel axis. First, the center of each vessel cross section was identified, then the vessel axis was obtained by connecting all the centers. The vessel length between two adjacent bifurcation points of the selected trunk vessel segment was measured along the vessel axis. Vessel tortuosity was defined as the ratio of the vessel length to the linear distance between the two bifurcation points. As shown in Figure 7.2c, the same artery–vein pair segment was chosen for the quantification of vessel volume, diameter, length, and tortuosity throughout the 60-day monitoring.

7.2.2 Lesion Thickness

In dermatology, particularly for the diagnosis of skin melanoma, a key staging parameter is lesion thickness. However, photoacoustic measurements at single optical wavelengths cannot guarantee an accurate measurement, given the possibility that light has not fully penetrated the lesion.

To solve this issue, a multiwavelength method was developed [6]. In studying a blue nevus, a melanin-rich skin lesion, a dual-wavelength (570 and 700 nm) photoacoustic measurement was utilized to determine its actual thickness (Figure 7.3). As is well known, the optical absorption of melanin decays exponentially with increasing optical wavelength. Its absorption coefficient at 700 nm is ~2.5 times lower than that at 570 nm [7], which leads to a 2.5-fold deeper penetration in nevus tissue at 700 nm, given the same incident optical fluence. However, similar lesion thicknesses were measured at these two wavelengths (Figure 7.3e and f), which indicated that light had fully penetrated the lesion at both wavelengths. Therefore, both measurements accurately reflected the true thickness of the nevus. If the thickness measurements had been different, higher fluence or a longer wavelength would have been needed to confirm full penetration of the lesion.

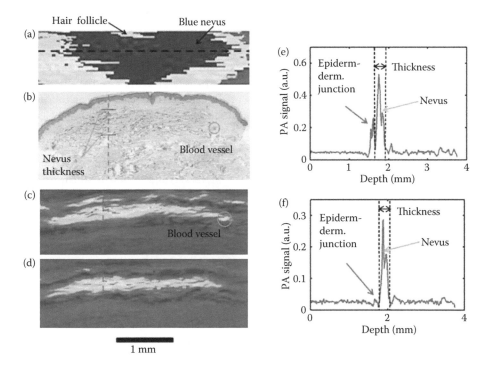

FIGURE 7.3 (**See color insert.**) (a) Maximum amplitude projection of a nevus generated from serial histology sections (top view, that is, *en face* view, of the nevus) used to match histology with photoacoustic data. The nevus is shown in red, and a large hair follicle near the top was used to orient the image with the photoacoustic MAP image. (b) Histology section along the dashed line in (a), in which the location of the nevus thickness is indicated by the dashed line and two hash marks. A blood vessel used for orientation is circled in green. (c) Cross-sectional photoacoustic image of the nevus acquired at 570 nm. The circled blood vessel matches with the vessels outlined in (b). (d) Cross-sectional photoacoustic image of the nevus acquired at 700 nm. The red dashed lines in (c) and (d) indicate the depth profiles shown in (e) and (f), respectively.

7.2.3 Hemoglobin Oxygen Concentration

Taking advantage of the differential spectra of oxyhemoglobin (HbO_2) and deoxyhemoglobin (HbR), PAT can quantify the relative concentrations of HbO_2 and HbR within single microvessels, from which relative HbT and absolute sO_2 can be computed.

Based on the above assumption, the blood absorption coefficient can be expressed as

$$\mu_a(\lambda_i, x, y, z) = \varepsilon_{HbR}(\lambda_i) \cdot [HbR] + \varepsilon_{HbO_2}(\lambda_i) \cdot [HbO_2], \quad (7.1)$$

where $\mu_a(\lambda_i, x, y, z)$ is the local blood absorption coefficient at the wavelength λ_i. Further, $\varepsilon_{HbO_2}(\lambda_i)$ and $\varepsilon_{HbR}(\lambda_i)$ are the molar extinction coefficients of HbO_2 and HbR, respectively, and $[HbO_2]$ and $[HbR]$ are their relative concentrations at the local position (x, y, z). Knowing that the photoacoustic signal amplitude $\Phi(\lambda_i, x, y, z)$ is proportional to the product of $\mu_a(\lambda_i, x, y, z)$ and the local optical fluence $F(\lambda_i, x, y, z)$, we can replace $\mu_a(\lambda_i, x, y, z)$ by $\Phi(\lambda_i, x, y, z)$ in Equation 7.1 and calculate $[HbO_2]$ and $[HbR]$ in relative values. Consequently, the sO_2 can be computed as

$$sO_2 = \frac{[HbO_2]}{[HbR] + [HbO_2]}. \quad (7.2)$$

HbT can be computed in relative values as the summation of $[HbO_2]$ and $[HbR]$. Alternatively, relative HbT can be directly measured using PAT at an isosbestic wavelength, where the molar extinction coefficients of HbR and HbO_2 are equal.

If we simply replace $\mu_a(\lambda_i, x, y, z)$ by $\Phi(\lambda_i, x, y, z)$ in Equation 7.1, $F(\lambda_i, x, y, z)$ is assumed to be wavelength independent, which is just an approximation neglecting the wavelength dependence of tissue absorption and scattering. To recover accurate sO_2 values without this approximation, fluence compensation is necessary. Fluence compensation may be implemented in an invasive manner [8,9], where solid absorbers or liquid contrast agents with known absorption spectra are administered into the tissue for spectral calibration. Although effective, the invasive methods disturb the intrinsic physiology and impede clinical translation.

Recently, a self-calibrating fluence compensation method based on the acoustic spectrum of the received photoacoustic signal was proposed to quantify optical absorption and chromophore concentration [10]. The acoustic spectrum of the received photoacoustic signal $[S(\omega)]$ depends on three factors. First is the real object spectrum $[O(\omega)]$ measured with unit fluence, which is determined by the shape, size, and optical properties of the object, as well as the direction of the incident fluence. Second is the system-dependent acoustic response $[H(\omega)]$, which is the Fourier transform of the detected photoacoustic signal from an ideal point absorber, with no acoustic attenuation in the tissue. Third is the acoustic attenuation effect $[a(\omega)]$, which is related to the acoustic properties of the tissue that lies between the object and the detector. Assuming system linearity, we have $S(\omega) = F(\lambda_i) \cdot O(\omega) \cdot H(\omega) \cdot a(\omega)$. Note that the last two terms are independent of optical wavelength and can be readily cancelled out in the dual-wavelength measurement of sO_2.

Specifically in OR-PAM, the optically defined lateral resolution is much smaller than the diameter of trunk vessels. Thus, a trunk vessel's surface can be roughly treated as flat. In this case, the fluence inside the blood vessel follows Beer's law, and the acoustic spectrum of the trunk vessel excited with unit fluence is

$$O(t) = \Gamma \cdot \mu_a \cdot \exp(-\mu_a v t), \quad (7.3)$$

where v is the speed of sound in the biological tissue and Γ is the Grüneisen coefficient.

If the photoacoustic signals of the blood vessel are measured at two optical wavelengths, the ratio of the detected acoustic spectra at the two optical wavelengths can be written as

$$\frac{S_1(\omega)}{S_2(\omega)} = \frac{F(\lambda_1) \cdot O_1(\omega) \cdot H(\omega) \cdot a(\omega)}{F(\lambda_2) \cdot O_2(\omega) \cdot H(\omega) \cdot a(\omega)} = \frac{F(\lambda_1) \cdot \sqrt{(\omega/\mu_{a2})^2 + v^2}}{F(\lambda_2) \cdot \sqrt{(\omega/\mu_{a1})^2 + v^2}},$$

$$(7.4)$$

where $O(\omega)$ is the Fourier transform of Equation 7.3. By fitting the acoustic spectrum, we can derive the absolute values of μ_{a1} and μ_{a2} and the ratio of $F(\lambda_1)/F(\lambda_2)$.

As an *in vivo* demonstration, a 1×1 mm^2 region of interest in a nude mouse ear was imaged at two optical wavelengths (561 and 570 nm). An artery–vein pair was selected from the MAP image acquired at 570 nm (labeled as V_1 and V_2 in the inset of Figure 7.4) for the quantification of sO_2 and absolute μ_a. The A-scans acquired within these two vessels were properly aligned by their maximum values and then averaged. The absorption coefficients and the incident fluence ratios were quantified from the acoustic spectral ratios (Figure 7.4). The HbT and sO_2 values were sequentially calculated based on the quantified optical absorption coefficients at the two optical wavelengths (Table 7.1). Note that the computed sO_2 values were off by ~8% and 11% in the artery and vein, respectively, if the optical fluence was assumed to be wavelength independent (i.e., $F(\lambda_1)/F(\lambda_2) = 1$).

Moreover, and promisingly, since $H(\omega)$ and $a(\omega)$ are roughly the same for both vessels, the optical absorption coefficients of both vessels can be quantified with a single optical wavelength (561 nm) measurement. The μ_a values measured at 561 nm for both blood vessels were quantified from the acoustic spectral ratio (Figure 7.4), and the fitted μ_a agree with the values in Table 7.1.

Recently, another single-wavelength sO_2 measurement method was developed based on relaxation photoacoustic microscopy (rPAM) [11,12]. The significant difference in the relaxation times of HbO_2 and HbR, rather than the spectral difference, was utilized to quantify the relative concentration of each molecule *in vivo*. Thus, rPAM is able to measure sO_2 with a single optical wavelength, avoiding the compensation for wavelength-dependent optical fluence.

In the nonsaturation regime of optical absorption, the photoacoustic signal Φ is proportional to the local optical fluence:

$$\Phi(F) \propto K \cdot \mu_a \cdot F, \quad (7.5)$$

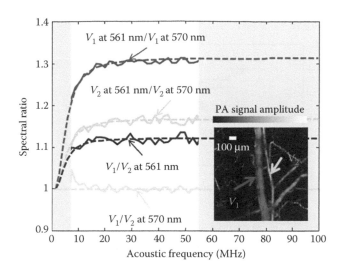

FIGURE 7.4 OR-PAM quantification of the optical absorption coefficients of arterial and venous blood in a living mouse ear. Structural image acquired at 570 nm (inset), and ratios of the acoustic spectral amplitudes of photoacoustic signals measured in an artery–vein pair *in vivo* with two optical wavelengths (570 and 561 nm).

where K is a proportionality coefficient determined by the ultrasonic parameters, the Grüneisen coefficient, and the heat conversion efficiency; μ_a is the optical absorption coefficient of the imaging target; and F the local optical fluence, which is the product of the local laser intensity I and the laser pulse width τ_{laser}.

In contrast, in the absorption saturation regime of hemoglobin, Equation 7.5 needs to be rewritten as

$$\Phi(F) \propto K \cdot F \cdot \left[\varepsilon_{HbO_2} \cdot \frac{[HbO_2]}{1 + F / \left(\tau_{laser} \cdot I_{sat}^{HbO_2} \right)} \right.$$
$$\left. + \varepsilon_{HbR} \cdot \frac{[HbR]}{1 + F / \left(\tau_{laser} \cdot I_{sat}^{HbR} \right)} \right], \quad (7.6)$$

where $I_{sat}^{HbO_2}$ and I_{sat}^{HbR} are the saturation intensities of HbO_2 and HbR.

By measuring the photoacoustic signal at various laser fluences at a single wavelength, the relative concentration of $[HbO_2]$ and $[HbR]$ can be extracted by solving

$$K \cdot \begin{bmatrix} [HbO_2] \\ [HbR] \end{bmatrix} = (H^T \cdot H)^{-1} \cdot H^T \cdot \underline{y}, \quad (7.7)$$

where $\underline{y}$ is the photoacoustic measurement vector, in which each line corresponds to a different incident laser intensity. The transformation matrix is

$$H = \begin{bmatrix} a_{11} & a_{12} \\ \vdots & \vdots \\ a_{i1} & a_{i2} \\ \vdots & \vdots \\ a_{n1} & n2 \end{bmatrix}, \quad \text{in which } a_{i1} = \frac{\varepsilon_{HbO_2}}{1 + F_i / \left(\tau_{laser} \cdot I_{sat}^{HbO_2} \right)} \cdot F_i$$

and
$$a_{i2} = \frac{\varepsilon_{HbR}}{1 + F_i / \left(\tau_{laser} \cdot I_{sat}^{HbR} \right)} \cdot F_i.$$

As an *in vivo* demonstration, a mouse ear was imaged with 4 ns laser pulses. The optical focal spot size of 3.9 μm was set approximately 120 μm below the skin surface. To measure the relaxation time of hemoglobin *in vivo*, the local fluence F_i inside the tissue had to be estimated. Assuming the illumination beam was Gaussian in shape, we had $F_i = E_i/(\pi \cdot w^2)$, where E_i was the local pulse energy and w was the Gaussian beam radius at the corresponding depth. Within the optical diffusion limit, the local pulse energy of OR-PAM could be estimated using Beer's law as $E_i = E_0 \cdot e^{-\alpha \cdot z}$, where E_0 is the incident energy on the skin surface, α is the average optical attenuation of the tissue, and z is the depth of the absorber. According to Equation 7.5, the averaged photoacoustic signal generated from a vessel at a given depth is proportional to the local fluence in the nonsaturation regime. Thus, the photoacoustic signals of three representative points in a vessel segment and their corresponding depths were experimentally measured to estimate α (Figure 7.5).

After obtaining α, the same area of the mouse ear was re-imaged at four different laser intensities. At each point in the MAP image (Figure 7.5a), the photoacoustic amplitude was experimentally measured and the normalized local pulse energy was computed based on the aforementioned model. Figure 7.6a and b shows the average photoacoustic amplitudes versus the local fluence in a vein and an artery, respectively. The photoacoustic signal in the artery shows more significant saturation, due to the longer absorption relaxation time of HbO_2 at 576 nm. The sO_2 map was computed using Equations 7.2 and 7.7 (Figure 7.6c). The average sO_2 values in the artery and the vein were measured to be 94% ± 2% and 74% ± 3%, respectively. As a mutual validation, a dual-wavelength measurement (at 576 and 592 nm) was performed sequentially (Figure 7.6d). The average sO_2 in the artery and in the vein were measured to be 96% ± 3% and 69% ± 3%, respectively, and agreed with the single-wavelength measurement.

TABLE 7.1 Absorption Coefficients, Fluence Ratio, Hemoglobin Concentrations, and sO_2 Quantified from Two Optical Wavelength Measurements

	$\mu_a(\lambda_1)$ (cm^{-1})	$\mu_a(\lambda_2)$ (cm^{-1})	$F(\lambda_2)/F(\lambda_1)$	[HbT] (g/L)	[HbO$_2$] (g/L)	[HbR] (g/L)	sO$_2$	sO$_2$[a]
V_1 (artery)	143 ± 3	188 ± 4	0.96 ± 0.01	110.6 ± 8.1	106.2 ± 4.3	4.4 ± 3.8	0.96 ± 0.04	0.88
V_2 (vein)	159 ± 4	186 ± 5	0.96 ± 0.01	110.2 ± 9.2	77.1 ± 4.9	33.1 ± 4.3	0.70 ± 0.07	0.62

[a] With $F(\lambda_2)/F(\lambda_1) = 1$.

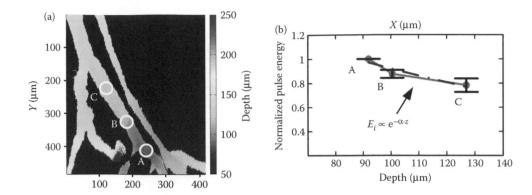

FIGURE 7.5 **(See color insert.)** (a) Depth-encoded MAP image of a mouse ear. (b) The exponential fitting (blue-dashed curve) and the normalized pulse energy as a function of depth (solid red dots).

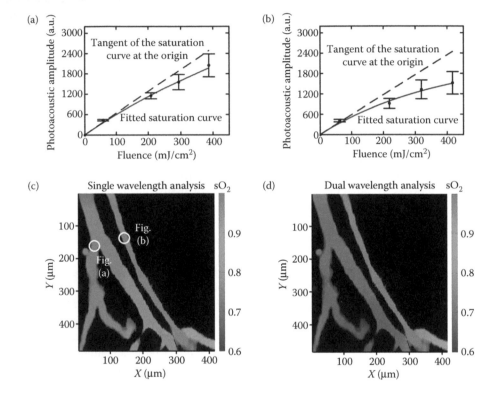

FIGURE 7.6 **(See color insert.)** Saturation profiles of photoacoustic signals and sO_2 mapping in a mouse ear. Average photoacoustic amplitude as a function of the fluence (a) at a low sO_2 location (a vein) and (b) at a high sO_2 location (an artery). Error bars represent the standard deviations in 15 adjacent measurements. (c) Single-wavelength (576 nm) and (d) dual-wavelength (576 and 592 nm) measurements of sO_2.

7.2.4 Flow Dynamics

With excellent spatial scalability, Doppler photoacoustics bridges the spatial gap between Doppler optics and ultrasound. More attractively, the high optical absorption contrast between the intravascular blood and extravascular background makes Doppler PAT an ideal technology to measure blood flow *in vivo*.

The photoacoustic Doppler effect was first observed in 2007 in flowing light-absorbing particles [13]. Recently, a photoacoustic Doppler bandwidth broadening method was proposed for *in vivo* blood flow measurement [14,15].

In the Doppler OR-PAM system shown in Figure 7.7a [15], the acoustic paths L1 and L2, respectively, subtend an obtuse angle and an acute angle with respect to the flow direction. Thus, they contribute to the photoacoustic Doppler shift with opposite signs and induce a bandwidth broadening of

$$\text{PDB}_{\text{OR-PAM}} = 2f_0 \cdot \frac{v_f}{v_s} \cdot \sin\theta \cdot \sin\phi, \quad (7.8)$$

where v_f is the flow speed of red blood cells, v_s is the propagation speed of sound, θ is the angle subtended by the flow direction

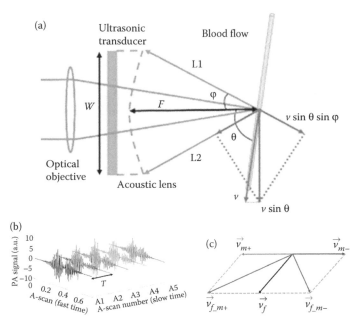

FIGURE 7.7 **(See color insert.)** (a) Schematic of the Doppler OR-PAM system. (b) Sequential A scans used to estimate the bandwidth broadening. (c) Bidirectional scanning to determine the flow direction.

and the acoustic axis, φ is the aperture angle of the acoustic lens, and f_0 is the center frequency of the ultrasonic transducer.

Typically, five sequential A-scan signals are acquired along the acoustic axis per pulse excitation (Figure 7.7b), and are then bandpass-filtered (center frequency: $f_0 = 75$ MHz; 3-dB bandwidth: 1 MHz). By autocorrelating the filtered sequential A-scans, the bandwidth broadening is estimated as the standard deviation of the photoacoustic Doppler spectrum.

In Doppler OR-PAM, the measured flow speed is actually a vector combination of the flow speed and the motor scanning speed (Figure 7.7c). The measured flow speeds under the two opposite scanning directions are

$$v_{f\_m+} = \sqrt{v_f^2 + v_m^2 + 2v_f \cdot v_m \cdot \cos \varphi} \qquad (7.9)$$

and

$$v_{f\_m-} = \sqrt{v_f^2 + v_m^2 - 2v_f \cdot v_m \cdot \cos \varphi}, \qquad (7.10)$$

where v_m is the motor scanning speed, and φ is the angle subtended by $\vec{v}_{m+}$ and $\vec{v}_f$.

Thus, the flow speed can be computed as

$$v_f = \sqrt{\frac{v_{f\_m+}^2 + v_{f\_m-}^2 - 2v_m^2}{2}},$$

and the flow direction can be determined by comparing the absolute values of v_{f_m+} and v_{f_m-}.

In a mouse ear, a region of interest that contained an artery–vein pair was selected for an *in vivo* demonstration of label-free photoacoustic imaging of blood flow (Figure 7.8) [15]. A dual-wavelength measurement was used to image the vascular anatomy (Figure 7.8a) and sO_2 (Figure 7.8b). Further, the flow speed (Figure 7.5c) and direction (Figure 7.5d) were determined using the aforementioned bidirectional scanning, with a motor speed of 0.75 mm/s and a step size of 0.625 µm. The flow directions in the artery and vein were opposite, which agreed with the known vascular physiology.

7.2.5 Thermodynamics

In the photoacoustic effect, the fractional volume expansion due to optical absorption is

$$\frac{dV}{V} = -\kappa p + \beta T, \qquad (7.11)$$

where κ is the isothermal compressibility (~5×10^{-10} Pa^{-1} for water or soft tissue), β is the thermal coefficient of volume expansion (~4×10^{-4} K^{-1} for muscle), and p and T are the optical-absorption induced rises in pressure and temperature, respectively.

If the laser excitation is in both thermal and stress confinement (i.e., the laser pulse duration is much shorter than required time for appreciable heat conduction or stress propagation), the fractional volume expansion is negligible. Thus, the local pressure rise p_0 immediately after the laser excitation can be derived from Equation 7.11 as

$$p_0 = \beta T / \kappa. \qquad (7.12)$$

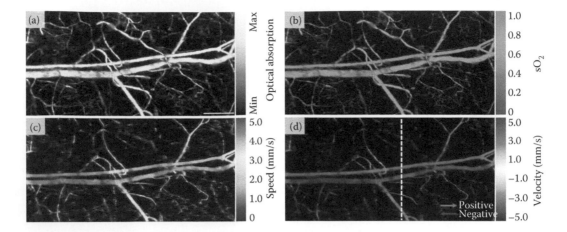

FIGURE 7.8 (**See color insert.**) Doppler OR-PAM of sO_2 and blood flow in a living mouse ear, showing MAP images of (a) vascular anatomy (scale bar: 250 μm), (b) sO_2, (c) blood flow speed, and (d) blood flow velocity with directions.

Assuming that conversion efficiency of the absorbed optical energy to heat is η_{th}, we can estimate the temperature rise generated by the short laser pulse as

$$T = \eta_{th}A_e/\rho C_V,\tag{7.13}$$

where A_e is the specific or volumetric optical absorption, ρ is the mass density, and C_V is the specific heat capacity at constant volume.

From Equations 7.12 and 7.13, we have

$$p_0 = \beta\eta_{th}A_e/\kappa\rho C_V.\tag{7.14}$$

Defining the Grüneisen parameter (dimensionless) as

$$\Gamma = \beta/\kappa\rho C_V,\tag{7.15}$$

we can rewrite Equation 7.14 as

$$p_0 = \Gamma\eta_{th}A_e.\tag{7.16}$$

Thus, the initial pressure rise p_0 is proportional to the Grüneisen parameter. Equations 7.13 and 7.15 show that the Grüneisen parameter is temperature dependent and can be utilized for local temperature measurements.

Note that a 1-mK temperature rise in biological tissues results in an 800-Pa pressure rise [16], which is above the noise level of a typical ultrasonic transducer. Usually, the instantaneous temperature increase of the imaging target due to photoacoustic excitation is on the order of millikelvin, which provides sufficient signal but negligible heating, an important feature for photoacoustic-based temperature measurements.

Figure 7.9 shows the photoacoustic temperature measurements of an ink solution [17]. As shown in Figure 7.9a, the photoacoustic signal decreased along with the temperature of the solution, and followed the actual temperature profile very well. The photoacoustic signal and the actual temperature showed a linear relationship, with an R^2 of 0.98 (Figure 7.9b). Increasing the temperature of the ink solution resulted in a corresponding elevation of the photoacoustic signal (Figure 7.9c), and the linear relationship was well maintained, with an R^2 of 0.98 (Figure 9d).

7.3 Biomedical Applications

7.3.1 Hemodynamic Monitoring

Hemodynamic monitoring is essential for understanding blood perfusion in biological tissues in normal and disease states. Capable of label-free imaging of vascular anatomies (e.g., length, diameter, and tortuosity) and functions (i.e., HbT, sO_2, and blood flow), PAT is well suited for noninvasive hemodynamic monitoring *in vivo*.

Vasodilation and vasomotion [18,19], two important regulation mechanisms of vessel diameter, have been associated with tissue oxygen delivery. OR-PAM was recently applied to explore vasodilation and vasomotion in response to tissue oxygen variation *in vivo* [20]. As shown in Figure 7.10, to acquire anatomical and sO_2 images a 1×1 mm² region of interest in a nude mouse ear was imaged with two optical wavelengths (570 and 578 nm). Then, an arteriole–venule pair (labeled A1 and V1 in Figure 7.10b) was selected for cross-sectional (B-scan) monitoring along the fast scanning direction (the yellow dashed line in Figure 7.10b), from which the actual diameters of both vessels could be quantified using the aforementioned method.

The cross sections of the vessel pair were continuously monitored for 70 min, during which the physiological state of the mouse was switched between systemic hyperoxia and hypoxia by alternating the inspired gas between pure oxygen and hypoxic gas (5% O_2, 5% CO_2, and 90% N_2). The red and blue time segments in Figure 7.11a and b, respectively, indicate hyperoxia and hypoxia. Vasodilation and vasomotion in response to the physiological state changes were clearly observed as the expansion and

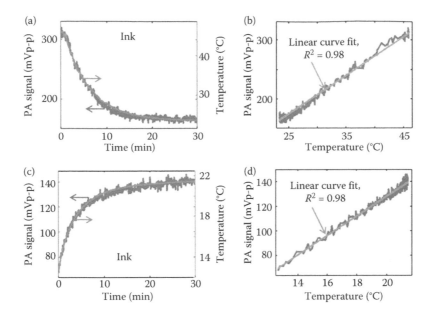

FIGURE 7.9 (a) Time courses of the photoacoustic signal and actual temperature of a diluted black ink solution ($\mu_a = 30$ cm^{-1}), as the solution was cooled to room temperature. (b) The photoacoustic signal and temperature show a linear relationship. Straight line is the linear curve fitting, with an R^2 of 0.98. (c) Time courses of the photoacoustic signal and the actual temperature of ink solution, as the solution was warmed to room temperature. (d) The photoacoustic signal and temperature show a linear relationship. Straight line is the linear curve fitting, with an R^2 of 0.98.

oscillation of the vessel diameter, respectively (Figure 7.11c). To study vasodilation, the vasomotion-induced diameter oscillation was smoothed out by a 60-point moving average. The smoothed curve showed a significant arteriolar dilation of 96% ± 3% and a moderate venous dilation of 26% ± 5% (Figure 7.11c). The 10–90% full-scale response time of arteriolar vasodilation to hypoxia was measured to be ~3 min. To study vasomotion, a Fourier analysis was performed to extract the diameter-oscillation tone of both the arteriole and venule. A strong arteriolar vasomotion tone of ~1.6 cycles/min was observed under hyperoxia (Figure 7.11d), which is in agreement with the observation from a previous invasive study [21]. In contrast, the venular vasomotion tone was much weaker, but with a similar oscillation frequency (Figure 7.11e).

Besides the oxygen-induced vascular dynamics, blood-pressure-induced vascular dynamics were also observed in a human palm by acoustic-resolution photoacoustic microscopy (AR-PAM) [22]. In this study, a region of interest (8 × 4 mm^2) in a human volunteer was imaged using AR-PAM at 570 nm (Figure 7.12a). Then, a B-scan position in the MAP image (the dashed line in Figure 7.12a) was selected to monitor the hemodynamic response to arm cuffing-induced arterial occlusion (Figure 7.12b–d). Clear changes were observed during and after the release of the occlusion, as compared to the baseline before occlusion. Specifically, there was a drastic reduction in capillary perfusion during the occlusion and a slight hyperreperfusion after the release of the occlusion.

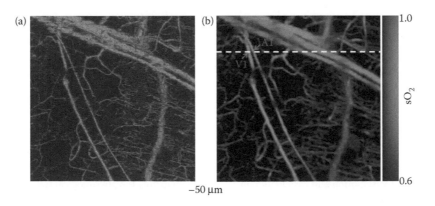

FIGURE 7.10 (**See color insert.**) Anatomical and functional OR-PAM imaging of a nude mouse ear *in vivo*. (a) Anatomical image acquired at 570 nm. (b) Vessel-by-vessel sO$_2$ mapping based on dual-wavelength (570 and 578 nm) measurements. The calculated sO$_2$ values are shown in the color bar. A1: a representative arteriole; V1: a representative venule.

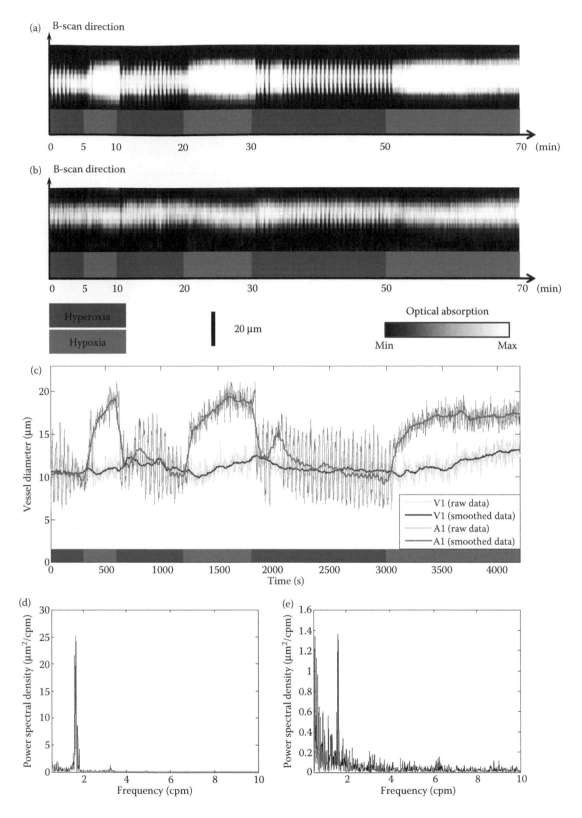

FIGURE 7.11 (**See color insert.**) Vasomotion and vasodilation in response to switching the physiological state between systemic hyperoxia and hypoxia. (a) B-scan monitoring of the cross section of arteriole A1. (b) B-scan monitoring of the cross section of venule V1. (c) Changes in arteriolar and venular diameters in response to changes in physiological state (raw data were smoothed via 60-point moving averaging to isolate the effect of vasodilation). (d) Power spectrum of the arteriolar vasomotion tone. (e) Power spectrum of the venular vasomotion tone.

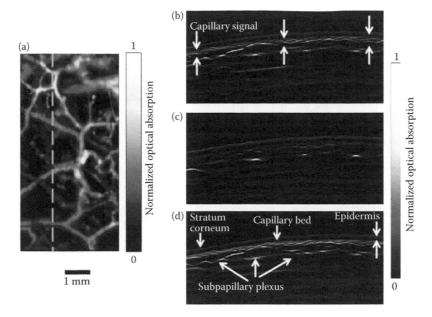

FIGURE 7.12 (a) MAP image of a human palm, acquired by AR-PAM. The dashed line indicates the B-scan position selected for hemodynamic monitoring shown in Figure 7.12b–d. (b) B-scan image before the blood flow was occluded. (c) B-scan image during the occlusion. (d) B-scan image after the occlusion was released. The scale bar applies to all panels.

To visualize the time course of the hemodynamic response to periodic arterial occlusions, a 15-min dual-wavelength (561 and 570 nm) B-scan monitoring was performed in a healthy volunteer, who was subjected to periodic arterial occlusions (Figure 7.13). These MAPs are projections of single B-scans acquired at different times. The left-hand edge of the MAP shows the first acquired B-scan and the right-hand side shows the last B-scan collected 15 min later. In the images acquired at the isosbestic point of 570 nm (Figure 7.13a and b), there is an obvious decrease in signal strength during the occlusions, indicating a reduction in blood perfusion. In contrast, a slight increase in signal strength was observed after releasing the occlusion, indicating hyperperfusion in the ischemic tissue. Figure 7.13c shows the corresponding B-scan monitoring of blood oxygenation, as revealed by the ratio between the signals acquired at 561 and 570 nm. In the absorption spectra of hemoglobin [7], 561 nm is a HbR-absorption-dominant wavelength, while 570 nm is an isosbestic wavelength. Thus, the ratio indicates the relative change in [HbR] or the degree of ischemia.

For quantitative analysis, photoacoustic signals from the capillary vessels in the palms were integrated for each B-scan. Figure 7.14 shows the relative changes in the ratio between the

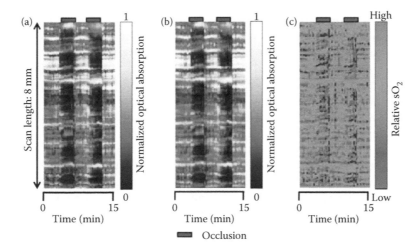

FIGURE 7.13 (**See color insert.**) MAP images of each B-scan at every time point during monitoring of periodic arterial occlusion. (a) The MAP image acquired at 570 nm. (b) MAP image acquired at 561 nm. Both (a) and (b) show a reduction in the photoacoustic signal during the occlusion periods, as indicated by the blue boxes on the top of the images. (c) The ratio between the MAP images acquired at 561 and 570 nm. The increase of the blue regions during the occlusion shows the relative increase in [HbR].

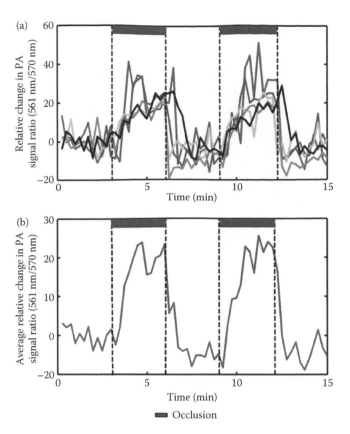

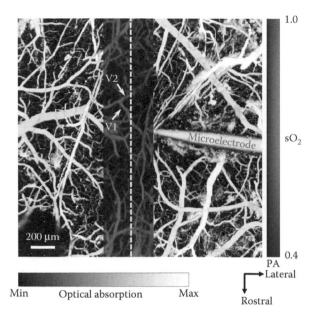

FIGURE 7.15 (**See color insert.**) Superimposed open-skull photoacoustic images of the mouse cortical microvasculature. The maximum-amplitude projection image acquired at 570 nm is shown in gray scale, and the vessel-by-vessel hemoglobin oxygen saturation mapping of a smaller region calculated from dual-wavelength measurements is shown in color scale. B-scan monitoring of the vascular response was performed along the yellow dashed line. V1 and V2 are the two microvessels studied in Figure 7.16.

FIGURE 7.14 The relative change in the photoacoustic signal generated from the capillary bed of each of five volunteers. The ratio between the signals acquired at 561 and 570 nm indicates the relative change in [HbR]. The relative [HbR] increased during the occlusions (marked by the boxes at the top of the plots) and dropped below the resting level immediately following the occlusion. (a) Individual data from each volunteer. (b) Averaged data for all five volunteers.

photoacoustic signals acquired at 561 and 570 nm in five volunteers. A baseline measurement was determined by averaging the measurements acquired before each occlusion, and the relative changes were calculated based on the baseline. All five volunteers consistently showed a strong response to the arterial occlusion, in which the relative [HbR] rose after the onset of ischemia (Figure 7.14a). The average maximum relative increases in [HbR] were 20.3% ± 3.8% and 21.3% ± 8.1% for the first and second occlusions, respectively, indicating the occlusion-induced local hypoxia. In contrast, there was a relative decrease in [HbR] immediately following the release of the occlusion, indicating the reperfusion-induced renormalization of blood oxygenation. These results are consistent with the literature [23].

7.3.2 Neurovascular Coupling

Neurovascular coupling [24,25], describing the relationship between neural activities and cerebral hemodynamics, suggests the use of cerebral hemodynamics as a surrogate for functional brain study.

Recently, OR-PAM was used to explore neurovascular coupling at the microscopic level in mice under cortical electrical stimulations [26]. As shown in Figure 7.15, the vascular anatomy (gray scale) and sO$_2$ (color scale) in a 2×2 mm^2 cortical region were imaged through a cranial window. Based on the sO$_2$ map, a B-scan crossing multiple arterioles and venules (indicated by the yellow dashed line) was selected for hemodynamic monitoring. Each monitoring trial lasted for 600 s, during which two identical electrical stimulation sequences were executed, at 200 and 400 s after the starting point. Each stimulus consisted of a train of four identical square pulses (duration: 0.3 ms; repetition rate: 300 Hz; current intensity: 100–400 µA), which were generated from a stimulator (A365, World Precision Instruments) triggered by a function generator (DS345, Stanford Research Systems). The time interval between two adjacent trials was 10 min.

Figure 7.16 demonstrates the vasoconstriction and vasodilation of microvessel V1 in response to electrical stimulations at different current levels. At 100 µA, a distinct vasoconstriction appeared after each of the two stimuli (Figure 7.16a). At 150 µA, the first stimulus produced a pronounced and prolonged vasodilation, which prevented the vessel from recovering to the resting state before the arrival of the second stimulus (Figure 7.16b). Note that the baseline diameters of the microvessel—defined as the average transverse dimension of the vessel prior to the first stimulation in each monitoring trial—were different in the two trials, possibly due to the isoflurane anesthesia and open-skull condition [27].

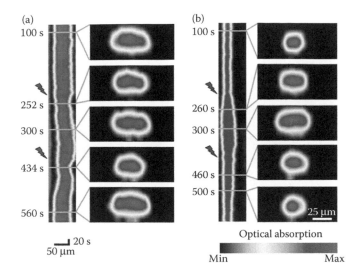

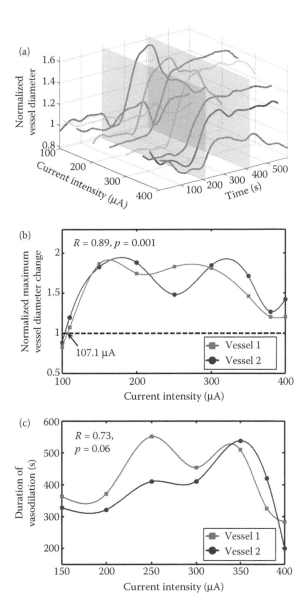

FIGURE 7.16 B-scan monitoring of vasoconstriction and vasodilation induced by direct electrical stimulations at (a) 100 μA and (b) 150 μA (B-scan rate is 0.5 Hz). In each panel, the left column is the time course of the change in vessel diameter (represented by the projection of the vessel cross section). The right column is the vessel cross-sectional image at different time points, which are indicated by the green lines. The red lightning symbol indicates the onset of the stimulation.

Figure 7.17 further shows the current-dependent responses of microvessel V1. At 100 μA (Figure 7.17a), a rapid vasoconstriction was induced right after each stimulus, with as much as an 18% decrease in vessel diameter. However, higher stimulation currents (up to 400 μA) led to prolonged vasodilation, with a diameter increase of as high as 87%. The normalized maximum diameter changes of V1 at a variety of stimulation currents are plotted in Figure 7.17b (red squares) and fitted with the smoothing spline function (red curve). According to the fitting curve, the critical stimulation intensity corresponding to the transition from vasoconstriction to vasodilation was estimated to be ~107 μA. Interestingly, the two sister vessels (V1 and V2 in Figure 7.15) showed very similar responses in terms of both the relative diameter change and the vasodilation duration (Figure 7.17b and c).

The large field of view of OR-PAM enabled studying the spatial features of the microvascular response. A total of 27 microvessels in three mice were studied (Figure 7.18). Statistically, when the microvessels were within 300 μm from the electrode tip, increasing stimulation intensity induced a clear transition from vasoconstriction to vasodilation (dark blue bars in Figure 7.18a). During this transition, both the response amplitude and the number of responding vessels were diminished (blue bars in Figure 7.18a and b). However, for vessels that were farther from the electrode tip (>300 μm), vasodilation became dominant (Figure 7.18a). Another interesting observation was that the total number of responsive vessels decreased with increased distance from the electrode tip, which was likely due to the current-diffusion-induced decay in stimulation intensity (Figure 7.18b).

FIGURE 7.17 (**See color insert.**) Current-dependent vascular responses studied in individual microvessels. (a) Time courses of the diameter change in V1 under various stimulation intensities. Correlation of the current-dependent vascular responses of two sister vessels (V1 and V2) is shown in (b), which plots the normalized maximum diameter change, and (c), which plots the duration of vasodilation. The two semitransparent planes in (a) indicate the time points of the two stimuli.

In a further study, the temporal features of vasodilation and vasoconstriction occurring at the lowest stimulation current for each vessel were examined. Statistically, the response and recovery times of vasodilation were significantly longer than those of vasoconstriction (Figure 7.19). Although both the response phase of vasodilation and the recovery phase of vasoconstriction corresponded to an increase in vessel diameter, their time scales were quite different (the same behavior as for the recovery phase of vasodilation and the response phase of vasoconstriction). These observations suggest different underlying mechanisms for the two types of vascular response.

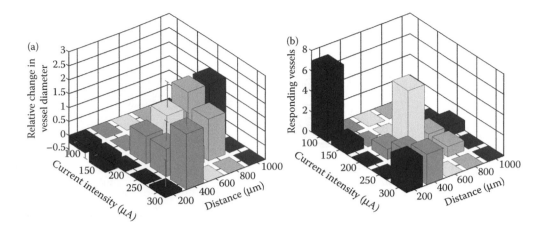

FIGURE 7.18 Spatial and current dependence of vasoconstriction and vasodilation studied in 27 microvessels. (a) Relative change in vessel diameter versus distance from the electrode tip. (b) Number of responding vessels versus distance from the electrode tip.

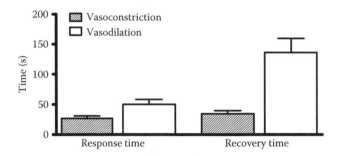

FIGURE 7.19 Temporal characteristics of vasoconstriction (in nine vessels) and vasodilation (in 15 vessels). The response (rising) time and recovery (falling) time of vasodilation are compared with those of vasoconstriction, using the 1-tailed Wilcoxon signed rank test. The response and recovery times of vasodilation are statistically significantly longer than those of vasoconstriction (*p* values for both response and recovery cases are 0.005).

7.3.3 Neovascularization

Neovascularization is crucial for tumor growth and metastasis, wound healing, and tissue preservation and recovery following ischemia. PAT, providing a comprehensive characterization of vascular anatomy, function, and metabolism, is ideally suited for *in vivo* noninvasive study of neovascularization.

Recently, OR-PAM was applied to test the role of vascular endothelial growth factor (VEGF) during epithelial HIF-1-induced neovascularization in a doxycycline-regulated HIF-1 transgenic mouse model (VEGF$^{f/f}$) [4,5]. Sixty-day monitoring of an individual mouse under continuous HIF-1 activation revealed significant increases in the capillary density, as well as the diameter and tortuosity of trunk vessels (Figure 7.20a, day 60 data is not shown). In contrast, Cre-mediated VEGF deletion (VEGF$^\Delta$) in the same type of transgenic mouse completely abrogated the HIF-1-mediated capillary angiogenesis and trunk vessel remodeling (Figure 7.20b), indicating that VEGF is essential for HIF-1-mediated neovascularization.

Further, the resolving power of OR-PAM and aforementioned segmentation analysis enabled multiparameter quantification of the neovascularization. As shown in Figure 7.20c, 60-day HIF-1 activation resulted in up to an eightfold increase in the capillary volume and a maximal 3.5-fold elevation in the trunk vessel volume. Moreover, the remarkable increases in trunk vessel diameter and tortuosity indicated the undergoing vascular remodeling.

7.3.4 Oxygen Metabolism

PAT, capable of measuring vessel diameter, HbT, sO$_2$, blood flow, and the tissue volume of interest, is the only existing technology capable of imaging oxygen metabolism *in vivo* based on endogenous contrasts. For a region of interest with clearly defined feeding and draining vessels, the oxygen extraction fraction (OEF) and the absolute metabolic rate of oxygen (MRO$_2$) can be computed as

$$\text{OEF} = \frac{sO_2^{in} - sO_2^{out}}{sO_2^{in}} \tag{7.17}$$

and

$$\text{MRO}_2 = \frac{\varepsilon \cdot [\text{HbT}] \cdot (sO_2^{in} \cdot A_{in} \cdot v_{in} - sO_2^{out} \cdot A_{out} \cdot v_{out})}{W}, \tag{7.18}$$

respectively. sO$_2^{in/out}$, $A_{in/out}$, and $v_{in/out}$ are the sO$_2$ values, the cross-sectional areas, and the flow velocities of the feeding and draining vessels, respectively [28].

Recently, label-free metabolic photoacoustic imaging was demonstrated in a living mouse ear with a B16 melanoma xenograft [29]. As shown in Figure 7.21, 7 days after the xenotransplantation, there was significant dilation of the trunk vessels that supported the growth of the tumor, along with a 1.5-fold increase in the volumetric blood flow rate. Moreover, the OEF of the tumor region decreased by 43% (Figure 7.21c), likely due

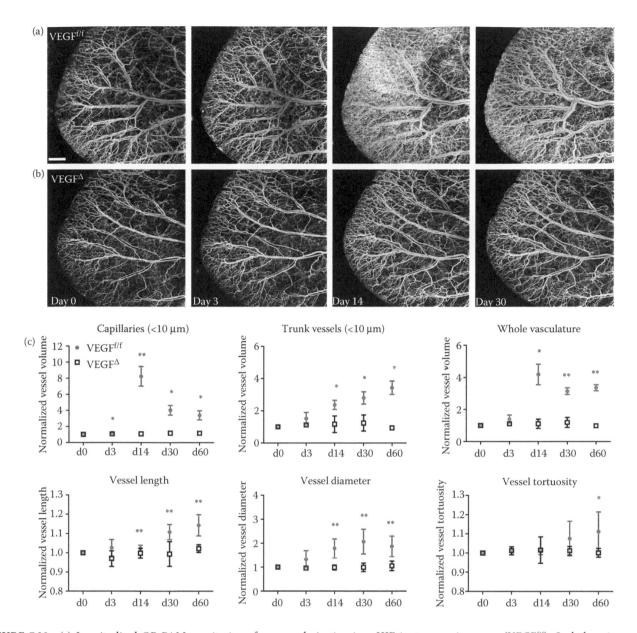

FIGURE 7.20 (a) Longitudinal OR-PAM monitoring of neovascularization in a HIF-1α transgenic mouse (VEGF[f/f]). Scale bar: 1 mm. (b) Longitudinal OR-PAM revealing the lack of neovascularization in a HIF-1α transgenic mouse with Cre-mediated VEGF deletion (VEGF[Δ]). (c) Multiparameter quantification of the neovascularization. Three mice were measured per time point, and the statistical significance was quantified using the unpaired Student's *t* test (*$p < 0.05$, **$p < 0.01$, and ***$p < 0.001$).

to the increased blood flow [30]. The vasculature and melanoma were spectroscopically differentiated according to their different absorption spectra, and the tumor volume was estimated. Based on the experimentally measured tumor volume and vascular parameters, the MRO_2 time course was quantified in Figure 7.21c, where the 36% increase in MRO_2 at day 7 indicated the hypermetabolism of the melanoma.

7.4 Perspectives

To date, PAT has demonstrated multiparameter imaging of anatomical, functional, metabolic, molecular [31], and genetic [32]

contrasts *in vivo*. However, each individual contrast can hardly provide a comprehensive characterization of complex biological systems. In cancer research, for instance, details of the vascular anatomy, blood oxygenation, and metabolic rate of oxygen are highly desired to study neovascularization, tumor hypoxia, and tumor hypermetabolism, respectively. In some disease models, exogenous molecular biomarkers are needed to visualize non-absorbing molecules [33]. Integrating the multiple photoacoustic contrasts discussed in this chapter into a single imaging platform that can operate at multiple spatial scales will greatly enhance the impact of photoacoustics in basic and translational biomedicine.

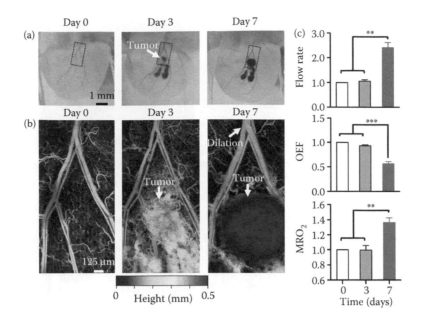

FIGURE 7.21 (**See color insert.**) (a) White-light photographs of a mouse ear before (day 0), 3 days, and 7 days after the xenotransplantation of B16 melanoma tumor cells. (b) PAM images of the tumor region [dashed boxes in (a)] color coded by the tumor height. (c) PAM quantification of blood flow rate, OEF and MRO_2, all of which were normalized by the corresponding values of day 0 (i.e., flow rate: 1.77 ± 0.50 μL/min; OEF: 0.31 ± 0.04; MRO_2: 0.38 ± 0.03 mL/100 g/min). Statistics: paired Student's t test. $**p < 0.01$, $***p < 0.001$, $n = 5$. Data are presented as means ± standard error of the mean.

Animal Ethics

All experimental animal procedures were carried out in conformance with the laboratory animal protocol approved by the School of Medicine Animal Studies Committee of Washington University in St. Louis.

Acknowledgments

The authors appreciate Prof. James Ballard's close reading of the manuscript. This work was sponsored by National Institutes of Health Grants R01 EB000712, R01 EB008085, R01 CA134539, U54 CA136398, R01 CA157277, R01 EB010049, and R01 CA159959. L.V.W. has financial interests in Microphotoacoustics, Inc. and Endra, Inc., which, however, did not support this work.

References

1. Wang, L. V. 2009. Multiscale photoacoustic microscopy and computed tomography. *Nat. Photon.* 3:503–509.
2. Maslov, K., H. F. Zhang, S. Hu, and L. V. Wang. 2008. Optical-resolution photoacoustic microscopy for *in vivo* imaging of single capillaries. *Opt. Lett.* 33:929–931.
3. Hu, S., K. Maslov, and L. V. Wang. 2011. Second-generation optical-resolution photoacoustic microscopy with improved sensitivity and speed. *Opt. Lett.* 36:1134–1136.
4. Oladipupo, S., S. Hu, J. Kovalski, J. Yao, A. Santeford, R. E. Sohn, R. Shohet, K. Maslov, L. V. Wang, and J. M. Arbeit. 2011. VEGF is essential for hypoxia-inducible factor-mediated neovascularization but dispensable for endothelial sprouting. *Proc. Natl. Acad. Sci. USA* 108:13264–13269.
5. Oladipupo, S. S., S. Hu, A. C. Santeford, J. Yao, J. R. Kovalski, R. V. Shohet, K. Maslov, L. V. Wang, and J. M. Arbeit. 2011. Conditional HIF-1 induction produces multistage neovascularization with stage-specific sensitivity to VEGFR inhibitors and myeloid cell independence. *Blood* 117:4142–4153.
6. Favazza, C. P., O. Jassim, L. A. Cornelius, and L. V. Wang. 2011. *In vivo* photoacoustic microscopy of human cutaneous microvasculature and a nevus. *J. Biomed. Opt.* 16:016015.
7. Jacques, S. L., and S. A. Prahl. http://omlc.ogi.edu/spectra/hemoglobin/index.html.
8. Zhang, H. F., K. Maslov, M. Sivaramakrishnan, G. Stoica, and L. V. Wang. 2007. Imaging of hemoglobin oxygen saturation variations in single vessels *in vivo* using photoacoustic microscopy. *Appl. Phys. Lett.* 90:3.
9. Rajian, J. R., P. L. Carson, and X. Wang. 2009. Quantitative photoacoustic measurement of tissue optical absorption spectrum aided by an optical contrast agent. *Opt. Express* 17:4879–4889.
10. Guo, Z., S. Hu, and L. V. Wang. 2010. Calibration-free absolute quantification of optical absorption coefficients using acoustic spectra in 3D photoacoustic microscopy of biological tissue. *Opt. Lett.* 35:2067–2069.
11. Danielli, A., C. P. Favazza, K. Maslov, and L. V. Wang. 2010. Picosecond absorption relaxation measured with nanosecond laser photoacoustics. *Appl. Phys. Lett.* 97:163701.
12. Danielli, A., C. P. Favazza, K. Maslov, and L. V. Wang. 2011. Single-wavelength functional photoacoustic microscopy in biological tissue. *Opt. Lett.* 36:769–771.

13. Fang, H., K. Maslov, and L. V. Wang. 2007. Photoacoustic Doppler effect from flowing small light-absorbing particles. *Phys. Rev. Lett.* 99:184501.

14. Yao, J., and L. V. Wang. 2010. Transverse flow imaging based on photoacoustic Doppler bandwidth broadening. *J. Biomed. Opt.* 15:021303.

15. Yao, J., K. Maslov, Y. Shi, L. A. Taber, and L. V. Wang. 2010. *In vivo* photoacoustic imaging of transverse blood flow by using Doppler broadening of bandwidth. *Opt. Lett.* 35:1419–1421.

16. Wang, L. V., and H. Wu. 2007. *Biomedical Optics: Principles and Imaging*. Wiley, Hoboken, NJ.

17. Pramanik, M., and L. V. Wang. 2009. Thermoacoustic and photoacoustic sensing of temperature. *J. Biomed. Opt.* 14:054024.

18. Nilsson, H., and C. Aalkjaer. 2003. Vasomotion: Mechanisms and physiological importance. *Mol. Interv.* 3:79–89, 51.

19. Aalkjer, C., and H. Nilsson. 2005. Vasomotion: Cellular background for the oscillator and for the synchronization of smooth muscle cells. *Br. J. Pharmacol.* 144:605–616.

20. Hu, S., K. Maslov, and L. V. Wang. 2009. Noninvasive label-free imaging of microhemodynamics by optical-resolution photoacoustic microscopy. *Opt. Express* 17:7688–7693.

21. Bertuglia, S., A. Colantuoni, G. Coppini, and M. Intaglietta. 1991. Hypoxia- or hyperoxia-induced changes in arteriolar vasomotion in skeletal muscle microcirculation. *Am. J. Physiol. Heart Circ. Physiol.* 260:H362–372.

22. Favazza, C. P., L. A. Cornelius, and L. V. Wang. 2011. *In vivo* functional photoacoustic microscopy of cutaneous microvasculature in human skin. *J. Biomed. Opt.* 16:026004.

23. de Mul, F. F. M., F. Morales, A. J. Smit, and R. Graaff. 2005. A model for post-occlusive reactive hyperemia as measured with laser-Doppler perfusion monitoring. *IEEE Trans. Biomed. Eng.* 52:184–190.

24. Allen, N. J., and B. A. Barres. 2009. Neuroscience: Glia—More than just brain glue. *Nature* 457:675–677.

25. Inyushin, M. Y., A. B. Vol'nova, and D. N. Lenkov. 2001. Use of a simplified method of optical recording to identify foci of maximal neuron activity in the somatosensory cortex of white rats. *Neurosci. Behav. Physiol.* 31:201–205.

26. Tsytsarev, V., S. Hu, J. Yao, K. Maslov, D. L. Barbour, and L. V. Wang. 2011. Photoacoustic microscopy of microvascular responses to cortical electrical stimulation. *J. Biomed. Opt.* 16:076002.

27. Schwinn, D. A., R. W. McIntyre, and J. G. Reves. 1990. Isoflurane-induced vasodilation: Role of the alpha-adrenergic nervous system. *Anesth. Analg.* 71:451–459.

28. Wang, L. V. 2008. Prospects of photoacoustic tomography. *Med. Phys.* 35:5758–5767.

29. Yao, J., K. I. Maslov, Y. Zhang, Y. Xia, and L. V. Wang. 2011. Label-free oxygen-metabolic photoacoustic microscopy *in vivo*. *J. Biomed. Opt.* 16:076003.

30. Buxton, R. B., and L. R. Frank. 1997. A model for the coupling between cerebral blood flow and oxygen metabolism during neural stimulation. *J. Cereb. Blood Flow Metab.* 17:64–72.

31. Ntziachristos, V., and D. Razansky. 2010. Molecular imaging by means of multispectral optoacoustic tomography (MSOT). *Chem. Rev.* 110:2783–2794.

32. Li, L., R. J. Zemp, G. Lungu, G. Stoica, and L. V. Wang. 2007. Photoacoustic imaging of lacZ gene expression *in vivo*. *J. Biomed. Opt.* 12:020504.

33. Hu, S., P. Yan, K. Maslov, J.-M. Lee, and L. V. Wang. 2009. Intravital imaging of amyloid plaques in a transgenic mouse model using optical-resolution photoacoustic microscopy. *Opt. Lett.* 34:3899–3901.

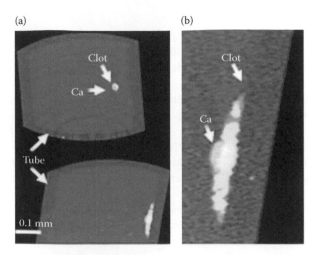

FIGURE 2.22 Bismuth nanoparticle contrast agent highlighting soft plaque in a calcified clot. (Pan, D. et al. *Angewandte Chemie (International Ed. in English)* 9635–9639. 2010. Copyright Wiley-VCH Verlag GmbH & Co. KGaA. Reproduced with permission.)

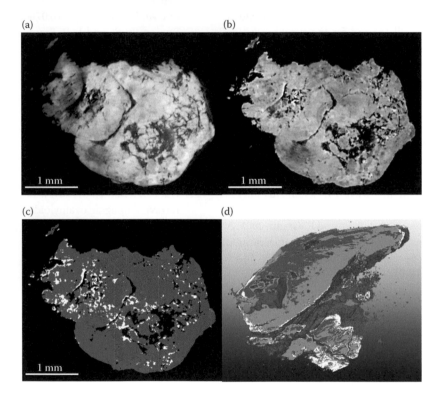

FIGURE 3.10 Microstructural analysis of a calcified nodule derived from a bicuspid aortic heart valve: (a) electron density map, (b) effective atomic number map, (c) segmented tissue of the same slice, and (d) 3D visualization of the entire segmented sample. Blue: myxoid tissue, red: dense tissue, and white: regions of elevated calcium concentration.

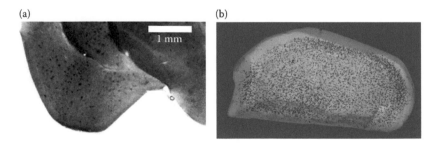

FIGURE 3.11 Amyloid plaque deposit in the cerebral cortex of a mouse model as revealed by GI: (a) zoom into a phase reconstruction slice and (b) a 3D rendering of the segmented mouse brain with the cerebral cortex in gray and amyloid plaques in red.

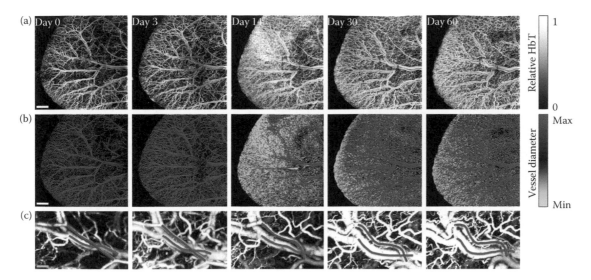

FIGURE 7.2 Vessel-segmentation-enabled multiparameter OR-PAM of neovascularization in a transgenic mouse model. (a) Representative OR-PAM maximum-amplitude-projection images of the entire ear vasculature of a longitudinally imaged vascular endothelial growth factor floxed (VEGF$^{f/f}$) knock-in mouse treated with doxycycline for 60 days. (b) Differentiation between capillaries and trunk vessels via vessel segmentation. Capillaries are false-colored green and trunk vessels are red. (c) Quantification of trunk vessel remodeling (i.e., length, vasodilation, and tortuosity) in a selected pair of arterial-venous segments. Red and blue solid curves indicate the vessel axes of the arterial and venous segments, respectively. Scale bars in Panels (a)–(c) are 1 mm, 1 mm, and 200 μm, respectively.

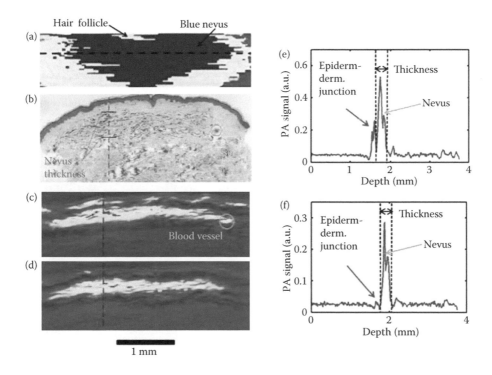

FIGURE 7.3 (a) Maximum amplitude projection of a nevus generated from serial histology sections (top view, that is, *en face* view, of the nevus) used to match histology with photoacoustic data. The nevus is shown in red, and a large hair follicle near the top was used to orient the image with the photoacoustic MAP image. (b) Histology section along the dashed line in (a), in which the location of the nevus thickness is indicated by the dashed line and two hash marks. A blood vessel used for orientation is circled in green. (c) Cross-sectional photoacoustic image of the nevus acquired at 570 nm. The circled blood vessel matches with the vessels outlined in (b). (d) Cross-sectional photoacoustic image of the nevus acquired at 700 nm. The red dashed lines in (c) and (d) indicate the depth profiles shown in (e) and (f), respectively.

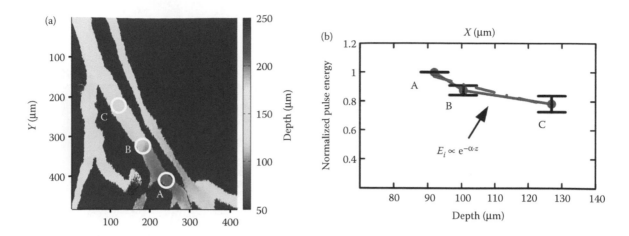

FIGURE 7.5 (a) Depth-encoded MAP image of a mouse ear. (b) The exponential fitting (blue-dashed curve) and the normalized pulse energy as a function of depth (solid red dots).

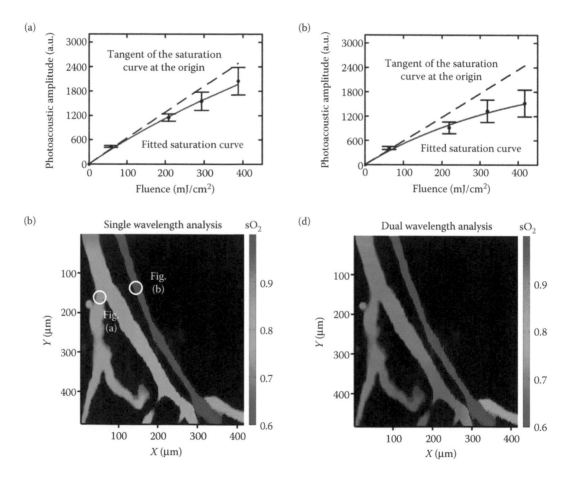

FIGURE 7.6 Saturation profiles of photoacoustic signals and sO$_2$ mapping in a mouse ear. Average photoacoustic amplitude as a function of the fluence (a) at a low sO$_2$ location (a vein) and (b) at a high sO$_2$ location (an artery). Error bars represent the standard deviations in 15 adjacent measurements. (c) Single-wavelength (576 nm) and (d) dual-wavelength (576 and 592 nm) measurements of sO$_2$.

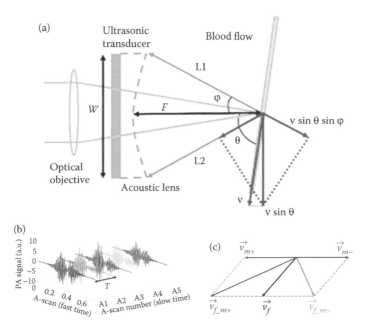

FIGURE 7.7 (a) Schematic of the Doppler OR-PAM system. (b) Sequential A scans used to estimate the bandwidth broadening. (c) Bidirectional scanning to determine the flow direction.

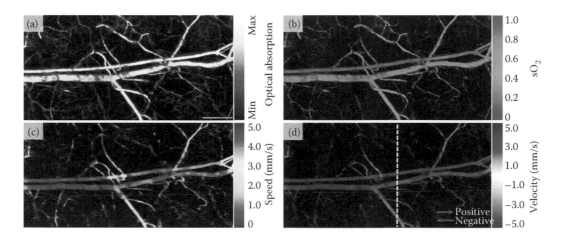

FIGURE 7.8 Doppler OR-PAM of sO$_2$ and blood flow in a living mouse ear, showing MAP images of (a) vascular anatomy (scale bar: 250 μm), (b) sO$_2$, (c) blood flow speed, and (d) blood flow velocity with directions.

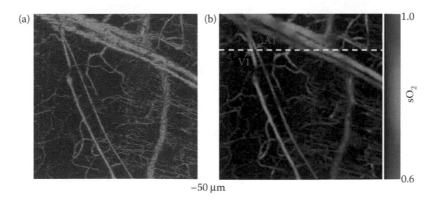

FIGURE 7.10 Anatomical and functional OR-PAM imaging of a nude mouse ear *in vivo*. (a) Anatomical image acquired at 570 nm. (b) Vessel-by-vessel sO$_2$ mapping based on dual-wavelength (570 and 578 nm) measurements. The calculated sO$_2$ values are shown in the color bar. A1: a representative arteriole; V1: a representative venule.

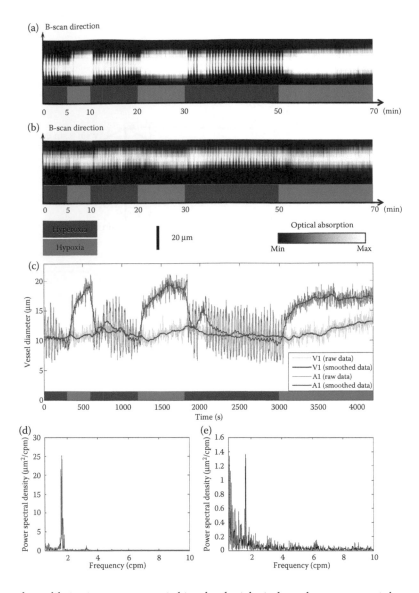

FIGURE 7.11 Vasomotion and vasodilation in response to switching the physiological state between systemic hyperoxia and hypoxia. (a) B-scan monitoring of the cross section of arteriole A1. (b) B-scan monitoring of the cross section of venule V1. (c) Changes in arteriolar and venular diameters in response to changes in physiological state (raw data were smoothed via 60-point moving averaging to isolate the effect of vasodilation). (d) Power spectrum of the arteriolar vasomotion tone. (e) Power spectrum of the venular vasomotion tone.

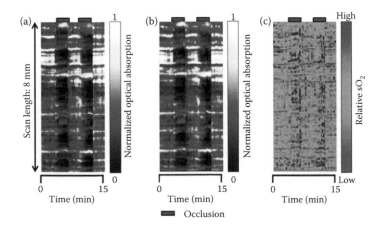

FIGURE 7.13 MAP images of each B-scan at every time point during monitoring of periodic arterial occlusion. (a) The MAP image acquired at 570 nm. (b) MAP image acquired at 561 nm. Both (a) and (b) show a reduction in the photoacoustic signal during the occlusion periods, as indicated by the blue boxes on the top of the images. (c) The ratio between the MAP images acquired at 561 and 570 nm. The increase of the blue regions during the occlusion shows the relative increase in [HbR].

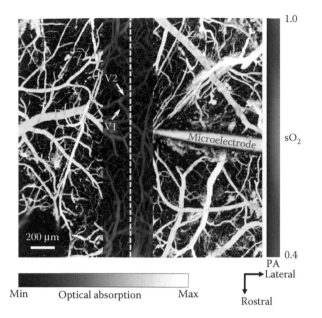

FIGURE 7.15 Superimposed open-skull photoacoustic images of the mouse cortical microvasculature. The maximum-amplitude projection image acquired at 570 nm is shown in gray scale, and the vessel-by-vessel hemoglobin oxygen saturation mapping of a smaller region calculated from dual-wavelength measurements is shown in color scale. B-scan monitoring of the vascular response was performed along the yellow dashed line. V1 and V2 are the two microvessels studied in Figure 7.16.

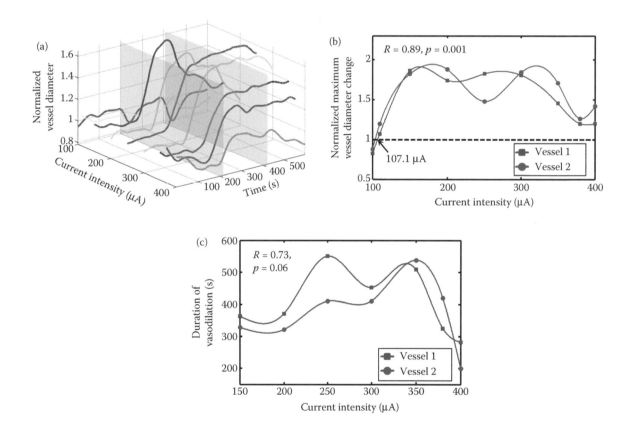

FIGURE 7.17 Current-dependent vascular responses studied in individual microvessels. (a) Time courses of the diameter change in V1 under various stimulation intensities. Correlation of the current-dependent vascular responses of two sister vessels (V1 and V2) is shown in (b), which plots the normalized maximum diameter change, and (c), which plots the duration of vasodilation. The two semitransparent planes in (a) indicate the time points of the two stimuli.

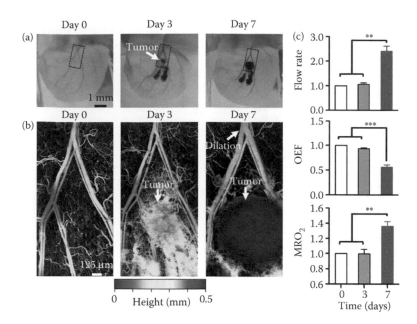

FIGURE 7.21 (a) White-light photographs of a mouse ear before (day 0), 3 days, and 7 days after the xenotransplantation of B16 melanoma tumor cells. (b) PAM images of the tumor region [dashed boxes in (a)] color coded by the tumor height. (c) PAM quantification of blood flow rate, OEF and MRO_2, all of which were normalized by the corresponding values of day 0 (i.e., flow rate: 1.77 ± 0.50 μL/min; OEF: 0.31 ± 0.04; MRO_2: 0.38 ± 0.03 mL/100 g/min). Statistics: paired Student's t test. $^{**}p < 0.01$, $^{***}p < 0.001$, $n = 5$. Data are presented as means $\pm$ standard error of the mean.

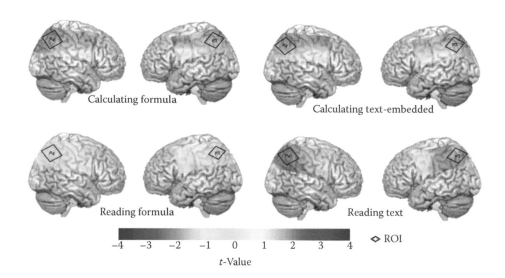

FIGURE 10.3 Changes in $[HbO_2]$ during arithmetical tasks presented as either formulae (left-hand side) or as a sentence of text (right-hand side). The top row shows the response when solving the tasks, and the bottom row shows the response when just reading. Data is displayed as t-values resulting from a statistical t-test analysis, and superimposed on a generic brain. (Reproduced with permission from Richter, M. M. et al. 2009. *Journal of Neural Transmission* 116:267–273.)

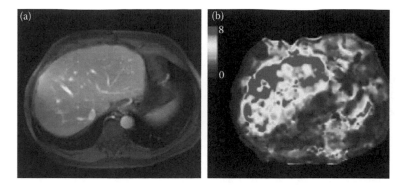

FIGURE 12.7 MRI of liver (a) and MRE analysis (b) demonstrating stiffness consistent with fibrosis. Image courtesy of Russell N. Low, M.D., Sharp and Children's MRI Center, San Diego, CA. Image acquired on GE Healthcare Signa HDxt 1.5T MRI.

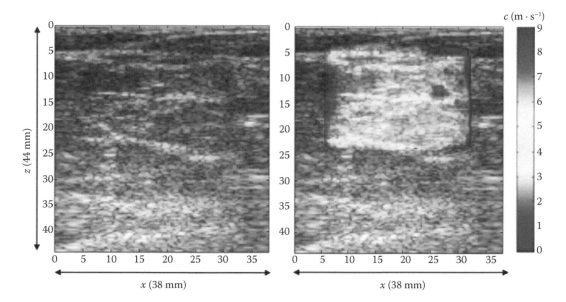

FIGURE 13.7 Invasive ductal carcinoma, grade III, evaluated with shear wave imaging. The lesion is undetectable in the B-mode image but is shown to be detectable in the shear wave image. This cancerous lesion is showing to be significantly stiffer than the surrounding parenchymal tissue. (Reprinted from *Ultrasound Med Biol 34*, Tanter, M. et al., Quantitative assessment of breast lesion viscoelasticity: Initial clinical results using supersonic shear imaging, 1373–1386, Figure 7, Copyright 2008, with permission from Elsevier.)

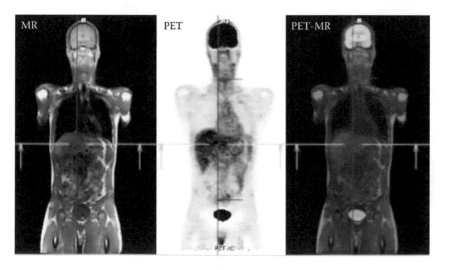

FIGURE 14.2 Coronal slice of a simultaneous [18]FDG-PET-MR acquisition. (Courtesy of C. Catana and B. Rosen, MGH.)

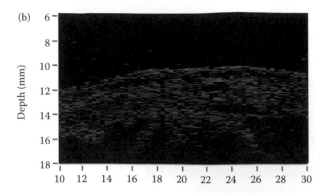

FIGURE 15.4 (b) Combined US (gray scale) and photoacoustic (red) image showing absorption in blood vessels and the skin. Acquisition time was 128 ms at a fluence of 1.9 mJ/cm².

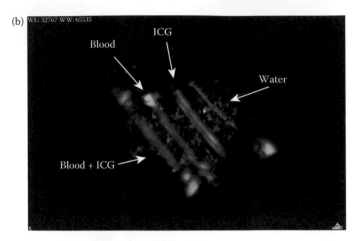

FIGURE 15.11 (b) 3D-rendered photoacoustic (hot-metal color scale) image overlaid on a pulse-echo image (gray scale) displayed at 70 dB dynamic range. The image shows four tubes filled with water, ICG, blood, blood + ICG. (Reprinted from Vaithilingam, S. et al. 2009. *IEEE Transactions on Ultrasonics, Ferroelectrics and Frequency Control* 56(11):2411–19. With permission from IEEE, Courtesy of S. Vaithilingam, Stanford University, Stanford.)

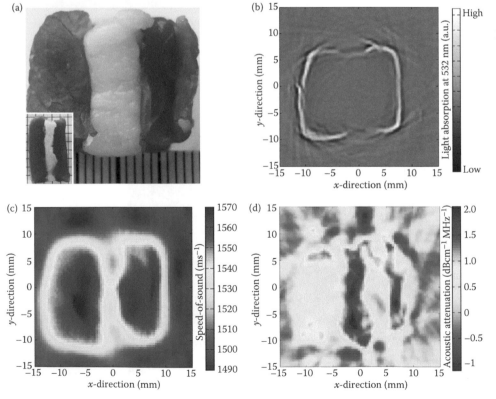

FIGURE 15.18 (a) Photograph of four layers of pork tissue showing from left to right: muscle, fat, muscle and a mixture of fat and muscle. (b) The photoacoustic image shows the boundaries of the phantom. (c) The AS distribution shows the presence of the muscle tissue with high AS (red) and the fat with low AS (blue). (d) The AA image shows the presence of the highest attenuators in red which are the second layer (fat) and the fourth layer (mixture of fat and muscle). (Reprinted from Jose, J. et al. 2011b. Passive element enriched photoacoustic computed tomography (PER PACT) for simultaneous imaging of acoustic propagation properties and light absorption. *Optics Express* 19(3):2093–104. With permission from Optical Society of America.)

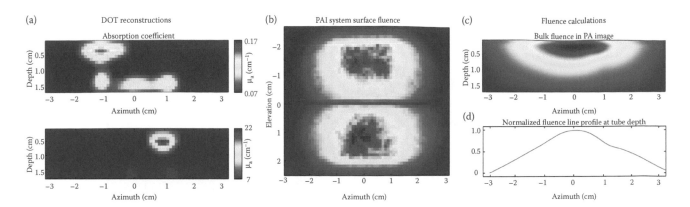

FIGURE 16.3 DOT reconstruction data and fluence calculations: (a) DOT reconstructions of the absorption and scattering properties of the phantom. (b) Fluence pattern on the surface of the phantom during PA data collection. This source distribution and DOT reconstructions are used as the input to the finite difference solution of Equation 16.11. (c) Cross-sectional slice of the bulk fluence distribution calculated using the finite-difference method. (d) Fluence line profile at the depth of the capillary tubes (1.2 cm) showing azimuthal inhomogeneity. Note the reduction in fluence magnitude of ~33% at the location of the capillary tubes (±1 cm azimuth).

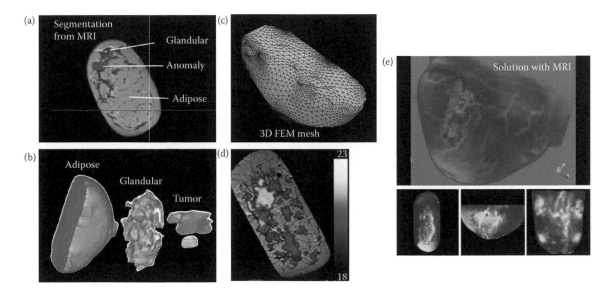

FIGURE 17.5 (a) Optical reconstruction is guided by MRI information generated from segmented DICOM images. (b) 3D regions of glandular, adipose, and tumor regions are identified and assigned. (c) A finite element mesh is generated. (d) Optical images are generated and projected onto slices of coregistered MR. (e) Optical solutions are presented in volumetric form within the coregistered MR image volume.

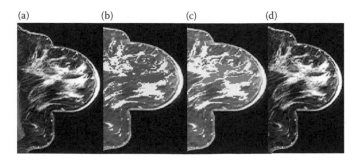

FIGURE 17.6 (a) Methods can be used to segment suspicious regions from breast MRI. (b) Segmentation based on subtraction of pre- and post-contrast scans (suspicion for tumor in red). (c) Segmentation based on curve fitting to contrast washout kinetics from several postcontrast scans (suspicion for tumor in pink). (d) Differences in region classifications between (b) and (c) with suspicion of tumor differences shown in green.

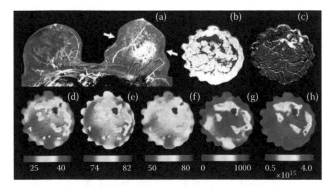

FIGURE 17.11 Example from Carpenter et al. of using MRI/NIRS to differentiate malignant from benign lesions. (a) MRI maximum intensity projection image. (b) T_1-weighted anatomically coronal cross section in the plane of the optical fibers. (c) Same as (b) but after Gd contrast injection. (d) NIRS total hemoglobin image. (e) NIRS oxygen saturation image. (f) NIRS water content image. (g) NIRS scatter amplitude image. (h) NIRS scatter power image. (From Carpenter, C. M. et al. 2007. *Opt Lett*, 32, 933–935.)

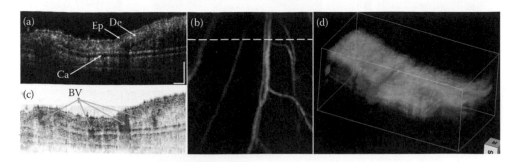

FIGURE 18.5 Simultaneously acquired LSOR-PAM and SD-OCT images of a mouse ear. (a) SD-OCT B-scan at the location in panel b highlighted by the dashed line; (b) LSOR-PAM projecting image; (c) fused LSOR-PAM and SD-OCT B-scan image; and (d) fused LSOR-PAM and SD-OCT volumetric image. Ep: epidermis; De: dermis; Ca: cartilaginous backbone; BV: blood vessels; bar: 200 µm. (Reprinted with permission from Jiao, S. et al. 2009a. *Opt Lett*, 34, 2961–3.)

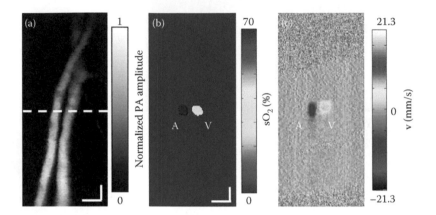

FIGURE 18.6 Multimodal functional imaging using multiwavelength PAM and Doppler OCT. (a) Imaged sO$_2$ using LSOR-PAM; (b) imaged blood flow velocity using SD-OCT. Bar: 100 µm. A: artery; V: vein. (Reprinted with permission from Liu, T. et al. 2011a. *Biomed Opt Express*, 2, 1359–65.)

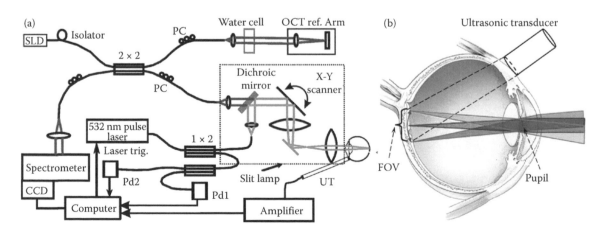

FIGURE 18.7 Experimental system of the OCT-guided PAOM. (a) Schematic diagram of the optical setup; and (b) geometry of the optical illumination and ultrasonic detection in PAOM. SLD: superluminescent diode; PC: polarization controller; Pd: photodiode; FOV: field of view; UT: ultrasonic transducer. (Reprinted with permission from Jiao, S. et al. 2010a. *Opt Express,* 18, 3967–72.)

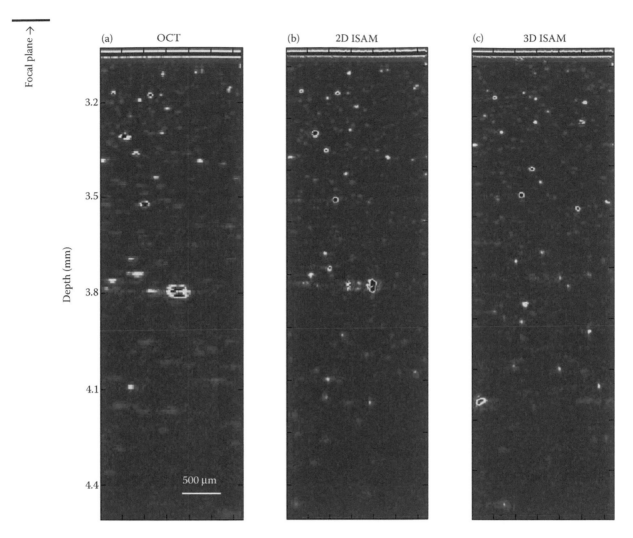

FIGURE 19.2 ISAM in a silicone phantom with 1 μm TiO₂ particles. OCT cross-sectional image of a phantom (a), in-plane 2D ISAM (b), and 3D ISAM (c). The data were acquired at an A-scan rate of 8 kHz on a 1300 nm swept-source OCT system with NA = 0.07. (Courtesy of Diagnostic Photonics, Inc.)

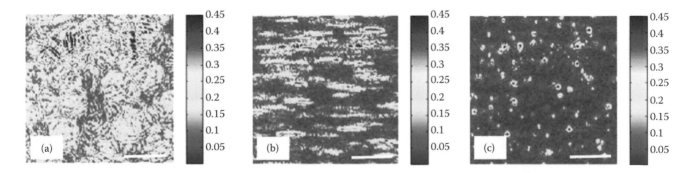

FIGURE 19.3 ISAM in a silicone phantom with 1 μm TiO₂ particles. *En face* planes 17 Rayleigh ranges above the focus with (a) OCT processing, (b) 2D ISAM processing, and (c) 3D ISAM processing. The scale bar denotes 100 μm. Gamma correction (signal $S_{Corr} = S^{\gamma}$, $\gamma = 0.3$) was utilized for dynamic range compression. The data were acquired at an A-scan rate of 92 kHz on a 1300 nm spectral-domain OCT system with NA = 0.1.

IV

Diffuse Optical Imaging

Diffuse Optical Tomography: Imaging Physics and Image Formation Principles

Soren D. Konecky
Caliper—A PerkinElmer
Company

8.1 Introduction

Biomedical optics is a rapidly expanding field, which is providing new ways to detect, diagnose, and study the disease (Wang 2007). Optical methods allow one to study intrinsic signals associated with endogenous chromophores such as hemoglobin, as well as extrinsic signals associated with targeted exogenous probes. They use nonionizing radiation, are low cost, and are extremely sensitive to variations in chemical concentrations and environments that allow both biologists and physicians to assay tissue function and monitor dynamic biological processes. There are many optical imaging techniques commonly used in biology and medicine. Laser scanning confocal microscopy provides high-resolution, depth-resolved, three-dimensional images. Techniques such as two-photon fluorescence microscopy and second harmonic generation microscopy, whose dependence on the applied light intensity is nonlinear, achieve even higher resolution and deeper penetration into tissues. Optical coherence tomography (OCT) forms cross-sectional images of tissue scattering by detecting the interference between backscattered light from the sample and light from a reference beam.

While these techniques are all extremely useful, they have a common limitation. They can only be used to image tissue near the surface. More than a few hundred microns beneath the surface, optical methods which rely on ballistic light transport cannot be used. This is because detected light that has interrogated tissue at these depths has been scattered multiple times due to the index-of-refraction changes at the interfaces between microscopic tissue structures. Confocal microscopy penetrates up to depths of about 50 μm. Two-photon microscopy allows one to look slightly deeper, up to hundreds of micrometers. OCT goes even deeper, but only to depths of at most 2–3 mm in highly scattering tissue.

A second obstacle to looking deep into tissue with visible light is the high degree of light absorption by tissue. There exists, however, a spectral region in the near-infrared (NIR) where the absorption of light by tissue is relatively low. As can be seen in Figure 8.1, light absorption by oxy- and deoxyhemoglobin drops dramatically at around 600 nm. Likewise, the absorption of light by water is very low through wavelengths up to around 900 nm. As a result, there is a window in the NIR from about 650–950 nm, where light can penetrate more deeply into tissue. Work has been done using NIR light with the techniques mentioned in the previous paragraph in order to maximize their penetration depth. For example, two-photon microscopy is able to image deeper in tissue than standard single-photon fluorescence confocal microscopy because it uses NIR excitation light. However, even with NIR light, the high degree of light scattering in tissue limits the use of any of the above techniques to a few millimeters. The mean free path of NIR light in tissue is only about 100 μm, and multiple scattering events will cause the direction of the average photon to be randomized after about 1 mm. Thus, techniques which rely on ballistic or quasi-ballistic light are inherently limited in depth.

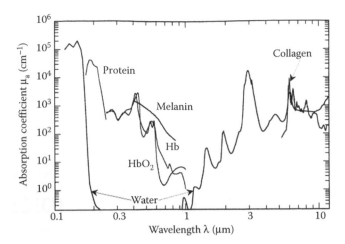

FIGURE 8.1 Optical absorption coefficients of principal tissue chromophores in the 0.1–12 μm spectral region. (Reprinted with permission from Vogel, A., and V. Venugopalan. 2003. Mechanisms of pulsed laser ablation of biological tissues, *Chem. Rev.*, 103: 577–644. Copyright 2003, American Chemical Society.)

The problem is not that NIR light cannot penetrate deeply into the tissue. It does. For example, NIR light transmitted through ~10 cm of human breast tissue can be readily detected. The problem is that beyond a few millimeters deep, almost all of the remaining photons will have been scattered multiple times, and their directions will be random. Thus, in order to image tissue deep below the surface, a method is needed that permits information to be extracted from detected photons, which have been scattered many times. This is the goal of diffuse optical tomography (DOT).

Attempts to see into the body with light date back to the 1930s when Max Cutler attempted to detect cancer by shining light through the human breast (Cutler 1931). Further research looking at the human breast and brain was conducted in the 1980s (Alveryd et al. 1990, Jobsis 1977). However, all of these studies were limited by the lack of a rigorous model to describe how light propagates through biological tissue. More recently, researchers have recognized that light propagation through tissue can be accurately modeled as a transport process. In particular, if the distance between where light is injected into tissue and where it is detected is much greater than the distance over which the paths of the individual photons become randomized, then the paths of photons in tissue can be described as a random walk. The random walk step length is equal to the distance over which the photon directions become randomized. As mentioned above, this distance is about 1 mm in many human tissues. As a result, if experiments are performed over distances much greater than the step length, the propagation of the light can be modeled by a diffusion equation (Ishimaru 1997). With an accurate model of how the light propagates, experimenters can then use the light they detect to gain information about the tissue through which it has passed, including the concentration of absorbing chromophores such as oxy- and deoxyhemoglobin, the amount of light scattering, and the concentration and lifetime of exogenous fluorophores.

Furthermore, by adapting methods originally developed for x-ray computed tomography (CT), one can also use diffusing light to reconstruct images of the spatially varying optical properties of tissue. In CT, x-rays are shot through the sample along multiple lines. Almost all the x-ray photons will either travel along a straight line to a detector or be absorbed by the sample. Detectors on the opposite side of the sample detect the x-rays that are not absorbed. If all the lines are parallel, the result will only reveal the total absorption of the sample along one direction. However, by detecting x-rays from multiple directions, a cross-sectional image of x-ray absorption can be derived. DOT uses this same idea of shooting photons through the sample from multiple directions. However, unlike x-rays, NIR photons do not travel in straight lines through the tissue. Instead they diffuse through the tissue. By using the diffusion model, one can determine the most likely volume through which detected photons have traveled. If many measurements are made corresponding to different volumes within the tissue, a three-dimensional image of optical properties can be reconstructed (Gonatas et al. 1995, O'Leary et al. 1995).

The number of applications for diffuse optical tomography is continually increasing. As described in Chapter 9, many researchers are now using diffuse optical tomography to detect, characterize, and monitor the human breast lesions (Choe et al. 2009, Fang et al. 2011, Grosenick et al. 2004, Gu et al. 2004, Ntziachristos and Chance 2001, Pera et al. 2003, Pifferi et al. 2003, Pogue et al. 2001, Schmitz et al. 2005). Chapter 10 gives an overview of the significant portion of DOT research devoted to brain imaging (Hebden and Austin 2007, Joseph et al. 2006, White and Culver 2010). Chapters 16 and 17 discuss the multimodality use of DOT in combination with photoacoustic tomography and magnetic resonance imaging (MRI), respectively. Other applications include whole-body imaging of small animals (Leblond et al. 2010, Nothdurft et al. 2009, Ntziachristos et al. 2005) and the imaging of finger joints (Netz et al. 2008).

In this chapter, we introduce the reader to the basic methods and underlying physical principles of DOT. Section 8.2 describes the underlying equations and solutions which govern light transport in tissues. Section 8.3 contains an introduction to the image reconstruction methods used in DOT. Finally, Section 8.4 gives an overview of future directions of DOT research, and describes two recent extensions of DOT: fluorescence DOT and spatial-frequency domain imaging (SFDI).

8.2 Light Transport in Biological Tissues

8.2.1 Governing Equations

The propagation of incoherent NIR light in biological tissue is well described by the rules of geometric optics. The light can be thought of as being composed of particles (i.e., photons), which travel in straight lines until they scatter at the interfaces between microscopic tissue structures where the tissue's index of refraction changes. As the particles of light undergo multiple scattering events, their directions become randomized. While we cannot keep track of the individual paths of the photons, the electromagnetic energy density due to all of the photons can be modeled as a transport process governed by the radiative transport equation (RTE):

$$\frac{\partial I(\mathbf{r},\hat{\mathbf{s}},t)}{\partial t} = -\hat{\mathbf{s}} \cdot \nabla I(\mathbf{r},\hat{\mathbf{s}},t) - (\mu_a + \mu_s)I(\mathbf{r},\hat{\mathbf{s}},t)$$

$$+ \mu_s \int d^2 s' P(\hat{\mathbf{s}},\hat{\mathbf{s}}')I(\mathbf{r},\hat{\mathbf{s}}',t) + S(\mathbf{r},\hat{\mathbf{s}},t). \quad (8.1)$$

Here, the power over a narrow wavelength range, at the point **r**, traveling in the direction of the unit vector $\hat{\mathbf{s}}$, per unit area normal to $\hat{\mathbf{s}}$ per unit solid angle is denoted by the radiance $I(\mathbf{r},\hat{\mathbf{s}},t)$. The absorption coefficient μ_a equals the inverse mean distance a photon travels before it is absorbed, while the scattering coefficient μ_s equals the inverse mean distance a photon travels before being scattered. $S(\mathbf{r},\hat{\mathbf{s}},t)$ describes the light source, and $P(\hat{\mathbf{s}},\hat{\mathbf{s}}')$ is the phase function. The phase function gives the probability density that in a scattering event a photon traveling in the direction $\hat{\mathbf{s}}'$ is scattered into the direction $\hat{\mathbf{s}}$. It is normalized to unity (i.e., $\int P(\hat{\mathbf{s}},\hat{\mathbf{s}}')d^2 s = 1$). The term on the left-hand side of the RTE gives the rate of change in time of the radiance in a differential volume and solid angle. The four terms on the right account for this rate of change and correspond to the energy leaving the volume, the energy being absorbed or scattered out of the solid angle, the energy being scattering into the solid angle, and the energy being added into the volume/solid angle by a light source.

The RTE is difficult to solve both analytically and numerically. There are researchers in the DOT community who are developing faster and more accurate solutions to the RTE (Kim and Schotland 2006, Markel 2004, Montejo et al. 2010). However, most research in DOT has focused on using the diffusion approximation to the RTE (Ishimaru 1997, Van Rossum and Nieuwenhuizen 1999):

$$\frac{\partial u(\mathbf{r},t)}{\partial t} = \nabla \cdot D(\mathbf{r})\nabla u(\mathbf{r},t) - c\mu_a(\mathbf{r})u(\mathbf{r},t) + S(\mathbf{r},t). \quad (8.2)$$

This equation governs the energy density $u(\mathbf{r},t)$, which is proportional to the radiance integrated over all angles. In this equation, $D(\mathbf{r})$ is the diffusion coefficient and c is the speed of light in the medium. The diffusion coefficient equals

$$D = \frac{c}{3(\mu_a + \mu_s')}, \quad (8.3)$$

where μ_s' is the reduced scattering coefficient. The reduced scattering coefficient represents the inverse of the random walk step length ℓ' for photons in the medium. It is related to the scattering coefficient μ_s appearing in the RTE by the equation

$$\mu_s' = \mu_s(1 - g) = (1 - g)/\ell, \quad (8.4)$$

where ℓ is the average distance photons travel between scattering events and g is the anisotropy factor. The anisotropy factor is the average cosine of the angle through which photons are scattered during a single scattering event. It is defined as

$$g = \int_{4\pi} (\hat{\mathbf{s}} \cdot \hat{\mathbf{s}}')P(\hat{\mathbf{s}} \cdot \hat{\mathbf{s}}') \, d^2 s, \quad (8.5)$$

where $\hat{\mathbf{s}}'$ and $\hat{\mathbf{s}}$ are the unit vectors found in the RTE, which point in the incoming and outgoing photon directions, and $P(\hat{\mathbf{s}} \cdot \hat{\mathbf{s}}')$ is the phase function found in the RTE. By writing the phase function as a function of the angle between $\hat{\mathbf{s}}'$ and $\hat{\mathbf{s}}$, we are assuming that there is no preferred scattering direction. This assumption corresponds to spherically symmetric (or randomly orientated) particles. When the phase function is constant (i.e., isotropic scattering), the anisotropy factor is zero and $\mu_s = \mu_s'$. The anisotropy factor is also zero when the phase function is symmetric about $\pi/2$. However, most biological tissue is highly forward scattering such that g is positive and close to unity (e.g., $g \sim 0.9$ is typical). Thus, it takes many scattering events before the direction of a photon is randomized, and μ_s' is much less than μ_s. A detailed derivation of the diffusion approximation to the RTE in which all the mathematical steps are explicitly written can be found in Wang and Wu (2007).

The diffusion approximation is valid when $\mu_a \ll \mu_s'$ and the distance between sources and detectors is much greater than the random walk step length. This approximation is valid for many applications in breast and brain imaging. For example, for NIR light in human breast, typical optical properties are $\mu_a = 0.005$ mm^{-1}, $l = 100$ μm, $g = 0.9$, $l' = 1$ mm, and $\mu_s' = 1$ mm^{-1} (Durduran et al. 2002). Depending on the experimental geometry, the distances between sources and detectors typically range from 1 to 10 cm. Thus, we expect the diffusion approximation to be valid in breast imaging applications.

8.2.2 Relationship of Absorption and Scattering to Biological Quantities

The parameters of interest in the diffusion equation are the optical absorption and reduced scattering coefficients μ_a and μ_s'. By themselves these parameters are not extremely relevant to physicians. However, by measuring the absorption coefficient at many wavelengths of light, one can use the known absorption spectra of the absorbing chromophores to relate the wavelength-dependent absorption to the chromophore concentrations according to

$$\mu_a(\lambda) = \sum_{i}^{N} \varepsilon_i(\lambda) \, c_i. \quad (8.6)$$

Here, $\varepsilon_i(\lambda)$ is the extinction coefficient that specifies how absorbing the i'th chromophore is at wavelength λ, and c_i is the concentration of that chromophore. Some researchers use Equation 8.6 in order to solve directly for the chromophore concentrations without ever explicitly determining the absorption coefficient. In the DOT literature, such an approach is referred to as spectrally constrained (Corlu et al. 2005, Srinivasan et al. 2005).

The two chromophores that are most absorbing in the NIR are oxyhemoglobin and deoxyhemoglobin. Diffuse optics has provided strong evidence for links between hemoglobin concentrations and breast tumors, and diffuse optical brain studies have associated hemoglobin changes with brain activity similar to what is more commonly done with functional MRI. Knowledge of the tissue concentrations of oxyhemoglobin ($ctHbO_2$) and dexoyhemoglobin (ctHb) can be used to determine the total hemoglobin concentration ($ctHbT = ctHbO_2 + ctHb$) as well as the oxygen saturation ($S_tO_2 = ctHbO_2/ctHbT$) of the tissue. Note the tissue oxygen saturation is primarily a measure of the hemoglobin in capillaries and should not be confused with the arterial saturation (S_aO_2) typically measured using pulse oximetry, which is generally close to 100%. Diffuse optical researchers have also measured water and lipid concentrations (Cerussi et al. 2006, Chung et al. 2010).

Although less well studied, changes in light scattering (μ_s') may also be associated with both breast cancer and brain activity. It is common place in the diffuse optics community to model the spectral dependence of scattering with a power law of the form (Bevilacqua et al. 2000, Mourant et al. 1997)

$$\mu_s' = A\,\lambda^{-b}, \tag{8.7}$$

where the scattering prefactor A depends primarily on the number density and size of the scatterers, and the scattering exponent b depends on the size of the scatterers.

8.2.3 Measurement Types

There are three types of standard measurements made in diffuse optics (see Figure 8.2). The most straightforward approach is to use a continuous wave (CW) light source and measure the average intensity of the detected light over some finite period of time (Figure 8.2a). The light can be detected with a photomultiplier tube (PMT), (avalanche) photo diode (APD), or charge coupled device (CCD). CW is the simplest and least expensive type of measurement to implement. It typically has the highest signal-to-noise ratio. However, the CW approach has a major limitation that has led to the development of frequency domain (FD) and time domain (TD) measurements: it cannot readily distinguish between absorption and scattering effects.

In general, both the absorption and scattering properties of the tissue, governed by μ_a and μ_s', respectively, affect the integrated intensity of the detected light. As long as the light source and the detector are far apart (a condition for the diffusion approximation to be valid), increasing either μ_a or μ_s' will lead to a decrease in the measured intensity. It is straightforward to see that increasing μ_a leads to a decrease in measured intensity, since μ_a is a measure of how absorbing the tissue is. However, increasing μ_s' causes photons to take a more indirect path through the tissue thereby increasing the distance they must travel, and thus the probability that they are absorbed. Thus, increasing μ_s' also leads to a decrease in the measured intensity. As a result, it is difficult to disentangle the effects of absorption and scattering with an intensity only measurement, limiting one's ability to accurately determine the concentrations of the absorbing chromophores.

In order to remedy this difficulty, many researchers have added a measurement related to the time it takes the photons to traverse the tissue. By measuring both the intensity of the detected light and the time it takes for light to travel through the tissue, it is possible to simultaneously measure both the absorption and scattering properties of the tissue. In frequency-domain measurements, the light source is intensity modulated with typical frequencies in the 10–100 MHz range. As the light propagates through the tissue, the modulated light intensity decreases in amplitude and acquires a phase delay directly related to the time it takes the photons to traverse the tissue (Figure 8.2b). In time domain measurements, a pulsed light source is used, and the time of flights of the individual photons can be measured. The mean time of flight, as well as the dispersion of the photon

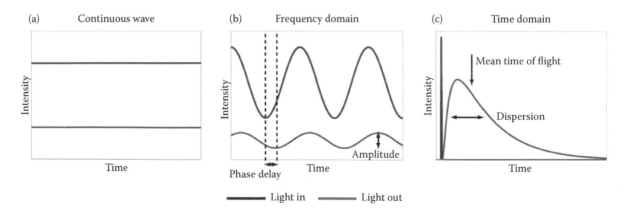

FIGURE 8.2 Types of measurement made in DOT. One can measure (a) the mean intensity of a continuous wave source, (b) the amplitude and phase delay of an intensity-modulated source, or (c) the mean intensity, mean time of flight, and dispersion of a pulsed light source.

arrival times, can be used in addition to integrated intensity to determine the absorption and scattering properties of the tissue (Figure 8.2c). Some researchers take the Fourier transform of their time domain measurements to acquire the amplitude and phase at many modulations frequencies.

8.2.4 Boundary Conditions

To derive unique solutions, the diffusion equation (Equation 8.2) must be supplemented by a boundary condition. In an infinite medium, we only require that the electromagnetic energy density goes to zero as we move far from any sources. However, in all cases of interest (i.e., in tissue), there will be a boundary satisfying the following condition:

$$u(\mathbf{r}) + \ell \, \nabla u(\mathbf{r}) \cdot \hat{\mathbf{n}} = 0. \qquad (8.8)$$

Here, ℓ is the extrapolation distance (Haskell et al. 1994), and $\hat{\mathbf{n}}$ is the outward unit vector normal to the surface of the medium. If there is an index of refraction mismatch at the boundary, the extrapolation distance is modeled as $\ell = D/(c\,\alpha(n))$. Here, D is the diffusion coefficient, c is the speed of light, and α (which is of order 1) depends on the index of refraction mismatch $n = n_{\text{tissue}}/n_{\text{outside}}$ on the boundary. It accounts for the internal reflection of diffusing photons at the boundary. (See Haskell et al. (1994) for a derivation of the both boundary condition Equation 8.8 and $\alpha(n)$.)

Once the diffuse equation has been solved, one can predict the electromagnetic energy leaving the medium along the outward normal at a given point on the surface using (Arridge 1999).

$$J(\mathbf{r}, \hat{\mathbf{n}}) = -D\nabla u(\mathbf{r}) \cdot \hat{\mathbf{n}} = \frac{D}{\ell} u(\mathbf{r}) = c\alpha(n)u(\mathbf{r}). \qquad (8.9)$$

Here, $J(\mathbf{r}, \hat{\mathbf{n}})$ is the component of the electromagnetic current density flowing in the direction of the outward normal $\hat{\mathbf{n}}$ at a position $\mathbf{r}$ on the surface. This is what is measured by a detector at that location. The first equality is Fick's rule, which states the current density is proportional to the concentration gradient. The second equality comes from Equation 8.8, and the third equality comes from the definition of ℓ stated above.

8.2.5 Solutions to the Diffusion Equation

The remainder of this section consists of a review of the methods used to solve the diffusion equation. For simple geometries and homogeneous optical properties, analytic solutions are available and can be derived either by working in the spatial Fourier domain or by using an extrapolated boundary condition. We review both these methods here. In Table 8.1, solutions in the time domain, the temporal-frequency domain, and the spatial-frequency domain are listed for the three most common geometries: the infinite medium, the semi-infinite medium, and the slab. The semi-infinite geometry accurately describes measurements of light remitted from tissue when the thickness of the

sample is large compared to all other relevant lengths. The slab geometry is used for transmission experiments, such as optical mammography measurements taken when the human breast is compressed between two plates (Choe et al. 2009).

8.2.5.1 Extrapolated Boundary Solutions

The most common approach to solving the diffusion equation (Equation 8.2) for simple geometries and homogeneous optical properties is to use an extrapolated boundary condition and the method of images to derive an approximate solution (Aronson 1995, Haskell et al. 1994, Patterson et al. 1989). According to this approach, one starts with the boundary condition (Equation 8.8), which specifies the gradient of the electromagnetic energy density at the surface. One then makes the approximation that the rate at which the energy density decreases remains constant outside of the scattering medium. With this approximation, the electromagnetic energy density $u(\mathbf{r})$ becomes zero at an extrapolated boundary that is a distance ℓ from the actual boundary of the scattering medium (see Figure 8.3a).

The solution for an infinite homogeneous medium can be calculated exactly, and is shown in Table 8.1. Using the method of images, one adds a linear combination of the infinite medium solutions together such that the electromagnetic energy density is zero at the extrapolated boundary. In order to be consistent with the diffusion approximation, the laser source directed at a point on the medium surface is modeled as an isotropic point source located one random walk step length beneath this point on the surface ($z_s = 1/\mu_s'$). This requires an isotropic sink of equal magnitude to be placed a distance $2\ell + 1/\mu_s'$ outside the surface. For the semi-infinite geometry, this is the only sink required to satisfy the boundary condition. For the slab geometry, an infinite number of sources and sinks are required to satisfy the boundary conditions at both surfaces (Figure 8.3b). Solutions using this method in the time and temporal-frequency domain are found in Table 8.1.

8.5.2.2 Spatial Fourier Domain Solutions

It is also possible to derive exact solutions to the diffusion equation in the semi-infinite and slab geometries if one works in the spatial Fourier domain. While these solutions are exact, they are in the form of an integral which must be evaluated numerically. They form the foundation of Fourier domain image reconstruction algorithms (Section 8.3.5). They are also important in (SFDI), a variation of DOT in which data is collected directly in the spatial Fourier domain. Here, we only outline how these solutions are derived. For a more complete treatment, see Markel and Schotland (2002).

The Fourier domain solutions take advantage of the translation invariance of the semi-infinite and slab geometries. The solutions to the diffusion equation in these geometries depend only on the distance along the surface between the source and the detector, and not on their actual positions. That is, they depend on the quantity $|\boldsymbol{\rho}_s - \boldsymbol{\rho}_d|$, where $\boldsymbol{\rho}_s$ and $\boldsymbol{\rho}_d$ denote the location on the tissue surface of the source and detector. This

TABLE 8.1   Solutions to the Diffusion Equation for Simple Geometries

Time Domain—Extrapolated Boundary Condition

Infinite

$$G(\mathbf{r}_s, t_s = 0; \mathbf{r}, t) = c(4\pi Dt)^{-3/2} \exp\left(-\frac{|\mathbf{r} - \mathbf{r}_s|^2}{4Dt} - \mu_a ct\right)$$

Semi-infinite

$$G(\mathbf{r}_s, t_s = 0; \mathbf{r}, t) = c(4\pi Dt)^{-3/2} \exp(-\mu_a ct)\left\{\exp\left(-\frac{r_+^2}{4Dt}\right) - \exp\left(-\frac{r_-^2}{4Dt}\right)\right\}$$

Slab

$$G(\mathbf{r}_s, t_s = 0; \mathbf{r}, t) = c(4\pi Dt)^{-3/2} \exp(-\mu_a ct) \sum_{m=-\infty}^{m=\infty} \left\{\exp\left(-\frac{r_{+,m}^2}{4Dt}\right) - \exp\left(-\frac{r_{-,m}^2}{4Dt}\right)\right\}$$

Temporal-Frequency Domain—Extrapolated Boundary Condition

Infinite

$$G(\mathbf{r}_s, \mathbf{r}) = \frac{\exp\left(-k_0 |\mathbf{r} - \mathbf{r}_s|\right)}{4\pi D |\mathbf{r} - \mathbf{r}_s|}$$

Semi-infinite

$$G(\mathbf{r}_s, \mathbf{r}) = \frac{1}{4\pi D}\left\{\frac{\exp(-k_0 r_+)}{r_+} - \frac{\exp(-k_0 r_-)}{r_-}\right\}$$

Slab

$$G(\mathbf{r}_s, \mathbf{r}) = \frac{1}{4\pi D} \sum_{m=-\infty}^{m=\infty} \left\{\frac{\exp(-k_0 r_{+,m})}{r_{+,m}} - \frac{\exp(-k_0 r_{-,m})}{r_{-,m}}\right\}$$

Spatial- and Temporal-Frequency Domain—Exact Boundary Condition

Infinite

$$G(\mathbf{r}_s, \mathbf{r}) = \int \frac{d^3k}{(2\pi)^3} g(k) \exp(i\mathbf{k} \cdot (\boldsymbol{\rho}_s - \boldsymbol{\rho})), \text{ where } g(k) = \frac{1}{D}\frac{1}{k^2 - k_0^2}$$

Semi-infinite

$$G(\mathbf{r}_s, \mathbf{r}) = \int \frac{d^2q}{(2\pi)^2} g(q, z_s, z) \exp(i\mathbf{q} \cdot (\boldsymbol{\rho}_s - \boldsymbol{\rho})), \text{ where}$$

$$g(q, z_s, z) = \frac{1}{2QD}\left\{\frac{Q\ell - 1}{Q\ell + 1}\exp\left[-Q|z_s + z|\right] + \exp\left[-Q|z_s - z|\right]\right\}$$

Slab

$$G(\mathbf{r}_s, \mathbf{r}) = \int \frac{d^2q}{(2\pi)^2} g(q, z_s, z) \exp(i\mathbf{q} \cdot (\boldsymbol{\rho}_s - \boldsymbol{\rho})), \text{ where}$$

$$g(q, z_s, z) = \frac{\left[1 + (Q\ell)^2\right]\cosh\left[Q\left(L - |z_s - z|\right)\right] - \left[1 - (Q\ell)^2\right]\cosh\left[Q\left(L - |z_s + z|\right)\right] + 2Q\ell \sinh\left[\left(L - |z_s - z|\right)\right]}{2DQ\left[\sinh(QL) + 2Q\ell \cosh(QL) + (Q\ell)^2 \sinh(QL)\right]}$$

Note: $r_+ = \sqrt{(\rho - \rho_s)^2 + (z - z_s)^2}$ $z_s = \dfrac{1}{\mu_s'}$

$r_- = \sqrt{(\rho - \rho_s)^2 + (z + z_s + 2\ell)^2}$ $D = \dfrac{c}{3(\mu_a + \mu_s')}$

$r_{+,m} = \sqrt{(\rho - \rho_s)^2 + (z - 2m(L + 2\ell) - z_s)^2}$ $k_0^2 = \dfrac{c\mu_a - i\omega}{D}$

$r_{-,m} = \sqrt{(\rho - \rho_s)^2 + (z - 2m(L + 2\ell) + 2\ell + z_s)^2}$ $Q = \sqrt{q^2 + k_0^2}$

$\mathbf{r} = (\rho, z).$

symmetry allows us to expand the Green's functions for the diffusion equation as a summation of plane waves according to

$$G(\mathbf{r}_s, \mathbf{r}) = \int \frac{d^2q}{(2\pi)^2} g(\mathbf{q}, z_s, z) \exp(i\mathbf{q} \cdot (\boldsymbol{\rho}_s - \boldsymbol{\rho})). \quad (8.10)$$

Here, $\mathbf{r} = (\rho, z)$, where z represents the depth in the medium, $\boldsymbol{\rho}$ points parallel to the surface(s), and the spatial Fourier variable

is $\mathbf{k} = (\mathbf{q}, k_z)$. The function $g(\mathbf{q}, z_s, z)$ gives the amplitude and phase of each plane wave. Substituting Equation 8.10 into Equation 8.2 yields a one-dimensional differential equation for each value of the continuous variable $\mathbf{q}$:

$$\left[\frac{\partial^2}{\partial z^2} - \left(q^2 - \frac{c\mu_a - i\omega}{D}\right)\right] g(q, z_s, z) = \frac{-1}{D}\delta(z - z_s). \quad (8.11)$$

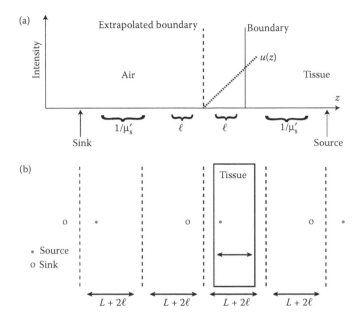

FIGURE 8.3 Schematics of the method of images using the extrapolated boundary condition. (a) The electromagnetic energy density $u(z)$ is plotted as a function of distance from the surface of a semi-infinite medium. The extrapolated boundary is located halfway between a source and sink such that $u(z) = 0$ at the extrapolated boundary. (b) For the slab geometry, an infinite number of sources and sinks are needed to ensure $u(z) = 0$ at the extrapolated boundaries on both sides of the slab.

Using the boundary condition Equation 8.8, the Green's function of this equation can be determined using the standard methods for ordinary differential equations found in most mathematical methods text books for physicists and engineers (Arfken and Weber 2005). The solutions are summarized in Table 8.1.

8.5.2.3 Finite Element Method Solutions

The analytic solutions to the diffusion equation presented in the preceding two sections are accurate and can be calculated extremely quickly. However, they are only valid for homogeneous media with simple boundaries. The finite element method (FEM) has become popular in the optical tomography community because it allows one to solve for the solution of the diffusion equation for arbitrary geometries and optical property distributions (Arridge et al. 1993). While it requires the inversion of a large matrix, and thus is slower than analytical methods, it is significantly faster than the Monte Carlo methods. FEMs are important when using nonlinear model-based reconstructions, where the expected electromagnetic energy density must be computed multiple times with heterogeneous optical properties (see Section 8.3.6). Here, we briefly outline how the method can be used to solve the diffusion equation.

In the spatial Fourier domain, the function specifying the electromagnetic energy density is written as a sum of plane waves that are smooth and have infinite support. In contrast, with FEMs, the electromagnetic energy density is approximated as a sum of piecewise linear functions with finite support such that

$$u(\mathbf{r}) \approx \tilde{u}(\mathbf{r}) = \sum_{i=1}^{N} u_i \psi_i(\mathbf{r}), \qquad (8.12)$$

where N is the number of nodes in the finite element mesh. The FEM method provides a systematic way to find the coefficients u_i such that the difference between $u(\mathbf{r})$ and $\tilde{u}(\mathbf{r})$ is minimized. As explained in Arridge et al. (1993), this leads to the matrix equation

$$\left[\mathbf{K}(D) + \mathbf{C}(c\mu_a) + c\alpha(n)\mathbf{A} + i\omega\mathbf{B}\right]\mathbf{u} = \mathbf{S}, \qquad (8.13)$$

where

$$K_{i,j} = \int_V D(\mathbf{r}) \, \nabla\psi_i(\mathbf{r}) \, \nabla\psi_j(\mathbf{r}) \, dV \qquad (8.14)$$

$$C_{i,j} = \int_V c\mu_a(\mathbf{r}) \, \psi_i(\mathbf{r}) \, \psi_j(\mathbf{r}) \, dV \qquad (8.15)$$

$$A_{i,j} = \int_S \psi_i(\mathbf{r}) \, \psi_j(\mathbf{r}) \, dS \qquad (8.16)$$

$$B_{i,j} = \int_V \psi_i(\mathbf{r}) \, \psi_j(\mathbf{r}) \, dV \qquad (8.17)$$

$$S_j = \int_V \psi_i(\mathbf{r}) \, S(\mathbf{r}) \, dV. \qquad (8.18)$$

For large 3D problems, this matrix equation becomes very large, and is difficult to solve with direct matrix-inversion methods. However, the matrices are sparse, and can be efficiently inverted with iterative matrix solvers such as linear conjugate gradients and biconjugate gradients (Schweiger et al. 2005).

8.3 Image Reconstruction Methods

8.3.1 Overview of Reconstruction Methods

Reconstruction methods in optical tomography are similar to those used in x-ray CT and PET, but there are also major differences. In x-ray CT and PET, each measurement is sensitive to a narrow volume that connects the source position to the detector (or connects the pair of detectors in PET). This is because the detected x-ray and gamma ray photons travel primarily in straight lines with very little scattering. In contrast, for the NIR light used in DOT, the average distance between scattering events is only ~100 μm. As a result, the multiply-scattered photons diffuse through the tissue.

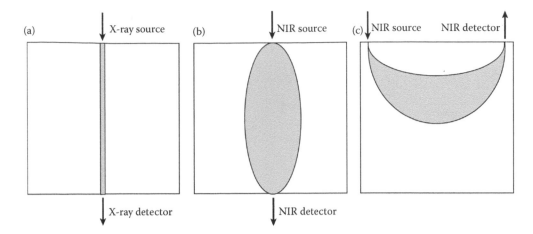

FIGURE 8.4 Schematic of sensitivity functions. (a) A collimated beam of x-rays is only sensitive to a narrow volume of tissue connecting the source and the detector. (b) Diffusing NIR light transmitted through tissue is sensitive to a large volume of the tissue. (c) When the source and the detector are located on the same side of the tissue, the NIR sensitivity function looks like a banana.

Photon diffusion has two major consequences on image reconstruction. First, each measurement is sensitive to a very large volume. Unlike with x-ray CT and PET where the volumes sensitive to a measurement are narrow and well approximated by straight lines, DOT measurements are sensitive to large "banana"-shaped regions referred to as sensitivity functions (see Figure 8.4). The large and overlapping sensitivity functions act as low-pass filters that reduce the spatial resolution. Second, the shape of the sensitivity functions depends not only on the scanner geometry, but also on the optical properties of the tissue. This makes the DOT image reconstruction problem inherently nonlinear. In order to reconstruct an image, it is necessary to start with an estimate of the absorption and scattering properties of the tissue, which allows for computation of the sensitivity functions. Usually, this estimate is made by fitting the measured data to a solution of the diffusion equation that assumes that the tissue is homogeneous (e.g., the solutions in Table 8.1). Then, the sensitivity functions are used to calculate the spatially varying corrections to the homogeneous absorption and the scattering estimates. One can stop at this point to obtain an image in the linear approximation. However, the new absorption and scattering values can also be used to calculate more accurate sensitivity functions, and the process can be iterated until some convergence criterion is met.

8.3.2 Reference Scans

As described above, image reconstruction in DOT usually begins with an estimate of the average optical properties of the sample. Image reconstruction algorithms are then used to determine the spatially dependent perturbations to the average properties. Typically a reference scan of a homogeneous tissue-simulating phantom is performed in addition to the scan of the sample of interest. The optical properties of the phantom are chosen to be as close as possible to the sample of interest. As will be shown below, it is the change in data between these two scans that is used for image reconstruction. Tissue-simulating phantoms can be made in liquid form using fatty emulsions such as Intralipid® and Lyposin® to scatter light, and India ink as an absorber. Solid phantoms are typically constructed from silicon rubber in which titanium dioxide is added to scatter light, and ink or carbon is added for absorption (Cooper et al. 2010, Pogue and Patterson 2006). It is also possible to do imaging without a reference scan, in which case the scan is compared directly with simulated data. However, the use of an experimentally obtained reference scan reduces the effects of systematic errors such as boundary effects and the variable strengths and sensitivities of sources and detectors. In dynamic experiments, the sample itself can be used as the reference. For example, in functional brain imaging, the brain is scanned both before and after a stimulus is presented to a subject. Likewise, in breast imaging, the breast can be measured both before and after the administration of an exogenous contrast agent.

8.3.3 Sensitivity Functions

In the temporal-frequency domain, the diffusion equation with a point source (e.g., a collimated laser beam) appears as

$$[-\nabla \cdot D(\mathbf{r})\nabla + c\mu_a(\mathbf{r}) - i\omega]G(\mathbf{r}_s, \mathbf{r}) = \delta(\mathbf{r} - \mathbf{r}_s). \quad (8.19)$$

Perturbation theory is used to relate the changes in the data to the spatially varying optical properties. The absorption coefficient $\mu_a(\mathbf{r})$ is decomposed as $\mu_a(\mathbf{r}) = \mu_a^{(0)} + \delta\mu_a(\mathbf{r})$, where $\mu_a^{(0)}$ is the constant value of the absorption coefficient in the reference, while $\delta\mu_a(\mathbf{r})$ represents the spatial fluctuations. Likewise, the diffusion coefficient $D(\mathbf{r})$ is decomposed as $D(\mathbf{r}) = D^{(0)} + \delta D(\mathbf{r})$. One then substitutes into the diffusion equation (Equation 8.2), and pulls the $\delta\mu_a(\mathbf{r})$ and $\delta D(\mathbf{r})$ terms to the right-hand side giving

$$[-D^{(0)}\nabla^2 + c\mu_a^{(0)} - i\omega]G(\mathbf{r}_s, \mathbf{r}) = \delta(\mathbf{r} - \mathbf{r}_s)$$
$$- [c\delta\mu_a(\mathbf{r}) - \nabla \cdot \delta D(\mathbf{r})\nabla]G(\mathbf{r}_s, \mathbf{r}). \quad (8.20)$$

Equation 8.20 is the diffusion equation for homogeneous media with an additional source term. This extra term can be thought of as an additional source distribution caused by the interaction of the diffusing photons with the perturbations in the optical properties. Convolving with the Green's function for homogeneous media results in

$$G(\mathbf{r}_s, \mathbf{r}_d) - G_0(\mathbf{r}_s, \mathbf{r}_d)$$
$$= -\int_V G_0(\mathbf{r}, \mathbf{r}_d)\, [c\delta\mu_a(\mathbf{r}) - \nabla \cdot \delta D(\mathbf{r})\nabla]\, G(\mathbf{r}_s, \mathbf{r})\, d^3 r$$
$$= -\int_V G_0(\mathbf{r}, \mathbf{r}_d) G(\mathbf{r}_s, \mathbf{r})\; c\delta\mu_a(\mathbf{r})\, d^3 r$$
$$\quad - \int_V \nabla G_0(\mathbf{r}, \mathbf{r}_d) \cdot \nabla G(\mathbf{r}_s, \mathbf{r})\; \delta D(\mathbf{r})\, d^3 r, \qquad (8.21)$$

where the integral is taken over the volume of the sample and G_0 is Green's function in the homogeneous reference medium with $\mu_a(\mathbf{r}) = \mu_a^{(0)}$ and $D(\mathbf{r}) = D^{(0)}$. It can be calculated using any of the methods described in Section 8.2. The left-hand side of Equation 8.21 is the difference between the sample and reference measurements. A readable derivation of the second equality of Equation 8.21 can be found in O'Leary (1996). The terms

$$G_0(\mathbf{r}, \mathbf{r}_d) G(\mathbf{r}_s, \mathbf{r}) \qquad (8.22)$$

and

$$\nabla G_0(\mathbf{r}, \mathbf{r}_d) \cdot \nabla G(\mathbf{r}_s, \mathbf{r}) \qquad (8.23)$$

which appear in Equation 8.21 are the sensitivity functions for absorption and scattering, respectively. These functions govern how sensitive each measurement is to the optical properties at different point in the tissue. According to Equation 8.21, the change in the measured data due to the perturbations $\delta\mu_a(\mathbf{r})$ and $\delta D(\mathbf{r})$ is calculated by multiplying $\delta\mu_a(\mathbf{r})$ and $\delta D(\mathbf{r})$ by their respective sensitivity function at every point in space, and adding all the products together.

Note, since $G(\mathbf{r}_s, \mathbf{r})$ depends on the optical properties for which we are solving (i.e., $\delta\mu_a(\mathbf{r})$ and $\delta D(\mathbf{r})$), Equation 8.21 is nonlinear. In the linear approximation (also called the Born approximation), $G(\mathbf{r}_s, \mathbf{r})$ is replaced by Green's function for the homogeneous diffusion equation $G_0(\mathbf{r}_s, \mathbf{r})$. In an analogous formulation called the Rytov approximation, the left-hand side of Equation 8.21 is replaced by $G_0(\mathbf{r}_s, \mathbf{r}_d) \log [G(\mathbf{r}_s, \mathbf{r}_d)/G_0(\mathbf{r}_s, \mathbf{r}_d)]$. The Rytov approximation has the advantage of allowing the sample measurement $G(\mathbf{r}_s, \mathbf{r}_d)$ to be divided by the reference measurement $G_0(\mathbf{r}_s, \mathbf{r}_d)$. This division automatically calibrates the measurement to systematic errors such as differences in efficiency between different source and detector positions.

8.3.4 Numerical Reconstructions

Equation 8.21 gives the relationship in DOT between the optical properties of the sample, and the measured data. If one makes many measurements corresponding to different source and detector positions $\mathbf{r}_s$ and $\mathbf{r}_d$, one obtains a version of Equation 8.21 for each measurement. The most straightforward approach to inverting this system of integral equations is to discretize the volume into small cubes of volume h^3, where h is the width of each cube. The system of integral equations then becomes the matrix equation $\mathbf{A}\mathbf{x} = \mathbf{b}$, where $\mathbf{A}$ is composed of two blocks such that $\mathbf{A} = [\mathbf{A}^\mu \mid \mathbf{A}^D]$. Letting the i's index source/detector pairs numbered 1 to N, and the j's index volume elements numbered 1 to M results in (O'Leary et al. 1995)

$$A_{i,j}^\mu = G_0(\mathbf{r}_{si}, \mathbf{r}_j) G(\mathbf{r}_j, \mathbf{r}_{di})\, c h^3 \qquad (8.24)$$

$$A_{i,j}^D = \nabla G_0(\mathbf{r}_{si}, \mathbf{r}_j) \cdot \nabla G(\mathbf{r}_j, \mathbf{r}_{di})\, c h^3 \qquad (8.25)$$

$$x_j = \begin{cases} \delta\mu_a(\mathbf{r}_j) & j = 1 : M \\ \delta D(\mathbf{r}_j) & j = M + 1 : 2M \end{cases} \qquad (8.26)$$

$$b_i = G(\mathbf{r}_{si}, \mathbf{r}_{di}) - G_0(\mathbf{r}_{si}, \mathbf{r}_{di}). \qquad (8.27)$$

This matrix equation can be inverted by a number of methods, including singular value decomposition and the algebraic reconstruction technique.

8.3.5 Fourier Domain Image Reconstruction

For three-dimensional problems, the matrix equation described above can become quite large. The number of rows of $\mathbf{A}$ is equal to the number of measurements N. Some prototype scanners use a lens-coupled CCD as a detector and galvanometer mirrors to scan a collimated laser source, and can easily collect data sets with 10^7 source/detector pairs. Likewise, the number of columns of $\mathbf{A}$ is $M = $ (number of voxels) × (number of parameters). For example, in breast imaging experiments in the slab geometry, one might reconstruct a volume with typical dimensions of $20 \times 20 \times 6$ cm. Dividing this volume into volume elements with dimension of 3 mm results in 10^5 volume elements for both absorption and scattering. Since the matrix is dense, methods for solving large sets of sparse equations will not be helpful.

In order to make the image reconstruction with large data sets feasible, some researchers have developed Fourier domain image reconstruction algorithms for DOT (Konecky et al. 2008b, Markel et al. 2003a, Markel and Schotland 2004, Schotland and Markel 2001). These algorithms can process data sets of up to 10^7 source–detector pairs with <1 min of CPU time on a 1.3 GHz workstation. An example of such a reconstruction appears in Figure 8.5, where the letters "PENN" and "DOT" were suspended in a 6-cm-thick highly scattering liquid with optical properties similar to breast tissue (Konecky et al. 2008b).

Similar to the central slice theorem which forms the basis for many x-ray CT reconstruction algorithms, a relationship can be found for DOT which relates the Fourier transform of the measured data to the Fourier transform of the reconstructed image. However, due to the boundary conditions for diffuse

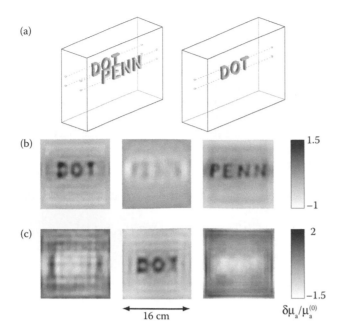

FIGURE 8.5 Slices from a three-dimensional image reconstructions of the relative absorption coefficient ($\delta\mu_a/\mu_a^{(0)}$) for targets suspended in a 6-cm-thick slab filled with highly scattering fluid. The three slices shown for each reconstruction correspond to depths of 1 cm (left), 3 cm (middle), and 5 cm (right) from the source plane. The field of view of each slice is 16×16 cm. (a) Schematics of the positions of the letters during the experiment. Left: The target consists of the letters "DOT" and "PENN," suspended 1 and 5 cm from the source plane, respectively. Right: The target consists only of the letters "DOT" suspended 3 cm from the source plane, that is, in the center of the slab. (b) Reconstructed image of the letters "DOT" and "PENN." (c) Reconstructed image of the letters "DOT." (From Konecky, S. D., et al. 2008b. Imaging complex structures with diffuse light. *Opt Express* 16:5048–5060. Reprinted with permission of Optical Society of America.)

light (Equation 8.8), the relationship in DOT is more complicated than in the x-ray case. The relationship can be derived from Equation 8.21 using the spatial-frequency domain Green's functions listed in Table 8.1. We will not perform the derivation here (see Markel et al. (2003a) and Markel and Schotland (2004) for derivations). For the semi-infinite and slab geometries, the spatial-frequency domain analog to Equation 8.21 reads

$$\tilde{\varphi}(\mathbf{q}_s,\mathbf{q}_d) = \int [\kappa_A(\mathbf{q}_s,\mathbf{q}_d;z)\,\delta\tilde{\mu}_a(\mathbf{q},z) + \kappa_D(\mathbf{q}_s,\mathbf{q}_d;z)\,\delta\tilde{D}(\mathbf{q},z)]\,dz,$$

(8.28)

where

$$\kappa_A(\mathbf{q}_s,\mathbf{q}_d;z) = c\,g(\mathbf{q}_s;z_s,z)\,g(\mathbf{q}_d;z,z_d)$$

(8.29)

$$\kappa_D(\mathbf{q}_s,\mathbf{q}_d;z)$$
$$= c\left[\frac{\partial\,g(\mathbf{q}_s;z_s,z)}{\partial z}\frac{\partial\,g(\mathbf{q}_d;z,z_d)}{\partial z} - \mathbf{q}_s\cdot\mathbf{q}_d\,g(\mathbf{q}_s;z_s,z)\,g(\mathbf{q}_d;z,z_d)\right]$$

(8.30)

$$\tilde{\varphi}(\mathbf{q}_s,\mathbf{q}_d)$$
$$= \int\left[G(\mathbf{r}_s,\mathbf{r}_d) - G_0(\mathbf{r}_s,\mathbf{r}_d)\right]\exp\left[i\left(\mathbf{q}_s\cdot\boldsymbol{\rho}_s+\mathbf{q}_d\cdot\boldsymbol{\rho}_d\right)\right]d^2\rho_s\,d^2\rho_d$$

(8.31)

$$\delta\tilde{\mu}_a(\mathbf{q},z) = \int\delta\mu_a(\boldsymbol{\rho},z)\exp\left[i\mathbf{q}\cdot\boldsymbol{\rho}\right]d^2\rho$$

(8.32)

$$\delta\tilde{D}(\mathbf{q},z) = \int\delta D(\boldsymbol{\rho},z)\exp\left[i\mathbf{q}\cdot\boldsymbol{\rho}\right]d^2\rho.$$

(8.33)

Equations 8.28 through 8.33 actually have a straight forward explanation. The term on the left-hand side, $\tilde{\varphi}(\mathbf{q}_s,\mathbf{q}_d)$, is the Fourier transform of the measured data with respect to the source and detector locations $\mathbf{r}_s$ and $\mathbf{r}_d$ (see Equation 8.31). The functions $\kappa_A(\mathbf{q}_s,\mathbf{q}_d;z)$ and $\kappa_D(\mathbf{q}_s,\mathbf{q}_d;z)$ are sensitivity functions in the Fourier domain. The unknown functions $\delta\tilde{\mu}_a(\mathbf{q},z)$ and $\delta\tilde{D}(\mathbf{q},z)$ are the two-dimensional Fourier transforms of the absorption and scattering images we wish to reconstruct. The procedure for image reconstruction goes as follows:

1. Fourier transform the data according to Equation 8.31.
2. For each Fourier component $\mathbf{q}$, solve the one-dimensional integral Equation 8.28 to find $\delta\tilde{\mu}_a(\mathbf{q},z)$ and $\delta\tilde{D}(\mathbf{q},z)$.
3. Take the inverse Fourier transform of the functions $\delta\tilde{\mu}_a(\mathbf{q},z)$ and $\delta\tilde{D}(\mathbf{q},z)$ to get the real space solution.

The principal advantages of the Fourier domain reconstruction method are speed and feasibility. Equation 8.28 is one-dimensional integral equation, whereas in real space, the integral equation is three-dimensional (see Equation 8.21). As described in Section 8.3.4, the three-dimensional equation is often too large to be easily inverted. Solving the problem in the Fourier domain allows one to substitute the massive three-dimensional problem with a series of small one-dimensional problems (one for each value of $\mathbf{q}$). Each integral equation can be discretized in the z direction, and solved numerically in a similar fashion to what was shown in Section 8.3.4. However, the integral equations can also be inverted analytically, without any need to discretize in space at all. The result is analytic inversion formulas that give the change in optical properties at specified points in space (Markel et al. 2003a).

8.3.6 Nonlinear Model-Based Reconstructions

While the Fourier domain inversion strategy presented in the previous section is fast and allows for large data sets, it has several limitations. Most notably it cannot be performed in arbitrary geometries, and it only solves the linear approximations (although nonlinear corrections to the Fourier domain algorithms have been proposed (Markel et al. 2003b)). In model-based reconstructions, the experimental data is compared to data predicted by the diffusion model calculated with the FEM (see Section 8.2.5). The optical properties of the medium are then adjusted in order to minimize a function representing the difference between predicted and measured data. This process is iterated until some criterion is met. Freely available software for model-based DOT reconstruction includes TOAST (time-resolved optical absorption and scattering tomography) (UCL) and NIRFAST (near-infrared and spectral tomography) (Dehghani et al. 2009).

In the model-based reconstruction, one needs to minimize a function $\chi^2(\mathbf{x})$, where $\mathbf{x} = (x_1, x_2, \ldots, x_N)$ is a vector containing the unknown values of $\mu_a(\mathbf{r})$ and $\mu_s'(\mathbf{r})$ in each voxel in the sample. If N measurements are made, and the tissue is divided into M voxels, $\chi^2(\mathbf{x})$ typically takes the form

$$\chi^2(\mathbf{x}) = \frac{1}{2} \sum_i^N \left(\frac{r_i^{(m)} - r_i^{(C)}(\mathbf{x})}{\sigma_i} \right)^2 + \frac{1}{2} \sum_{i,j}^M (L^T L)_{i,j} (x_i - x_i^{(0)})(x_j - x_j^{(0)}).$$

(8.34)

The first summation gives the difference between the measured and calculated data. Here, $r_i^{(m)}$ and $r_i^{(c)}(\mathbf{x})$ are the measured and calculated data for the measurement i, σ_i is the uncertainty for measurement i, and $\mathbf{x}$ is the vector of unknowns. Unfortunately, the minimization problem is ill-posed in the sense that there are many vectors $\mathbf{x}$, which will cause the first term to have a value near its minimum, and most of these solutions yield noisy images with many artifacts. For this reason, a second "regularization" term is added to $\chi^2(\mathbf{x})$. In the second term, L is a regularization matrix and $\mathbf{x}^{(0)}$ is vector representing *a priori* information about the solution vector $\mathbf{x}$. The second term penalizes solutions in which $\mathbf{x}$ differs from $\mathbf{x}^{(0)}$. The simplest choice for L, called Tikhonov regularization, is to make it a multiple of the identity matrix. If in addition, $\mathbf{x}^{(0)}$ is a constant equal to the average optical properties of the sample, then the effect of the second term is to make the solution $\mathbf{x}$ smooth. This occurs because solutions with large variations from the average optical properties are penalized. The larger L is, the bigger the penalty, and the smoother the resulting image.

There are other choices for L and $\mathbf{x}^{(0)}$ as well. In particular, many researchers have begun performing multimodality imaging studies in which DOT is combined with another imaging modality such as MRI. The MRI images can then be segmented into different tissue types, and the regularization term can be used to penalize the large optical fluctuations within a given tissue type. Combined DOT/MRI methods are described in more detail in Chapter 17.

The minimization of $\chi^2(\mathbf{x})$ is typically accomplished using standard optimization methods such as the Gauss–Newton, Levenberg–Marquardt, and nonlinear conjugant gradient methods (Fletcher 1980, Nocedal and Wright 1999). There is, however, a trick which is necessary for any practical implementation of these methods for DOT. A common requirement for all of these optimization methods is that one needs to calculate the partial derivatives of $\chi^2(\mathbf{x})$ with respect to a change in the optical properties in a given voxel of tissue. For example, for the kth voxel, one needs to calculate

$$\frac{\partial}{\partial x_k} \chi^2(\mathbf{x}) = \sum_i^N \frac{\partial r_i^{(c)}(\mathbf{x})}{\partial x_k} \left(\frac{r_i^{(m)} - r_i^{(c)}(\mathbf{x})}{\sigma_i} \right) + \sum_i^M (L^T L)_{k,i} (x_i - x_i^{(0)}).$$

(8.35)

The problem with Equation 8.35 is that it requires knowing $\partial r_i^{(c)}(\mathbf{x})/\partial x_k$ for every voxel and measurement combination. Using a brute force approach would require numerically calculating $M \times N$ partial derivatives (i.e., one for each voxel/measurement combination). Determining each partial derivative would require solving the diffusion equation multiple times using the FEM method (see Section 8.2.5). A much more efficient approach is to calculate the partial derivative using the sensitivity functions Equations 8.22 and 8.23. These functions give the change in the data for a given source–detector pair with respect to a small change in the optical properties at each location. Nonlinear optimization in DOT amounts to iteratively solving the Born (Rytov) approximation (see Equation 8.21), and updating their kernels at every step. For example, the partial derivative with respect to a change in μ_a in the kth voxel is

$$\frac{\partial r_i^{(\iota)}(\mathbf{x})}{\partial \mu_{a(k)}} = -G(r_{s(i)}, r_k) \, G^+(r_k, r_{d(i)}).$$

(8.36)

In order to calculate all the needed partial derivatives using this approach, one only needs to solve the diffusion equation $N_S + N_D$ times, where N_S and N_D are the number of source and detector positions. It also turns out that the entire gradient can actually be calculated by solving the forward problem $2N_S$ times. This is helpful when there are many more detector positions than source positions, as is common when a CCD array is used for detection (Culver et al. 2003). For more details see Arridge and Schweiger (1995, 1998).

8.3.7 Multispectral DOT

As mentioned in Section 8.2.2, our goal is not to know merely the optical properties (i.e., μ_a and μ_s') of the tissue we are imaging, but to derive the underlying composition of the tissue which determines these optical properties. To this end, researchers use multiple wavelengths of light to determine the tissue concentrations of different chromophores such as ctHb, ctHbO$_2$, water, and lipids. The simplest approach is to reconstruct images of the absorption coefficient at each wavelength, and then use Equation 8.6 to solve for the chromophore concentrations in each voxel. However, it has been shown that solving directly for the chromophore concentrations yields more accurate images with fewer artifacts (Corlu et al. 2005, Srinivasan et al. 2005). The modifications needed to make any of the reconstructions methods described in this chapter multispectral are straightforward. The primary modification is that the partial derivatives (i.e., sensitivity functions) are now with respect to the chromophore concentrations. They can be calculated using Equations 8.6 and 8.7 along with the chain rule for partial derivatives. For example, the derivative of the ith measurement with respect to a change in the jth chromophore in voxel number k is

$$\frac{\partial r_i^{(c)}(\mathbf{x},\lambda)}{\partial c_{j(k)}} = \frac{\partial r_i^{(c)}(\mathbf{x},\lambda)}{\partial \mu_a(k)} \frac{\partial \mu_a(\lambda)}{\partial c_j} = -\varepsilon_j(\lambda)\, G(r_{s(i)},r_k)\, G^+(r_k,r_{d(i)}),$$

(8.37)

where $\varepsilon_j(\lambda)$ is the extinction coefficient of the jth chromophore (see Equation 8.6).

8.4 New Approaches to DOT

8.4.1 Overview

The field of DOT is constantly evolving as researchers look for new applications and develop new methods. Advances in the application of DOT to problems in breast cancer imaging and functional brain imaging are ongoing and will be described in Chapters 9 and 10, respectively. Another major theme in current DOT research is multimodality imaging in which DOT is performed in conjunction with another imaging modality. DOT has been used in conjunction with MRI (Brooksby et al. 2006, Hsiang et al. 2005, Ntziachristos et al. 2000), ultrasound (Zhu et al. 2003), x-ray mammography (Li et al. 2003), and PET (Konecky et al. 2008a). Typically, the other modality has higher spatial resolution and can provide structural information to complement functional DOT images. Chapters 16 and 17 of this volume cover combined DOT/MRI and combined DOT/PAT imaging. Finally, research in DOT is also focusing on using new types of contrast, and new ways of delivering light to tissue through the development of fluorescent DOT (FDOT), and SFDI, respectively. The remainder of this chapter consists of a brief introduction to these two approaches.

8.4.2 Fluorescence DOT

Fluorescent DOT is a major theme in current diffuse optics research. Targeted fluorescent agents allow one to visualize the location of molecular events *in vivo*. They have been used to image tumor-specific receptors (Ballou et al. 1995, Bloch et al. 2005, Foster et al. 2005, Ke et al. 2003, Ntziachristos et al. 2002, Weissleder et al. 1999), osteoclast activity (Bloch et al. 2005), pH (Kuwana and Sevick-Muraca 2003), and intracellular calcium concentrations (Lakowicz et al. 1994). DOT research using fluorescence to image deep inside tissue has been pioneered by many groups (Hull et al. 1998, O'Leary et al. 1996, Wu et al. 1997). The contrast agent indocyanine green (ICG) is approved by the FDA, and has therefore received considerable interest in the DOT community for use in humans. Several DOT studies of breast cancer indicate that the leaky vasculature of tumors delays the washout of ICG, thereby increasing the relative concentration of ICG between normal and cancerous tissue (Corlu et al. 2007, Intes et al. 2003, Ntziachristos et al. 2000).

All of the reconstruction methods discussed in Section 8.3 can be modified for fluorescence DOT. To predict the intensity

of CW light exiting the tissue at the fluorescence emission wavelength, the source of emission light is convolved with Green's function for the diffusion equation according to

$$u(\mathbf{r}_d,\lambda_{em}) = \int G(\mathbf{r}_d,\mathbf{r},\lambda_{em})\, S(\mathbf{r})\, d^3r.$$

(8.38)

Here, λ_{em} is the emission wavelength, $u(\mathbf{r}_d,\lambda_{em})$ is the amount of emission light at the detector location, and $S(\mathbf{r})$ describes the spatial distribution of the source of emitted photons. $S(\mathbf{r})$ is related to the amount of excitation light in the tissue by

$$S(\mathbf{r}) = C(\mathbf{r})\, \varepsilon(\lambda_{ex})\, \eta(\lambda_{ex},\lambda_{em})\, G(\mathbf{r}_s,\mathbf{r},\lambda_{ex}).$$

(8.39)

Here, $C(\mathbf{r})$ is the unknown concentration of the fluorophore in the tissue, ε and η are the extinction coefficient and quantum fluorescence yield of the fluorophore, and $G(\mathbf{r}_s,\mathbf{r},\lambda_{ex})$ is the amount of excitation light in the tissue due to a light source at $\mathbf{r}_s$. Combined, these two equations give a relationship between the amount of detected emission light and the three-dimensional distribution of the fluorophore in the tissue:

$$u(\mathbf{r}_d,\lambda_{em}) = \int G(\mathbf{r}_d,\mathbf{r},\lambda_{em})\, G(\mathbf{r}_s,\mathbf{r},\lambda_{ex})\, \varepsilon(\lambda_{ex})\, \eta(\lambda_{ex},\lambda_{em}) C(\mathbf{r})\, d^3r.$$

(8.40)

This equation is almost identical to the absorption part of Equation 8.21. The only major differences are that the absorption term $\delta\mu_a$ has been replaced by the concentration of the fluorophore $C(\mathbf{r})$, and the two Green's functions should be calculated using the absorption and scattering properties at the excitation and emission wavelengths. With these two modifications, the strategies for solving the linear approximation presented in Sections 8.3.4 and 8.3.5 can be used for fluorescence DOT. Many researchers use the homogeneous Green's functions presented in Table 8.1. However, others reconstruct the spatially varying absorption and scattering properties of the tissue first, and then use these values to more accurately calculate the Green function with the FEM method (Section 8.2.5).

8.4.3 Spatial-Frequency Domain Imaging

SFDI is a relatively new variant of DOT which combines the benefits of camera-based imaging, with the quantitative accuracy and tomographic capabilities of diffuse optics (Cuccia et al. 2005, 2009b). In this modality, sinusoidal patterns of light intensity are projected on a sample, and the remitted light is detected with a CCD array (Figure 8.6). As the photons propagate through the tissue, the sinusoidal intensity patterns become blurred. If at least two spatial frequencies are sequentially projected onto the tissue, the degree to which the patterns are blurred can be used to determine the spatially varying optical properties of the tissue. Furthermore, high-frequency patterns decay more rapidly as they penetrate the tissue, and are therefore less sensitive to deeper structures in the tissue. As a result, a combination of low

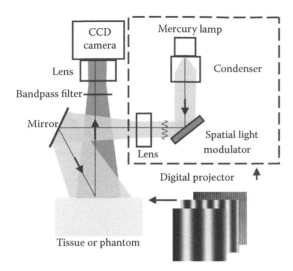

FIGURE 8.6 Schematic of a spatial-frequency domain imaging device. (Reprinted with permission from Cuccia, D. J. et al. 2009. Quantitation and mapping of tissue optical properties using modulated imaging. *Journal of Biomedical Optics* 14:024012. Copyright 2009. SPIE.)

and high projected frequencies can be used to localize the depth of objects in the tissue.

SFDI is the Fourier analog of conventional DOT. In conventional DOT, optical fibers or collimated laser beams are used to inject light into the tissue. Using the Fourier domain reconstruction algorithms of Section 8.3.5 requires taking the Fourier transform of the data with respect to the source (and detector) positions. With SFDI, sinusoidal patterns of light intensity are projected onto the tissue, and the measured data is already in the Fourier domain with respect to the source positions. The amplitudes of the projected patterns are governed by the spatial Fourier domain solutions in Table 8.1, and the Fourier domain reconstruction methods of Section 8.3.5 can be used with little modification (Konecky et al. 2009).

SFDI enables rapid, wide-field, noncontact imaging of tissue absorption, scattering, and fluorescence. SFDI has been used to quantitatively image stroke (Abookasis et al. 2009), brain injury (Weber et al. 2011), cortical spreading depression (Cuccia 2009a), layered structures in skin (Weber et al. 2009), and depth-resolved fluorescent signals (Mazhar et al. 2010). It has also been used to perform tomographic imaging (Belanger et al. 2010, Konecky et al. 2009), detect inhomogeneities (Bassi et al. 2009), and determine optical properties over a broad range (Erickson et al. 2010). It may have a large impact in medical imaging applications such as surgical guidance, reconstructive flap monitoring, and burn wound assessment, in which clinicians need a rapid, portable, low-cost method for quantitatively imaging subsurface tissue properties over a wide field of view.

Acknowledgments

This work was supported by a postdoctoral fellowship from the Hewitt Foundation for Medical Research. Additional support for this work was provided by the NIH Laser Microbeam and Medical Program (LAMMP, P41-RR01192). Beckman Laser Institute programmatic support from the Beckman Foundation and Air Force Research Laboratory Agreement Number FA9550-04-1-0101 is acknowledged. I also thank Bruce J. Tromberg, Arjun G. Yodh, and John C. Schotland for their continued guidance.

References

Abookasis, D., C. C. Lay, M. S. Mathews, M. E. Linskey, R. D. Frostig, and B. J. Tromberg. 2009. Imaging cortical absorption, scattering, and hemodynamic response during ischemic stroke using spatially modulated near-infrared illumination. *J Biomed Opt* 14:024033.

Alveryd, A., I. Andersson, K. Aspegren, G. Balldin, N. Bjurstam, G. Edstrom, G. Fagerberg et al. 1990. Lightscanning versus mammography for the detection of breast cancer in screening and clinical practice. A Swedish multicenter study. *Cancer* 65:1671–1677.

Arfken, G. B., and Weber, H. J. 2005. *Mathematical Methods for Physicists*. Academic Press, Burlington, MA.

Aronson, R. 1995. Boundary conditions for diffusion of light. *J Opt Soc Am A Opt Image Sci Vis* 12:2532–2539.

Arridge, S. 1999. Optical tomography in medical imaging. *Inv Prob* 15:R41–R93.

Arridge, S., and M. Schweiger. 1998. A gradient-based optimisation scheme for optical tomography. *Opt Express* 2:213–226.

Arridge, S. R., and M. Schweiger. 1995. Photon-measurement density functions. Part 2: Finite-element-method calculations. *Appl Opt* 34:8026–8037.

Arridge, S. R., M. Schweiger, M. Hiraoka, and D. T. Delpy. 1993. A finite element approach for modeling photon transport in tissue. *Med Phys* 20:299–309.

Ballou, B., G. W. Fisher, A. S. Waggoner, D. L. Farkas, J. M. Reiland, R. Jaffe, R. B. Mujumdar, S. R. Mujumdar, and T. R. Hakala. 1995. Tumor labeling *in vivo* using cyanine-conjugated monoclonal antibodies. *Cancer Immunol Immunother* 41:257–263.

Bassi, A., C. D'Andrea, G. Valentini, R. Cubeddu, and S. Arridge. 2009. Detection of inhomogeneities in diffusive media using spatially modulated light. *Opt Lett* 34:2156–2158.

Belanger, S., M. Abran, X. Intes, C. Casanova, and F. Lesage. 2010. Real-time diffuse optical tomography based on structured illumination. *J Biomed Opt* 15:016006.

Bevilacqua, F., A. J. Berger, A. E. Cerussi, D. Jakubowski, and B. J. Tromberg. 2000. Broadband absorption spectroscopy in turbid media by combined frequency-domain and steady-state methods. *Appl Opt* 39:6498–6507.

Bloch, S., F. Lesage, L. McIntosh, A. Gandjbakhche, K. Liang, and S. Achilefu. 2005. Whole-body fluorescence lifetime imaging of a tumor-targeted near-infrared molecular probe in mice. *J Biomed Opt* 10:054003.

Brooksby, B., B. W. Pogue, S. Jiang, H. Dehghani, S. Srinivasan, C. Kogel, T. D. Tosteson, J. Weaver, S. P. Poplack, and K. D. Paulsen. 2006. Imaging breast adipose and fibroglandular tissue molecular signatures by using hybrid MRI-guided near-infrared spectral tomography. *Proc Natl Acad Sci USA* 103:8828–8833.

Cerussi, A., N. Shah, D. Hsiang, A. Durkin, J. Butler, and B. J. Tromberg. 2006. *In vivo* absorption, scattering, and physiologic properties of 58 malignant breast tumors determined by broadband diffuse optical spectroscopy. *J Biomed Opt* 11:044005.

Choe, R., S. D. Konecky, A. Corlu, K. Lee, T. Durduran, D. R. Busch, S. Pathak et al. 2009. Differentiation of benign and malignant breast tumors by in-vivo three-dimensional parallel-plate diffuse optical tomography. *J Biomed Opt* 14:024020.

Chung, S. H., A. E. Cerussi, S. I. Merritt, J. Ruth, and B. J. Tromberg. 2010. Non-invasive tissue temperature measurements based on quantitative diffuse optical spectroscopy (DOS) of water. *Phys Med Biol* 55:3753–3765.

Cooper, R. J., R. Eames, J. Brunker, L. C. Enfield, A. P. Gibson, and J. C. Hebden. 2010. A tissue equivalent phantom for simultaneous near-infrared optical tomography and EEG. *Biomed Opt Express* 1:425–430.

Corlu, A., R. Choe, T. Durduran, K. Lee, M. Schweiger, S. R. Arridge, E. M. Hillman, and A. G. Yodh. 2005. Diffuse optical tomography with spectral constraints and wavelength optimization. *Appl Opt* 44:2082–2093.

Corlu, A., R. Choe, T. Durduran, M. A. Rosen, M. Schweiger, S. R. Arridge, M. D. Schnall, and A. G. Yodh. 2007. Three-dimensional *in vivo* fluorescence diffuse optical tomography of breast cancer in humans. *Opt Express* 15:6696–6716.

Cuccia, D. J., D. Abookasis, R. D. Frostig, and B. J. Tromberg 2009a. Quantitative *in vivo* imaging of tissue absorption, scattering, and hemoglobin concentration in rat cortex using spatially-modulated structured light. In *In Vivo Optical Imaging of Brain Function*, 2nd ed. R. D. Frostig, editor. CRC, Boca Raton, FL.

Cuccia, D. J., F. Bevilacqua, A. J. Durkin, F. R. Ayers, and B. J. Tromberg. 2009b. Quantitation and mapping of tissue optical properties using modulated imaging. *J Biomed Optics* 14:024012.

Cuccia, D. J., F. Bevilacqua, A. J. Durkin, and B. J. Tromberg. 2005. Modulated imaging: Quantitative analysis and tomography of turbid media in the spatial-frequency domain. *Opt Lett* 30:1354–1356.

Culver, J. P., R. Choe, M. J. Holboke, L. Zubkov, T. Durduran, A. Slemp, V. Ntziachristos, B. Chance, and A. G. Yodh. 2003. Three-dimensional diffuse optical tomography in the parallel plane transmission geometry: Evaluation of a hybrid frequency domain/continuous wave clinical system for breast imaging. *Med Phys* 30:235–247.

Cutler, M. 1931. Transillumination of the breast. *Ann Surg* 93:223–234.

Dehghani, H., M. E. Eames, P. K. Yalavarthy, S. C. Davis, S. Srinivasan, C. M. Carpenter, B. W. Pogue, and K. D. Paulsen. 2009. Near infrared optical tomography using NIRFAST: Algorithm for numerical model image reconstruction. *Commun Numer Methods Eng* 25:711–732.

Durduran, T., R. Choe, J. P. Culver, L. Zubkov, M. J. Holboke, J. Giammarco, B. Chance, and A. G. Yodh. 2002. Bulk optical properties of healthy female breast tissue. *Phys Med Biol* 47:2847–2861.

Erickson, T. A., A. Mazhar, D. Cuccia, A. J. Durkin, and J. W. Tunnell. 2010. Lookup-table method for imaging optical properties with structured illumination beyond the diffusion theory regime. *J Biomed Opt* 15:036013.

Fang, Q., J. Selb, S. A. Carp, G. Boverman, E. L. Miller, D. H. Brooks, R. H. Moore, D. B. Kopans, and D. A. Boas. 2011. Combined optical and x-ray tomosynthesis breast imaging. *Radiology* 258:89–97.

Fletcher, R. 1980. *Practical Methods of Optimization*. John Wiley and Sons, Hoboken, NJ.

Foster, T. H., B. D. Pearson, S. Mitra, and C. E. Bigelow. 2005. Fluorescence anisotropy imaging reveals localization of meso-tetrahydroxyphenyl chlorin in the nuclear envelope. *Photochem Photobiol* 81:1544–1547.

Gonatas, C. P., M. Ishii, J. S. Leigh, and J. C. Schotland. 1995. Optical diffusion imaging using a direct inversion method. *Phys Rev E Stat Phys Plasmas Fluids Relat Interdisc Topics* 52:4361–4365.

Grosenick, D., H. Wabnitz, K. T. Moesta, J. Mucke, M. Moller, C. Stroszczynski, J. Stossel, B. Wassermann, P. M. Schlag, and H. Rinneberg. 2004. Concentration and oxygen saturation of haemoglobin of 50 breast tumours determined by time-domain optical mammography. *Phys Med Biol* 49:1165–1181.

Gu, X., Q. Zhang, M. Bartlett, L. Schutz, L. L. Fajardo, and H. Jiang. 2004. Differentiation of cysts from solid tumors in the breast with diffuse optical tomography. *Acad Radiol* 11:53–60.

Haskell, R. C., L. O. Svaasand, T. T. Tsay, T. C. Feng, M. S. McAdams, and B. J. Tromberg. 1994. Boundary conditions for the diffusion equation in radiative transfer. *J Opt Soc Am A Opt Image Sci Vis* 11:2727–2741.

Hebden, J. C., and T. Austin. 2007. Optical tomography of the neonatal brain. *Eur Radiol* 17:2926–2933.

Hsiang, D., N. Shah, H. Yu, M. Y. Su, A. Cerussi, J. Butler, C. Baick, R. Mehta, O. Nalcioglu, and B. Tromberg. 2005. Coregistration of dynamic contrast enhanced MRI and broadband diffuse optical spectroscopy for characterizing breast cancer. *Technol Cancer Res Treat* 4:549–558.

Hull, E. L., M. G. Nichols, and T. H. Foster. 1998. Localization of luminescent inhomogeneities in turbid media with spatially resolved measurements of cw diffuse luminescence emittance. *Appl Opt* 37:2755–2765.

Intes, X., J. Ripoll, Y. Chen, S. Nioka, A. G. Yodh, and B. Chance. 2003. *In vivo* continuous-wave optical breast imaging enhanced with Indocyanine Green. *Med Phys* 30:1039–1047.

Ishimaru, A. 1997. *Wave Propagation and Scattering in Random Media*. I.E.E.E. Press, Piscataway, NJ.

Jobsis, F. F. 1977. Noninvasive, infrared monitoring of cerebral and myocardial oxygen sufficiency and circulatory parameters. *Science* 198:1264–1267.

Joseph, D. K., T. J. Huppert, M. A. Franceschini, and D. A. Boas. 2006. Diffuse optical tomography system to image brain activation with improved spatial resolution and validation with functional magnetic resonance imaging. *Appl Opt* 45:8142–8151.

Ke, S., X. Wen, M. Gurfinkel, C. Charnsangavej, S. Wallace, E. M. Sevick-Muraca, and C. Li. 2003. Near-infrared optical imaging of epidermal growth factor receptor in breast cancer xenografts. *Cancer Res* 63:7870–7875.

Kim, A. D., and J. C. Schotland. 2006. Self-consistent scattering theory for the radiative transport equation. *J Opt Soc Am A Opt Image Sci Vis* 23:596–602.

Konecky, S. D., R. Choe, A. Corlu, K. Lee, R. Wiener, S. M. Srinivas, J. R. Saffer et al. 2008a. Comparison of diffuse optical tomography of human breast with whole-body and breast-only positron emission tomography. *Med Phys* 35:446–455.

Konecky, S. D., G. Y. Panasyuk, K. Lee, V. Markel, A. G. Yodh, and J. C. Schotland. 2008b. Imaging complex structures with diffuse light. *Opt Express* 16:5048–5060.

Konecky, S. D., A. Mazhar, D. Cuccia, A. J. Durkin, J. C. Schotland, and B. J. Tromberg. 2009. Quantitative optical tomography of sub-surface heterogeneities using spatially modulated structured light. *Opt Express* 17:14780–14790.

Kuwana, E., and E. M. Sevick-Muraca. 2003. Fluorescence lifetime spectroscopy for pH sensing in scattering media. *Anal Chem* 75:4325–4329.

Lakowicz, J. R., H. Szmacinski, K. Nowaczyk, W. J. Lederer, M. S. Kirby, and M. L. Johnson. 1994. Fluorescence lifetime imaging of intracellular calcium in COS cells using Quin-2. *Cell Calcium* 15:7–27.

Leblond, F., S. C. Davis, P. A. Valdes, and B. W. Pogue. 2010. Pre-clinical whole-body fluorescence imaging: Review of instruments, methods and applications. *J Photochem Photobiol B* 98:77–94.

Li, A., E. L. Miller, M. E. Kilmer, T. J. Brukilacchio, T. Chaves, J. Stott, Q. Zhang, T. Wu, M. Chorlton, R. H. Moore, D. B. Kopans, and D. A. Boas. 2003. Tomographic optical breast imaging guided by three-dimensional mammography. *Appl Opt* 42:5181–5190.

Markel, V. A. 2004. Modified spherical hamonics method for solving the radiative transport equation. *Waves Random Media* 14:L13–L19.

Markel, V. A., V. Mital, and J. C. Schotland. 2003a. Inverse problem in optical diffusion tomography. III. Inversion formulas and singular-value decomposition. *J Opt Soc Am A Opt Image Sci Vis* 20:890–902.

Markel, V. A., J. A. O'Sullivan, and J. C. Schotland. 2003b. Inverse problem in optical diffusion tomography. IV. Nonlinear inversion formulas. *J Opt Soc Am A Opt Image Sci Vis* 20:903–912.

Markel, V. A., and J. C. Schotland. 2002. Inverse problem in optical diffusion tomography. II. Role of boundary conditions. *J Opt Soc Am A Opt Image Sci Vis* 19:558–566.

Markel, V. A., and J. C. Schotland. 2004. Symmetries, inversion formulas, and image reconstruction for optical tomography. *Phys Rev E Stat Nonlin Soft Matter Phys* 70:056616.

Mazhar, A., D. J. Cuccia, S. Gioux, A. J. Durkin, J. V. Frangioni, and B. J. Tromberg. 2010. Structured illumination enhances resolution and contrast in thick tissue fluorescence imaging. *J Biomed Opt* 15:010506.

Montejo, L. D., A. D. Klose, and A. H. Hielscher. 2010. Implementation of the equation of radiative transfer on block-structured grids for modeling light propagation in tissue. *Biomed Opt Express* 1:861–878.

Mourant, J. R., T. Fuselier, J. Boyer, T. M. Johnson, and I. J. Bigio. 1997. Predictions and measurements of scattering and absorption over broad wavelength ranges in tissue phantoms. *Appl Opt* 36:949–957.

Netz, U. J., J. Beuthan, and A. H. Hielscher. 2008. Multipixel system for gigahertz frequency-domain optical imaging of finger joints. *Rev Sci Instrum* 79:034301.

Nocedal, J., and S. Wright. 1999. *Numerical Optimization.* Springer, New York, NY.

Nothdurft, R. E., S. V. Patwardhan, W. Akers, Y. Ye, S. Achilefu, and J. P. Culver. 2009. *In vivo* fluorescence lifetime tomography. *J Biomed Opt* 14:024004.

Ntziachristos, V., and B. Chance. 2001. Probing physiology and molecular function using optical imaging: applications to breast cancer. *Breast Cancer Res* 3:41–46.

Ntziachristos, V., J. Ripoll, L. V. Wang, and R. Weissleder. 2005. Looking and listening to light: the evolution of whole-body photonic imaging. *Nat Biotechnol* 23:313–320.

Ntziachristos, V., C. H. Tung, C. Bremer, and R. Weissleder. 2002. Fluorescence molecular tomography resolves protease activity in vivo. *Nat Med* 8:757–760.

Ntziachristos, V., A. G. Yodh, M. Schnall, and B. Chance. 2000. Concurrent MRI and diffuse optical tomography of breast after indocyanine green enhancement. *Proc Natl Acad Sci USA* 97:2767–2772.

O'Leary, M. A. 1996. *Imaging with Diffuse Photon Density Waves.* Department of Physics and Astronomy. University of Pennsylvania, Philadelphia.

O'Leary, M. A., D. A. Boas, B. Chance, and A. G. Yodh. 1995. Experimental images of heterogeneous turbid media by frequency-domain diffusing-photon tomography. *Opt Lett* 20:426–428.

O'Leary, M. A., D. A. Boas, X. D. Li, B. Chance, and A. G. Yodh. 1996. Fluorescence lifetime imaging in turbid media. *Opt Lett* 21:158–160.

Patterson, M. S., B. Chance, and B. C. Wilson. 1989. Time resolved reflectance and transmittance for the non-invasive measurement of tissue optical properties. *Appl Opt* 28:2331–2336.

Pera, V. E., E. L. Heffer, H. Siebold, O. Schutz, S. Heywang-Kobrunner, L. Gotz, A. Heinig, and S. Fantini. 2003. Spatial second-derivative image processing: An application to optical mammography to enhance the detection of breast tumors. *J Biomed Opt* 8:517–524.

Pifferi, A., P. Taroni, A. Torricelli, F. Messina, R. Cubeddu, and G. Danesini. 2003. Four-wavelength time-resolved optical mammography in the 680–980-nm range. *Opt Lett* 28:1138–1140.

Pogue, B. W., and M. S. Patterson. 2006. Review of tissue simulating phantoms for optical spectroscopy, imaging and dosimetry. *J Biomed Opt* 11:041102.

Pogue, B. W., S. P. Poplack, T. O. McBride, W. A. Wells, K. S. Osterman, U. L. Osterberg, and K. D. Paulsen. 2001.

Quantitative hemoglobin tomography with diffuse near-infrared spectroscopy: pilot results in the breast. *Radiology* 218:261–266.

Schmitz, C. H., D. P. Klemer, R. Hardin, M. S. Katz, Y. Pei, H. L. Graber, M. B. Levin et al. 2005. Design and implementation of dynamic near-infrared optical tomographic imaging instrumentation for simultaneous dual-breast measurements. *Appl Opt* 44:2140–2153.

Schotland, J. C., and V. A. Markel. 2001. Inverse scattering with diffusing waves. *J Opt Soc Am A Opt Image Sci Vis* 18:2767–2777.

Schweiger, M., S. R. Arridge, and I. Nissila. 2005. Gauss-Newton method for image reconstruction in diffuse optical tomography. *Phys Med Biol* 50:2365–2386.

Srinivasan, S., B. W. Pogue, S. Jiang, H. Dehghani, and K. D. Paulsen. 2005. Spectrally constrained chromophore and scattering near-infrared tomography provides quantitative and robust reconstruction. *Appl Opt* 44:1858–1869.

UCL. TOAST Optical Tomography, Home Page: http://web4.cs.ucl.ac.uk/research/vis/toast/.

Van Rossum, M. C. W., and T. M. Nieuwenhuizen. 1999. Multiple scattering of classical waves: microscopy, mesoscopy and diffusion. *Rev. Mod. Phys.* 71:313–371.

Vogel, A., and V. Venugopalan. 2003. Mechanisms of pulsed laser ablation of biological tissues, *Chem. Rev.*, 103:577–644.

Wang, L. V., and H. Wu 2007. *Biomedical Optics: Principles and Imaging.* John Wiley & Sons, Inc, Hoboken, NJ.

Weber, J. R., D. J. Cuccia, A. J. Durkin, and B. J. Tromberg. 2009. Noncontact imaging of absorption and scattering in layered tissue using spatially modulated structured light. *J App Phys*: 105:102028.

Weber, J. R., D. J. Cuccia, W. R. Johnson, G. H. Bearman, A. J. Durkin, M. Hsu, A. Lin, D. K. Binder, D. Wilson, and B. J. Tromberg. 2011. Multispectral imaging of tissue absorption and scattering using spatial frequency domain imaging and a computed-tomography imaging spectrometer. *J Biomed Opt* 16:011015.

Weissleder, R., C. H. Tung, U. Mahmood, and A. Bogdanov, Jr. 1999. *In vivo* imaging of tumors with protease-activated near-infrared fluorescent probes. *Nat Biotechnol* 17:375–378.

White, B. R., and J. P. Culver. 2010. Quantitative evaluation of high-density diffuse optical tomography: *In vivo* resolution and mapping performance. *J Biomed Opt* 15:026006.

Wu, J., L. Perelman, R. R. Dasari, and M. S. Feld. 1997. Fluorescence tomographic imaging in turbid media using early-arriving photons and Laplace transforms. *Proc Natl Acad Sci USA* 94:8783–8788.

Zhu, Q., M. Huang, N. Chen, K. Zarfos, B. Jagjivan, M. Kane, P. Hedge, and S. H. Kurtzman. 2003. Ultrasound-guided optical tomographic imaging of malignant and benign breast lesions: Initial clinical results of 19 cases. *Neoplasia* 5:379–388.

9

Paola Taroni
Politecnico di Milano
CNR—Istituto di Fotonica e
Nanotecnologie

Antonio Pifferi
Politecnico di Milano
CNR—Istituto di Fotonica e
Nanotecnologie

Giovanna Quarto
Politecnico di Milano

Lorenzo Spinelli
CNR—Istituto di Fotonica e
Nanotecnologie

Alessandro Torricelli
Politecnico di Milano

Rinaldo Cubeddu
Politecnico di Milano
CNR—Istituto di Fotonica e
Nanotecnologie

Diffuse Optical Imaging: Application to Breast Imaging

9.1 Breast Cancer and Breast Imaging

About one in eight women in the United States (12%) will develop invasive breast cancer over the course of her lifetime. It is by far the commonest cancer in women in the United States, accounting for approximately 30% of all cases in women. Breast cancer death rates are higher than those for any other cancer, besides lung cancer. The situation is similar in Northern Europe, while other EU countries (like Spain and Greece) have much lower incidence, although mortality rates are still comparable (Ferlay et al. 2010a, b). Less industrialized countries, like parts of Asia and Africa, are generally less affected, but the risk for women who migrate from low- to high-risk countries typically increases, suggesting a strong effect for lifestyle or environmental factors (Deapen et al. 2002).

Early diagnosis and subsequent therapy significantly reduce the mortality (Tabar et al. 2003). Even small improvements in diagnosis and therapy can have huge impact in terms of spared lives and quality of life. This is the reason why many countries (e.g., UK, Italy, Australia) offer screening programs to women, typically between 50 and 70 years of age. Breast screening essentially relies on x-ray mammography. However, false negatives and false positives hamper the efficacy of the screening and lead to unnecessary treatment (typically consisting of core biopsy), with significant physical and psychological morbidity, as well as nonnegligible social and economic costs. Mammography is also less accurate in patients with dense breasts, including young women (Marshall 2010), with reported sensitivity as low as 48% (Kolb et al. 2002).

Magnetic resonance imaging (MRI) is the most sensitive technique to diagnose invasive breast cancer, with good performance even in dense breast tissue, but the increased sensitivity can also lead to false-positive results (i.e., low specificity) (Hylton 2005, Lord et al. 2007). Moreover, MRI suffers from lower sensitivity for ductal carcinoma *in situ*. It is also characterized by high costs and long examination times, which hinder its routine use for screening purposes.

Breast imaging is also possible with positron emission tomography (PET). With a PET examination, women with suspicious mammograms may gain information on the nature of the detected lesion. PET is also effectively used in cancer staging. However, it is expensive and involves the administration of a radioactive tracer (Wu and Gambhir 2003).

Breast ultrasonography (US) is often applied as supplemental breast-imaging technique in the population of women with mammographically dense breast tissue, and permits detection of small, otherwise occult, breast cancers (Nothacker et al. 2009). Potential adverse impact for women in this group is associated with an increased biopsy rate. The technique is not particularly expensive and the equipment is generally available in the doctor's office. However, US is a real-time imaging method and data interpretation strongly relies on the operator's experience.

It is worth mentioning that breast imaging aims not only at diagnosing diseases, but also at monitoring therapy. While mammography is still the gold standard for diagnosing breast pathologies, MRI is often applied for the latter use. Screening is repeated periodically, typically every year or every other year, for decades, while therapy monitoring involves several imaging sessions concentrated in a short period. Therefore, in both cases, the use of ionizing radiation can have adverse effects that should be avoided.

It is evident that new breast-imaging methods are needed. Optical techniques are inherently noninvasive, suggesting their possible exploitation for the screening of wide populations and for therapy monitoring. They have the potential to provide functional information that can be of great importance for diagnostic purposes. An intrinsic limitation of optical methods is related to the nonnegligible attenuation of light in biological tissues: it is on the basis of the information that can be derived on tissue, but at the same time it limits the penetration depths and consequently the thickness of tissues that can be probed using light. However, breast imaging only involves soft tissue, with average thickness of 5–10 cm, depending on whether a compressed-breast geometry is adopted. Thus, light attenuation is high, but generally not prohibitive. Finally, optical instruments are portable and reasonably cheap as compared to other instrumentation that is conventionally used for medical diagnosis.

Based on these considerations, at least in principle, optical imaging methods have the potential for a successful application to breast screening and monitoring.

9.2 Light Propagation in Tissues

Light propagation in most biological tissues is severely affected by the fine variations of the dielectric constant that produce multiple refraction and diffraction perturbations to the propagating electromagnetic wave (Ishimaru 1978, Martelli et al. 2009). A meaningful picture can be derived following the corpuscular description of light propagation. Photons injected into a diffusive medium (e.g., biological tissue) can freely propagate along a straight path up to the occurrence of either absorption or scattering events. In the former case, the photon is lost. In the latter, its direction is changed, while—in the most common situation of elastic scattering—its energy (wavelength) is preserved. Thus, photon migration through a diffusive medium can be seen as a random path through multiple scattering interactions, up to the point where the photon is either absorbed or exits the medium at the boundaries (see Figure 9.1). From a probabilistic point of view, it is possible to define the absorption (μ_a), and the scattering (μ_s) coefficients as the inverse of the mean photon paths (free paths) before an absorption or a scattering interaction takes place, respectively.

After a scattering interaction, the new photon direction can be completely randomly oriented (isotropic scattering) or on average more aligned along the prescattering path (forward scattering). A useful parameter is the reduced scattering coefficient $\mu_s' = (1-g)\mu_s$, where g is the anisotropy factor and corresponds to the mean cosine of the scattering angle. Here, μ_s' can be interpreted as the equivalent scattering coefficient of the medium assuming a completely isotropic scattering ($g = 0$). After the first scattering interactions, where the scattering

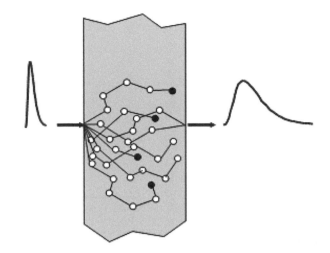

FIGURE 9.1 Photon migration: injected and remitted (transmittance geometry), photon paths with scattering centers (white circle), and absorption (black circle).

anisotropy can still preserve memory of the initial photon direction, the photon path can be efficiently described by μ_s' alone, meaning that tissues with different anisotropy properties still yield quite similar photon path distributions for equal values of μ_s'. For instance, light attenuation is ruled by the effective transport coefficient $\mu_{\mathrm{eff}} = \sqrt{3\mu_a\mu_s'}$, depending only on the absorption and reduced scattering coefficient. More generally, the photon distribution can be effectively described using the diffusion approximation of the radiative transport equation, which assumes a completely isotropic scattering and exploits μ_s' to account for the scattering properties (Martelli et al. 2009, see also Chapter 8).

In the 600–1000 nm range—often called the "therapeutic window"—$\mu_s' \gg \mu_a$, since—for most tissues—$\mu_a \approx 0.05$–0.5 cm^{-1}, and $\mu_s' \approx$5–15 cm^{-1}. This "diffusive regime" has some important features, like: (i) spatial information on photon-visited structures is rapidly smeared out due to random photon paths; (ii) light attenuation is much higher than for propagation through a nondiffusive medium with the same absorption properties, due to the much longer path (from 100 to 1000 times) required by photons to come across a given tissue; and (iii) absorption and scattering properties are strongly linked to determine the photon distribution in the medium, thus making it difficult to decouple absorption from scattering contributions (Yodh and Chance 1995, Durduran et al. 2010). Ultimately, it is due to the diffusive nature of biological tissue that one cannot see though the human body. If scattering were negligible, the hand, for instance, would appear transparent, with some feeble coloring for different structures. Thus, classical tools for optics in nonscattering media cannot be used, and a new wealth of methods, instrumentation, and theoretical models are needed to properly face the hard challenges of tissue optics. In Section 9.4, some possible experimental approaches are presented.

9.3 Optical Properties of the Breast

9.3.1 Absorption Properties of the Main Breast Constituents

A well-documented collection of the absorption spectra of some tissue constituents is found in the website of Prahl (2011). Figure 9.2 displays a selection of key absorbers in the 600–1100 nm range. For water (Kou et al. 1993) and lipids (van Veen et al. 2005, Taroni et al. 2009b), the spectrum refers to the pure substance, that is, a mass density of 1 g/cm^3 and 0.9 g/cm^3, respectively. Oxy- (O$_2$Hb) and deoxyhemoglobin (HHb) are referred to a molar concentration of 1 μM, while collagen was normalized to a reference density of 1 g/cm^3.

Water has a high peak around 980 nm, and a shoulder in the 750 nm region. Its relatively high and broad absorption band is the main cause of light attenuation in tissues beyond 900 nm. The absorption spectrum of lipid has a narrow peak at 930 nm that exactly equals water absorption at the same wavelength (water–lipid isosbestic point). HHb shows a generally decreasing absorption trend with increasing wavelength, yet with a relative maximum at 760 nm. O$_2$Hb, besides having a decreasing tail in the red, has a broad maximum around 900 nm. Finally, collagen (Taroni et al. 2007, 2009b) has a rather flat spectrum, with a decreasing tail in the red, a minor peak around 930 nm, and a main peak around 1020 nm. Indeed, collagen content in tissues is often low (<0.1 g/cm^3), thus its overall contribution to the absorption spectrum is limited. The knowledge of the spectral features of the key tissue constituents can be exploited to derive their abundance from a spectroscopy measurement. The recovered absorption spectrum of the investigated tissue can be fitted with a linear combination of the reference component spectra to retrieve their concentration, assuming a uniform distribution of the different absorbers within the probed volume (Torricelli et al. 2001). Conversely, to characterize a suspect lesion buried within the breast, inhomogeneous models must be enforced,

such as perturbative approaches (Martelli et al. 2009, Spinelli et al. 2003) or fully tomographic schemes (Durduran et al. 2010, see also Chapter 8).

9.3.2 Scattering Spectra

For the interpretation of the scattering spectra, an empirical macroscopic description is usually adopted. Simulations derived using the Mie scattering model and experimental measurements have shown that the scattering spectrum in most biological tissues in the red–NIR range roughly follows a power law of the form $\mu'_s(\lambda) \approx a(\lambda/\lambda_0)^{-b}$, where λ_0 is an arbitrary reference wavelength used for dimensional requirements, b is the scattering power, while a is an amplitude factor that represents the reduced scattering coefficient at λ_0 (Mourant et al. 1997, Nilsson et al. 1998). A useful physical meaning of b can be inferred by analyzing a set of Mie simulations for homogeneous spheres with different radii using the scattering power law. In that case, b can be related to the radius (dimension) of the sphere, and a to their volume concentration. It must be underlined that a and b derived from a tissue measurement are not the *real* average size and density of the tissue scatterers but rather are effective parameters for a sphere distribution yielding an equivalent scattering spectrum. Nonetheless, these empirical quantities can be effectively used to quantify microstructure-related features and possibly to classify the different tissue structures.

9.3.3 *In Vivo* Optical Spectroscopy of the Breast

The female breast can have quite diverse optical properties related to the different structure and composition of the constituent tissues (Leff et al. 2008). As an example, Figure 9.3 (Taroni et al. 2009b) displays the spectra obtained using a time-resolved broadband diffuse spectrometer operated continuously in the 600–1100 nm range (Pifferi et al. 2007, Bassi et al. 2007) for three subjects representative of the different spectral features observed *in vivo* on the breast. Data were obtained by averaging the optical properties measured on both breasts in transmittance geometry right in the central region, with the breast being slightly compressed between two black PVC plates. The absorption spectrum (Figure 9.3a) shows large variations both over wavelength and among subjects. The main spectral features are the decreasing tail of HHb and O$_2$Hb in the red, the lipid peak at 930 nm, and water absorption around 980 nm. The region around 760 nm is quite complex due to the overlapping of minor water, lipid, and HHb peaks. The broad absorption maximum of O$_2$Hb about 900 nm cannot be identified since it is overwhelmed by the water and lipid peaks; conversely, it contributes to the overall absorption in the NIR region. Also, collagen features are not visible since the intrinsic absorption of collagen is rather low—thus making its quantification challenging. The key differences among women can be easily seen in the 900–1000 nm region, where different abundance of lipids and water yield substantially different absorption properties.

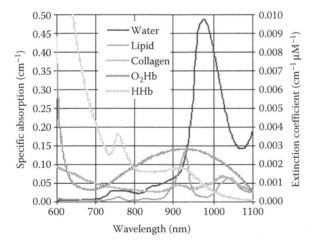

FIGURE 9.2 Absorption spectra of the main tissue absorbers. Water, lipid, and collagen are referred to the left axis as pure substances, while HHb and O$_2$Hb to the right axis as μM dilutions.

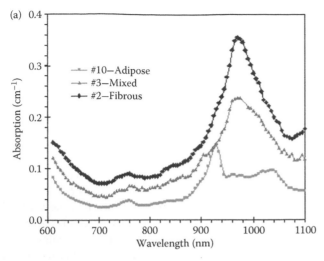

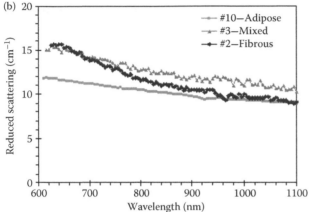

FIGURE 9.3 Spectra of the absorption (a) and the reduced scattering (b) coefficient of the female breast derived *in vivo* on three subjects. (Reproduced with permission from P. Taroni et al. 2009. *J. Biomed. Opt.* 14:054030.)

The scattering spectra (Figure 9.3b) show lower spectral and intersubject variations. The key differences are changes in amplitude and in slope, the latter being generally higher in water-rich breasts. The reason is that water-rich breasts have a prevalence of glandular and fibrous tissues that, due to the presence of collagen fibrils, exhibit a steeper declining scattering spectrum.

9.4 Techniques and Instrumentation for Breast Imaging

9.4.1 Imaging Geometries

Given an object embedded in a diffusive medium, the general problem of optical imaging, and specifically of breast imaging, is

- To detect the object, that is, to be sensitive to spatial and/or temporal changes of the optical properties of the object

- To localize the object in terms of depth and position
- To identify the object, that is, to quantify its geometrical properties (dimension, shape) and optical properties (absorption, scattering, fluorescence) (Yodh and Chance 1995)

In the literature, two distinct approaches have been reported (Arridge and Hebden 1997):

The *direct* approach constructs images by filtering the least scattered photons from the multiple scattered, diffusive photons.

The *indirect* approach is based on the hypothesis that, given a set of measurements at the boundary of an object, a unique 3D distribution of internal scatterers and absorbers exists that would yield that specific set. Therefore, optical imaging involves solving an inverse problem using a proper model of photon transport.

Both approaches have been implemented leading to two different data acquisition geometries:

- *Coaxial scanning* (see Figure 9.4a), typically with a single source and detector, thus yielding a projection image (shadowgram), where each individual measurement corresponds directly to a single pixel in the final image.
- *Tomographic sampling* (see Figure 9.4b), typically with multiple sources and detectors, thus yielding, through a reconstruction procedure, a slice section where each pixel is dependent on the neighboring ones.

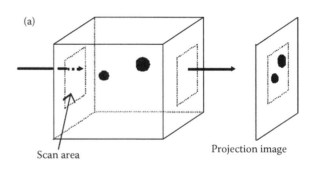

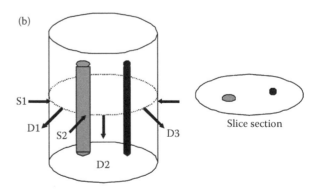

FIGURE 9.4 (a) Geometry for coaxial scanning and corresponding projection image. (b) Geometry for tomographic sampling and corresponding slice section.

9.4.2 Imaging Set-Ups

Diffuse optical imaging through a scattering medium suffers from low contrast and spatial resolution because of the blurring introduced by the spread of photon paths in the medium (as described in Section 9.2). The effect of multiple light scattering can be properly taken into account by direct measurements of photon pathlength. Since photon pathlength is directly related to the time-of-flight in the medium, the natural choice is to perform time-resolved measurements (Chance et al. 1998, Delpy et al. 1988, Jacques 1989, Wilson et al. 1992, Liu et al. 1993).

In the time-resolved or time domain (TD) approach, following the injection of a light pulse into a turbid medium, the temporal distribution (see Figure 9.1) of the remitted photons at a distance ρ from the injection point will be delayed (depending on the finite time light takes to travel the distance between source and detector), broadened (due to the many different paths that photons undergo because of multiple scattering), and attenuated (since absorption reduces the probability of detecting a photon, and diffusion into other directions within the medium decreases the number of detected photons in the considered direction).

It is interesting to note that in principle the investigation of diffusive media by means of the analysis of the remitted light can be also accomplished through different modalities such as frequency domain (FD) and continuous wave (CW) approaches.

FD photon migration (Lakowicz and Berndt 1990, Patterson et al. 1991, Sevick et al. 1991, 1994, Pogue and Patterson 1994, Fishkin et al. 1995, Kienle and Patterson) is the natural alternative to the TD approach. An intensity-modulated light wave is injected into the medium, and the remitted wave is detected. The optical properties of the traversed medium can be recovered by the analysis of the phase shift and of the demodulation signals.

CW optical methods have been used for a long time to investigate natural and biological media, thanks to the availability of low-cost, easy-to-use laser sources with different spectral and power characteristics. However, the main disadvantage of CW methods, when applied to diffusive media, is the impossibility to discriminate between the scattering and the absorbing contributions of the diffusive medium (Nomura et al. 1997). Measurements of the total intensity attenuation are, in fact, affected by both the absorbing and the scattering properties of the investigated sample. At least in principle, these properties can be assessed even by CW methods, if one is able to apply the measurement procedures strictly following the basic definition of the parameters to be determined. Nonetheless, this always requires the manipulation of the sample so as to produce a particular geometry, typically thin sections, so as to get rid of the multiple scattering. In many cases (e.g., *in vivo* measurements), this is not possible and, whenever possible, great care should be taken in order to avoid the artifacts in the measured optical properties due to the preparation procedures that the sample typically undergoes.

In recent years, CW optical methods have thus evolved to more complicated configurations so as to noninvasively determine the optical parameters of diffusive media. In the CW space-resolved (Farrell et al. 1997, Kienle et al. 1996) or CW angle-resolved (Wang and Jacques 1995, Lin et al. 1997) configurations, remitted photons are collected at different source–detector distances or at different angles with respect to the surface of the medium. In those ways, it is possible to discriminate photon paths, thus yielding the absorption and the scattering contributions separately.

It is not clear which method is the best. In principle, TD data should have the highest information content, as compared to FD or CW data. In the TD approach, in fact, the path length is determined for each detected photon, as opposed to the ensemble averages that are determined in the FD approach. On the other hand, CW and FD approaches are potentially easier to implement.

Table 9.1 reports the prototypes and the instruments developed and effectively used in the last decade to noninvasively

TABLE 9.1 Prototypes and Instruments for Optical Mammography

Group	Approach	Technique	Wavelength (nm)	Detectors	Reference
PTB, DE	Scanning	TD	785	1	Grosenick et al. (1999)
			670, 785, 843/884	1	Grosenick et al. (2005a)
			670, 785, 843/884	4	Möller et al. (2003)
PoliMi, IT	Scanning	TD	637, 785, 915, 985	1	Pifferi et al. (2003)
			637, 656, 683, 785, 916, 975, 985	1	Taroni et al. (2005)
			635, 680, 785, 905, 930, 975, 1060	1	Taroni et al. (2010)
Hamamatsu, KK	Tomographic	TD	765, 800, 835	48	Ueda et al. (2011)
ART, CA	Scanning	TD	760, 780, 830, 850	1	Intes (2005)
UCL, UK	Tomographic	TD	780, 815	32	Yates et al. (2005)
UPENN, MD	Tomographic	TD	780, 830	24	Ntziachristos et al. (1998)
Imaging Diagnostic, FL	Tomographic	CW	808	84	Floery et al. (2005)
UPENN, MD	Tomographic	FD + CW	690, 750, 786, 830	9 + CCD	Culver et al. (2003)
Dartmouth, NH	Tomographic	FD	761, 785, 808, 826	48	Dehghani et al. (2003)
Clemson, SC	Tomographic	FD	785	16	Jiang et al. (2002)
Siemens, DE	Scanning	FD	690, 750, 788, 856	1	Fantini et al. (1996)
				1	Götz et al. (1998)

detect the breast lesions by optical mammography. A detailed description of each device is beyond the scope of this work, therefore we refer the reader to the selected references listed in Table 9.1. More information can also be found in the paper by van de Ven (2011).

9.4.3 Data Analysis

Depending on the geometry (e.g., coaxial scanning or tomographic sampling) and technique (space/time/frequency domain), the theoretical models as well as laboratory instrumentation needed to produce images will be notably different. Therefore, the accuracy, precision, and robustness of measurements will be the characteristic of the specific system.

To analyze the TD data, the *time-gating* method has been perhaps the most widely used, thanks to its straightforwardness. In the coaxial scanning geometry, the time-gating technique involves illuminating the object with a picosecond pulse of light, and filtering to retain only the small fraction of transmitted light that follows a path closely approximating the straight line between the source and the detector (the so-called early photons), as shown in Figure 9.5. Both the experiments and theoretical models have shown that obtaining an image with the highest contrast and spatial resolution requires the rejection of photons with longer path within the medium (Andersson-Engels et al. 1992).

However, the significant drawback of time gating is that the vast majority of transmitted light is discarded, and the scarcity of detected photons with the shortest path lengths limits the signal-to-noise ratio. Moreover, since the time-gated intensity is influenced by both the absorbing and scattering properties of the medium, and since at early times, the effects of both

the optical parameters result in a reduction of the transmitted intensity, the time-gating technique is unable to identify the object. In some experimental conditions, the sensitivity can also be reduced. For example, since both absorption and scattering decrease the intensity of the remitted signal, if we consider an object with opposite scattering and absorbing contrast, it may so happen that these two effects tend to cancel out.

Therefore, the initial enthusiasm for time-gated optical imaging has been substantially dampened. Monte Carlo simulations (Zaccanti and Donelli 1994), theoretical arguments (Moon et al. 1996), and experimental results (Hall et al. 1997) have in fact indicated that this technique is unlikely to exceed a spatial resolution of 1 cm in a 5-cm-thick diffusive ($\mu'_s = 10$ cm^{-1}, $\mu_a = 0.1$ cm^{-1}) medium. When applied to optical mammography, this value falls short of the performance needed in a clinical system for routine screening of the breast disease. Moreover, since at early times, the effect of both absorption and scattering is a reduction of the remitted intensity, the time-gating technique can hardly discriminate the nature of the investigated medium.

Photons experience a probability to be absorbed that increases with their path in the medium. Thus, the absorption influences the tail of time-resolved curves, more than their leading edge. Consequently, by integrating the transmitted intensity over time windows positioned at late times, it is possible to enhance the information on the absorption properties (Cubeddu et al. 1996). Therefore, an improvement of the time-gating method involves constructing several time-gated intensity images by displaying the number of counts collected within different time windows of variable delay and fixed width. This differs from the conventional time-gating technique, which uses a fixed delay $D = 0$ ns (corresponding to the time the first nonscattered photons are expected), and a variable width W (typically set so as to achieve the best trade-off between signal-to-noise ratio and contrast, by collecting a significant number of photons, but only among the early less-scattered ones). In the dynamic time-gating approach, a time gate with fixed width W and variable delay D is used. The width W of the time gate should be large enough to collect a significant number of photons. The delay D varies from 0 ns up to the instant when the time-resolved curve falls to the noise level. Figure 9.5 shows a time-resolved transmittance curve with typical examples of an early time-gate interval ($D = 0$ ns and $W = 0.4$ ns) and a late time-gate interval ($D = 4.4$ ns and $W = 0.8$ ns).

A limitation of all the time-gating approaches is that no quantitative data can be obtained for the optical parameters (i.e., a lesion can be detected and localized, but not characterized). To improve lesion characterization, the estimate of absorption and scattering coefficients should be provided. In the simplest approach, μ_a *images* and μ'_s *images* are constructed by plotting the values of the absorption coefficient and of the reduced scattering coefficient as obtained from fitting the experimental data with a standard solution of the transport equation in the diffusion approximation for a homogeneous medium. When operating in the TD, typically the theoretical curve is convolved with the system transfer function and normalized to the area of the

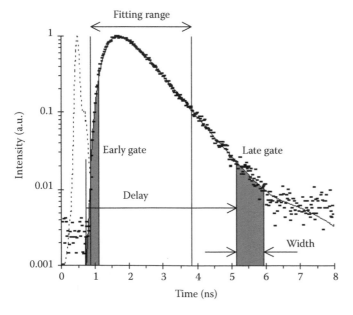

FIGURE 9.5 Typical time-resolved transmittance curve with an early time-gate interval ($D = 0$ ns and $W = 0.4$ ns), and a late time-gate interval ($D = 4.4$ ns and $W = 0.8$ ns).

experimental curve. The fitting range may also include points on the rising edge of the curve (e.g., points with a number of counts higher than 1% of the peak value), where the diffusion approximation fails, but where photon paths are straighter and provide better spatial resolution. It should be noted that the theoretical model used to fit the experimental data is not accurate to describe the photon time-distribution, since it is derived for a homogeneous slab. Therefore, the absolute value of the measured μ_a and μ_s' cannot be used to evaluate the exact optical properties of a heterogeneous medium. Nevertheless, the fitted parameters depend on the absorption and scattering coefficients experienced by the photons along their paths. Hence they vary in the presence of a local inhomogeneity and, at least in principle, can be used to make a rough characterization.

To improve the estimate of the spatial distribution of optical properties, the natural way forward is either to resort to advanced methods, which introduce complexity in data acquisition (multisource/detector configurations) and in data analysis like perturbative approaches (Martelli et al. 2009), or to introduce reconstructive inverse algorithms (as described in Chapter 10). The reconstructive approach can be applied to data acquired in any measurement regime (CW, TD, and FD).

9.5 Breast Lesion Detection and Identification

9.5.1 Historical Evolution

Different approaches to optical breast imaging, described in the previous sections of this chapter, have been exploited over the years, following the evolution of light sources and detection systems.

The first optical attempt to detect breast lesions was made by Cutler in the late 1920s with a very simple scheme, named "transillumination" or "diaphanography" (Cutler 1929). It involved shining light from a powerful lamp through the pendant breast and visually inspecting the transmitted light pattern. The "shadowgrams" allowed the localization of vascularized lesions that caused higher attenuation than surrounding tissue, but could not distinguish between malignant lesions and benign ones, which can also be highly vascularized.

In the 1970s and 1980s, the development of sensitive films and video cameras renewed the interest in transillumination, still using CW broadband (red–NIR) light. However, extensive clinical studies soon revealed the main limitations of the approach, mostly related to the strong scattering of NIR light in tissue (Geslien et al. 1985, Gisvold et al. 1986, Alveryd et al. 1990). Overall, the clinical performance was poor, with significantly lower sensitivity for deep lesions as compared to superficial ones, low specificity, and strong dependence of the results on the operator and on previous knowledge of the lesion. The simplest technical approach was exploited, but the informative content was necessarily limited. No attempt was conceivable to untangle absorption and scattering contributions to the measured attenuation, thus mixing different pieces of information on the tissue. Moreover, broadband light from filtered lamps was used and no spectral discrimination was performed on detection, thus averaging out the information that comes from the different behavior of light at different wavelengths.

In the meantime, technical advancements had made more sophisticated approaches feasible, operating either in the time or in the frequency domain so as to uncouple absorption from scattering. Moreover, with the use of laser sources, illumination at single wavelengths was made possible, which allowed for wavelength selection to investigate the tissue and derive diagnostically useful information.

At that time (1980s and 1990s), the instrumentation for phase-resolved studies was simpler than that for time-resolved measurements, and this allowed the realization of compact instruments for clinical use. Hence, initial clinical studies were carried out in the frequency domain. Usually point measurements were performed in transmittance geometry and the compressed breast was scanned to build images. Data were analyzed with the diffusion theory for a homogeneous infinite slab. The finite extent of the compressed breast could not be accounted for, and this caused artifacts that limited the sensitivity in breast areas close to tissue boundaries. To make up for this problem, an original algorithm was proposed, deriving information on breast thickness from phase data and exploiting it to correct amplitude data (Fantini et al. 1996). The resulting images, called "*N-images*" had reduced boundary artifacts, and higher contrast was achieved. However, a single set of corrected data was thus available, making it impossible in practice to untangle absorption from scattering. Moreover, in a clinical study performed with a phase-resolved instrument developed by Siemens AG (Medical Engineering, Erlangen, Germany), data were acquired at four wavelengths (690, 750, 788, and 856 nm), but analysis was published only on data at the shortest wavelength (Götz et al. 1998). Overall, *N-images* at 690 nm were sensitive to attenuation due to blood, that is, the main tissue absorber at that wavelength. This led to a promising sensitivity for the first clinical trial of a novel diagnostic technique, but the specificity was low. Later, the application of an image-processing method that relies on a spatial second-derivative operator allowed the improved detection of regions with low contrast, thus improving the sensitivity. At the same time, an attempt to increase the specificity was made through the analysis of the data at all four wavelengths in order to estimate the blood parameters, especially blood oxygenation (Heffer et al. 2004).

In recent years, due to the development of picosecond-pulsed diode lasers emitting in the red and NIR, and compact detectors with extended spectral sensitivity, time-resolved measurements no longer necessarily require complex and expensive laboratory systems, and can reliably be performed with portable instruments. From 2000 to 2004, a European project—"Optimamm: Imaging and characterization of breast lesions by pulsed near-infrared laser light"—investigated the potential of time-resolved transmittance imaging for optical mammography. Extensive clinical data were collected by partners in Germany (Grosenick et al. 2005a) and Italy (Taroni et al. 2005). Both research groups developed and tested time-domain scanning optical mammographs operating in transmittance geometry on the compressed breast and acquiring images from both breasts in two views (i.e., cranio-caudal and medio-lateral, or cranio-caudal and oblique), in agreement with routine clinical practice for x-ray mammography. The choice of acquiring projection images of the compressed breast allowed easier comparison between optical images and standard x-ray mammograms for the retrospective assessment of the optical technique.

In particular, the group in Berlin (Grosenick et al. 2005a) developed a triple-wavelength (670, 785, 843/884 nm) instrument and applied it to 154 patients with suspected breast cancer. Optical mammograms were obtained displaying (inverse) photon counts collected in selected time windows, as well as the absorption and reduced-scattering coefficients. Setting the percentage content of water and lipids to fixed values, total hemoglobin concentration and blood oxygen saturation were also estimated from two- or three-wavelength data. Seventy-two out of 102 histologically confirmed tumors (71%) were detected retrospectively in both optical projection mammograms, and 20 more lesions (20%) were detected in one projection only.

The group in Milan (Taroni et al. 2005) initially developed an instrument operating at four wavelengths in the range of 637–985 nm, and tested it on 194 patients with 225 malignant and benign lesions. In particular, two wavelengths were longer than 900 nm (i.e., 915 and 975/985 nm) to increase the sensitivity to lipids and water, which show major absorption peaks in that wavelength range. Forty-one out of 52 cancers (79%) were detected in both views and nine more (17%) in just one view. Concerning benign lesions, a significant number of cysts was analyzed ($n = 82$), achieving a detection rate of 83% in both views, which rose up to 90% if detection in a single view was accepted.

Both the German and Italian studies had similar detection rates for malignant breast lesions. On average, late-gated

intensity images at short wavelengths (i.e., <700 nm) provided the highest tumor-to-healthy contrast and also the highest contrast-to-noise ratio. At short wavelengths, blood absorption is dominant, so the contrast was attributed to high blood volume in the lesion. For the estimate of tissue composition and specifically to quantify the constituents in the detected lesions, simple perturbative approaches were used, considering a single spherical lesion in an otherwise homogeneous background. Both studies confirmed significantly higher hemoglobin content in the lesion, as compared to the surrounding healthy tissue, while blood oxygen saturation proved to be a poor discriminator (Grosenick et al. 2005b, Spinelli et al. 2005). Neoangiogenesis is a well-known aspect associated with cancer development. Thus, the high blood volume that is typically assessed through the optical measurements could be an effective indication of malignancy and prove useful for diagnostic purposes.

As mentioned earlier, when time-domain projection imaging is performed, time-gated intensity images can easily be produced. In particular, late gates opened on the tail of the transmitted pulses increase the sensitivity to the absorption properties, providing information on tissue absorption with better spatial resolution than achievable when the diffusion theory is applied to estimate the absorption properties. A proper selection of the imaging wavelengths may then allow one to investigate, at least qualitatively, the tissue composition. An example is reported in Figure 9.6, where cranio-caudal late-gated intensity images at different wavelengths are compared to the corresponding x-ray mammograms (Taroni et al. 2004). Complementary information is obtained from the scattering images that are also displayed in Figure 9.6. Two cysts in the external quadrants of the right breast are identified in the scattering images. One cyst, close to the breast edge, is partially obscured by the artificially low scattering values due to the edge effects. In this case, the absorption properties of the cysts at different wavelengths are

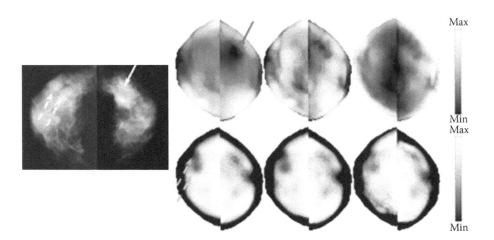

FIGURE 9.6 Cranio-caudal views of both breasts of a patient with two cysts [20 mm maximum diameter each (dashed arrows)] in the upper outer quadrant of her right breast and a carcinoma [30 mm maximum diameter (solid arrow)] in the upper outer quadrant of her left breast. From left to right: x-ray mammograms, late-gated intensity images (top), and scattering images (bottom) at 683, 785, and 913 nm. (Reproduced with permission from P. Taroni et al. 2004. *J. Biomed. Opt.* 9: 464–473.)

not significantly different from those of the surrounding tissue. Specifically, some reduction in absorption can be observed at 785 and 913 nm, corresponding to the cyst located deeper in the breast tissue. As mentioned earlier, malignant lesions are often identified based on their marked absorption at short wavelengths. This trend is confirmed for the carcinoma in the left breast shown in Figure 9.6.

Besides information useful for lesion detection and identification, gated intensity imaging performed at several wavelengths can also provide information on tissue physiology. As an example, strong absorption at short wavelengths (683 and 785 nm) due to high vascularization can be observed in the nipple area, especially in the left breast, where superficial blood vessels are also easily identified. At 913 nm, the absorption is dominated by lipids. Thus the corresponding image provides information on the abundance and localization of adipose tissue. Based on the x-ray images, here adipose tissue is mostly located close to the chest wall, in an otherwise dense breast. The gated intensity image at 913 nm provides the same information: strong lipid absorption is essentially observed close to the chest wall only.

In parallel with the development of projection imaging in the time domain, tomography has also recently gained a lot of interest in the biomedical optics community. Several research groups now focus on tomographic approaches, exploring various techniques and geometries. They generally apply CW or frequency-domain methods or a combination of the two, where some wavelengths are modulated, while others are not (Nielsen et al. 2009, Culver et al. 2003, Dehghani et al. 2003, Jiang et al. 2002). However, the time-resolved approach has also been investigated (Hebden et al. 2005, Intes 2005). In some systems, the patient lies prone with her breast pendant through a hole, while sources and detectors are distributed along one or more rings around the breast (Nielsen et al. 2009, Jiang et al. 2002, Dehghani et al. 2003, Hebden et al. 2005, Intes 2005). In other set-ups, the pendant breast is compressed between parallel plates (Culver et al. 2003). A matching fluid is often used to improve the light coupling between injection/collection devices and breast. Moreover, hand-held probes for reflectance imaging have also been developed and applied with the patient lying supine, similar to that done for US imaging (Xu et al. 2007).

Using recently upgraded set-ups, images have been collected even at six wavelengths and above 900 nm (Wang et al. 2010). However, tomographic systems generally operate at 2–4 wavelengths in the range of 650–850 nm. Thus, for lesion detection and identification, they rely on blood parameters and scattering properties. The clinical results achieved so far substantially agree with those obtained using projection imaging: high hemoglobin content is typically chosen as the main signature of malignancy. Angiogenesis is a well-known aspect associated with the cancer development. It has to be noted that some studies also observed scattering contrast between lesions and healthy breast tissue, where lesions are characterized by stronger scattering (Choe et al. 2009). Such observation can be explained if we consider that rapidly growing tumor cells may be characterized by the altered density and size of the subcellular organelles (e.g., mitochondria) that are at the origin of tissue scattering.

9.5.2 New Trends in Optical Mammography

9.5.2.1 Multimodality Imaging

Low spatial resolution, inherently due to the long wavelengths that are employed, is a major limitation of optical imaging. The choice of the spectral range is motivated by the need to penetrate through several centimeters of tissue, but the dominant scattering strongly affects the achievable spatial resolution. Moreover, the resolution depends on the position of the inhomogeneity/lesion (i.e., on its distance from sources and detectors) and on the optical contrast (smaller lesions can be detected if the contrast is high enough).

To overcome such limitations and to help in solving the underdetermined inverse problem of optical tomography, several groups have recently started to combine optical imaging with other imaging modalities that can provide detailed knowledge of tissue morphology, which is used as *a priori* information for optical imaging. Co-registration or subsequent registration in the same geometry has been performed with MRI (Ntziachristos et al. 2002, Azar et al. 2007, Srinivasan et al. 2010), x-ray tomosynthesis (Fang et al. 2011), and ultrasound imaging (Zhu et al. 2003). Only initial results have been obtained thus far, but they suggest that multimodal approaches might not only improve the spatial resolution of optical techniques by providing *a priori* morphologic information, but also contribute to increasing the overall diagnostic efficacy, with different techniques contributing the distinct and complementary physiological data.

9.5.2.2 Extended Spectral Information

Major constituents of soft tissues, such as water and lipids, are characterized by strong absorption peaks between 900 and 1000 nm. However, over the years, most breast imaging studies have been performed at 2–4 wavelengths within the range of 650–850 nm. At wavelengths shorter than 650 nm, blood attenuation increases rapidly, strongly reducing the signal level, while the use of wavelengths longer than 850 nm was often precluded by the limitations of available commercial detectors. Thus in the past, only very limited attempts were made to perform imaging at wavelengths longer than 850 nm (Taroni et al. 2005). A few other groups included long wavelengths in their imaging systems, but they did not fully exploit those wavelengths in the estimate of tissue composition, either by excluding lipid from the tissue absorbers they accounted for (Li et al. 2006) or by fixing water and lipid content and estimating only blood parameters (Konecky et al. 2008).

In principle, based on Beer's law, if n constituents of known absorption properties are considered: (i) n wavelengths are strictly needed to estimate their concentrations in tissue, and (ii) any n wavelengths could be chosen. However, in practice, measurements are not ideal, they are affected by noise. Thus the availability of information at a higher number of wavelengths provides an improved solution. Moreover, better results are achieved if optimized wavelengths are selected, taking advantage of the different spectral behavior of distinct tissue chromophores. Specifically, wavelengths should be chosen so that (ideally) each constituent is the dominant absorber at a different wavelength. This is intuitive, because it reduces the coupling between the equations to be solved, and was also proved with simulations and phantom data (Corlu et al. 2005, Eames et al. 2008). It should also be taken into account that projection imaging often relies on gated intensity images for lesion detection. In that case, the choice of the imaging wavelength is of key importance, as different wavelengths are used to highlight the absorption of distinct tissue constituents and provide information on their spatial distribution, as shown in Figure 9.6.

Recent spectroscopic studies (that involve single-point measurements over a broad wavelength range) have started to investigate the spectral changes in the absorption properties of water and lipids that may occur as a consequence of cancer onset and progression (Chung et al. 2008, Kukreti et al. 2007). As for imaging, a clinical study is going on with an instrument operating at seven wavelengths in the range of 635–1060 nm (Taroni et al. 2009a). In particular, a wavelength longer than 1000 nm is used to increase the sensitivity of the measurement to collagen. Collagen is a structural protein and a major constituent of breast tissue. Notwithstanding, this has always been neglected in the interpretation of optical data. Specifically, collagen is perhaps the most important extracellular matrix protein, and fibrillar collagens are essential in determining the stromal architecture. In turn, alterations of stromal architecture and composition are a well-known aspect of both benign and malignant pathologies, and may play an initial role in breast carcinogenesis (Guo et al. 2001). So the *in vivo* assessment of collagen could have significant relevance for breast-cancer detection.

9.5.2.3 Imaging with Exogenous Contrast Agents

Low spatial resolution is not the only aspect that negatively affects lesion detectability. When endogenous contrast is considered, as in the common case of hemoglobin absorption, lesions and healthy tissue are characterized by absolute concentrations of hemoglobin that are often much bigger than their difference. Thus, lesion detection relies on the sensitivity to a small change relative to a high background value, and the optical contrast is often low. Moreover, changes in endogenous contrast is frequently associated with normal physiological conditions and also consequently observed when no pathology is present. An example is the strong vascularization that is typical of the nipple area, which leads to high levels of hemoglobin-associated absorption.

The aforementioned limitations would be overcome if imaging were performed using a selective contrast agent that accumulates in the tumor, but not in healthy tissue.

If an absorbing exogenous contrast agent is considered, it should obviously absorb strongly and its absorption properties should ideally match a spectral region of low tissue absorption in order to increase the optical contrast between lesion (where the absorbing agent accumulates) and healthy tissue (that absorbs weakly at the imaging wavelength). An even more favorable situation is envisaged if a fluorescence contrast agent is considered. Obviously, good lesion targeting is again a key requirement. Moreover, to allow deep penetration into breast tissue, NIR excitation and emission wavelengths are needed. This also allows one to take advantage of negligible endogenous fluorescence from tissue. If these conditions are fulfilled, the fluorescence signal coming from the contrast agent localized in the lesion can be detected against a negligible background, with beneficial effects on the contrast value.

Unfortunately, until now no specific contrast agent has been developed, either for absorption or fluorescence imaging. The only contrast agent that is approved for clinical use and suitable for breast imaging is indocyanine green (ICG). It was originally introduced for fluoroangiography and is characterized by strong absorption and fluorescence in the NIR range. Thus, it can be used in both modalities. However, it is a blood pool agent, not selective for malignant lesions. Tumor/healthy contrast is achieved by exploiting the slower kinetics in the tumor due to an increased resistance of the chaotic tumor vasculature to blood flow.

While waiting for an optimal contrast agent, various research groups have started trials on humans with ICG (or related compounds), exploiting its absorption (Ntziachristos et al. 2000, Intes et al. 2003) and/or its fluorescence properties (Corlu et al. 2007, van de Ven et al. 2010, Poellinger et al. 2011). Figure 9.7 reports an example of the results that were achieved performing the imaging at late times, several minutes after the end of the systemic administration of ICG to the patient (i.e., during the extravascular phase). As shown in the figure, absorption and fluorescence mammograms are strongly affected by hemoglobin absorption, and in some situations, hemoglobin absorption is even dominant. Much better conditions for lesion detection are achieved with ratio images, obtained through ratioing point-by-point fluorescence and absorption data (Hagen et al. 2009).

9.5.3 Monitoring of Cancer Therapy

Besides cancer detection and screening, other applications of breast imaging are being investigated. Among them, monitoring of cancer therapy deserves mention. Specifically, presurgical neoadjuvant chemotherapy is often administered to reduce the tumor size and allow more conservative surgery. Direct and immediate knowledge about the individual response to specific drugs would be highly beneficial as it would allow one to take corrective actions with nonresponders, improving the cancer therapy and reducing the useless exposure to ineffective drugs with heavy side effects.

In initial monitoring studies, diffuse optical techniques were applied by simply performing linescans across the tumor area. Recently, more complete spatial information has been acquired, collecting the data over a bidimensional grid. In the case of good therapeutic response, significant changes are observed in optically derived parameters, including a reduction in blood

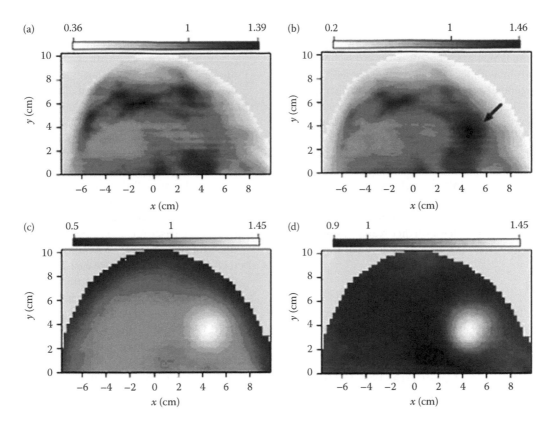

FIGURE 9.7 Cranio-caudal mammograms of the breast of a patient with a carcinoma: (a) time-integrated fluorescence intensity image; (b) absorption image; (c) ratio image of (a) and (b) without edge correction; and (d) with edge correction. The carcinoma (indicated by a black arrow) is visible in absorption and ratio images, while it cannot be identified in the fluorescence image due to the cancellation effects. (Reproduced with permission from A. Hagen et al. 2009. *Opt. Expr.* 17: 17016–17033.)

content (especially deoxyhemoglobin), water content, and scattering power, and an increase in lipid content (Cerussi et al. 2007, Soliman et al. 2010, Zhu et al. 2008, Roblyer et al. 2011). Moreover, it seems possible to distinguish responders from nonresponders based on optical measurements 1 day after the first treatment (Roblyer et al. 2011). Similar information can be obtained with contrast-enhanced MRI, but optical techniques are cost-effective, do not require the administration of exogenous contrast agents and may allow shorter measurement times. Thus, they seem to be good candidates to monitor the effectiveness of neoadjuvant chemotherapy, and even to predict it at early stages of treatment.

9.5.4 Estimate of Cancer Risk: Assessment of Breast Density

The most significant risk factors for breast cancer are gender (being a woman) and age (growing older). Besides those, breast density is a recognized strong independent risk factor (Boyd et al. 2001). Optical techniques could contribute to risk assessment by providing a noninvasive means for the assessment of breast density, which presently requires the use of ionizing radiation (x-ray mammography). In one study, CW transmittance data were collected over a broad spectral range (550–1300 nm) in

four selected positions per breast and interpreted using principal component analysis. This allowed the researchers to achieve very good correlation with quantitative mammographic features (Blackmore et al. 2008).

Changes in density are expected to reflect stromal changes (Alowami et al. 2003) and, as mentioned earlier, collagen is related to stromal changes. Thus, the quantification of collagen content based on optical measurements could allow the direct assessment of breast density. In this vein, an ongoing clinical study performed with a seven-wavelength (635–1060 nm) time-domain instrument has given promising preliminary results. Excellent agreement was achieved between breast density estimated using conventional mammography and an optical index based on tissue composition and scattering parameters, as derived from optical measurements. Moreover, direct correlation was achieved between mammographic density and collagen content, and could show promise for the direct assessment of cancer risk (Taroni et al. 2010).

9.6 Conclusion and Perspectives of Breast Imaging

Innovative optical instruments for breast imaging and modalities for related data analysis have been developed in recent decades. The results achieved are promising, but not yet conclusive. As described earlier, new approaches are being proposed to overcome limitations that are inherent in optical techniques or that became apparent with the development of the field.

Even though general agreement exists on some key aspects (e.g., the diagnostic potential of optically derived hemoglobin content for cancer detection), the clinical results obtained by different research groups can hardly be compared, due to differences (in some cases, major ones) between instruments, measurement protocols, procedures for data analysis, and so on. This is acceptable in the initial development phase of a new research field, but now standards are needed to allow the reliable comparisons at all stages, from instrument performance to clinical results and to guarantee a further real advancement of the field.

References

Alowami, S., S. Troup, S. Al-Haddad, I. Kirkpatrick, and P.H. Watson. 2003. Mammographic density is related to stroma and stromal proteoglycan expression. *Breast Cancer Res.* 5:R129–R135.

Alveryd, A., L. Andersson, K. Aspegren et al. 1990. Lightscanning versus mammography for the detection of breast cancer in screening and clinical practice. A Swedish multicenter study. *Cancer* 65:1671–1677.

Andersson-Engels S., R. Berg, and S. Svanberg. 1992. Effects of optical constants on time-gated transillumination of tissue and tissue-like media. *J. Photochem. Photobiol. B* 16:155–167.

Arridge, S.R. and J.C. Hebden. 1997. Optical imaging in medicine: II. Modelling and reconstruction. *Phys. Med. Biol.* 42:841–853.

Azar, F.S., K. Lee, A. Khamene et al. 2007. Standardized platform for coregistration of nonconcurrent diffuse optical and magnetic resonance breast images obtained in different geometries. *J. Biomed. Opt.* 12:051902.

Bassi, A., A. Farina, C. D'Andrea, A. Pifferi, G. Valentini, and R. Cubeddu. 2007. Portable, large-bandwidth time-resolved system for diffuse optical spectroscopy. *Opt. Expr.* 15:14482–14487.

Blackmore, K.M., J.A. Knight, and L. Lilge. 2008. Association between transillumination breast spectroscopy and quantitative mammographic features of the breast. *Cancer Epidemiol. Biomarkers Prevent.* 17:1043–1050.

Boyd, N.F., L.J. Martin, J. Stone, C. Greenberg, S. Minkin, and M.J. Yaffe. 2001. Mammographic densities as a marker of human breast cancer risk and their use in chemoprevention. *Curr. Oncol. Rep.* 3:314–321.

Cerussi, A., D. Hsiang, N. Shah et al. 2007. Predicting response to breast cancer neoadjuvant chemotherapy using diffuse optical spectroscopy. *PNAS* 104:4014–4019.

Chance, B., J.S. Leigh, H. Miyake et al. 1988. Comparison of time-resolved and -unresolved measurements of deoxyhemoglobin in brain. *Proc. Natl. Acad. Sci. USA* 85:4971–4975.

Choe, R., S.D. Konecky, A. Corlu et al. 2009. Differentiation of benign and malignant breast tumors by *in-vivo*

three-dimensional parallel-plate diffuse optical tomography. *J. Biomed. Opt.* 14:024020.

Chung, S.H., A.E. Cerussi, C. Klifa et al. 2008. *In vivo* water state measurements in breast cancer using broadband diffuse optical spectroscopy. *Phys. Med. Biol.* 53:6713–6727.

Corlu, A., R. Choe, T. Durduran et al. 2005. Diffuse optical tomography with spectral constraints and wavelength optimization. *Appl. Opt.* 44: 2082–2093.

Corlu, A., R. Choe, T. Durduran et al. 2007. Three-dimensional *in-vivo* fluorescence diffuse optical tomography of breast cancer in humans. *Opt. Expr.* 15:6696–6716.

Cubeddu, R., A. Pifferi, P. Taroni, A. Torricelli, and G. Valentini. 1996. Imaging of optical inhomogeneities in highly diffusive media: Discrimination between scattering and absorption contributions. *Appl. Phys. Lett.* 69: 4162–4164.

Culver, J.P., R. Choe, M.J. Holboke et al. 2003. Three-dimensional diffuse optical tomography in the parallel plane transmission geometry: Evaluation of a hybrid frequency domain/continuous wave clinical system for breast imaging. *Med. Phys.* 30:235–247.

Cutler, M. 1929. Transillumination of the breast, *Surg. Gynecol. Obstet.* 48:721–727.

Deapen, D., L. Liu, C. Perkins, L. Bernstein, and R.K. Ross. 2002. Rapidly rising breast cancer incidence rates among Asian-American women. *Int. J. Cancer* 99:747–750.

Dehghani, H., B.W. Pogue, P.P. Poplack, and K.D. Paulsen. 2003. Multiwavelength three-dimensional near-infrared tomography of the breast: Initial simulation, phantom and clinical results. *Appl. Opt.* 42:135–145.

Delpy D.T., M. Cope, P. van der Zee, S.R. Arridge, S. Wray, and J.S. Wyatt. 1988. Estimation of optical pathlength through tissue from direct time of flight measurement. *Phys. Med. Biol.* 33:1433–1442.

Durduran, T., R. Choe, W.B. Baker, and A.G. Yodh. 2010. Diffuse optics for tissue monitoring and tomography. *Rep. Prog. Phys.* 73:076701.

Eames, M.E., J. Wang, B.W. Pogue, and H. Dehghani. 2008. Wavelength band optimization in spectral near-infrared optical tomography improves accuracy while reducing data acquisition and computational burden. *J. Biomed. Opt.* 13:054037.

Fang, Q., J. Selb, S.A. Carp et al. 2011. Combined optical and x-ray tomosynthesis breast imaging. *Radiology* 258:89–97.

Fantini, S., M.A. Franceschini, G. Gaida et al. 1996. Frequency-domain optical mammography: Edge effect corrections. *Med. Phys.* 23:149–157.

Farrell T.J., M.S. Patterson, and B.C. Wilson. 1997. A diffusion theory model of spatially resolved, steady-state diffuse reflectance for the non-invasive determination of tissue optical properties *in vivo*. *Med. Phys.* 19:879–888.

Ferlay, J., D.M. Parkin, and E. Steliarova-Foucher. 2010a. Estimates of cancer incidence and mortality in Europe in 2008. *Eur. J. Cancer*, 46:765–781.

Ferlay, J., H.R. Shin, F. Bray, D. Forman, C. Mathers, and D.M. Parkin. 2010b. GLOBOCAN2008v1.2, Cancer Incidence and Mortality Worldwide: IARC Cancer Base No. 10 [Internet] Lyon, France: International Agency for Research on Cancer. Available from: http://globocan.iarc.fr.

Fishkin J.B., P.T.C. So, A.E. Cerussi, S. Fantini, M.A. Franceschini, and E. Gratton. 1995. Frequency-domain method measuring spectral properties in multiple-scattering media: Methemoglobin absorption spectrum in a tissue-like phantom. *Appl. Opt.* 34:1143–1155.

Floery D., T.H. Helbich, C.C. Riedl et al. 2005. Characterization of benign and malignant breast lesions with computed tomography laser mammography (CTLM): Initial experience. *Invest. Radiol.* 40:328–335.

Geslien, G.E., J.R. Fisher, and C. De Laney. 1985. Transillumination in breast cancer detection: Screening failures and potential. *AJR* 144:619–622.

Gisvold, J.J., L.R. Brown, R.G. Swee, D.J. Raygor, N. Dickerson, and M.K. Ranfranz. 1986. Comparison of mammography and transillumination light scanning in the detection of breast lesions. *AJR* 147:191–194.

Götz, L., S.H. Heywang-Köbrunner, O. Schütz, and H. Siebold, 1998. Optical mammography on preoperative patients (Optische Mammographie an präoperativen Patientinnen). *Akt. Radiol.* 8:31–33.

Grosenick, D., K.T. Moesta, M. Möller et al. 2005a. Time-domain scanning optical mammography: I. Recording and assessment of mammograms of 154 patients. *Phys. Med. Biol.* 50:2429–2449.

Grosenick, D., H. Wabnitz, K.T. Moesta, J. Mucke, P.M. Schlag, and H. Rinneberg. 2005b. Time-domain scanning optical mammography: II. Optical properties and tissue parameters of 87 carcinomas. *Phys. Med. Biol.* 50:2451–2468.

Grosenick, D., H. Wabnitz, H.H. Rinneberg, K.T. Moesta, and P.M. Schlag. 1999. Development of a time-domain optical mammograph and first *in vivo* applications. *Appl. Opt.* 38:2927–2943.

Guo, Y.P., L.J. Martin, W. Hanna et al. 2001. Growth factors and stromal matrix protein associated with mammographic densities. *Cancer Epidemiol. Biomarkers Prev.* 10:243–248.

Hagen, A., D. Grosenick, R. Macdonald et al. 2009. Late-fluorescence mammography assesses tumor capillary permeability and differentiates malignant from benign lesions. *Opt. Expr.* 17:17016–17033.

Hall D.J., J.C. Hebden, and D.T. Delpy. 1997. Evaluation of spatial resolution as a function of thickness for time-resolved optical imaging of highly scattering media. *Med. Phys.* 24:361–368.

Hebden, J.C., T.D. Yates, A. Gibson et al. 2005. Monitoring recovery after laser surgery of the breast with optical tomography: A case study. *Appl. Opt.* 44:1898–1904.

Heffer, E., V. Pera, O. Schütz et al. 2004. Near-infrared imaging of the human breast: Complementing hemoglobin concentration maps with a color-coded display of hypoxic areas. *J. Biomed. Opt.* 9:1152–1160.

Hylton, N. 2005. Magnetic resonance imaging of the breast: Opportunities to improve breast cancer management. *J. Clin. Oncol.* 23:1678–1684.

Intes, X. 2005. Time-domain optical mammography SoftScan: Initial results. *Acad. Radiol.* 12:934–947.

Intes X., J. Ripoll, Y. Chen, S. Nioka, B. Chance, and A.G. Yodh. 2003. *In vivo* continuous-wave optical breast imaging enhanced with Indocyanine Green. *Med. Phys.* 30:1039–1047.

Ishimaru, A. 1978. *Wave Propagation and Scattering in Random Media.* New York, Academic Press, 175–190.

Jacques, S.L. 1989. Time resolved propagation of ultrashort laser pulses within turbid tissues. *Appl. Opt.* 28:2223–2229.

Jiang, H., N.V. Iftimia, Y. Xu, J.A. Eggert, L.L. Fajardo, and K.L. Klove. 2002. Near-infrared optical imaging of the breast with model-based reconstruction. *Acad. Radiol.* 9:186–194.

Kienle A., L. Lilge, A. Vitkin et al. 1996. Why do veins appear blue? A new look at an old question. *Appl. Opt.* 35:1151–1160.

Kienle A. and M.S. Patterson. 1997. Determination of the optical properties of semi-infinite turbid media from frequency domain reflectance close to the source. *Phys. Med. Biol.* 42:1801–1819.

Kolb, T.M., J. Lichy, and J.H. Newhouse. 2002. Comparison of the performance of screening mammography, physical examination and breast US and evaluation of factors that influence them: An analysis of 27, 825 patient evaluations. *Radiology* 225:165–175.

Konecky, S.D., R. Choe, A. Corlu et al. 2008. Comparison of diffuse optical tomography of human breast with whole-body and breast-only positron emission tomography. *Med. Phys.* 35:446–455.

Kou, L., D. Labrie, and P. Chylek. 1993. Refractive indices of water and ice in the 0.65–2.5 μm spectral range. *Appl. Opt.* 32:3531–3540.

Kukreti, S., A. Cerussi, B. Tromberg, and E. Gratton. 2007. Intrinsic tumor biomarkers revealed by novel double differential spectroscopic analysis of near-infrared spectra. *J. Biomed. Opt.* 12:020509.

Lakowicz J.R. and K.W. Berndt. 1990. Frequency-domain measurements of photon migration in tissues. *Chem. Phys. Lett.* 166:246–252.

Leff, D.R., O.J. Warren, L.C. Enfield et al. 2008. Diffuse optical imaging of the healthy and diseased breast: A systematic review. *Breast Cancer Res. Treat.* 108:9–22.

Li, C., H. Zhao, B. Anderson, and H. Jiang. 2006. Multispectral breast imaging using a ten-wavelength, 64×64 source/detector channels silicon photodiode-based diffuse optical tomography system. *Med. Phys.* 33:627–636.

Lin, S.-P., L. Wang, S.L. Jacques, and F.K. Tittel. 1997. Measurement of tissue optical properties by the use of oblique-incidence optical fiber reflectometry. *Appl. Opt.* 36:136–143.

Liu H., M. Miwa, B. Beauvoit, G.N. Wang, and B. Chance. 1993. Characterization of absorption and scattering properties of small-volume biological samples using time-resolved spectroscopy. *Anal. Biochem.* 213:378–385.

Lord, S.J., W. Lei, P. Craft et al. 2007. A systematic review of the effectiveness of magnetic resonance imaging (MRI) as an addition to mammography and ultrasound in screening young women at high risk of breast cancer. *Eur. J. Cancer* 43:1905–1917.

Möller, M., H. Wabnitz, A. Kummrow et al. 2003. A four-wavelength multi-channel scanning time-resolved optical mammograph. *Proc. SPIE* 5138:290–297.

Marshall, E. 2010. Brawling over mammography. *Science* 327:936–938.

Martelli, F., S. del Bianco, A. Ismaelli, and G. Zaccanti. 2009. *Light Propagation through Biological Tissue and Other Diffusive Media: Theory, Solutions, and Software.* Bellingham: SPIE Press.

Moon, J.A., P.R. Battle, M. Bashkansky, R. Mahon, M.D. Duncan, and J. Reintjes. 1996. Achievable spatial resolution of time-resolved transillumination imaging systems which utilize multiple scattered light. *Phys. Rev. E* 53:1142–1155.

Mourant, J.R., T. Fuselier, J. Boyer, and I.J. Bigio. 1997. Predictions and measurements of scattering and absorption over broad wavelength ranges in tissue phantoms. *Appl. Opt.* 36:949–957.

Nielsen, T., B. Brendel, R. Ziegler et al. 2009. Linear image reconstruction for a diffuse optical mammography system in a noncompressed geometry using scattering fluid. *Appl. Opt.* 10:D1–D13.

Nilsson, A.M., K.C. Sturesson, D.L. Liu, and S. Andersson-Engels. 1998. Changes in spectral shape of tissue optical properties in conjuction with laser-induced thermotheraphy. *Appl. Opt.* 37:1256–1267.

Nomura, Y., O. Hazeki, and M. Tamura. 1997. Relationship between time-resolved and non-time-resolved Beer-Lambert law in turbid media. *Phys. Med. Biol.* 42:1009–1023.

Nothacker, M., V. Duda, M. Hahn et al. 2009. Early detection of breast cancer: Benefits and risks of supplemental breast ultrasound in asymptomatic women with mammographically dense breast tissue. A systematic review. *BMC Cancer* 9:335.

Ntziachristos, V., X.H. Ma, and B. Chance. 1998. Time-correlated single photon counting imager for simultaneous magnetic resonance and near-infrared mammography *Rev. Sci. Instrum.* 69: 4221–4233.

Ntziachristos, V., A.G. Yodh, M. Schnall, and B. Chance. 2000. Concurrent MRI and diffuse optical tomography of breast after indocyanine green enhancement. *Proc. Natl. Acad. Sci. USA.* 97:2767–2772.

Ntziachristos, V., A.G. Yodh, M.D. Schnall, and B. Chance. 2002. MRI-guided diffuse optical spectroscopy of malignant and benign breast lesions. *Neoplasia* 4:347–354.

Patterson, M.S., J.D. Moulton, B.C. Wilson, K.W. Berndt, and J.R. Lakowicz. 1991. Frequency-domain reflectance for the determination of the scattering and absorption properties of tissue. *Appl. Opt.* 30:4474–4476.

Pifferi, A., P. Taroni, A. Torricelli, F. Messina, R. Cubeddu, and G. Danesini. 2003. Four-wavelength time-resolved optical mammography in the 680–980 nm range. *Opt. Lett.* 28:1138–1140.

Pifferi, A., A.Torricelli, P. Taroni, D. Comelli, A. Bassi, and R. Cubeddu. 2007. Fully automated time domain spectrometer for the absorption and scattering characterization of diffusive media. *Rev. Sci. Instrum.* 78:053103.

Poellinger, A., S. Burock, D. Grosenick et al. 2011. Breast cancer: Early- and late-fluorescence near-infrared imaging with indocyanine green—A preliminary study. *Radiology* 258:409–416.

Pogue, B.W. and E. Patterson. 1994. Frequency-domain optical absorption spectroscopy of finite tissue volumes using diffusion theory. *Phys. Med. Biol.* 39:1157–1180.

Prahl, S., Oregon Medical Laser Center website, 2011, (http://omlc.ogi.edu/spectra/water/index.html).

Roblyer, D., S. Ueda, A. Cerussi et al. 2011. Optical imaging of breast cancer oxyhemoglobin flare correlates with neoadjuvant chemotherapy response one day after starting treatment. *Proc. Natl. Acad. Sci. USA* 108:14626–14631.

Sevick, E.M., B. Chance, J.S. Leigh, S. Nioka, and M. Maris. 1991. Quantitation of time- and frequency-resolved optical spectra for the determination of tissue oxygenation. *Anal. Biochem.* 195:330–351.

Sevick E.M., J.K. Frisoli, C.L. Burch, and J.R. Lakowicz. 1994. Localization of absorbers in scattering media by use of frequency-domain measurements of time-dependent photon migration. *Appl. Opt.* 33:3562–3570.

Soliman, H., A. Gunasekara, M. Rycroft et al. 2010. Functional imaging using diffuse optical spectroscopy of neoadjuvant chemotherapy response in women with locally advanced breast cancer. *Clin. Cancer Res.* 16:2605–2614.

Spinelli, L., A. Torricelli, A. Pifferi, P. Taroni, G. Danesini, and R. Cubeddu. 2003. Experimental test of a perturbation model for time-resolved imaging in diffusive media. *Appl. Opt.* 42:3145–3153.

Spinelli, L., A. Torricelli, A. Pifferi, P. Taroni, G. Danesini, and R. Cubeddu. 2005. Characterisation of female breast lesions from multi-wavelength time-resolved optical mammography. *Phys. Med. Biol.* 50:2489–2502.

Srinivasan, S., C.M. Carpenter, H.R. Ghadyani et al. 2010. Image guided near-infrared spectroscopy of breast tissue *in vivo* using boundary element method. *J. Biomed. Opt.* 15:061703.

Tabar, L., M.-F. Yen, B. Vitak et al. 2003. Mammography service screening and mortality in breast cancer patients: 20-year follow-up before and after introduction of screening. *Lancet* 361:1405–1410.

Taroni, P., A. Bassi, D. Comelli, A. Farina, R. Cubeddu, and A. Pifferi. 2009a. Diffuse optical spectroscopy of breast tissue extended to 1100 nm. *J. Biomed. Opt.* 14:054030.

Taroni, P., D. Comelli, A. Pifferi, A. Torricelli, and R. Cubeddu. 2007. Absorption of collagen: Effects on the estimate of breast composition and related diagnostic implications. *J. Biomed. Opt.* 12:014021.

Taroni, P., G. Danesini, A. Torricelli, A. Pifferi, L. Spinelli, and R. Cubeddu. 2004. Clinical trial of time-resolved scanning optical mammography at 4 wavelengths between 683 and 975 nm. *J. Biomed. Opt.* 9:464–473.

Taroni, P., A. Pifferi, E. Salvagnini, L. Spinelli, A. Torricelli, and R. Cubeddu. 2009b. Seven-wavelength time-resolved optical mammography extending beyond 1000 nm for breast collagen quantification. *Opt. Expr.* 17:15932–15946.

Taroni, P., A. Pifferi, G. Quarto et al. 2010. Non-invasive assessment of breast cancer risk using time-resolved diffuse optical spectroscopy. *J. Biomed. Opt.* 15:060501.

Taroni, P., A. Torricelli, L. Spinelli et al. 2005. Time-resolved optical mammography between 637 and 985 nm: Clinical study on the detection and identification of breast lesions. *Phys. Med. Biol.* 50:2469–2488.

Torricelli, A., A. Pifferi, P. Taroni, E. Giambattistelli, and R. Cubeddu. 2001. In vivo optical characterisation of human tissues from 610 to 1010 nm by time-resolved reflectance spectroscopy. *Phys. Med. Biol.* 46:2227–2237.

Ueda, Y., K. Yoshimoto, E. Ohmae et al. 2011. Time-resolved optical mammography and its preliminary clinical results. *Technol. Cancer Res. Treat.* 10:393–401.

van de Ven, S.M.W.Y., Optical Breast Imaging. 2011. Thesis, Utrecht University, ISBN: 978-90-393-5509-1.

van Veen, R.L.P., H.J.C.M. Sterenborg, A. Pifferi, A. Torricelli, E. Chikoidze, and R. Cubeddu. 2005. Determination of visible near-IR absorption coefficients of mammalian fat using time- and spatially resolved diffuse reflectance and transmission spectroscopy. *J. Biomed. Opt.* 10:054004.

van de Ven, S., A. Wiethoff, T. Nielsen et al. 2010. A novel fluorescent imaging agent for diffuse optical tomography of the breast: First clinical experience in patients. *Mol. Imaging Biol.* 12:343–348.

Wang, L. and S.L. Jacques. 1995. Use of a laser beam with an oblique angle of incidence to measure the reduced scattering coefficient of a turbid medium. *Appl. Opt.* 34:2362–2366.

Wang, J., S. Jiang, Z. Li, R.M. diFlorio-Alexander, R.J. Barth, P.A. Kaufman, B.W. Pogue, and K.D. Paulsen. 2010. *In vivo* quantitative imaging of normal and cancerous breast tissue using broadband diffuse optical tomography. *Med. Phys.* 37:3715–3724.

Wilson, B.C., E.M. Sevick, M.S. Patterson, and K.P. Chan. 1992. Time-dependent optical spectroscopy and imaging for biomedical applications. *Proc. IEEE* 80:918–930.

Wu, D. and S.S. Gambhir. 2003. Positron emission tomography in diagnosis and management of invasive breast cancer: Current status and future perspectives. *Clin. Breast Cancer* 4:S55–S63.

Xu, R.X., D.C. Young, J.J. Mao, and S.P. Povoski. 2007. A prospective pilot clinical trial evaluating the utility of a dynamic near-infrared imaging device for characterizing suspicious breast lesions. *Breast Cancer Res.* 9:R88.

Yates, T., J.C. Hebden A. Gibson, N. Everdell, S.R. Arridge, and M. Douek 2005. Optical tomography of the breast using a multi-channel time-resolved imager. *Phys. Med. Biol.* 50:2503–2517.

Yodh, A.G. and Chance B. 1995. Spectroscopy and imaging with diffusing light. *Phys. Today* 48:34–40.

Zaccanti, G. and P. Donelli. 1994. Attenuation of energy in time-gated transillumination imaging: Numerical results. *Appl. Opt.* 33:7023–7030.

Zhu, Q., M. Huang, N. Chen, K. Zarfos, B. Jagjivan, M. Kane, P. Hedge, and S.H. Kurtzman. 2003. Ultrasound-guided optical tomographic imaging of malignant and benign breast lesions: Initial clinical results of 19 cases. *Neoplasia* 5:379–388.

Zhu, Q., S. Tannenbaum, P. Hegde, M. Kane, C. Xu, and S.H. Kurtzman. 2008. Noninvasive monitoring of breast cancer during neoadjuvant chemotherapy using optical tomography with ultrasound localization. *Neoplasia* 10:1028–1040.

10
Diffuse Optical Imaging: Application to Brain Imaging

Jeremy C. Hebden
University College London

10.1 Introduction

The development of near-infrared spectroscopy (NIRS) as a tool for continuous monitoring of cerebral oxygenation and hemodynamics noninvasively was prompted by the pioneering work of Jöbsis who showed that the oxygenated status of cytochrome aa$_3$, an enzyme in the oxidative metabolic pathway, could be determined using measurements of diffusely transmitted light [1]. This is due to the fact that cytochrome aa$_3$, like hemoglobin, has oxygenated and deoxygenated forms with different characteristic absorption spectra in the visible and near-infrared wavelength range. While hemoglobin, which is only present in red blood cells, provides an indicator of blood oxygenation, cytochrome aa$_3$ provides an indicator of tissue oxygenation. Although NIRS is often still used to monitor cytochrome, its most effective and popular application is as a tool for measuring the changes in the cerebral concentrations of oxyhemoglobin [HbO$_2$] and deoxyhemoglobin [Hb].

Although the relatively low attenuation of near-infrared light in most tissues facilitates transmittance measurements over distances of several centimeters, the profound scatter of light renders those measurements sensitive to a much larger volume of tissue than that occupying a direct line between the source and the detector. The light traveling between two points on the surface of a uniform diffusing medium separated by a few centimeters has typically explored a banana-shaped volume of tissue, often referred to as the photon measurement density function (PMDF; see Figure 10.1) [2]. The breadth of the PMDF limits the spatial resolution achievable using the diffuse optical

imaging (DOI) methods, although as discussed later, gains can be achieved by measuring more than just the intensity of the transmitted light, and by a judicious combination of multiple measurements with overlapping PMDFs. Note that due to the dominance of scatter, the terms "transmitted" and "reflected" are somewhat arbitrary when applied to optical measurements on the head, although the latter term is generally used when the separation between the source and the detector is less than about one-quarter of the circumference.

During the past 20 years, DOI has evolved as a tool for imaging changes in blood volume and oxygenation in the brain as a result of developments in two areas of technology. First is the availability of near-infrared sources and sensitive detectors, including low-cost portable devices. Second is the development

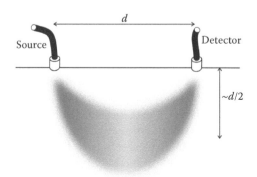

FIGURE 10.1 The photon measurement density function (PMDF): The volume of tissue sampled by detected photons.

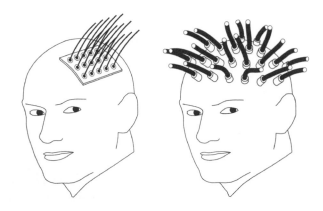

FIGURE 10.2 Typical arrangements of optodes for performing (a) optical topography and (b) optical tomography of the brain.

of sophisticated algorithms for reconstructing images from measurements of diffuse light through tissue, associated with the availability of powerful computers able to model photon migration through large thicknesses of biological tissues. Subsequently, two DOI approaches have emerged, known as optical topography and optical tomography, illustrated in Figure 10.2. Although the distinction between them is somewhat arbitrary, the term "optical topography" is used to describe the techniques, which provide maps of changes in hemodynamic parameters (such as blood volume) close to the surface, with little or no depth resolution, while optical tomography involves reconstructing an image representing the three-dimensional (3D) distribution of optical properties, or a transverse slice across the 3D volume. The distinction has evolved because of the differences in the complexity of the imaging problem and of the associated instrumentation.

The directly measured quantities employed in DOI are commonly known as datatypes, of which transmitted (or reflected) intensity is generally the easiest to acquire. Changes in intensity, for example, at two or more wavelengths can be converted into the estimates of changes in [HbO$_2$] and [Hb], usually invoking some assumptions about the optical properties of the sampled tissue volume. However, an accurate assessment of absolute values of [HbO$_2$] and [Hb] requires knowledge of the average pathlength taken by the detected light, which can be measured using time-domain or frequency-domain instruments. Time-domain systems measure flight times of individual photons, while frequency-domain systems measure the modulation depth and phase delay of light transmitted through the tissue illuminated by an intensity-modulated source. The combination of intensity with additional datatypes such as mean flight time also enables the effects of absorption to be distinguished from those of scatter [3].

DOI methods often present images representing physical parameters such as absorption or scatter coefficient. However, with appropriate datatypes at sufficient wavelengths, DOI methods can in principle generate images representing spatial variation in physiological parameters such as [HbO$_2$] and [Hb], concentration of water and lipids, and tissue oxygenation = [HbO$_2$]/([HbO$_2$] + [Hb]).

As presented in Chapter 8, reconstructing datatypes into images represents a challenging inverse problem that is nonlinear,

severely ill-posed, and generally underdetermined. Finding reliable and unique solutions for a given set of data involves considerable theoretical and practical difficulties, and various compromises and approximations are unavoidable. For many DOI applications, it is common to record changes in datatypes over time. For example, intensity or mean flight time can be measured before and during an invoked or naturally evolving alteration in the physiological state of the tissue. There are two very considerable advantages stemming from the use of datatype differences. First, it avoids the requirement to fully calibrate the measuring system. For example, absolute values of transmitted intensity are notoriously dependent on coupling of light into and out of the tissue, which in many practical circumstances is highly uncontrollable and variable. However, such coupling can usually be assumed to be constant over the duration of a measurement-rendering intensity ratios (or differences in log intensity) as dependent only on absorption or scatter changes occurring within the interrogated tissue. Second, the image reconstruction can usually be reduced to a relatively straightforward linear problem, yielding images representing the change in a physical or physiological parameter. This approximation is illustrated below, by representing the forward problem by the following equation:

$$y = F(x), \tag{10.1}$$

where y is the data, x is the set of optical properties to be reconstructed, and F is an operator representing the model of light propagation in the medium. The inverse problem may be expressed as

$$x = F^{-1}(y). \tag{10.2}$$

Equation 10.1 can be expanded about a fixed value x_0 in a Taylor series:

$$y = y_0 + \frac{dF(x_0)}{dx} \cdot (x - x_0) + \frac{d^2F(x_0)}{dx^2} \cdot (x - x_0)^2 + \cdots, \tag{10.3}$$

where $y_0 = F(x_0)$. If the true optical properties x are close to an initial estimate x_0 and the measured data y are close to the modeled data generated from that estimate, y_0, then the forward problem can be linearized by neglecting the higher terms. This may be expressed as

$$\Delta y = \mathbf{J}\,\Delta x, \tag{10.4}$$

where $\Delta y = y - y_0$, $\Delta x = x - x_0$, and $\mathbf{J}$ is the matrix form of the first derivative of $F(x_0)$, known as the Jacobian. Image reconstruction therefore involves inverting the matrix $\mathbf{J}$. As described in Chapter 8, regularization is required to stabilize the inversion against ill-conditioning caused by the ill-posedness of the inverse problem. The matrix $\mathbf{J}$ can be calculated from the forward model based on a "best guess" of the distributions of optical properties.

The linear "difference imaging" approach is readily applied to the study of the hemodynamic response of the brain due to sensory stimuli and mental processing, and many of the research applications of DOI are currently focused on such studies. The activated areas of the brain are generally quite superficial, and therefore easily interrogated optically with sources and detectors coupled to the scalp. Nevertheless, the linear approach is clearly inappropriate if the measured data does not represent a difference between two similar states, or if the "best guess" is insufficiently representative of the true optical properties. In this case, a nonlinear approach must be used, which involves iteratively updating the estimate of **J** by replacing the "best guess" with the distributions derived from the previous iteration. This is described in detail in Chapter 8. Detailed reviews of DOI theory and its applications have been published by Arridge and Schotland [4], Gibson et al. [5], and Hillman [6].

10.2 Optical Topography

The imaging technique known as optical topography involves measuring diffusely reflected light between multiple pairs of sources and detectors at discrete locations spread over the area of the tissue of interest. The principal motivation for the development of optical topography has been the display of hemodynamic activity in the cerebral cortex. The source–detector separations are deliberately small enough (a few cm or less) to ensure that measurements with good signal-to-noise ratio (SNR) can be acquired quickly, enabling images to be displayed in real time at a rate of a few Hertz or faster.

The earliest attempts at optical topography of the brain were performed using several single-channel near-infrared spectrometers. For example, Gratton et al. [7] employed an NIRS instrument to generate two-dimensional (2D) maps of the adult head by acquiring the measurements at a series of points on the head sequentially. This enabled them to study the response of the adult brain to repeated visual stimulation with an effective temporal resolution of 50 ms. The first demonstration of *simultaneous* optical mapping of human brain activity was presented by Hoshi and Tamura [8], who employed five NIRS systems in parallel to detect the region-specific changes in hemoglobin saturation and blood volume in the adult brain during visual and auditory stimulation and in response to various mental tasks. More recently, dedicated instruments have been built for brain imaging that consist of an array of distinct sources and detectors coupled to the head, usually via optical fibers or fiber bundles often known as "optodes." To acquire the data quickly, sources are activated simultaneously. This requires that the detected signal from each source is uniquely identifiable, which is achieved by modulating the intensity of each source at a different frequency. The detectors then operate in parallel, employing either lock-in amplifiers or digital processing methods to isolate the signal from each source. Since the addition of a source commonly involves little more than the cost of an extra fiber, topographic systems often employ more source fibers than detectors. Several commercial optical topography systems are available, such as the ETG-4000 system developed by Hitachi Medical Corporation (Japan). This system samples data from up to 52 discrete source–detector pairs at two wavelengths (695 and 830 nm) at a rate of 10 Hz.

A data-sampling rate of 10 Hz is typical of most systems, and determines the temporal resolution. Although this is more than sufficient for most applications that attempt to measure the hemodynamic responses characterized by time courses extending over several seconds (Section 10.4), some optical techniques have been used to record a so-called fast event-related optical signal, where changes observed over a few hundred milliseconds are considered to correspond directly with the neuronal activity associated with information processing [9].

The probes developed for optical topography systems typically consist of a flexible pad that supports an array of optodes, each coupled to either a source (e.g., a laser diode) or a detector (e.g., avalanche photodiode). Securing a probe safely, securely, and comfortably to the head, while minimizing the adverse effects of hair, has always been one of the greatest technical challenges in the implementation of DOI techniques, and a broad variety of designs have been implemented. Not surprisingly, much of the pioneering development work has been performed on bald male volunteers.

The physical separations between sources and detectors on a probe largely dictate the depth sensitivity, or "penetration depth" of the measurement. As a rough guide, the mean depth sampled by a given PMDF in biological tissue is about half the source–detector distance (see Figure 10.1). Thus, to sample brain tissues 15 mm below the scalp requires the measurements of diffusely reflected light using sources and detectors separated by approximately 30 mm. However, the larger the separation, the weaker the signal, and the longer the period of time over which measurement needs to be integrated in order to achieve sufficient SNR. The ideal probe employs an arrangement of sources and detectors whose combination of PMDFs covers as much of the underlying volume of tissue of interest as possible, down to a desired penetration depth. For studies of the adult cortex, the sampled volume will include the tissue superficial to the brain, such as scalp and skull, and sensitivity to these regions cannot be eliminated.

The spatial resolution achievable using optical topography is difficult to characterize in general terms because of its strong dependence on the shape and size of PMDFs for each active source–detector combination, and on their degree of overlap. Immediately below a source or detector, the PMDFs are the narrowest, and spatial resolution is the highest. To achieve good lateral resolution close to the surface, a high density of sources and detectors is required. Further below the surface, the PMDFs are broader, and spatial resolution is worse. However, the resolution at a given location within the sampled volume can be enhanced by sampling using multiple, overlapping PMDFs. An imaging algorithm that utilizes signals from overlapping PMDFs can identify the locations of regions of changing optical properties with a precision better than the width of a single PMDF. It is important to note that PMDFs are dependent on the datatype and on the optical properties of the medium. For example, PMDFs for mean-time measurements are narrower than those for intensity measurements [2]. Furthermore, PMDFs are generally narrower and interrogate less deeply in higher absorbing media, leading to smaller penetration depth but higher spatial resolution.

For brain imaging, one must also consider the influence of the layer of the cerebrospinal fluid (CSF) that occupies the subarachnoid space between the skull and the brain. CSF is predominantly water, and therefore low-absorbing and minimally scattering. Computer models of photon migration in the vicinity of the CSF layer have indicated that it has the effect of reducing the overall penetration of light into the brain, particularly for adult studies where the layer is thicker [10].

Optical topography has evolved into a particularly powerful research tool for brain studies that are not feasible or ethical using more established techniques such as functional magnetic resonance imaging (fMRI) or positron emission tomography (PET). An example described in Section 10.5 is the investigation of the developing brain. Babies and younger children are unlikely to remain sufficiently motionless in an MRI magnet for extended periods without sedation, whereas it is relatively easy to affix an optical probe to their scalp. Thus, it is important that optical topography measurements are as insensitive as possible to head motion, and that means are used to identify and reject data that are contaminated by movement artifact. In practice, motion of the probe relative to the head is usually characterized by very sudden jumps in intensity that are easily distinguished from comparatively slower signals with a physiological origin. Nevertheless, some investigators employ independent means of monitoring head motion to ensure that contaminated optical data are identified [11].

10.3 Optical Tomography

Optical tomography, or 3D imaging of tissue using measurements of diffusely transmitted light, is a very challenging problem. Breast imaging (see Chapter 9) rather than brain imaging has arguably made more progress toward clinical applications, where coupling of sources and detectors to the surface is easier, one can access a greater portion of the surface area, and the tissue is less severely attenuating. The challenge of optical tomography is to generate the spatial maps representing either the intrinsic optical properties (i.e., the absorption and scattering coefficients) at one or more wavelengths, chromophore concentrations, or physiological parameters.

Optical tomography of the brain requires acquiring measurements of light transmitted between discrete sources and detectors located over as much of the available surface of the tissue as realistically achievable, which in practice is limited to the scalp and forehead. As for optical topography, hair represents a significant obstacle. Imaging deep below the surface requires that systems employ large source–detector separations, and the consequent high attenuation of light requires powerful sources (consistent with patient safety) and sensitive (e.g., photon-counting) detectors.

The attenuation of near-infrared light transmitted across adult brain tissues is too high to access the center of the brain, and optical tomography is limited to partial imaging of superficial tissues, which is sufficient for functional imaging studies of the cortex (Section 10.4). However, it is feasible to obtain the measurements across the full thickness of a newborn (especially premature) infant brain, and optical tomography images of the entire infant brain have been obtained (Sections 10.5 and 10.7). This is possible because of the lower intrinsic scattering by infant brain matter (due to lower myelination) as well as the smaller head size.

The deeper one attempts to image using optical tomography, the more time is required to acquire the data with a sufficient SNR. Consequently, optical tomography has a lower temporal resolution than the optical topography. Thus, the former is less suitable for imaging hemodynamic changes in brain tissue, which typically occur over timescales of a few seconds. Although 3D images of functional changes have been generated for both the adult and infant brain (Sections 10.4 and 10.5), the development of optical tomography has largely focused on the significantly more difficult problem of generating images that represent the 3D distribution of absolute values of optical properties. Such an approach is strongly reliant on robust image reconstruction algorithms to convert the measured datatypes into images [4]. Although various ad hoc backprojection methods have been proposed, such approaches are not applicable to imaging arbitrary-shaped volumes. Instead, it is necessary to utilize a model of the propagation of light within the tissue that allows a set of predicted measurements on the surface to be derived for any given distribution of internal optical properties. Optical tomography is an inverse problem that is nonlinear, severely ill-posed, and generally underdetermined. Consequently, finding reliable and unique solutions for a given set of data involves considerable theoretical and practical difficulties, and various compromises and approximations are unavoidable. The reconstruction problem can be significantly alleviated by utilizing prior information, such as knowledge of the external geometry and/or the locations of internal boundaries between different types of tissues. For this reason, the combination of optical tomography with anatomical imaging techniques such as MRI has been explored, particularly for breast imaging (see Chapter 17).

The highly sensitive (photon-counting) systems typically employed for optical tomography of the brain enable penetration depths to exceed 5 or 6 cm. The spatial resolution inevitably decreases with depth, although for measurements between one side of an infant's head and the other, the relatively high absorption by blood suppresses the width of the PMDF, enabling a spatial resolution of about 2 cm to be achieved for features near the center of the neonatal brain.

10.4 Imaging Evoked Hemodynamic Response to Sensory Stimuli and Cognitive Activity

The so-called blood oxygen level-dependent (BOLD) fMRI exploits the fact that the magnetic susceptibility of blood depends on the oxygenation status of the hemoglobin. The stimulation of a region of the brain results in additional metabolic

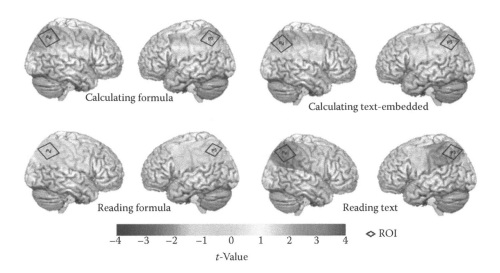

Calculating formula

Calculating text-embedded

Reading formula

Reading text

$-4 \quad -3 \quad -2 \quad -1 \quad 0 \quad 1 \quad 2 \quad 3 \quad 4$ ◇ ROI

t-Value

FIGURE 10.3 (**See color insert.**) Changes in [HbO$_2$] during arithmetical tasks presented as either formulae (left-hand side) or as a sentence of text (right-hand side). The top row shows the response when solving the tasks, and the bottom row shows the response when just reading. Data is displayed as *t*-values resulting from a statistical *t*-test analysis, and superimposed on a generic brain. (Reproduced with permission from Richter, M. M. et al. 2009. *Journal of Neural Transmission* 116:267–273.)

demand, which produces a measurable increase in local blood volume and/or a change in blood oxygenation. BOLD contrast results from a localized change in [Hb]. The potential of DOI techniques as functional imaging tools depends on several significant advantages over fMRI: they offer greater temporal resolution and sensitivity, are portable, less expensive, less influenced by moderate head movement, and are sensitive to changes in both [Hb] and [HbO$_2$].

The regions of the brain associated with processing of sensory information are generally cortical areas located within a couple of centimeters of the scalp in the adult head, and assessment of the evoked response of the brain to sensory stimuli has been a major focus of research with optical topography. Imaging probes typically consist of arrays of sources and detectors covering the area of the scalp immediately over the region of the cortex of interest. If the subject is not bald, then it is normally necessary to part the hair beneath each source/detector fiber to expose a small area of scalp. The imaging protocol typically involves recording the data during consecutive periods of stimulus and rest, each lasting a few tens of seconds, and then repeating and averaging the data as necessary to increase the SNR. Differences in the measured datatypes (typically just intensity) are used to display images that reveal the underlying changes in blood volume and/or oxygenation.

Although most systems could in principle display images in real time, it is more common to perform a significant amount of postprocessing of data to remove the artifacts due to movement, and unwanted signals such as those due to systematic variability associated with the cardiac and respiratory cycles [12].

DOI studies have been published which reveal the response of the healthy brain to a variety of visual, tactile, and motor stimuli, and to certain forms of cognitive activity, such as mental calculations and anagram solving. For example, Figure 10.3

shows changes in [HbO$_2$] recorded using the Hitachi ETG-4000 optical topography system during a study of the brain response to arithmetical tasks [13]. Simple two-digit addition tasks were presented to healthy adult volunteers as either numerical formulae or as a sentence of the text. A statistical *t*-test analysis of the recorded data was performed to display changes in [HbO$_2$] as *t*-values. Figure 10.3 shows the responses in the left and right hemispheres when reading the tasks and when solving the tasks. The response is greater when solving compared to just reading, and greater still during text-embedded tasks than during numeric tasks. The recorded regions of interest are shown superimposed on images of a generic brain surface.

Only a small minority of studies attempt a 3D reconstruction of optical properties below the probe, most investigators being content to display simple 2D maps of the underlying activity, or even just time courses sampled at specific locations on the scalp. However, some 3D images of hemodynamic changes in the adult brain have been attempted, such as changes in the contralateral motor cortex during hand movement [14], and within the frontal regions during the Valsalva maneuver [15]. In addition, there has been one published study of a full-brain imaging of evoked response in the newborn infant [16].

10.5 Developmental/Psychological Studies on Infants and Children

So far, the most successful and widespread medical translation of DOI technology has been in the field of quantitative psychology. Optical topography has evolved into a popular tool for the study of cognitive and psychological processes, and especially for developmental studies on infants and children. The

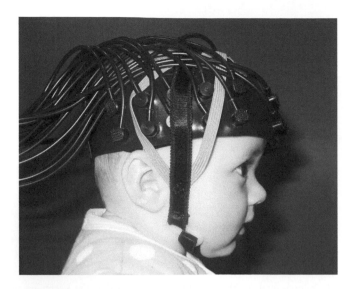

FIGURE 10.4 A probe for optical topography studies of the infant brain. (Reproduced with permission from Sarah Lloyd-Fox, Birkbeck, University of London.)

portability and silent function of the hardware make the technique particularly suitable for psychological studies on awake infants. The latter group are particularly difficult to study using fMRI where sustained periods without movement in an MRI magnet are required. Published DOI results include studies of face perception, language processing, memory, and object recognition; the application of DOI to studies of neurocognitive development in infants is reviewed by Minagawa-Kawai et al. [17]. Like for studies of sensory stimuli, the measurement of the hemodynamic response to mental processing typically requires extensive averaging from repeated episodes. The overwhelming technical challenge in developmental studies is coupling large numbers of optodes to a child's scalp safely and comfortably, in a way that does not distract the child's attention and which is not prone to significant artifact due to moderate head motion. Figure 10.4 shows a successful probe developed for psychological studies on young infants at Birkbeck College and University College London (UCL) [18].

10.6 Imaging Epilepsy

An epileptic seizure is a pathological transient neurological event that involves spontaneous and synchronous activation of many thousands of neurons. The accepted clinical tool for routinely diagnosing and monitoring abnormal electrical activity in the brain, such as epileptic seizures, is electroencephalography (EEG), which involves coupling an array of passive electrodes to the scalp and measuring the relative changes in electrical potential. While EEG is sensitive to the electromagnetic dynamics of groups of active neurons, optical imaging provides a means of measuring the associated changes in hemodynamic activity as a result of the neurons' increased metabolic demand. The changes in blood volume are expected to be significantly greater than those associated

with evoked response to sensory stimuli or mental processing, and thus data averaging may not be necessary, potentially enabling the hemodynamic activity associated with isolated seizures to be displayed in real time. The clinical potential of DOI techniques for imaging seizures has yet to be adequately explored, although optical topography has already been used to display the continuous changes in cerebral blood volume during seizures [19,20], and has been combined simultaneously with EEG [21,22].

10.7 Imaging Neonatal Brain Injury

One of the principal clinical motivations for the earliest work in NIRS was the diagnosis and monitoring of hypoxic–ischemic injury in newborn infants. This is caused by insufficient oxygen reaching the brain at or near the time of birth, and often leads to permanent brain injury, especially without appropriate intervention during the first few hours or days of life. Infants born very prematurely are particularly vulnerable to this condition, which results from insufficient development of their respiratory system. The need for a means of diagnosing hypoxic–ischemic injury and other neonatal cerebral pathologies noninvasively at the crib-side also motivated some of the earliest attempts at optical tomography of the brain. A research team at Stanford University acquired the first optical tomography images that revealed intracranial hemorrhage [23], and focal regions of low oxygenation after acute stroke [14]. Later, a faster system was employed at University College London to reconstruct 3D images of the whole infant brain that have revealed the incidence of intraventricular hemorrhage [24,25], and changes in blood volume and oxygenation in the brain-injured infant, induced by small alterations to ventilator settings [26]. This system was used to generate the images displayed in Figure 10.5, which shows coronal slices from 3D images of blood volume and oxygenation from a 34-week gestational age infant with left-sided intraventricular hemorrhage, and a cranial ultrasound scan for comparison [25]. A distinct area of desaturated hemoglobin is exhibited on the left-hand side. This is located more superficially than the hemorrhage indicated in the ultrasound scan and blood-volume image, suggesting the evidence of an ischemic penumbra surrounding the hemorrhage.

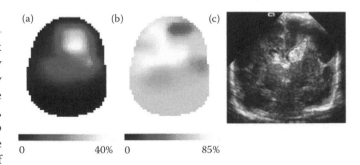

FIGURE 10.5 Coronal sections of infant brain images of (a) regional blood volume, (b) regional oxygen saturation, and (c) the corresponding ultrasound scan. (Reproduced with permission from Hebden, J. C., Austin, T. 2007. *European Radiology* 17:2926–2933.)

Unlike some other potential medical applications of DOI, such as optical mammography (Chapter 9), neonatal brain imaging has no directly competing imaging technology. Ultrasound is routinely used to identify brain hemorrhage in newborns, but cannot report on oxygenation. Although MRI can provide limited functional information, such as brain perfusion and diffusion, it is unsuitable for long-term monitoring, and transfer of the most critically ill infants into a magnet is difficult and involves risk. Nuclear imaging techniques, which involve the use of radioactive isotopes, are typically not approved for infant studies.

However, the application of DOI techniques to this most vulnerable of patient groups is an exceptional challenge for two very significant reasons. First, obtaining data is difficult. It requires large numbers of optodes to be coupled to a small and fragile head safely and without interfering with the routine clinical care and monitoring of the infant. Figure 10.6 shows the adjustable helmet developed at UCL to couple up to 32 optodes to the infant head [25]. The infants that are most likely to benefit from the diagnosis offered by an effective DOI system are those that have the smallest heads (due to their prematurity) and are surrounded by various life-support and monitoring equipment. To be effective at diagnosing hypoxic–ischemic injury at a stage where intervention can make a positive difference, the DOI method needs to be applied during the first few hours after birth. So far, however, studies have focused on older, more stable infants, where there is no risk of the study jeopardizing clinical care. The second significant difficulty is the demand made on the content and quality of the images. Generating

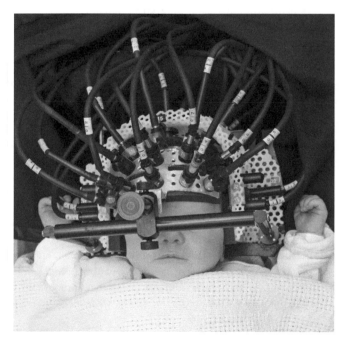

FIGURE 10.6   A probe for optical tomography studies of the infant brain. (Reproduced with permission from Hebden, J. C., Austin, T. 2007. *European Radiology* 17:2926–2933.)

static images of cerebral blood volume and oxygenation relies on single sets of datatypes rather than the measurements of changes. Reconstructing 3D images representing absolute optical properties requires algorithms that are inherently sensitive to tissue geometry, and deriving the external infant head geometry in the NICU environment is also a very challenging task. Although various approximations can be invoked, these inevitably reduce the quantitative accuracy of the images. Note that this problem is often overcome for breast imaging by conforming the breast to a known geometry or placing it within a coupling fluid (see Chapter 9).

10.8   The Future

As described in Section 10.5, optical topography has become a popular research tool for mapping the cerebral hemodynamic activity, particularly for psychological studies of the developing brain where alternative imaging modalities have limited use. The prospects are good that this trend will increase, especially as optical systems become more portable and user friendly. It is inevitable that progress in the development of optoelectronic devices will lead to smaller and lighter imaging systems that employ wireless probes. Wireless technology opens up a new realm of medical applications, especially psychological studies where freedom to walk and move around is an essential component. The facility of DOI technology to detect and map mental processing within the cortex could eventually be exploited as a brain–computer interface, such as a neural prosthetic device to assist the people with impaired motor or language skills.

Over the next few years, DOI methods would appear to have greater clinical potential as an adjunct modality to established tools such as EEG and MRI, with which they can be easily integrated (DOI-MRI is discussed further in Chapter 17), rather than as a stand-alone technique. This stems from the ability of optical imaging to provide complementary information on blood flow and oxygenation at relatively low cost.

One of the problems still inhibiting the medical utility of optical tomography is the reliance on numerical models of photon migration in the brain. An inevitable progress toward faster computing power will enable increasingly more complex models to be routinely implemented, including those that incorporate nondiffusing regions. Models based on notoriously computationally expensive Monte Carlo methods may eventually become the most effective solution. Note that the combination of DOI with MRI also provides a direct means of acquiring head and brain geometries to inform the models used for image reconstruction (Chapter 17).

Effective coupling of optodes to the scalp remains a challenge, particularly in the presence of thick dark hair, and is an awkward obstacle to routine application. However, attachment of EEG optodes is also a time-consuming procedure, but this has not prevented EEG being adopted as a common and effective clinical modality. DOI probe design must also take into account that the most sensitive optical systems are vulnerable to exposure by background sources of light, and consequently many

DOI studies are currently performed in conditions of low room lighting if not near-total darkness.

Finally, the future implementation of DOI for brain imaging is likely to be influenced by the ongoing development of extrinsic contrast agents. So far, indocyanine green (ICG) is the only contrast agent approved for optical studies, and has been shown to improve the spatial resolution for functional brain imaging applications through both its absorbing [27] and fluorescing properties [28]. However, it is possible that more specific optical markers being developed for preclinical applications may eventually become accepted as agents for clinical studies.

References

1. Jobsis, F. F. 1977. Noninvasive, infrared monitoring of cerebral and myocardial oxygen sufficiency and circulatory parameters. *Science* 198:1264–1267.
2. Arridge, S. R. 1995. Photon-measurement density functions. Part 1: Analytical forms. *Applied Optics* 34:7395–7409.
3. Arridge, S. R., Lionheart W. R. B. 1998. Nonuniqueness in diffusion-based optical tomography. *Optics Letters* 23:882–884.
4. Arridge, S. R., Schotland, J. C. 2009. Optical tomography: Forward and inverse problems. *Inverse Problems* 25:123010.
5. Gibson, A. P., Hebden, J. C., Arridge, S. R. 2005. Recent advances in diffuse optical imaging. *Physics in Medicine and Biology* 50:R1–R43.
6. Hillman, E. M. C. 2007. Optical brain imaging in vivo: Techniques and applications from animal to man. *Journal of Biomedical Optics* 12:051402.
7. Gratton, G., Corballis P. M., Cho E., Fabiani M., Hood D. C. 1995. Shades of gray matter: Noninvasive optical images of human brain responses during visual stimulation. *Psychophysiology* 32:505–509.
8. Hoshi, Y., Tamura, M. 1993: Dynamic multichannel near-infrared optical imaging of human brain activity. *Journal of Applied Physiology* 75:1842–1846.
9. Gratton, G., Fabiani, M. 2003. The event related optical signal (EROS) in visual cortex: Replicability, consistency, localization, and resolution. *Psychophysiology* 40:561–571.
10. Okada, E., Delpy, D. T. 2003. Near-infrared light propagation in an adult head model. I. Modeling of low-level scattering in the cerebrospinal fluid layer. *Applied Optics* 42:2906–2914.
11. Blasi, A., Phillips, D., Lloyd-Fox, S., Koh, P. H., Elwell, C. E. 2010. Automatic detection of motion artifacts in infant functional optical topography studies. *Advances in Experimental Medicine and Biology* 662:279–284.
12. Boas, D. A., Dale, A. M., Franceschini, M. A. 2004. Diffuse optical imaging of brain activation: Approaches to optimizing image sensitivity, resolution, and accuracy. *Neuroimage* 23:S275–S288.
13. Richter, M. M., Zierhut, K. C, Dresler, T., Plichta, M. M., Ehlis, A-C., Reiss, K., Pekrun, R., Fallgatter, A. J. 2009. Changes in cortical blood oxygenation during arithmetical tasks measured by near-infrared spectroscopy. *Journal of Neural Transmission* 116:267–273.
14. Benaron, D. A., Hintz, S. R., Villringer, A., Boas, D., Kleinschmidt, A., Frahm, J., Hirth, C. et al. 2000. Noninvasive functional imaging of human brain using light. *Journal of Cerebral Blood Blow and Metabolism* 20:469–477.
15. Bluestone, A., Abdoulaev, G., Schmitz, C., Barbour, R., Hielscher, A. 2001. Three-dimensional optical tomography of hemodynamics in the human head. *Optics Express* 9:272–286.
16. Gibson, A.P., Austin, T., Everdell, N. L., Schweiger, M., Arridge, S. R., Meek, J. H., Wyatt, J. S., Delpy, D. T., Hebden, J. C. 2006. Three-dimensional whole-head optical tomography of passive motor evoked responses in the neonate. *Neuroimage* 30:521–528.
17. Minagawa-Kawai, Y., Mori, K., Hebden, J. C., Dupoux, E. 2008. Optical imaging of infants' neurocognitive development: Recent advances and perspectives. *Developmental Neurobiology* 68, 712–728.
18. Blasi, A., Fox, S., Everdell, N., Volein, A., Tucker, L., Csibra, G., Gibson, A. P., Hebden, J. C., Johnson, M. H., Elwell, C. E. 2007. Investigation of depth dependent changes in cerebral haemodynamics during face perception in infants. *Physics in Medicine and Biology* 52:6849–6864.
19. Watanabe, E., Maki, A., Kawaguchi, F., Yamashita, Y., Koizumi, M. Y. 1998. Noninvasive cerebral blood volume measurement during seizures using multichannel near infrared spectroscopic topography. *Journal of Epilepsy* 11:335–340.
20. Gallagher, A., Lassonde, M., Bastien, D., Vannasing, P., Lesage, F., Grova, C., Bouthillier, A. et al. 2008. Non-invasive pre-surgical investigation of a 10 year-old epileptic boy using simultaneous EEG-NIRS. *Seizure* 17:576–582.
21. Cooper, R. J., Hebden, J. C., O'Reilly, H., Mitra, S., Michell, A. W., Everdell, N. L., Gibson, A. P., Austin, T. 2011. Transient haemodynamic events in neurologically compromised infants: A simultaneous EEG and diffuse optical imaging study. *Neuroimage* 55:1610–1616.
22. Wallois, F., Patil, A., Kongolo, G., Goudjil, S., Grebe, R. 2009. Haemodynamic changes during seizure-like activity in a neonate: A simultaneous AC EEG-SPIR and high-resolution DC EEG recording. *Clinical Neurophysiology* 39:217–227.
23. Hintz, S. R., Cheong, W-F., Van Houten, J. P., Stevenson, D. K., Benaron, D. A. 1999. Bedside imaging of intracranial hemorrhage in the neonate using light: Comparison with ultrasound, computed tomography, and magnetic resonance imaging. *Pediatric Research* 45:54–59.
24. Hebden, J. C., Gibson, A., Yusof, R., Everdell, N., Hillman, E. M. C., Delpy, D. T., Arridge, S. R., Austin, T., Meek, J. H., Wyatt, J. S. 2002. Three-dimensional optical tomography of the premature infant brain. *Physics in Medicine and Biology* 47:4155–4166.
25. Hebden, J. C., Austin, T. 2007. Optical tomography of the neonatal brain. *European Radiology* 17:2926–2933.
26. Hebden, J. C., Gibson, A., Austin, T., Yusof, R., Everdell, N., Delpy, D. T., Arridge, S. R., Meek, J. H., Wyatt, J. S. 2004.

Imaging changes in blood volume and oxygenation in the newborn infant brain using three-dimensional optical tomography. *Physics in Medicine and Biology* 49:1117–1130.

27. Habermehl, C., Schmitz, C. H., Steinbrink, J. 2011. Contrast enhanced high-resolution diffuse optical tomography of the human brain using ICG. *Optics Express* 19:18636–18644.

28. Steinbrink, J., Liebert, A., Wabnitz, H., Macdonald, R., Obrig, H., Wunder, A., Bourayou, R. et al. 2008. Towards noninvasive molecular fluorescence imaging of the human brain. *Neuro-Generative Diseases* 5:296–303.

V

Acoustical Imaging

Neb Duric
Wayne State University
Delphinus Medical Technologies

Olivier Roy
Wayne State University
Delphinus Medical Technologies

Cuiping Li
Delphinus Medical Technologies
Wayne State University

Steve Schmidt
Delphinus Medical Technologies

Xiaoyang Cheng
Delphinus Medical Technologies

Jeffrey Goll
Delphinus Medical Technologies

Dave Kunz
Delphinus Medical Technologies

Ken Bates
Applied Concepts, LLC

Roman Janer
Delphinus Medical Technologies

Peter Littrup
Wayne State University
Delphinus Medical Technologies

11

Ultrasound Tomography Systems for Medical Imaging

11.1 Introduction to Ultrasound Tomography

Ultrasound tomography (UST) is an emerging technique that uses methods similar to computed tomography (CT) to generate the images using sound signals. It is well suited for inferring the acoustic properties of a volume of tissue from measurements made along a surface surrounding the tissue. Unlike the x-rays used in conventional CT, sound is purely wave-like. Tomographic techniques that process sound data must, therefore, take into account wave propagation phenomena such as reflection, refraction, and diffraction. In an inhomogeneous medium, ultrasound pulses do not travel in straight lines, thereby complicating the tomographic inversion and placing additional burden on the computational requirements. The need for a high level of computing power and associated data-processing capability has been a major historical factor in limiting the development of UST compared to CT and other tomographic methods. For this reason, UST has not completely lived up to its early promise. However, in recent years, thanks to the increasing processing power of both computers and electronics, the landscape has changed dramatically.

11.1.1 History of UST Research

The idea of solving acoustic inverse problems in medicine can be traced back to the work of Wild and Reid (1952) and Howry and Bliss (1952) in the 1950s. The systems they built were crude mechanical scanners utilizing a single transducer that rotated on an arm and collected reflected signals using the pulse-echo technique. The first cross-sectional breast-tissue images were made at that time. However, the lack of computational power, combined with the slow rotation, made it impossible to apply this technique clinically. These early methods did give birth to what is now known as B-mode clinical ultrasound. The tomographic aspect of the early work was set aside. It took nearly 30 years for the concept of UST to be seriously revisited.

A number of investigators developed operator-independent ultrasound scanners, based on the principles of UST, in an attempt to perform *in vivo* imaging. Clinical examples include the work of Carson et al. at the University of Michigan (1981), Andre et al. at the University of California at San Diego (1997), Johnson et al. at TechniScan Medical Systems (1999), Marmarelis et al. at the University of South California (2003), Liu and Waag at the University of Rochester (1997), and Duric and Littrup et al. at the Karmanos Cancer Institute, Wayne State University (2005, 2007). More recently, Ruiter et al. (2011) have reported progress on a true 3D scanner utilizing a hemispherical array of transducers. Although no clinical results have yet been reported with this system, clinical studies are currently in the planning stages. The clinical systems developed by these groups employed similar patient-positioning systems. Patients were positioned in the prone position on a flat table with the breast suspended through

FIGURE 11.1 An ultrasound tomography scanner located at the Karmanos Cancer Institute in Detroit, MI, USA.

a hole in the table into a water bath lying just below the table surface. The water bath ensures minimal distortion of the breast while providing strong coupling of acoustic waves to the tissue.

In our laboratories, at the Karmanos Cancer Institute (KCI), our group has also focused on the development of UST for breast imaging. To that end, we have been developing and testing a clinical prototype in KCI's breast center (Figure 11.1).

The continuing development of the prototype and the associated UST methodology has been guided by feedback from clinical studies conducted at KCI. This development has led to continuing evolution in imaging performance, which has in turn led to increasingly greater clinical relevance. The prototype utilizes a solid-state ring-array transducer consisting of 256 elements that encircle the breast. It uses a 256-channel data-acquisition system that yields single-slice acquisitions in about 30 ms and whole-breast scans of 1 min or less. Furthermore, the prototype utilizes bent-ray image reconstructions. Although such images have lower spatial resolution compared to wave-based images, they can be reconstructed quickly, in keeping with the goal of a clinically fast system. The spin-off company, Delphinus Medical Technologies, is currently developing a commercial UST system based on experience with the clinical prototype. The commercial system will have 2048 active elements and a 512-channel data-acquisition system.

Clinical imaging with the UST prototype at KCI has been carried out with previously described tomographic reconstruction algorithms that yield (i) sound speed, (ii) attenuation, and (iii) reflection (Duric et al., 2007; Schmidt et al., 2011). Sound speed images are derived from the arrival times of acoustic signals. Previous studies have shown that cancerous tumors have increased sound speed relative to normal breast tissue (Duric et al., 2009, 2010; Ranger et al., 2009, 2010), a characteristic that can aid in the differentiation of masses, normal tissue, and fat. Attenuation images are derived from changes in acoustic wave amplitude. Cancerous masses show higher attenuation because they induce greater scattering of the ultrasound (US) wave. Attenuation data, in conjunction with sound speed, therefore provides a potentially effective means for determining the malignancy as illustrated visually in Figure 11.2. The mass at 7 o'clock appears distinct in both the UST and MRI images, indicating reliable visualization of breast lesions from UST data.

11.1.2 Current Trends

Over the past 30–40 years, computing power has increased by a factor of 10 million, as shown in Figure 11.3. This development has enabled sophisticated physics-based inversion algorithms such as waveform tomography to become practical. Inversions based on wave-form tomography have been successfully applied to clinical data by the Techniscan group (Johnson et al., 1999). At the same time, the processing power of electronics has also increased about 10 million-fold, leading to capabilities for massive parallelization of data acquisition and the corresponding ability to process large amounts of data. In fact, when framed in terms of processing power relative to energy consumed, the improvement in computational efficiency has been close to 100 billion, as shown in Figure 11.4. These two parallel trends have enabled the development of tomographic systems containing large numbers of sensors.

In the area of breast imaging, large transducer arrays are bringing about the realization of UST systems that are gaining clinical acceptance. UST is riding Moore's law into clinical relevancy. The remainder of this chapter will provide a technical overview of UST technology and will attempt to capture its current and future status as a viable imaging technology.

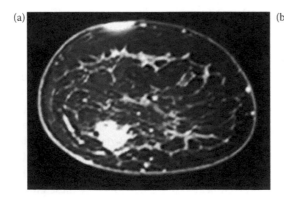
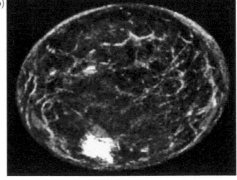

FIGURE 11.2 (a) MRI cross-sectional image showing a cancer at 7 o'clock. (b) A fused UST image (thresholded sound speed and attenuation images overlaid on a reflection image) showing the same region. The UST images were reconstructed using bent-ray tomography.

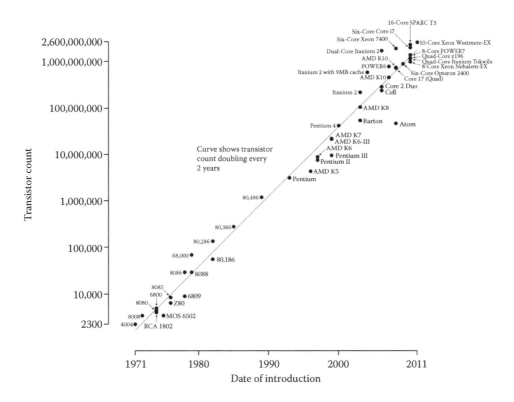

FIGURE 11.3 Transistor counts versus time. (By permission from Free Software Foundation.)

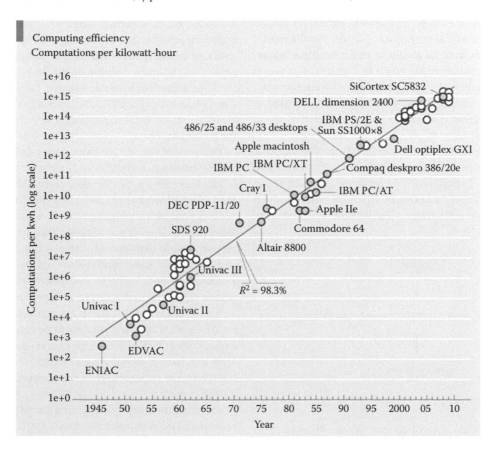

FIGURE 11.4 Computing efficiency over time. (Attribution: http://www.koomey.com. The Koomey.com website by Jonathan Koomey is licensed under a Creative Commons Attribution-NonCommercial-NoDerivs 3.0 Unported License.)

11.2 UST as an Inverse Problem

In a UST inverse problem, one wishes to find the properties of a tissue volume by illuminating it with known signals and then measuring the changes in the signals induced by their interaction with the object (Figure 11.5). In most practical applications, the object of interest can be confined either to a 2D plane or a 3D volume while the measurements are made in 1D or 2D space, respectively. In the case of UST, a typical setup involves either surrounding the object with an array of transducer elements or rotating a transducer around the object to probe the object with sound waves and to measure the resulting interaction between the sound waves and the object. The measurements are then recorded and used to reconstruct an image tomographically. Generally, the transducer elements are highly efficient at converting electrical energy into mechanical energy and vice versa. Often, the transducer elements are used either as receivers or emitters and the same element can be switched between the two modes.

11.2.1 Characterization of Lesions through Their Biomechanical Properties

Conventional reflection-mode ultrasound exploits differences in acoustic impedance between tissue types to generate anatomical images of breast tumors. However, reflection is just one aspect of a multifaceted set of acoustic signatures associated with the biomechanical properties of tissue. The diagnostic value of UST derives from its ability to probe multiple biomechanical properties of tissue. As shown in Figure 11.6, a sound wave interacts strongly with tissue, creating a scattered acoustic field. Sampling the scattered field yields data that are used to reconstruct the images of parameters such as reflectivity, sound speed, and attenuation.

Reflectivity is determined by gradients in the acoustic impedance of the medium being probed and is therefore well suited for imaging boundaries that separate different tissue types. Currently practiced ultrasound BIRADS criteria, which are used to aid in differentiating malignant and benign masses,

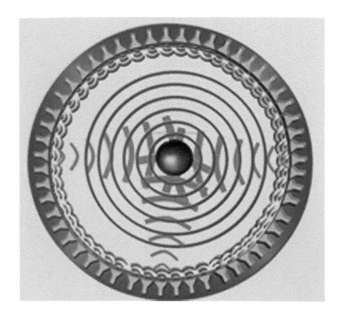

FIGURE 11.6 A stable solution that yields high-quality images requires (1) strong and uniform illumination of the object of interest from a variety of directions (shown here are acoustic sources along the periphery of the ring with three firings represented by the thick arcs) and (2) a set of dense measurements of the scattered field (shown as concentric circles), collected by the same array of sensors at the periphery of the ring.

are based on how tumor shape and margins are visualized, including echo-texture differences. The use of the BIRADS criteria can be easily extended to UST images where they can be applied to the entire volume of the breast. Recent results with the U-Systems whole-breast scanner have already shown that the application of the BIRADS criteria in the coronal plane can identify architectural distortion, a key element in diagnosing the breast cancer.

The speed of sound, c, in tissue varies with the physical parameters of density, ρ, and stiffness, K, according to equation $c^2 = K/\rho$. It has been shown in the laboratory (Masugata et al., 1999; Mast 2000; Weiwad et al., 2000) that the stiffness and density are not independent variables and that $K \propto \rho^3$ in human tissue. This approximate dependence suggests that sound speed is linearly proportional to density and varies as the cube root of the stiffness, on at least some spatial scales of the tissue. It is therefore a measure of both stiffness and density but with much greater relative sensitivity to density. Sound speed is an absolute quantity with units of km/s which allows for measurements taken at different times or on different patients to be compared. It is therefore a quantitative, virtual analog of manual palpation which recognizes that hard masses have a different "feel" relative to normal soft tissue. The typical range of sound speed in soft human tissue is ~1.35–1.6 km/s.

The amplitude of a propagating longitudinal wave changes because of two dominant processes: scattering and absorption. The former is an elastic process that redistributes the acoustic energy throughout the tissue while the latter converts sound energy into heat. The two processes act together to define the

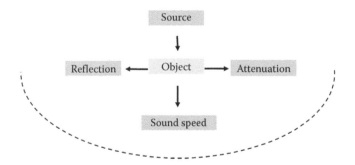

FIGURE 11.5 When a source of acoustic energy illuminates the object of interest, the sound waves interact strongly with it, causing a scattered acoustic field (dashed line). Measurements of that field yield parameters such as reflectivity, sound speed, and attenuation.

attenuation coefficient. They have strong but distinct dependencies on frequency such that absorption is relatively important at high frequencies (>10 MHz), while scattering contributes significantly at lower frequencies. Most UST scanners being developed today use frequencies below 10 MHz. Attenuation measurements can be therefore interpreted as a combination of absorption and scatter effects. Scattering depends strongly on the cellular structure of tissue as well as on macroscopic structures smaller than a wavelength in size. It is the scattering of such structures (nonspecular reflections) that gives rise to speckle in conventional ultrasound images. Cancer is known to alter tissue in its immediate environment, on cellular and macroscopic scales, which yields an attenuation signature. As Greenleaf et al. (1977) showed, cancerous tissues tend to have higher levels of attenuation relative to normal tissue. This observation is consistent with the shadowing artifact seen on conventional ultrasound images of cancerous tissues. Attenuation is strongly frequency dependent, and in normal breast tissue has a value of ~0.5 dB/cm/MHz.

The tissue parameters measured by UST can be combined to characterize the breast tissue and predict the presence of cancer, thereby providing significant diagnostic value. The inverse problem, diagrammed in Figure 11.5, can be solved in order to recover the images of these tissue parameters. The classes of algorithms required to perform the tomographic inversions are described in the following sections.

11.2.2 Solving the Inverse Problem

The quality of the reconstructed image depends on the quality of the signals acquired by the UST system and by the sophistication of the reconstruction algorithm. The latter is defined by how well the physics of the sound propagation are modeled. Generally, the simpler the assumptions for wave-based reconstructions, the faster an algorithm can run but the lower the quality of the final image. Therefore, a trade-off exists between reconstruction speeds and image quality.

11.2.2.1 Ray Tomography

The simplest approach to transmission tomography uses the straight-ray approximation. This approximation is analogous to the assumption made in x-ray CT reconstruction that x-rays travel in straight lines. The spatial resolution of the reconstructed UST images is poor under the straight-ray approximation because it does not take into account the refraction of the waves as they pass through an inhomogeneous medium. Time-of-flight measurements are used to reconstruct the images of sound speed while changes in signal amplitude are used to construct the attenuation images (analogous to x-ray CT).

Bent-ray tomography relies on the knowledge that refraction is governed by changes in sound speed. The initial model of sound speed used for the reconstruction can be homogeneous or heterogeneous. The sound speed model is used by the reconstruction algorithm to bend the rays as they propagate from one pixel to the next. The traced ray path and predicted arrival times

are used to generate the next sound speed model, which allows for more accurate ray bending. The process is repeated until convergence is achieved. The net effect of bending the rays is to compensate for the refractive effects, which reduce the image artifacts and improves the spatial resolution of the images. Bent-ray tomography continues to be an active field of research (Schomberg 1978; Andersen 1987; Norton 1987; Andersen 1990; Li et al., 2009; Hormati et al., 2010).

11.2.2.2 Reflection Tomography

In contrast to the use of transmitted signals, reflection tomography relies on the reflected echoes to construct the images of relative echo amplitudes. Since most UST systems do not employ beam-forming on the front end, the often-used Kirchhoff's migration technique does so "after the fact," that is, after all the data have been gathered. Consequently, there is a great deal of flexibility in how the data are used. In particular, it is possible to use apertures of arbitrary size to reconstruct a reflection image. Unlike standard B-mode imaging, this method allows receive aperture sizes that are limited only by the overall locus of the receiving transducer elements. Finally, since the Kirchhoff migration method operates on raw data, there is more flexibility in terms of how the signals are processed before and after the migration process is initiated. Figure 11.7 shows a reflection image of a clinical phantom, made with the Kirchoff migration method.

Generally, reflection tomography methods rely on the assumptions that every point in the object is a scatterer, is independent of other points, and that only one scatter occurs on a path that connects that point to an emitter–receiver pair. Consequently, these traditional reconstruction methods do not take into account diffraction and other wave properties. Some migration methods have been adapted to correct for refraction and for attenuation. Using the reconstructed datasets of sound speed and attenuation, the proper time delays can be calculated for signal alignment and variable signal amplification can be employed to correct for inhomogeneous energy loss. The resulting effects on the image include sharper boundaries, higher contrast, less background noise, and, in some cases, the detection of objects otherwise lost due to homogeneity assumptions (Schmidt et al., 2011).

11.2.2.3 Waveform Tomography

The ever-increasing computational power opens the possibility of solving the full wave equation, for both sound speed and attenuation (Pratt et al., 2007). The computational cost associated with waveform tomography buys a much better accounting of diffractive and refractive effects. Furthermore, by utilizing all of the recorded wave information (as opposed to the arrival time of the signal alone, as in ray tomography), waveform tomography has the potential to increase the image contrast while suppressing the artifacts. Waveform tomography's limiting resolution of λ is up to an order of magnitude better than that of ray tomography.

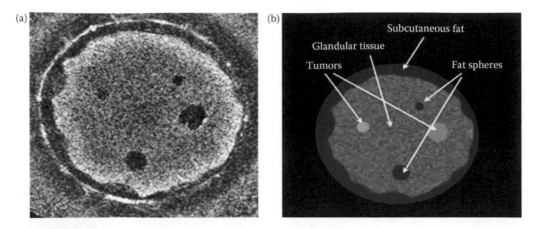

FIGURE 11.7 A reflection image of an anthropomorphic breast phantom constructed using reflection tomography (a). The truth image (b) is represented by a CT scan of the same phantom.

Waveform tomography reconstruction methods have been formulated in the time domain (Natterer and Wûbbeling 2001; Tarantola 2005) and in the frequency domain (Pratt 1999). The latter usually allows for a simpler formulation of the problem since convolution and differential operators are mapped to multiplications in Fourier's space. The reconstruction process is similar to that used in ray tomography. One starts from an initial model of the unknown parameters (sound speed, attenuation) and solves a forward problem. The solution of this forward problem is a set of simulated waveforms recorded at the transducer locations. The residual between the recorded waveforms and the measured ones is then used to update iteratively the unknown parameters until convergence is achieved.

For a given accuracy, the size of the model typically scales linearly with the frequency of the probing pulse. Complexity can be lowered using approximations. It can also be addressed by means of efficient parallel implementations (Micikevicius 2009; Roy et al., 2010). However, the complexity involved in waveform tomography remains a challenging issue, especially in medical imaging applications where reconstruction time must be kept at a minimum to maintain a high patient throughput.

11.2.3 Image Quality

UST inverse problems are generally ill-posed but high-quality images can be achieved if (i) the object being imaged is strongly and uniformly illuminated by the ultrasound sources, (ii) the measurement space is densely sampled, and (iii) robust algorithms are used to reconstruct the images. The first two conditions can be satisfied with a dense array of transducers (Figure 11.5) that are supported by highly parallel data-acquisition electronics. The third condition can be satisfied with highly parallel computing engines running the physics-based inversion algorithms. Fortunately, computers and electronics have followed Moore's law since the introduction of UST. Over the past few

decades, the number of transistors per unit area and the number of computations per dollar spent have risen many millionfold (Figures 11.3 and 11.4), to the point where the conditions for reconstructing high-quality images can now be met. The following sections will specifically focus on the basic hardware and embedded software components that allow practical implementation of UST, as determined by clinical performance constraints.

11.3 Clinical Performance Constraints

For a UST system to be clinically relevant, its design must be motivated by a compelling clinical need and constrained by practical clinical requirements.

11.3.1 Clinical Problem

According to SEER statistics, breast cancer incidence varies with the stage of the disease (National Cancer Institute, 2011) and the 5-year survival rate for women with localized cancer is 98% versus 23% for women with metastasized cancer. These statistics, applied to the population of the United States, results in an estimate of ~190,000 new cancer cases diagnosed each year, with a corresponding mortality rate of 40,000 per year. The mortality numbers suggest that many cancers are not detected at the earliest stages when they are most treatable. There are many reasons why cancers are not detected early enough but some of the major factors are the limited participation in breast screening and the performance of screening mammography.

National cancer-screening statistics indicate that only 51% of eligible women undergo annual screening mammograms (American Cancer Society, 2009). Limited access, fear of radiation, and discomfort are among the factors often cited to explain the low participation rate. Greater participation would lead to the detection of cancer at an earlier stage, which would lead to a greater survival rate. Increased participation and improved breast cancer detection would have the greatest impact on the nearly one in three women who are diagnosed each year with later stage (regional or greater) breast cancer, totaling ~60,000 women per year in the United States. The net effect would be an increase in survival time and a corresponding decrease in mortality rates, particularly among women who have been reluctant to accept the screening mammography.

For women with dense breast tissue, who are at the highest risk for developing breast cancer (Ursin et al., 2005; Chen et al., 2006; Armstrong et al., 2007; Martin and Boyd 2008), the performance of mammography is at its worst (Turnbull 2009). Consequently, up to 50% of cancers are missed at their earliest stages when they are the most treatable. Improved cancer detection for women with denser breasts would decrease the percentage of breast cancer diagnosed at later stages, which would result in lower mortality rates.

Although tomosynthesis may improve upon some of the limitations of standard mammography, it is unlikely to create a paradigm shift in performance and it uses at least as much radiation as mammography. MR, on the other hand, provides volumetric, radiation-free imaging with high sensitivity. Studies have shown that magnetic resonance (MR) can impact a large swath of the breast management continuum ranging from risk assessment to diagnosis to treatment monitoring (Partridge et al., 2002; Kuhl et al., 2007; Saslow et al., 2007; Bando et al., 2008; Bhattacharyya et al., 2008; Chen et al., 2008; Jansen et al., 2008; Sharma et al., 2009; Tozaki 2008). However, MR requires relatively long exam times and the use of contrast agents. Furthermore, MR has long been prohibitively expensive for routine use and there is a need for a low-cost alternative. The clinical utility of positron emission tomography is also limited by cost. Conventional sonography, which is inexpensive, comfortable, and radiation-free, continues to only play an adjunctive role in breast imaging because of its operator dependence and the time needed to scan the whole breast. The lack of an alternative that balances between the cost effectiveness of mammography and the imaging performance of MR is a barrier to dramatically impacting the mortality and morbidity through the improved screening.

UST has the potential to eliminate this trade-off by combining the low-cost advantage of mammography with the superior imaging performance of MR, thereby lowering barriers to participation. When fully realized, UST may have a positive societal impact because the accurate identification of cancer, at relatively low cost, would enable widespread access in clinics, and a reduction in death rates from breast cancer.

11.3.2 Safety

A new imaging modality is more likely to be adopted and used if it is considered to be low risk. No significant risk is associated with the clinical ultrasound imaging modality. It is interesting to note that UST utilizes acoustic energy levels that can be lower than conventional ultrasound. Typically, UST systems use unfocused small-aperture transducer elements (i.e., no electronic beam-forming) that are generally fired sequentially or they are beam-formed to launch plane waves. In either case, the instantaneous energy density and peak power level are well below those of B-mode ultrasound (i.e., <100 mW/cm² peak power). Furthermore, the paradigm of breast compression and the use of ionizing radiation are directly averted. The former is obviated by the use of a water bath, which eliminates the probe contact with the breast while the latter is eliminated by using sonic energy instead of electromagnetic energy.

11.3.3 Cost Effectiveness

For a system to be readily adopted and considered for screening, it must be cost effective compared to MRI and mammography. Ultrasound is inexpensive by the nature of its components and by the fact that no room shielding or special installation is required. The transducers used in ultrasound equipment are much cheaper than x-ray tubes and are far cheaper than the large magnets required by MRI. Furthermore, the electronics in ultrasound systems handles the low-power signals, which add to the cost advantage.

11.3.4 Scan Time

Imaging is the most effective when it minimizes the artifacts caused by patient motion. At a minimum, the scan time for one slice must be comparable to a typical breath-hold time. Unlike MRI, UST can be configured to scan the breast quickly, particularly in the case where mechanical rotation is eliminated and a highly parallelized set of data-acquisition channels is used. Some UST designs allow for slice scans of <0.03 s and whole-breast scans of the order of 1 min or less, thereby minimizing the motion artifacts (Duric et al., 2007).

11.3.5 Patient Throughput

Ideally, high-volume clinics would like images to be ready before the patient leaves the clinic. This requirement imposes a constraint that each patient's images are ready in about 15 min. High-volume clinics would also like an imaging system to be high-throughput, which generally means the system must image at least four patients per hour. Thus, the scan time, combined with the patient setup time and the image reconstruction time, should be no greater than 15 min. This constraint mandates a compromise between the cost of high-performance computing and the sophistication of the inversion algorithms. As shown below, this compromise can be optimized through parallelization of low-cost, low-power consumption electronics, and high-performance computing utilizing the multicore CPUs and GPUs.

11.3.6 Spatial Resolution

To be competitive with MRI and mammography, UST should achieve a spatial resolution of ~1 mm in all three spatial dimensions and should yield high-contrast images. Such constraints can be met with the use of dense arrays and robust algorithms. Figure 11.8 illustrates the ring array used in KCI's clinical prototype. The prototype's 256-element array allows the object to be illuminated from 256 distinct directions. Future arrays will have many more elements, increasing the illumination and sampling density further.

11.3.7 Penetration

Conventional US imaging trades off resolution, which increases with increasing frequency, and penetration, which

declines with increasing frequency. For a whole-breast imaging system such as UST, penetration takes on even greater importance. Whole-breast imaging systems must therefore operate at frequencies that are sufficiently low to allow adequate penetration while retaining spatial resolution of ~1 mm. Most UST systems are therefore designed to operate in the 1–4 MHz range; this is significantly lower than conventional US imaging devices.

11.3.8 Image Display and PACS Connectivity

To minimize the impact on a busy clinic's workflow and the radiologist's time, UST systems must provide easy-to-use workstations with connectivity with a clinic's or hospital's PACS (picture archiving and communications system). The UST images must therefore adhere to radiologic standards so that it can be fielded in any clinical setting. The next section describes how currently available hardware can be configured to satisfy the clinical constraints described above.

11.4 Technical Description of Ultrasound Tomography

The basic components of a generic UST system are shown in Figure 11.9. The front end consists of the sensor array, which is capable of providing the necessary illumination along with the resulting measurements. The sensor array feeds the analog front end which in turn feeds the back end where the digitized signals are processed. The digitized signals are pushed to an image reconstruction computer (the image-processing unit), which, in turn, pushes the images to a display (usually a stand-alone workstation with PACS connectivity). Each of these components is briefly discussed in the following sections.

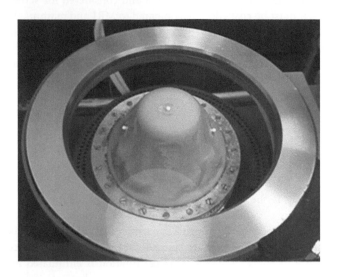

FIGURE 11.8 The ring transducer used at the Karmanos Cancer Institute (KCI). The ring's 256 elements surround the breast (a breast phantom is shown).

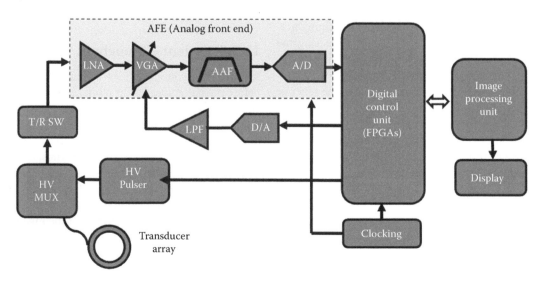

FIGURE 11.9 Block diagram of a generic UST system.

11.4.1 Transducer Array

Clinically useful tomographic imaging requires high-resolution time and amplitude measurements, made at numerous closely spaced locations, covering a wide range of paths through a medium with substantial attenuation. At KCI, a system with a ring-shaped transducer illuminates and acquires data in the plane of the ring, with subsequent movement parallel to the ring axis yielding three-dimensional coverage. In this case, transducer materials must be sufficiently pliable to allow the bending into the final ring geometry, under appropriate thermal and mechanical conditions.

These requirements place great demands on the design and fabrication of the transducer. Such a transducer should ideally have many elements with subwavelength spacing, each providing a wide field of view in the plane of the ring but a narrow focus in the perpendicular plane. The clinical constraints described in Section 11.2 translate into technical transducer requirements of high sensitivity, broad bandwidth, short impulse response, and excellent isolation of elements from each other. The techniques developed to enhance some of these properties may detract from the others. Nevertheless, materials and manufacturing techniques developed over decades have made it possible to achieve all of these demanding objectives. In general, one starts with a piezoelectric ceramic material that has strong coupling between mechanical pressure and electric fields. This strong coupling contributes to the required high sensitivity over a broad frequency bandwidth. Piezoelectric composite materials that properly combine the base ceramic material with epoxy-based materials develop even stronger piezoelectric coupling while providing a better acoustic match to water or to human tissue.

Wide angular response (element beam-width) results from physically narrow, subwavelength elements with excellent isolation from neighboring elements, which are typically tens of microns away. This is generally achieved by sawing between the elements and inserting highly attenuating materials in the gaps. The wide bandwidth and short impulse response are achieved using backing and acoustic impedance matching layers, both often based on epoxies or other polymers filled with selected materials. Highly absorptive backings reduce the reflections from the backside of the transducer and damp the natural oscillations. Carefully engineered acoustic matching layers similarly reduce the reflections at the front side and further damp and shorten the response. Acoustic lenses, molded from silicone rubber or similar materials, are used to focus the energy in the plane perpendicular to the ring. Sensitivity depends on the match between the electrical properties of the transducer and the driving and receiving circuits. The transducer's electrical properties can be varied somewhat, to improve the electrical match, by the final selection of materials. For example, the permittivity, which plays an important role in determining the electrical impedance of the transducer, varies substantially among available piezoelectric ceramic and composite materials. Finally, electrical connections must be made to each individual transducer element without generating the excessive crosstalk between elements. This is achieved using signal routing and shielding techniques based on those developed in the semiconductor and RF electronics industries.

Future transducer development can further improve ultrasonic tomographic imagers. Improvements in sensitivity, bandwidth, and angular response can improve upon current imaging performance and increase the quality of images obtained using alternative modalities, including harmonic and Doppler imaging. Harmonic and Doppler techniques respond to different physical tissue properties and have the potential to improve the diagnostic value of the images. Harmonic imaging requires bandwidths exceeding one octave. Doppler imaging is optimized at high frequencies—likely somewhat higher than are typically used in UST imaging—to quantify blood flow velocities based on scattering from individual blood cells. Finally, two-dimensional transducer arrays would allow comparably high resolution in both the azimuthal direction (in the plane of the ring) and the elevation direction (parallel to the ring axis). The current ring at KCI achieves this high resolution in the azimuthal direction, but has somewhat lower resolution in the elevation direction.

11.4.1.1 Need for Calibration

The characteristics of the acquisition hardware must be estimated in order to fully use the potential of the tomographic acquisition system. In fact, these characteristics have a direct effect on the shape of the acquired ultrasound signals. This effect on the signals translates directly into an effect on UST image quality. We now list a few of these characteristics.

11.4.1.1.1 Transducer Element Positions

It is difficult to manufacture a transducer ring with a perfectly known geometry. For example, a transducer ring typically consists of multiple segments (Figure 11.10). The relative offsets between these segments lead to uncertainty in the positions of the transducers. If the distances between transducer elements are inaccurate, the sound speed and attenuation image reconstructions will be inaccurate as well. Calibration can be achieved by propagating signals in a homogeneous environment with a

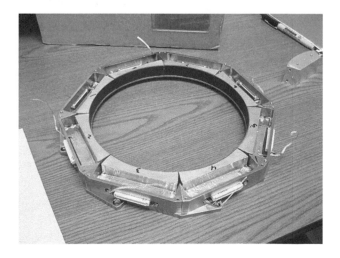

FIGURE 11.10 Transducer ring of the KCI prototype (eight segments).

known sound speed (e.g., a water bath with a known temperature). In this case, distances can be easily calculated from time-of-flight measurements (Roy et al., 2011).

11.4.1.1.2 Acquisition Delays

The data-acquisition electronics introduce some inherent electronic processing delays. These delays lead to inaccurate time-of-flight measurements which result in severe artifacts in the reconstructed images. These severe artifacts reduce resolution and contrast and produce out-of-focus effects (Figure 11.11). A water bath with known sound speed can be used as a controlled environment to estimate these delays.

11.4.1.1.3 Acquisition Gains

Variations in transmission power across transducers lead to an amplitude mismatch between different signals. This mismatch results in inaccurate attenuation image reconstruction and attenuation compensation, producing shadowing artifacts. In a similar manner to acquisition delay compensation, signals acquired in a water bath can be used to equalize the acquisition gains after compensation of the natural amplitude decay induced by geometrical spreading.

11.4.1.1.4 Transducer Beam Pattern

The response of a transducer in phase and amplitude, as a function of both angle and frequency, is referred to as the beam pattern—generally, the lower the central frequency of the beam, the larger the angle. The high-frequency components of the signals are thus greatly attenuated at large angles; this frequency shift stretches the signal in time and leads to inaccurate time-of-flight and attenuation measurements at large angles. The beam pattern can be measured using a hydrophone and signals can be characterized based on this information.

11.4.2 Data-Acquisition Electronics

In UST imaging, a transducer can be mechanically rotated around the object of interest to provide many points of illumination and measurement, or a transducer consisting of an array of elements can be placed around the object or over some portion of the object. An array of transducers is generally more expensive to build and field but it offers the possibility of either electronic multiplexing or a parallelized system that provides data channels for many or all elements, thereby greatly accelerating the data-acquisition process. There is therefore a trade-off between the cost and speed of any UST implementation. The exact amount of trade-off is governed by the application and in the case of clinical imaging favors the high-speed implementations as described in Section 11.3.

Figure 11.9 gives the block diagram of a data-acquisition system suitable for ultrasonic tomography. Current tomography systems may use transducer arrays with more than 2000 elements, which must all be individually addressable. With current IC technology, various blocks are most naturally implemented for smaller numbers of elements, such as 256. Multiple copies of these functional blocks are included to deal with the full complement of transducer elements. For example, multiple analog front end (AFE) circuits, multiple transmit/receive switches (T/R SW), high-voltage drivers (HV drivers), and high-voltage multiplexer (HV MUX) circuits are required for the simultaneous acquisition of a large amount of A-scan (full RF bandwidth) data by multiple elements. Today's large-scale integration and application-specific integrated circuits (ASICs) make possible the significant reductions in cost, space, and power which are required for clinical viability.

Many systems use a long interconnect, typically about 2 m in length, containing multiple micro-coaxial cables between the transducer array and the electronics. However, the cable capacitance loads the transducer elements, causes significant signal attenuation, and increases the noise figure (NF) of the receiver. With careful mechanical and electronic design, it is possible to replace this cable with short flex circuits with much lower capacitive loads. This helps to minimize the NF and optimize the dynamic range.

11.4.2.1 Analog Front End

As with most ultrasonic systems, the analog front end plays a major role in optimizing the signal-to-noise ratio (SNR) and

(a) 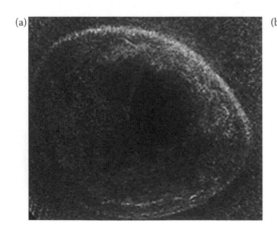 (b)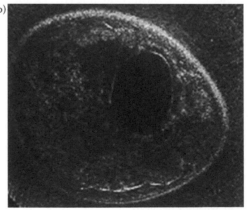

FIGURE 11.11 Tomographic reflection reconstruction with uncalibrated delays (a) and calibrated delays (b).

dynamic range, and in avoiding the image artifacts. The analog front end refers to a group of components in the signal-receiver chain following the transmit/receive switch. Typical components for the analog front end include the low-noise amplifier (LNA), variable-gain amplifier (VGA), antialiasing filter (AAF), and the analog-to-digital converter (ADC). The LNA is critical in minimizing the effects of noise in the subsequent stages such as the VGA and the AAF. The LNA is typically designed to be as physically close to the transducer as possible. Active LNA input impedance control can aid impedance matching to the transducer, thus optimizing the SNR. The VGA has a linear-in-dB gain control that is actually a time-dependent gain control. This allows compensation for attenuation of the signals returned by body tissues as a function of time. The VGA compresses a wide dynamic range input signal to fit within the fixed input range of the downstream ADC. The lower limit of this range is set in accordance with the quantization noise floor, which is determined by the ADC resolution. Modern ADCs use either 12 or 14 bits. AAF, which is placed before the ADC, removes unwanted frequency components beyond the ADC's Nyquist limit.

Several large vendors, such as Analog Devices (ADI) and Texas Instruments (TI), now offer multiple analog front end channels on a single die (Texas Instruments, 2011; Analog Devices, 2012). This development makes possible significant reductions in both size and power. Still higher density form factors for ultrasound analog front ends are under development and this will, in time, improve the capabilities of tomography systems and other medical ultrasound imaging systems.

11.4.2.2 Digital Back End

The digital back end performs final processing of the analog front end output, before providing the final information for display. The main digital back end functions include digital processing of the digitized image waveform, moving data to or from an external host system, logic control, and clock distribution. Modern digital components, such as field programmable gate arrays (FPGAs) and multicore digital signal processors (DSPs), play a major role in today's image-system architecture. For example, the digital back end platform of a fully three-dimensional ultrasound computed tomography (3D USCT) scanner being developed by Karlsruhe Institute in Germany uses 81 Altera Cyclone II FPGAs for controlling and computing purposes (Birk et al., 2011). The great quantity of data generated by UST imaging requires massive computing power and associated data-processing capacities in the digital back end. This also encourages rapid migration of key functions of traditional digital back ends from hardware to software, usually using PC-based software with multiple, powerful, central processing units (CPUs), and/or graphical processing units (GPUs).

11.4.2.3 Embedded Processing

The FPGAs which are part of the digital back end of the data-acquisition electronics are capable of performing many different processing tasks, from the simple to the complex. Depending on the complexity and sophistication of the UST system, the FPGAs

are used to package the received signals for processing and/or to perform some more complex data-processing tasks, shifting some of the processing load from the image-reconstruction hardware onto themselves. The latter use is particularly desirable in a UST system, as the processing load on the image reconstruction hardware is typically both heavy and important to the final image quality; all else being equal, the fewer processing tasks the image reconstruction hardware must perform, the better the quality of the image. The FPGAs therefore have the potential to improve the performance of a UST system without adding significant cost or power consumption to the system.

Most UST systems use FPGAs to perform at least some basic processing on the received signals. Typically, the FPGAs will pad and format the data for transmission to the image reconstruction hardware. In UST systems which decouple the image-reconstruction hardware from the data-acquisition electronics, the FPGAs may also add a header to the data that allows the data to be processed without any other information. In more sophisticated UST systems, the FPGAs may perform processing tasks that augment the performance of the system. Such processing tasks may improve the quality of the data, by performing signal averaging or decoding (in the case of encoded signals), for example. Processing tasks may also take on some of the data-processing burden, such as digital filtering, averaging reciprocal signals, or even receive beam-forming. As FPGA technology advances with time, the processing that can be done within the data-acquisition electronics will become more beneficial to a UST system's performance.

The FPGAs contain firmware that can be programmed after a UST system is built and assembled. The firmware adds flexibility to the system, allowing the system to be updated as new functionality is devised for the FPGAs. Most importantly, the programmability of the FPGA firmware allows the performance of a UST system's data-acquisition electronics to be upgraded without undergoing a long and costly hardware redesign. The FPGA firmware may even be updated in the field, allowing UST systems to be upgraded during routine service calls. The FPGA firmware contributes greatly to the flexibility and longevity of UST systems.

11.4.3 Image Reconstruction Hardware

The computational requirements of a UST scanner vary greatly depending on the amount of data being processed and the reconstruction algorithms being used. A typical tomographic reflection algorithm adds up the contribution of all reflected waveforms in the acquired dataset for each pixel in the final image. The computation complexity therefore scales linearly with the total number of pixels as well as with the square of the number of transducers. Sound speed and attenuation reconstruction based on the bent-ray approximation involve ray tracing and solvers for large, sparse, and possibly nonlinear systems of equations. While a wide range of methods exist to perform these tasks, their computational complexity follows a similar scaling behavior as that of reflection. Waveform tomography reconstruction methods require finite-difference modeling, which, despite the use of simplifying approximations, incurs a significant computational burden, especially at high-temporal frequencies.

To address this computational challenge, multithreaded routines running on multiple CPUs can be implemented. These routines may resort to optimized instruction sets (e.g., SSE instructions), which are processor dependent. A parallel scheme is harder to implement and debug, and is often more platform-dependent. However, parallelization is absolutely necessary to process the data quickly enough to ensure a reasonable image generation time. Another approach is to harness the computational power of graphics processing units (GPUs). While GPUs were originally designed to render complex graphical scenes, their use for general-purpose computation has recently attracted a lot of attention. Programming architectures, such as Nvidia's CUDA framework, allow for high level programming of image-reconstruction algorithms on GPUs. While GPUs offer tremendous computational power relative to their cost, global memory access is rather costly. Therefore, only algorithms with a relatively high computation-to-memory-access ratio are expected to be significantly faster on a GPU than on a CPU. Fortunately, many of the UST reconstruction schemes described above fall into this category. The use of multiple CPUs and GPUs on the same reconstruction platform is of course possible. In this case, particular attention has to be paid on how to distribute the burden of the computation across CPUs and GPUs to maximize the overall speed.

The basic floating-point operations required for tomographic reconstruction are addition, subtraction, multiplication, and division. A computer implementation of all these mathematical operations requires additional steps to properly control, sequence, and parallelize the operations. The total number of floating-point operations and computational time is scaled by the amount of data acquired by the UST system and the size of the grid model used in image reconstruction. On the commercial UST system being developed by Delphinus Medical Technologies, for example, the number of floating-point operations for sound speed and attenuation image reconstruction, based on a system of 1024 transducer elements and a 180×180 grid model, is ~1920 trillion for an average bilateral breast exam.

The total computational time for UST image reconstruction for an average patient can vary from minutes (with a sophisticated parallel computation scheme) to days (with a simple sequential computation scheme). For the average bilateral breast exam which is outlined above, the reconstruction takes ~3 h for a simple sequential implementation on an Intel Xeon E5654 Duo quad-core computer, while it only takes ~7 min for a sophisticated parallel implementation on a blade server with 8 NVidia Tesla M2060 GPUs. To meet the patient throughput requirements, sophisticated parallel implementations become a natural choice for UST image reconstruction.

The acquisition of ultrasound signals with a large array of transducers requires significant data storage capacity. The amount of data per acquired slice depends on the number of

TABLE 11.1 Size (GB) of a Single Slice for Different Configurations (Recording Length Set to 200 µs)

Sample Frequency (MHz)	Quantization (bits)	Number of Transducers		
		256	512	1024
10	10	0.15	0.61	2.44
	12	0.18	0.73	2.93
	14	0.21	0.85	3.42
15	10	0.23	0.92	3.66
	12	0.27	1.10	4.40
	14	0.32	1.28	5.13
20	10	0.30	1.22	4.88
	12	0.36	1.46	5.86
	14	0.42	1.70	6.84

transducers, the recording length, the sampling frequency, and the number of bits per sample. Table 11.1 provides numerical values for different configurations.

Volatile storage (e.g., RAM) and nonvolatile storage (e.g., solid state drives) can be used to store the data before it is processed for image reconstruction. If data can be written at a rate that is sufficiently high, then nonvolatile memory is usually the cheapest solution. However, in a clinical environment, the scan time must be reduced to a strict minimum to support a high patient throughput. In this case, volatile memory becomes the method of choice. The system should be designed with enough memory to compensate for the difference between the speed of data acquisition and the speed of image reconstruction and/or permanent data storage. Depending on the size of the breast, an entire scan can contain hundreds of slices. The storage task quickly becomes challenging and costly.

The hardware options for reconstruction are vast, but must be based on the reconstruction requirements for memory, operations, and parallelization. Computation can be done onsite or offsite at a remote dedicated reconstruction environment. A remote environment has the benefit of expandability and flexibility of climate controls. In a remote environment, regulated temperatures and loud noises are not as much of a concern as they are in a public or clinical environment. However, this concept of "cloud computing" does not lend itself well to the massive amount of data that would be required to be transferred over the Internet to the remote location. Even at large transfer rates, remote transfer is relatively slow when compared to a local computation environment; when reaching high remote transfer rates, the cost of hardware and high-speed networking access is very expensive. This leads to the necessity of an integrated reconstruction apparatus.

Three main factors weigh into the choice of the reconstruction hardware: cost, computational power, and form factor. Since tomographic ultrasound uses a combination of parallel and serial computations, the ratio of GPUs to CPUs for reconstruction is on the order of 1:1; however, this ratio can change depending on the algorithms. A number of vendors and computer configurations can satisfy this ratio. Options include clusters of networked workstations, rack servers, and blade servers.

Networked workstations would give the most flexible configuration in terms of computation power and hardware customization at a relatively low cost; however, the form factor is large due to all external switches, cables, and separate chassis. The networked workstation solution is typically not containable within the tomographic device. Rack servers provide a more compact solution than the networked workstations, but the configurability becomes limited to the options of the server vendors. Costs for rack servers tend to be higher due to the very specific designs of the devices. Finally, blade computing gives the most compact configuration of the options. All boards, switches, power supplies, and almost all cables are housed in one streamlined chassis. The hardware customization is the most limited in this case, but 1:1 GPU to CPU ratios are available. The cost of a blade server will vary depending on the vendor, but they are comparable in price to other rack server configurations. Because the form factor is the most condensed in comparison to other approaches, the reconstruction apparatus is most capable of being integrated inside the body of the tomographic device.

11.4.4 Workstations for Image Display

UST systems are often packaged with stand-alone workstations for image display and review. Stand-alone workstations are typically standard PCs with image-review software that allows the user to review the UST images. Workstations will often have multiple (typically 2–4) monitors to allow the user to select the images from a patient/study list on one monitor and to provide the user with as much image display area as possible. Image display area is especially important for UST-centered workstations, as UST systems produce multiple image stacks which are useful for diagnosis. Workstations will typically load and display non-UST images as well, which facilitates direct comparison of UST images with other breast imaging modalities such as mammography or MR.

UST-centered stand-alone workstations often incorporate UST-specific image review tools, which are not generally available on standard image-review workstations. For example, some UST-centered workstations allow the user to fuse the sound speed, attenuation, and reflection image stacks into a composite image stack which is particularly useful for diagnosis, as it presents much of the clinically relevant UST information on one image. UST-centered workstations may also calculate clinically useful parameters which are unique to UST, such as breast density, average sound speed, average attenuation, and the sound speed and/or attenuation of a user-defined region-of-interest. UST-specific workstations therefore often provide useful diagnostic information, which cannot be obtained at standard workstations.

Stand-alone UST workstations have the potential to introduce a workflow disadvantage, however, as they force the radiologist to review, UST images at the stand-alone workstation rather than at the workstation on which they typically review the images.

While the stand-alone workstation is necessary for any UST-specific image-viewing tools, the UST images can be viewed on any standard workstation using PACS connectivity. Most stand-alone UST workstations provide standard DICOM Storage and Query/Retrieve functionality as a Service Class User, which allows the workstation to interact with an installation's PACS and send-and-receive images to/from the PACS. Sending the UST images to the PACS allows them to be reviewed at any standard workstation that is connected to the PACS. A stand-alone UST workstation with PACS connectivity thus allows the user to review the UST images at the workstation with additional UST image-viewing tools, or to review the UST images at any standard workstation connected to the PACS.

11.4.5 Storing and Archiving UST Data

In the clinical setting, the UST data that is generated by the data-acquisition electronics and used to reconstruct the UST images can be safely discarded, as only the images are used for diagnosis. However, in research and developmental settings, it is often desirable to store the UST data in addition to the images. The storage and archival of UST data presents a significant challenge, as a single UST patient may reasonably result in 1–500 GB of data (even up to 1 TB in some cases), depending on the fine details of data acquisition and the patient's breast size. Given today's standard storage capacities, a typical PC or external hard drive can only store a handful of such data sets. In addition, most hard drives support a storage bandwidth of no more than about 100 MB per second. Copying the data for an average patient to a standard external hard drive may take more than 15 min, which would significantly impair the throughput of a UST system.

Most UST systems use external disk arrays for data storage and archival. External disk arrays can be bulky and expensive, but are the most practical solution offering the storage bandwidth and capacity necessary for UST data storage. With two external disk arrays in rotation, the data from a UST system operating at a four patients per hour throughput can be stored and archived by swapping the disk arrays every 2 weeks and transferring the contents of the disk array not in use into a data archival server. The disk array solution for UST data storage and archival is somewhat unwieldy; however, as the majority of UST systems do not require the data storage and archival, the disk array solution is adequate.

11.5 Ultrasound Tomography as a Clinical Product

As UST has matured, its clinical relevance has begun to be tested in actual clinical settings. In recent years, an increasing number of studies have tested the technology under real-world clinical conditions. Techniscan Inc. now has two active studies: in Freiberg, Germany, and at the University of California at San Diego. These studies have recruited more than 100 patients. For the past 6 years, our team at the Karmanos Cancer Institute in

Detroit has undertaken multiple studies in support of UST scanner development. To date, more than 600 patient scans have been completed. A spin-off company from this project, Delphinus Medical Technologies, is currently assembling a new generation of scanners which will also be used in international multicenter trials. The outcome of these multicenter trials will provide the first definitive assessment of UST efficacy by comparing its performance against mammography and MRI.

As of the writing of this paper, there are no FDA-approved true UST systems on the market. Although whole-breast US devices do exist, such as those sold by U-Systems and Siemens, they are not true UST devices because they do not perform transmission imaging. Most UST systems are either in the research stage or are in the process of commercialization (such as those described above). To the best of our knowledge, there are only two companies that are attempting commercialization of UST technology, at this time: Techniscan Medical Systems, based in Salt Lake City, Utah, USA and Delphinus Medical Technologies, based in Plymouth, Michigan, USA. In the United States, the key to any commercialization process is FDA approval, which is required to market any medical device in a clinical setting. The various levels of the approval process will govern the appearance of UST in the marketplace, both in the short term and in the long term.

The FDA offers a shorter term clearance option for emerging technologies, known as the 510(k) process. It requires a finding of substantial equivalence with a preexisting device that has been 510(k) cleared, or a device that was on the market prior to the Medical Device Amendments of 1976 (i.e., a predicate device). The 510(k) process can render a decision in as little as 3 months. Since whole-breast imaging scanners, such as the U-systems device, have been cleared in this manner, this is the most likely route by which UST devices will obtain clearance in the short term. The 510(k) process will allow UST devices to enter the diagnostic space in breast imaging. They will be used in conjunction with existing modalities and used mainly to follow up on mammographic findings. In this role, the potentially greater capabilities of UST compared to conventional US may lead to better differentiation of benign tumors from cancer, potentially reducing the number of biopsies.

The disadvantage of the 510(k) process is that it does not allow the applicant to make any claims. Consequently, the unique aspects of UST, such as transmission imaging and the associated claims of lesion characterization, cannot generally be approved in this way. The applicant must instead proceed with a premarket approval (PMA) process. The PMA process requires extensive clinical data to validate the claims that can be subsequently used for marketing purposes. The process is particularly time-consuming and expensive when claiming a screening indication. Multicenter studies, costing millions of dollars and enrolling many thousands of study subjects, need to be completed before a PMA submission for screening can be made.

In summary, the regulatory environment is likely to lead to a phased appearance of UST devices, some emerging quickly and entering the diagnostic space and others emerging in the years to come to enter the screening space.

11.6 Conclusions and Future Directions

Advances in electronics and computing have fueled the emergence of UST as a potentially viable technology for medical imaging. Ongoing clinical studies are aimed at defining the role of UST in breast imaging. The results from such studies have been highly encouraging. Studies have demonstrated images comparable, at least qualitatively, to MRI images. Furthermore, current prototype systems have been shown to be much faster than MRI, allowing much greater patient throughput. The combination of greater throughput and cheaper inherent cost promises much cheaper exams, which can be a great asset in both the diagnostic and screening settings. UST also has the advantages of being noninvasive and not requiring contrast injection, further reducing the costs and scan times.

Compared to mammography, UST offers the potential advantages of 3D versus 2D imaging. The absence of radiation eliminates the trade-off arising from radiation-based risk–benefit analysis. Finally, the UST approach does not require compression of the breast, which eliminates discomfort and pain and removes another barrier to patient acceptance and adherence to a regular screening regimen.

On the negative side, UST is not yet an FDA-approved technology. Ongoing commercialization efforts promise to remedy that situation in the near term through a 510(k) process aimed at clearing UST devices for diagnostic breast imaging. Studies supporting PMA applications for screening use are also in the planning stages. However, the latter is a longer-term process and is likely to take some years to complete. As the FDA approval process moves forward, the potential clinical impacts are noteworthy.

In the diagnostic arena, UST promises to reduce the biopsies through better differentiation of benign tumors and cancers. Conventional ultrasound is used routinely to differentiate the cysts from solid masses but rarely to differentiate solid benign masses, such as fibroadenomas, from solid malignant lesions. UST promises to differentiate the solid masses, which will greatly reduce the number of needed biopsies.

Mammography's performance is greatly curtailed for women with dense breast tissue. In patients with dense breast tissue, the false-negative rate can be as high as 50%. By virtue of its ability to penetrate tissue at low frequencies, UST has the potential to provide greater sensitivity, which will reduce the number of false-negatives.

In the diagnostic arena, UST also offers the possibility of treatment monitoring aimed at assessing the effectiveness of breast cancer therapy. The ability to image the whole breast frequently opens the door for early assessment of treatment response and rapid feedback to the clinician to inform their decision-making. A feedback loop centered on treatment response would allow more precise and personalized treatment that is better suited to the individual.

Since sound speed is a direct correlate of tissue density, UST mapping of the sound speed distribution in the breast offers a quantitative volume-based estimate of breast density. Such a measurement may prove more predictive of breast cancer risk and would enable the exploration of strategies aimed at preventing cancer from developing in high-risk women.

The potential for lower false-negative and false-positive rates combined with inexpensive, safe, and comfortable imaging makes UST an ideal technology for screening of the general population. The absence of radiation combined with low installation costs would allow easy placement in many clinical settings including not only hospitals, but also community clinics and even doctor's offices. Such a development would increase the access, leading to a larger proportion of cancer detected in early stages and therefore reducing the mortality from breast cancer.

Finally, UST is a platform technology. Combined with modular designs, such as those described in this chapter, it will be possible to readily adapt UST scanners to target other organs. Since UST relies primarily on electronics and computing, it benefits directly from Moore's law, as no limiting factors currently exist. It is therefore possible to imagine a future in which UST will have the capability to image most regions of the human body and possibly, one day, to image the whole body.

References

American Cancer Society, 2009. *Cancer Prevention & Early Detection Facts & Figures 2009,* Atlanta, GA: American Cancer Society, Inc.

Analog Devices, 2012. *Analog Devices | Semiconductors and Signal Processing ICs.* [Online] Available at: http://www.analog.com/en/index.html [Accessed December 12, 2012].

Andersen, A. H., 1987. Ray linking for computed tomography by rebinning of projection data. *Journal of the Acoustical Society of America,* 81(4), 1190–1192.

Andersen, A. H., 1990. A ray tracing approach to restoration and resolution enhancement in experimental ultrasound tomography. *Ultrasonic Imaging,* 12(4), 268–291.

Andre, M. P. et al., 1997. High-speed data acquisition in a diffraction tomography system employing large-scale toroidal arrays. *International Journal of Imaging Systems and Technology,* 8(1), 137–147.

Armstrong, K. et al., 2007. Screening mammography in women 40 to 49 years of age: A systematic review for the American College of Physicians. *Annals of Internal Medicine,* 146(7), 516–526.

Bando, H. et al., 2008. Imaging evaluation of pathological response in breast cancer after neoadjuvant chemotherapy by real-time sonoelastography and MRI. *European Journal of Cancer Supplements,* 6(7), 66–67.

Bhattacharyya, M. et al., 2008. Using MRI to plan breast-conserving surgery following neoadjuvant chemotherapy for early breast cancer. *British Journal of Cancer,* 98(2), 289–293.

Birk, M. et al., 2011. Evaluation of the reconfiguration of the data acquisition system for 3D USCT. *International Journal of Reconfigurable Computing,* 2011, 1–9.

Carson, P. L., Meyer, C. R., Scherzinger, A. L. and Oughton, T. V., 1981. Breast imaging in coronal planes with simultaneous pulse echo and transmission ultrasound. *Science,* 214(4525), 1141–1143.

Chen, J. et al., 2008. MRI evaluation of pathologically complete response and residual tumors in breast cancer after neoadjuvant chemotherapy. *Cancer,* 112(1), 17–26.

Chen, J. et al., 2006. Projecting absolute invasive breast cancer risk in white women with a model that includes mammographic density. *Journal of the National Cancer Institute,* 98(17), 1215–1226.

Duric, N. et al., 2010. *In-Vivo Imaging Results with Ultrasound Tomography: Report on an Ongoing Study at the Karmanos Cancer Institute.* San Diego, California, SPIE, p. 76290M.

Duric, N. et al., 2005. Development of ultrasound tomography for breast imaging: Technical assessment. *Medical Physics,* May, 32(5), 1375–1386.

Duric, N. et al., 2009. *Detection and Characterization of Breast Masses with Ultrasound Tomography: Clinical Results.* Lake Buena Vista, Florida, SPIE, pp. 72651G-1-8.

Duric, N. et al., 2007. Detection of breast cancer with ultrasound tomography: First results with the Computed Ultrasound Risk Evaluation (CURE) prototype. *Medical Physics,* 34(2), 773–785.

Greenleaf, J., Johnson, S., Bahn, R. and Rajagopalan, B., 1977. *Quantitative Cross-Sectional Imaging of Ultrasound Parameters.* Phoenix, Arizona: IEEE, pp. 989–995.

Hormati, A., Jovanović, I., Roy, O. and Vetterli, M., 2010. *Robust Ultrasound Travel-Time Tomography Using the Bent Ray Model.* San Diego, California: SPIE, p. 76290I.

Howry, D. H. and Bliss, W. R., 1952. Ultrasonic visualization of soft tissue structures of the body. *The Journal of Laboratory and Clinical Medicine,* 40(4), 579–592.

Jansen, S. et al., 2008. Differentiation between benign and malignant breast lesions detected by bilateral dynamic contrast-enhanced MRI: A sensitivity and specificity study. *Magnetic Resonance in Medicine,* 59(4), 747–754.

Johnson, S. A. et al., 1999. *Apparatus and Method for Imaging with Wavefields Using Inverse Scattering Techniques.* United States of America, Patent No. 6,005,916.

Kuhl, C. et al., 2007. MRI for diagnosis of pure ductal carcinoma in situ: A prospective observational study. *Lancet,* 370(9586), 485–492.

Li, C., Duric, N., Littrup, P. and Huang, L., 2009. *In vivo* breast sound-speed imaging with ultrasound tomograph. *Ultrasound in Medicine & Biology,* 35(10), 1615–1628.

Liu, D.-L. and Waag, R., 1997. Propagation and backpropagation for ultrasonic wavefront design. *IEEE Transactions on Ultrasonics, Ferroelectrics and Frequency Control,* 44(1), 1–13.

Marmarelis, V. Z., Kim, T. and Shehada, R. E., 2003. *High Resolution Ultrasonic Transmission Tomography.* San Diego, California: SPIE, pp. 33–40.

Martin, L. J. and Boyd, N. F., 2008. Potential mechanisms of breast cancer risk associated with mammographic density:

Hypotheses based on epidemiological evidence. *Breast Cancer Research,* 10(1), 201.

Mast, T. D., 2000. Empirical relationships between acoustic parameters in human soft tissues. *Acoustics Research Letters Online,* 1(2), 37–42.

Masugata, H. et al., 1999. Relationship between myocardial tissue density measured by microgravimetry and sound speed measured by acoustic microscopy. *Ultrasound in Medicine and Biology,* 25(9), 1459–1463.

Micikevicius, P., 2009. *3D Finite Difference Computation on GPUs Using CUDA.* Washington, D.C., Associatioin for Computing Machinery, pp. 79–84.

National Cancer Institute, 2011. *Surveillance Epidemiology and End Results.* [Online] Available at: http://seer.cancer.gov/ [Accessed November 11, 2011].

Natterer, F. and Wûbbeling, F., 2001. *Mathematical Methods in Image Reconstruction.* 1st ed. Philadelphia, PA: Society for Industrial and Applied Mathematics.

Norton, S. J., 1987. Computing ray trajectories between two points: A solution to the ray-linking problem. *Optical Society of America,* 4(10), 1919–1922.

Partridge, S. C. et al., 2002. Accuracy of MR imaging for revealing residual breast cancer in patients who have undergone neoadjuvant chemotherapy. *American Journal of Roentgenology,* 179(5), 1193–1199.

Pratt, R. G., 1999. Seismic waveform inversion in the frequency domain, part 1: Theory, and verification in a physical scale model. *Geophysics,* 64(3), 888–901.

Pratt, R. G., Huang, L., Duric, N. and Littrup, P., 2007. *Sound-Speed and Attenuation Imaging of Breast Tissue Using Waveform Tomography of Transmission Ultrasound Data.* San Diego, California: SPIE, p. 65104S.

Ranger, B. et al., 2009. *Breast Imaging with Acoustic Tomography: A Comparative Study with MRI.* Lake Buena Vista, Florida: SPIE, pp. 726510-1-8.

Ranger, B. et al., 2010. *Breast Imaging with Ultrasound Tomography: A Comparative Study with MRI.* San Diego, California: SPIE, pp. 76291C-1.

Roy, O. et al., 2011. *Robust Array Calibration Using Time Delays with Application to Ultrasound Tomography.* Lake Buena Vista, Florida: SPIE, p. 796806.

Roy, O. et al., 2010. *Sound Speed Estimation Using Wave-Based Ultrasound Tomography: Theory and GPU Implementation.* San Diego, California: SPIE, p. 76290J.

Ruiter, N. V. et al., 2011. *Realization of an Optimized 3D USCT.* Lake Buena Vista, Florida: SPIE, pp. 796805-1-8.

Saslow, D. et al., 2007. American Cancer Society guide-lines for breast screening with MRI as an adjunct to mammography. *CA: A Cancer Journal for Clinicians,* 57(2), 75–89.

Schmidt, S. et al., 2011. *Modification of Kirchhoff Migration with Variable Sound Speed and Attenuation for Tomographic Imaging of the Breast.* Lake Buena Vista, Florida: SPIE, pp. 796804-1-11.

Schomberg, H., 1978. An improved approach to reconstructive ultrasound tomography. *Journal of Physics D: Applied Physics,* 11(15), L181–L185.

Sharma, U., Danishad, K., Seenu, V. and Jagannathan, N., 2009. Longitudinal study of the assessment by MRI and diffusion-weighted imaging of tumor response in patients with locally advanced breast cancer undergoing neoadjuvant chemotherapy. *NMR in Biomedicine,* 22(1), 104–113.

Tarantola, A., 2005. *Inverse Problem Theory and Methods for Model Parameter Estimation.* Philadelphia: Society for Industrial and Applied Mathematics.

Texas Instruments, 2011. *Ultrasound Receive Chain Evaluation Module.* [Online] Available at: http://www.ti.com.cn/cn/lit/an/slaa320/slaa320.pdf [Accessed November 11, 2011].

Tozaki, M., 2008. Diagnosis of breast cancer: MDCT versus MRI. *Breast Cancer,* 15(3), 205–211.

Turnbull, L., 2009. Dynamic contrast-enhanced MRI in the diagnosis and management of breast cancer. *NMR Biomedicine,* 22(1), 28–39.

Ursin, G. et al., 2005. Greatly increased occurrence of breast cancers in areas of mammographically dense tissue. *Breast Cancer Research,* 7(5), 605–608.

Weiwad, W. et al., 2000. Direct measurement of sound velocity in various specimens of breast tissue. *Investigative Radiology,* 35(12), 721–726.

Wild, J. J. and Reid, J. M., 1952. Application of echo-ranging techniques to the determination of structure of biological tissues. *Science,* 115(2983), 226–230.

12

Imaging the Elastic Properties of Tissue

Kevin J. Parker
University of Rochester

12.1 Introduction and Overview of Elastography

From the earliest days of medicine, the palpable "feel" of tissues and organs has been used in diagnostic evaluations. Even today, screening exams include palpation of the liver, breast, prostate, thyroids, skin lesions, and other tissues to characterize their condition. However, manual palpation is limited to accessible surfaces and is not quantitative. By the 1980s, there existed a wide gap between the impressive imaging abilities within radiology and the important but limited information that was obtained from palpation. How could these two disparate domains be bridged to add hidden biomechanical properties into modern radiological imaging systems? Some early work took advantage of the availability of Doppler ultrasound devices to study the tissue motion and abnormalities in the 1970s and 1980s. Out of this background came a remarkable series of imaging approaches to map out the elastic properties of tissues. Initially, images were produced by vibration sonoelastography, where vibrational shear waves (typically between 50 and 1000 Hz continuous wave) are excited within tissue and the resulting vibrations are imaged and displayed using the Doppler-displacement or phase estimators. Other innovative approaches applied step compressions, transient forces, and a variety of additional imaging modalities, such as magnetic resonance elastography (MRE) and optical imaging, to uncover the biomechanical properties of tissue that were previously unobservable with conventional radiology.

While there are an impressive variety of approaches to imaging the elastic properties of tissue, it can be emphasized that all techniques lie on a continuum of biomechanical responses. In fact, the time-dependent terms can be seen as important drivers of the particular response of tissues, and this leads naturally to a continuum, or spectrum, of approaches ranging from very slow motion, through sinusoidal steady-state motion, to impulsive motion as depicted in Figure 12.1. An overview of the continuum analysis was given by Parker et al. (2005).

Consider a block of tissue-mimicking material as shown in Figure 12.2, constrained on the right side, free on top and bottom, but capable of being displaced from the left side, while being imaged so that internal displacements can be accurately estimated. The equation of motion that governs the displacements u given by

$$(\lambda + \mu)\frac{\partial^2 u_j}{\partial x_j \partial x_i} + \mu \frac{\partial^2 u_i}{\partial x_j \partial x_j} = \rho \frac{\partial^2 u}{\partial t^2}, \qquad (12.1)$$

or

$$(\lambda + \mu)\nabla(\nabla \cdot \mathbf{u}) + \mu\nabla^2\mathbf{u} = \rho\ddot{\mathbf{u}},$$

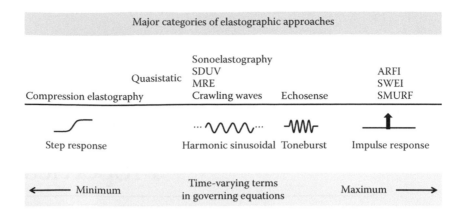

FIGURE 12.1   Some of the major approaches to imaging the elastic properties of tissue, arranged by the type and duration of the excitation. Broadly speaking the techniques can be organized into step-response, harmonic (sinusoidal steady state) applications, toneburst, and impulse-response excitations.

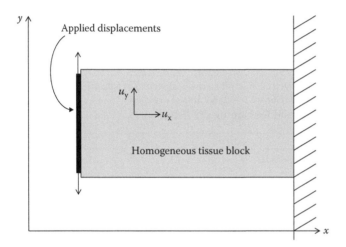

FIGURE 12.2   A homogeneous block of biomaterial is constrained on the right and is displaced on the left. As the applied displacements move from very slow shear or compression, to higher frequency harmonic shear in the *y*-direction, to broadband (impulsive) excitation, the form of the solution within the interior moves from linear with *x*, to sinusoidal in *x*, to a propagating pulse in *x*. However, all are solutions to the basic equation of motion. In this way, all the major elastographic imaging approaches can be seen as points along a common biomechanical spectrum.

where λ and μ are the Lamé constants, capturing the mechanical properties of the material, and where the body forces, such as gravity, have been assumed negligible. This equation, with given boundary and initial conditions, governs the general dynamic response of a homogeneous, isotropic, linearly elastic material to a force or displacement excitation.

Now if we very slowly compress the block, or slowly shear it in the *y*-direction, the time-varying terms are negligible and the solution to the displacement is simply linear with *x* (going to zero on the right side where the block is constrained). This is the basis for compression elastography. Furthermore, if we repeat the shear motion in the *y*-direction in a sinusoidal motion, as

the frequency increases, the time-varying terms of Equation 12.1 become nonnegligible and at some minimum frequency a modal pattern can be established. The displacement within the material is now a shear wave of a sinusoidal form, both in time and space. This is the basis for vibration elastography and MRE. Next, if we displace the left side with multiple frequencies (chords), or short tone bursts, then the displacements are comprised of multiple frequency components.

Finally, if we displace the left side with a sharp impulsive excitation, then a shear wave is produced that will propagate as a disturbance through the biomaterial, like a ripple on a taught string. This is the basis for impulsive shear-wave imaging. Thus, as the frequency content of the applied displacement moves from very low frequencies to sinusoidal harmonic frequencies above 40 Hz, to higher frequencies and wider bandwidths, and finally to an impulsive (broadband) excitation, the form of the solution varies, but all are solutions to Equation 12.1.

The organization of the chapter is as follows: first, we review some of the landmark studies of tissue motion leading up to the development of vibration elastography and compression elastography. Then, the historical development of some of the major imaging approaches is reviewed. A particular emphasis is given to harmonic shear-wave approaches. Transient shear-wave approaches and radiation-force excitations are treated in Chapter 13. The closing sections describe some major clinical trial results and future directions. The related field of inverse problems in elastography is a complex issue beyond the scope of this chapter. These are addressed in a recent review article by Doyley (2012).

12.2 Early Studies of Tissue Properties from Motion

12.2.1 Foundational Studies

In 1951, Oestreicher and colleagues published an early landmark study of the behavior of the human body when subjected to mechanical vibrations or sound fields (Oestreicher 1951;

Von Gierke et al. 1952). Photography and a strobe light were used to record the vibration waves on the skin. The resultant surface-wave propagation patterns gave wavelength and wavespeed data, which led to the formulation of a theory to correlate increased wave frequency with increased tissue impedance. From experimental data, the shear modulus was calculated.

12.2.2 Ultrasound A-Scan and B-Scan Studies

Wilson and Robinson presented a signal-processing technique to measure small liver-tissue displacements caused by vessel diameter differences during aortic pulsation in 1982 (Wilson and Robinson 1982). A coarse estimate of displacement was achieved by integrating velocity over time. In this time frame, Professor Kit Hill in the United Kingdom directed some of the landmark studies of tissue motion. Dickinson and Hill used the correlation coefficient between the consecutive A-scans to measure the tissue-motion frequency and amplitude (Dickinson and Hill 1982). In order to measure the changes of the interrogated region between two consecutive A-scans, they established a correlation parameter. The correlation parameter, which is unity for stationary tissue, decreases monotonically as tissue motion increases. Tristam et al. (1988, 1986) took this approach further to investigate the varied responses of normal liver and cancerous liver to cardiac movement. Various features on the correlation curves were found to distinguish normal liver from that with tumor. Livers with tumor were found to have lower maximum values, fewer peaks, and greater regularity. Modifying the correlation technique to measure tissue motion, De Jong et al. (1990) found the maximum cross-correlation using an interpolation algorithm.

Assessment of fetal pulmonary maturity and normal pulmonary development can be achieved by measurements of lung tissue elasticity. An attempt to qualitatively determine the stiffness of fetal lung was made by Birnholz and Farrell (1985). This was done by evaluating the ultrasound B-scans, which indicate lung compression due to cardiac pulsations. They proposed that soft lung tissue compresses (with maximal deformation immediately adjacent to the heart) while stiff lung tissue exhibits no regional deformation (it transmits cardiac pulsations moving as a block). More quantitative approximations were achieved by Adler et al. (1989, 1990).

Induced quasistatic compression was first investigated by Eisensher et al. (1983). This was accomplished by applying a 1.5 Hz vibration source to liver and breast tissue and using M-mode ultrasound for detection. The study found that benign lesions had characteristically sinusoidal quasistatic compression response, while malignant tumors tended to have a more nonlinear (flat) response.

12.2.3 Doppler Methods

Krouskop et al. reported one of the first quantitative measurements of tissue elasticity using gated pulsed Doppler (Krouskop et al. 1987). Assuming isotropy and incompressibility, they reduced the equations relating the tissue properties and tissue movements to linear forms. Measuring the tissue peak displacements and their gradients becomes the final reductive solution to finding the tissue elasticity. They measured actual tissue motion at various points of interest under external vibration created by one A-line pulsed Doppler instrument. A possible measurement of tissue stiffness in a very small region (i.e., 0.5×0.5 mm) was suggested. This paper is noteworthy in that it presages, with local Doppler vectors and external vibration, MRE and vibration sonoelastography.

In certain cases, the Doppler spectrum produced by the scattered ultrasound signal from a vibrating target can be similar to a pure-tone frequency modulation in that it has symmetric side harmonics in the area of the carrier frequency. The harmonic spacing is the same as the target vibrating frequency. The harmonic amplitudes are given by consecutive Bessel functions of the first kind. The expression for the signal (Taylor 1976) is

$$s_r(t) = A \sum_{-\infty}^{\infty} J_n(\beta) \cos[\omega_0 t + n(\omega_L t + \varphi)]. \qquad (12.2)$$

Using Doppler ultrasound to examine the abnormal oscillation of heart valves, Holen et al. (1985) first detected this characteristic Bessel-band Doppler spectrum. By counting the number of significant harmonics (as an estimation of the frequency modulated bandwidth), the amplitude of the vibration was estimated.

The Doppler ultrasound response to a low-frequency sound field in the auditory organs of fish was examined by Cox and Rogers (1987). By analyzing the ratio of the carrier to the first side band of the Doppler spectrum, they discovered the amplitude of the hearing organ's vibration. These studies set the foundation for the development of imaging techniques, which are described in the next sections.

12.3 Harmonic Shear Waves

12.3.1 Vibration Amplitude Sonoelastography: Early Results

In 1987, Lerner and Parker presented early work on the vibration amplitude sonoelastography (Lerner and Parker 1987; Lerner et al. 1988). Vibration amplitude sonoelastography entails the application of a continuous low-frequency vibration (50–1000 Hz) to excite the internal shear waves within the tissue of interest. A disruption in the normal vibration patterns will result if a stiff inhomogeneity is present in soft-tissue surroundings. A real-time vibration image can be created by the Doppler detection algorithms. Modal patterns can be created in certain organs with regular boundaries. The shear-wave speed of sound in the tissue of these organs can be ascertained with the information revealed by these patterns (Parker and Lerner 1992).

Figure 12.3 reproduces the first vibration-amplitude sonoelastography image (Lerner and Parker 1987; Lerner et al. 1988).

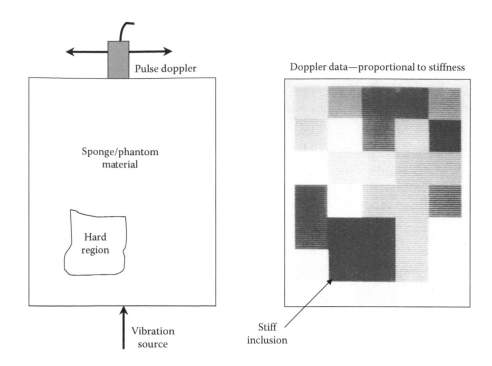

FIGURE 12.3 Original imaging data and schematic of the first known image of relative stiffness, derived from Doppler data in a phantom with applied vibration. (The original image was published in 1987 and 1988. Lerner, R. M. and K. J. Parker. 1987. Sonoelasticity images derived from ultrasound signals in mechanically vibrated targets. In Seventh European Communities Workshop. J. Tjissen, editor, Nijmegen, The Netherlands; Lerner, R. M. et al. 1988. *Acoust Imag* 16:317–327.)

The vibration within a sponge and saline phantom containing a harder area (the dark region) is depicted by this crude image. Range-gated Doppler was used to calculate the vibration amplitude of the interior of the phantom as it was vibrated from below. By 1990, a modified color Doppler instrument was used by the University of Rochester group to create real-time vibration-amplitude sonoelastography images. In these images, vibration above a certain threshold (in the 2 micron range) produced a saturated color (Figure 12.4).

Measurements of tissue elastic constants, finite-element modeling results, and phantom and *ex vivo* tissue sonoelastography were later reported (Lerner et al. 1990; Parker et al. 1993, 1990). Thus, by the end of 1990, the working elements of vibration elastography (sonoelastography and sonoelasticity were other names used at the time) were in place, including real-time imaging techniques and stress–strain analysis of tissues such as prostate, with finite-element models and experimental images demonstrating conclusively that small regions of elevated Young's modulus could be imaged and detected using conventional Doppler-imaging scanners.

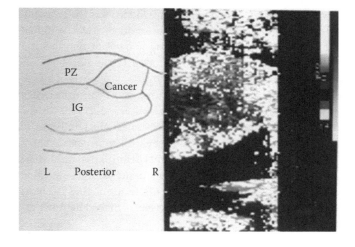

FIGURE 12.4 A representative first-generation image of vibration sonoelastography, circa 1990. Doppler spectral variance is employed as an estimator of vibration in the 1–10 micron range and displayed over the B-scan images. No color implies low vibrations below threshold. Shown is the fill-in of vibration within a whole prostate, with a growing cancerous region indicated by the deficit of color within the peripheral zone. (Courtesy of Dr. R. M. Lerner.)

12.3.2 Theory

Along with the useful finite-element approach of (Lerner et al. 1990; Parker et al. 1990), there were important theoretical questions to be answered by analytical or numerical techniques. How did vibration fields behave in the presence of elastic inhomogeneities? What is the image contrast of lesions in a vibration field,

and how does the contrast depend on the choice of parameters? How do we optimally detect sinusoidal vibration patterns and image them using the Doppler or related techniques? These important issues were addressed in a series of papers through the 1990s. One foundational result was published in 1992 under the title "Sonoelasticity of Organs, Shear Waves Ring a Bell" (Parker and Lerner 1992). This paper demonstrated experimental proofs that vibrational eigenmodes could in fact be created in whole organs, including the liver and kidneys, where extended surfaces create reflections of sinusoidal steady-state shear waves. Lesions would produce a localized perturbation of the eigenmode pattern, and the background Young's modulus could be calculated from the patterns at discrete eigenfrequencies. Thus, both quantitative and relative imaging contrast detection tasks could be completed with vibration elastography in a clinical setting, *in vivo*, by 1992. A later review of eigenmodes and a strategy for using multiple frequencies simultaneously (called "chords") were given in Taylor et al. (2000).

In 1994, a vibration-amplitude analytical model was created (Rubens et al. 1995). This model used a sonoelastic Born's approximation to solve the wave equations in an inhomogeneous, isotropic medium. The total wave field inside the medium can be expressed as

$$\Phi_{total} = \Phi_i + \Phi_s, \tag{12.3}$$

where Φ_i is the homogeneous or incident field and Φ_s is the field scattered by the inhomogeneity. They satisfy, respectively

$$(\nabla^2 + k)\Phi_i = 0 \tag{12.4}$$

$$(\nabla^2 + k)\Phi_s = \delta(x), \tag{12.5}$$

where $\delta(x)$ is a function of the properties of inhomogeneity. The theory accurately describes how a hard or soft lesion appears as disturbances in a vibration pattern. Figure 12.5 summarizes the theoretical and experimental trends.

Signal-processing estimators were also developed in the study of vibration amplitude sonoelastography. Huang et al. suggested a method to estimate a quantity denoted β, which is proportional to the vibration amplitude of the target, from the spectral spread

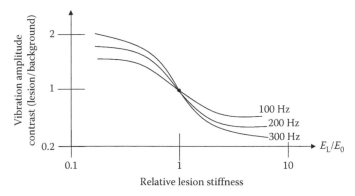

FIGURE 12.5 Theoretical results of the contrast of vibration sonoelastography for soft or hard lesions in a background medium. The image contrast increases with both increasing frequency and with increasing size of the lesion.

(Huang et al. 1990). They found a simple correlation between β and the Doppler spectral spread σ_ω:

$$\beta = \sqrt{2}(\sigma_\omega / \omega_L), \tag{12.6}$$

where ω_L is the vibration frequency of the vibrating target. This is an uncomplicated but very useful property of the Bessel Doppler function. The effect of noise, sampling, and nonlinearity on the estimation was also considered. In their later work, they studied real-time estimators of vibration amplitude, phase, and frequency that could be used for a variety of vibration sonoelastography techniques (Huang et al. 1992).

Finally, an overall theoretical approach that places vibration sonoelastography on a biomechanical spectrum with other techniques, including compression elastography, MRE, and the use of impulsive radiation force excitations, is found in "A Unified View of Imaging the Elastic Properties of Tissues" (Parker et al. 2005).

12.3.3 Vibration Phase Gradient Sonoelastography

In parallel to the early work at the University of Rochester, a vibration phase gradient approach to sonoelastography was developed by Sato and collaborators in the late 1980s at the University of Tokyo (Yamakoshi et al. 1990). They mapped the amplitude and the phase of low-frequency wave propagation inside the tissue. From this mapping, they were able to derive wave-propagation velocity and dispersion properties, which are directly linked to the elastic and viscous characteristics of tissue.

The phase-modulated (PM) Doppler spectrum of the signal returned from sinusoidally oscillating objects approximates that of a pure-tone frequency modulation (FM) process as given in Equation 12.2. This similarity indicates that the tissue vibration amplitude and phase of tissue motion may be estimated from the ratios of adjacent harmonics. The amplitude ratio between contiguous Bessel bands of the spectral signal is

$$A_{i+1} / A_i = |J_{i+1}(\beta) / J_i(\beta)|, \tag{12.7}$$

where A_i is the amplitude at the ith harmonic, β is the unknown amplitude of vibration in the tissue, and $J_i(\cdot)$ is an ith-order Bessel function. β can be estimated from the experimental data if A_{i+1}/A_i is calculated as a function of β beforehand. The phase of the vibration was calculated from the quadrature signals.

The display of wave propagation as a moving image is permitted by constructing phase and amplitude maps (Figure 12.6) as a function of time. The use of a minimum squared error algorithm to estimate the direction of wave propagation and to calculate phase and amplitude gradients in this direction allows images of amplitude and phase to be computed offline. By assuming that the shear viscosity effect is negligible at low frequencies, Sato's group obtained preliminary *in vivo* results (Yamakoshi et al. 1990).

Sato's technique (Yamakoshi et al. 1990) was used and refined by Levinson (Levinson et al. 1995), who developed a more general

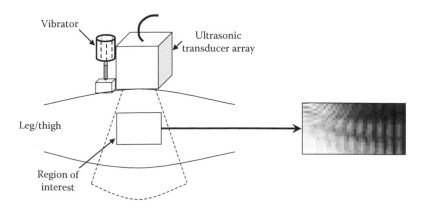

FIGURE 12.6 A depiction of the system developed by Professor Sato and colleagues at the University of Tokyo, including external vibration and an imaging array with signal processing for estimating the phase of the vibration in tissue. The rate of change of phase can be estimated to yield the tissue hardness.

model of tissue viscoelasticity and a linear recursive filtering algorithm based on cubic B-spline functions. Levinson took the Fourier transform of the wave equation and derived the frequency-domain displacement equation for a linear, homogeneous, and isotropic viscoelastic material. From this, equations that relate the shear modulus of elasticity and viscosity to the wave number and the attenuation coefficient of the wave can be derived.

Levinson et al. (1995) conducted a series of experiments on the quadriceps muscle group in human thighs. It was assumed that shear waves predominate and that viscosity at low frequencies is negligible. Phase gradient images of the subjects' thighs under conditions of active muscle contraction enabled the calculation of Young's modulus of elasticity. The tension applied to the muscle was controlled using a pulley device. The measured vibration propagation speed and the calculated values of Young's modulus increased with the increasing degrees of contraction needed to counteract the applied load.

12.3.4 Magnetic Resonance Elastography

Tissue motion estimates can be obtained using magnetic resonance imaging (MRI), and despite the comparatively long acquisition times (typically minutes per direction), MRE has also shown promise in the detection and differential diagnosis of breast (Kruse et al. 2000; Lorenzen et al. 2001; McKnight et al. 2002; Van Houten et al. 2003) and prostate (Kemper et al. 2004) cancer, monitoring minimally invasive therapeutic techniques (Wu et al. 2000), and characterizing the mechanical properties of brain (McCracken et al. 2005) and plantar (Weaver et al. 2005).

MRI is more expensive than diagnostic ultrasound, but it can measure all three spatial components of the induced tissue displacement with high accuracy and precision, and thus is more suited for quantitative elastography (Dresner et al. 2001; Manduca et al. 1998; Samani et al. 2003; Tikkakoski 2007; Van Houten et al. 2001; Weaver et al. 2001). Consequently, several groups have actively developed MRE; like its ultrasound counterpart, MRE visualizes the mechanical properties within soft tissues by employing either a quasistatic or dynamic mechanical

source (Kruse et al. 2000; Muthupillai et al. 1997; Plewes et al. 2000; Sack et al. 2004; Sinkus et al. 2007; Weaver et al. 2001). However, in MRE, the induced tissue motion is measured using either the saturation tagging (Axel and Dougherty 1989; Zerhouni et al. 1988) or the phase-contrast method.

12.3.4.1 Early MRE Approaches

Fowlkes et al. (1995) proposed a quasistatic approach to MRE that measures internal tissue strain using the saturation-tagging method. Although encouraging strain elastograms were produced with this technique, the spatial resolution of images is reduced. Plewes and colleagues (Bishop et al. 1998; Hardy et al. 2005; Plewes et al. 1995) proposed an MRE method that utilizes the phase-contrast imaging sequence to estimate the spatial strain incurred when the tissue or phantom under investigation is deformed quasistatically. They conducted phantom studies, which support the results observed in quasistatic ultrasound elastography (Bamber et al. 2002; Bamber and Bush 1996; O'Donnell et al. 1994; Ophir et al. 1997). More specifically, they confirmed that high-contrast focal lesions are discernable in strain elastograms. They also proposed to reconstruct the intrinsic tissue mechanical properties using an iterative inverse image reconstruction method (Bishop et al. 2000).

Muthupillai et al. (1995) proposed a dynamic approach to MRE based on the phase-contrast imaging method. They modified a standard spin-echo imaging sequence by introducing a motion-sensitizing, direction-specific MR gradient whose polarity was switched repeatedly at a defined frequency. They induced shear waves within the tissue under investigation at the same frequency as the motion-encoding gradient with a harmonic mechanical source that was coupled to the surface. The cyclic motion of nuclear spins produces a phase shift φ in the received NMR signal when the motion-sensitizing gradient is applied. The resulting phase shift is given by

$$\varphi(r,\theta) = \frac{2\gamma NT(\vec{G} \cdot \vec{\xi}_0)}{\pi} \sin(k \cdot \vec{r} + \theta), \qquad (12.8)$$

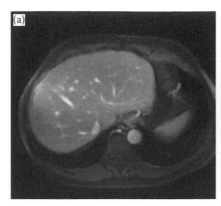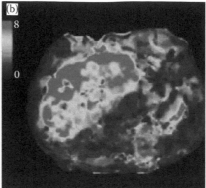

FIGURE 12.7 **(See color insert.)** MRI of liver (a) and MRE analysis (b) demonstrating stiffness consistent with fibrosis. Image courtesy of Russell N. Low, M.D., Sharp and Children's MRI Center, San Diego, CA. Image acquired on GE Healthcare Signa HDxt 1.5T MRI.

where $\vec{\xi}_0$ represents the displacement amplitude vector, $\vec{G}$ represents the motion-sensitizing magnetic gradient vector, θ represents the relative phase between the mechanical source and magnetic oscillations, γ represents the gyro magnetic ratio and is the spin position vector, $\vec{r}$ represents the mean position, k represents the wave vector, T represents the period of the mechanical excitation, and N represents the number of gradient cycles. Spins will have the maximum phase shifts of opposing polarities when their components of motion are in the direction of the gradient vector, and spins will have no net phase shift when their components of motion are 90° out of phase with the magnetic gradient vector. Muthupillai et al. (1995) demonstrated using phantoms and excised tissue samples that propagating shear waves can be visualized at high spatial resolution. Like sonoelasticity imaging, the computed shear modulus can be calculated directly from local estimates of wavelength. Although this technique is very elegant in terms of its simplicity, accurate quantification of wavelength in complex organs such as the breast can be difficult; therefore, Muthupillai and colleagues computed the shear modulus by applying the local frequency estimation (LFE) technique (Manduca et al. 1996) to MRE data, which has produced encouraging *in vivo* and *in vitro* results despite its origin as an imaging processing technique (Dresner et al. 1998; Kruse et al. 2008, 2000).

12.3.4.2 Steady-State Harmonic MRE

Two groups (Sinkus et al. 2000; Weaver et al. 2000) independently proposed an alternate approach to MRE that measured the mechanical properties of soft tissue under steady-state harmonic excitation. Like the MRE approach described by Muthupillai et al. (1995), this method also employed the phase-contrast imaging method. The main difference between the two approaches is the method used to reconstruct the shear modulus. Sinkus et al. (2000) proposed a direct inversion scheme where the elastic properties (e.g., shear modulus and Poisson's ratio) were calculated using a system of partial differential equations in which the spatial derivatives of the measured displacements appear as coefficients. This technique is very appealing because it is linear, efficient, and can be localized under the assumptions of

piecewise homogeneity in the mechanical properties. However, the technique is also very sensitive to the measurement noise, which is not surprising since significant noise amplification occurs when the measured data is differentiated. Consequently, the measured displacements and/or their spatial derivatives must be filtered to stabilize the performance in practice.

Weaver et al. (2001) compute the shear modulus from time-harmonic tissue displacements using an iterative inversion technique that does not require the data differentiation (Van Houten et al. 2001, 1999). While this method may be less sensitive to the measurement noise, its utility is hampered by the large computational overhead that is required to solve the three-dimensional (3D) inverse elastography problems on a highly resolved finite-element mesh, which the group overcame by implementing their overlapping subzone inversion technique (VanHouten et al. 1999) on a parallel computing form (Doyley et al. 2004). An example of MRE reconstruction is demonstrated in Figure 12.7.

12.3.5 Transient Shear Waves in Liver (from Mechanical Sources)

Catheline et al. (1999) devised a low-frequency (50 Hz) external mechanical vibrator integrated with an ultrasound M-mode system. This integrated applicator can be applied through the ribs to excite a short tone burst that propagates into the liver. The ultrasound tracking provides a displacement history along the axial center-line, and from this a global estimate of liver shear velocity and thus elasticity can be estimated. This technique is implemented in a stand-alone clinical device, the Fibroscan manufactured by Echosens in France, and is unique in not requiring the integration into a conventional imaging system.

12.3.6 Quasistatic Vibration

Salcudean and colleagues at the University of British Columbia have developed systems for periodic and low-frequency compression of tissues. The induced tissue motion at lower frequencies (e.g., 20 Hz) has reduced the inertial (or time derivative) effects compared to higher frequencies. Analysis at low frequencies can be done in a number of ways, analogous to compression elastography, or by deriving the transfer functions (Eskandari et al. 2008; Mahdavi et al. 2011; Salcudean et al. 2006). The advantages of this system include compatibility with conventional scanners, ease of use, and repeatability. Applications include cancer detection in the liver, prostate, and breast.

12.3.7 Crawling Waves

A fascinating extension of vibration sonoelastography is the use of an interference pattern formed by two parallel shear-wave sources. The interference patterns reveal the underlying local elastic modulus of the tissue. The term "crawling waves" comes from the useful fact that by implementing a slight frequency or progressive phase difference between the two parallel sources,

the interference pattern will move across the imaging plane at a speed controlled by the sources (Wu et al. 2004). Thus, the crawling waves are readily visualized by conventional Doppler imaging scanners at typical Doppler frame rates; there is no need for ultrafast scanning. Other advantages of crawling waves include (1) the region of interest excited between the two sources is large, (2) most of the energy in the crawling waves is aligned in the Doppler (axial) direction, and (3) a number of analysis or estimation schemes can be applied in a straightforward manner to derive the quantitative estimate of local shear-wave velocity and Young's modulus. Crawling waves can also be implemented by a number of techniques including mechanical line sources, surface applicators, or radiation force excitations of parallel beams.

The use of crawling waves was first described in 2004 by Wu et al. at the University of Rochester. It was shown that crawling waves could be used to accurately derive the Young modulus

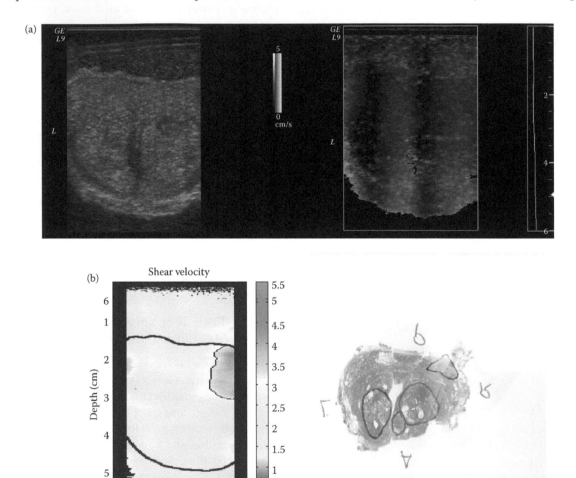

FIGURE 12.8 Crawling waves in the prostate. (a) Left: B-scan of whole excised prostate; right: crawling-waves frame. (b) Right: pathology with labels: upper right outlined area = cancer and lower central outlined areas = BPH; left: quantitative estimates of shear velocity from crawling-waves analysis, indicating the hard region as a red area (corresponding to the region with cancer). Estimates from an average of three frequencies are shown. All images are coregistered.

of materials and to delineate the stiff inclusions (Wu 2005; Wu et al. 2002, 2006, Zhang et al. 2007). Estimators of shear-wave speed and Young's modulus and the shear-wave attenuation are derived in Hoyt et al. (2007b,c, 2008a) and McLaughlin et al. (2007). Applications of crawling waves include the *ex vivo* prostate (Castaneda et al. 2009; Hoyt et al. 2006, 2008b), *in vivo* muscle (Hoyt et al. 2008c), and *ex vivo* liver (Zhang et al. 2007). An example from *ex vivo* prostate is given in Figure 12.8a and b.

Implementation into scanning probes can be accomplished by utilizing the radiation force excitation along the parallel beams (An et al. 2010; Cho et al. 2010; Hah et al. 2010).

12.4 Compression Elastography

Introduced in 1991 by Ophir et al. (1991), compression elastography utilizes a comparison of ultrasound B-scan RF information from tissue before and after a modest compression. Within 10 or so years, this approach was perhaps the most widely implemented elastography technique for breast-cancer studies. The concept can be explained by appeal to the stress–strain relations under simple uniaxial (one-dimensional) displacement. This was modeled by use of springs, as depicted in Figure 12.9.

In one-dimensional compression, a series of springs will displace according to their relative stiffness—the stiffest spring will compress the least. Thus, by calculating the strain (the derivative of displacement), areas of relative stiffness could be identified and imaged as low-strain regions, as demonstrated in Figure 12.10. Displacements are estimated by comparing the echoes before and after compression by correlation methods, or by other techniques. Compression elastography thus produces images of relative strain that are simple to interpret so

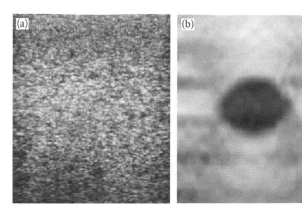

FIGURE 12.10 Representative compression elastography displays. (a) B-scan with embedded isoechoic lesion in a phantom. (b) Corresponding strain image demonstrating a dark region of low strain, indicating a hard lesion.

long as the applied stress is relatively uniform. The advantage of the technique is that the ultrasound scanning transducer, handheld in most applications, can be used to produce a localized compression near the region of interest in the breast and other applications, particularly for more superficial targets of interest.

The disadvantages include the relative nature of the strain image, the requirement for nearly uniform strain to interpret the image, the tendency of objects to move out-of-plane during compression, and the difficulty of compressing the deeper organs.

Major engineering efforts have improved the signal processing and systems implementation of compression elastography (Hall et al. 2003; Shiina et al. 2002; Sumi 2008; Varghese and Ophir 1997; Varghese et al. 1996). In addition, efforts have been made to solve the inverse problem and create quantitative estimates of the tissue elasticity from models applied to the raw displacement data (Barbone and Gokhale 2004; Fehrenbach 2007). However, the clinical applications currently under the investigation are predominantly conducted using the compression and interpretation of the resulting relative strain image.

Compression elastography can be implemented on B-scan imagers as real-time dedicated signal processing, or as postprocessing of the pre- and postcompression signals. The straightforward implementation has helped disseminate the approach to a number of platforms including Hitachi, Siemens, Ultrasonix, and other scanners (Varghese and Ophir 1997). Research applications include HIFU lesions in liver (Righetti et al. 1999), prostate (Kallel et al. 1999), breast (Garra et al. 1997), RF lesions in liver (Varghese et al. 2003), and HIFU lesions in the prostate (Curiel et al. 2005). Extensions of compression elastography include poroelastic imaging with time-dependent behavior of complex media (Berry et al. 2006) and shear-strain elastography (Thitaikumar et al. 2007). Clinical applications are covered in the next section.

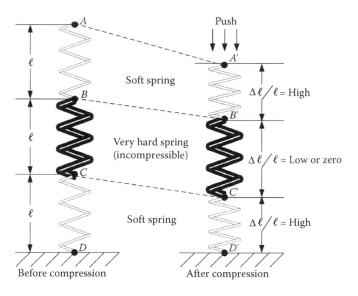

FIGURE 12.9 Schematic of conceptual argument used to explain the strain images produced by early compression elastography systems. (From Ophir, J. et al. 1991. *Ultrason Imaging* 13:111–134.)

12.5 Clinical Applications

Through the implementation of various time and frequency estimators, vibration sonoelastography has become a real-time diagnostic tool with applications for both deep and superficial organs. The liver, kidney, prostate, and breast can be imaged using the vibration sonoelastography to characterize the typical patterns and frequency ranges (Lee et al. 1991; Parker and Lerner 1992; Rubens et al. 1995; Zhang et al. 2008b). In a 1994 study (Rubens et al. 1995) of *ex vivo* prostate cancer, vibration amplitude sonoelastography was shown to have better sensitivity and predictive value than B-scan imaging alone.

Muscle tissue was studied to demonstrate the relationship between Young's modulus of *in vivo* muscle and applied load or exercise regimens (Hoyt et al. 2007a, 2008c; Levinson 1987; Levinson et al. 1995). An unusual application of sonoelasticity is the measurement of intraocular pressure to test for glaucoma, via the resonance of the human eye *in vivo* (Alam et al. 1994).

Another important application of elastography imaging is the precise delineation of thermal lesions produced intentionally by either microwaves, RF, or HIFU. A series of pig liver lesions created by multiple methods were imaged by sonoelastography and a high correlation was found between the lesion dimensions determined by imaging and later measurements from gross pathology (Taylor and Zhang 2002; Zhang et al. 2008a), both *in vivo* and *ex vivo*.

12.5.1 Three-Dimensional Prostate Cancer Detection

Given the high incidence of prostate cancer in older males, and the poor sensitivity of conventional imaging modalities for detecting the low-contrast cancers of the prostate, there is a pressing need for improved imaging, detection, and biopsy guidance. The earliest studies of vibration elastography appeared to be promising for prostate-cancer detection (Lee et al. 1991; Rubens et al. 1995). However, the most rigorous test of clinical utility would require comparison of 3D imaging of the prostate in B-scan-mode sonoelastography with, for a gold standard, sequential pathology slices demarcated by a pathologist for cancer, BPH (benign prostatic hyperplasia), and other abnormalities. We employed GE scanners with special Doppler's variance maps for imaging the vibration fields, along with an embedded 3D position sensor to recreate the images in both *in vivo* scans and *ex vivo* following the radical prostatectomy. A more recent example of sonoelasticity imaging of a whole prostate surgery specimen is given in Figure 12.11. Special coregistration techniques were developed to fuse the ultrasound and pathology images. This requires careful consideration due to the shrinkage and warping of the tissue during preparation for obtaining the sequential pathology slices (Figure 12.12).

The results demonstrated an increase in sensitivity and specificity on cancer detection into the mid 80% level (An et al. 2011;

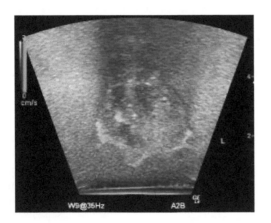
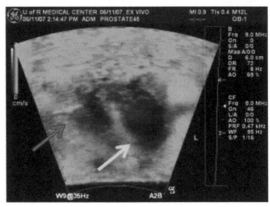

FIGURE 12.11 Prostate B-scan, sonelastography, and corresponding pathology. Larger circles on the pathology slides indicate BPH, black outline (on right side of prostate image) indicates cancer. The sonoelastography image shows a lower contrast deficit from the BPH (left arrow) and a higher contrast deficit, or dark region, from the cancer on the right border.

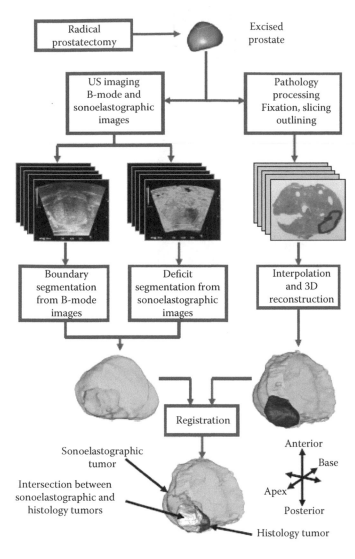

FIGURE 12.12 3D prostate imaging protocol. (Courtesy of Professor B. Castaneda.)

Castaneda et al. 2007; Hoyt et al. 2008b; Porter et al. 2001; Taylor et al. 2000, 2004, 2005). This represents an advance over conventional radiology and ultrasound, and increases the ability to localize the suspicious regions for biopsy. However, the results at this point are not accurate enough to eliminate the biopsy, and the role of BPH in creating false-positives requires further study. There are a number of improvements in techniques that could lead to a further increase in sensitivity and specificity of sonoelastography for prostate cancer.

12.5.2 Biomechanical Studies of Normal Tissue and Lesions

The entire field of imaging the elastic properties of tissues rests on the presumption that a meaningful elastic contrast exists between normal and diseased tissues, and this presumption is largely based on physical examination of tissues. Surprisingly,

there is a paucity of data where elastography studies are compared against a rigorous gold standard of stress–strain measurements. However, utilizing the available tools of vibration elastography and crawling waves, a number of studies have been conducted that include reference measurements. A four-way agreement of (i) shear-wave time-of-flight measurements, (ii) crawling-wave measurements, (iii) the Kelvin–Voight fractional derivative model (KVFD), and (iv) mechanical stress–strain measurement was described by Zhang et al. (2007). They demonstrated that the frequency-dependent properties of viscoelastic tissues, including *ex vivo* prostate and cancerous prostate tissues, liver, and others, could be derived from crawling-wave experiments at discrete frequencies, or alternatively by mechanical stress–strain measurements linked to the KVDF model of tissue viscoelasticity.

In other papers, utilizing these methods, the Rochester group measured an elastic contrast ratio of 2.6/1 between cancerous and noncancerous *ex vivo* prostate tissues, at 150 Hz (Zhang et al. 2008b). Much higher contrast was found for thermally ablated liver tissue compared to normal liver tissue (Zhang et al. 2008a). *In vivo* skeletal muscle properties under different conditions were also determined using these methods (Hoyt et al. 2008c). The change in contrast between two dissimilar media as a function of frequency was demonstrated and explained in Taylor et al. (2001). Overall, the combination of crawling-waves measurements over a frequency range, combined with the KVDF model, has proven to be a useful imaging method for estimating the viscoelastic properties of tissues, *in vivo* and *ex vivo*. Furthermore, the crawling-waves analysis closely corresponds to conventional mechanical measurements using stress and strain applied directly to cylindrical sections of tissue. Of course, the imaging techniques utilizing the crawling waves do not require any excisions, and can be performed across the overlying skin and fat.

12.5.3 Breast Cancer

Elastographic imaging of more superficial structures has been successful. The most widely adopted application in the United States has been breast-compression elastography. A direct compression technique has become commercially available from multiple manufacturers and the sensitivity and specificity for carcinoma versus benign disease is above 95% and 80%, respectively (Barr 2006; Ginat et al. 2009; Thomas et al. 2007). In a multicenter study from Wojcinski et al. (2010) assessing 779 patients, elastography demonstrated an improved specificity (89.5%) and an excellent positive predictive value (86.8%) compared to B-mode ultrasound (76.1% and 77.2%) for detecting the malignant lesions. In dense breasts, the specificity was even higher (92.8%). In breast lesions, malignancy creates a larger lesion on the elastogram than is seen on the routine grayscale B-scan image, presumably due to the desmoplastic reaction of the tumor and tethering in the adjacent tissue, as demonstrated in Figure 12.13. In contrast, benign lesions appear equal in size or smaller on the elastogram than they do on the grayscale image. Furthermore, a five-point

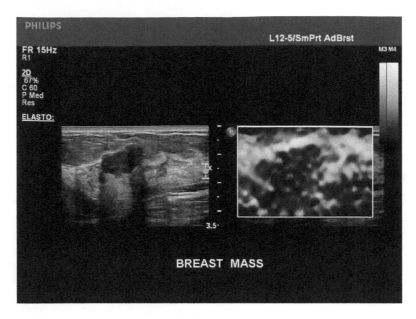

FIGURE 12.13 Breast nodule examined with B-scan (left) and compression elastography (right). The lesion appears larger in extent on elastography, a characteristic of malignancy. (Image courtesy of Philips.)

scale utilized in some studies using Hitachi scanners characterizes the strain patterns and has shown promise for breast-cancer diagnosis (Itoh et al. 2006; Zhi et al. 2010). Elastography therefore adds another imaging dimension, which can be used to help characterize the indeterminate lesions such as solid-appearing cysts or solid hypoechoic well-circumscribed nodules (Burnside et al. 2007; Svensson and Amiras 2006). Thus, it can potentially eliminate some unnecessary biopsies.

12.5.4 Liver Fibrosis

Over the past several years, elastography has moved out of the laboratory and into the hands of clinicians. The most rapidly adopted areas include liver, breast, prostate, and thyroid imaging. Liver elastography has become a mainstay for gastroenterologists who use a nonimaging instrument (Fibroscan) to assess the liver parenchymal stiffness in patients with hepatitis, at risk for developing cirrhosis. Calculations of liver stiffness based on transient elastography (Fibroscan) have shown better than 90% correlation with Metavir staging for hepatitis C, and have a better than 90% negative predictive value for cirrhosis (stage 4 disease) (Wang et al. 2009). When Fibroscan is combined with a blood test for liver fibrosis (fibrometer), sensitivity and specificity rise to over 90% for advanced (stage 4) disease (Boursier et al. 2009). One of the limitations of current liver elastography is that it has no imaging component. A second difficulty involves depth of penetration as well as liver contact. Very large patients and/or patients with ascites are not as effectively studied, with one in five obese patients having inadequate studies (Castera et al. 2008, 2010). New equipment and processing techniques are being developed to overcome these limitations.

12.5.5 Other Clinical Areas

A recent adaptation of compression elastography has been developed in the thyroid, using the gated pulsation of the carotid as the compressive force against the thyroid tissue. Dighe et al. (2010) have found excellent correlation between tissue stiffness and papillary carcinoma, the most prevalent type of thyroid cancer. Benign nodules are significantly softer, so that, in a cohort of 59 patients, if elastography had been used in addition to gray-scale features as a discriminator, as many as 60% of thyroid biopsies could have been safely deferred (Dighe et al. 2010). Given the low prevalence and the indolence of most thyroid cancer, avoiding the unnecessary biopsies may provide a great cost savings while permitting safe, conservative management of thyroid nodules.

A recent application of elastography to clinical disease states involves vascular pathology, including assessment of thrombus age and overall vascular compliance (Biswas et al. 2010; de Korte et al. 2000, 2002; Weitzel et al. 2009). Rubin et al. have shown marked changes in stiffness of common femoral clots as they age (Xie et al. 2004). This may potentially assist clinicians in deciding which thrombi require anticoagulation and for what time period. Direct measurement of vessel wall compliance would permit the monitoring of blood pressure and the efficacy of antihypertensive therapy.

A number of very specialized instruments have been developed to assess the elastic properties of accessible surfaces including the skin. Zheng and colleagues have devised a number of innovative approaches to the measurement of the effects of radiation or burns on skin, and the elastic properties of limbs, muscles, and tendons (Guo et al. 2008; Huang et al. 2007; Jun et al. 2008; Lau et al. 2005; Ling et al. 2007; Wang et al. 2007, 2008; Zheng et al. 2000). Mazza and colleagues have applied torsion waves and pressure/displacement tests to organs in surgery and to the assessment of uterine condition during late-term pregnancy (Nava et al. 2008; Valtorta and Mazza 2005).

12.6 Conclusion: Future Directions and Remaining Challenges

The previous section described some highly promising clinical applications that have been developed utilizing special technology for elastography. The near future will likely bring an expanding set of techniques, some integrated into scanners and others as stand-alone devices, with an expanding set of clinical diagnoses that benefit from an assessment of the biomechanical properties of tissues. However, some important areas have not received close attention and require dedicated effort to strengthen the foundations of elastography. Most importantly, we currently lack developmental models of diseases that link genetic, cellular, biochemical, and gross pathological changes with observations of the biomechanical properties. For example, as a cancer develops from under 1 mm in diameter to over 1 cm, what is the evolution of the biomechanical properties at a macroscale, and how do they depend on the cellular level dynamics? Similar questions can be asked for other important targets such as the development of arterial plaques, liver fibrosis, prostate BPH, benign lesions of the breast and thyroid, blood clots, brain injuries, and others. Some important foundational work has begun (Barnes et al. 2009; Carstensen et al. 2008; Insana et al. 2004; Kiss et al. 2004, 2006; Krouskop et al. 1998; Samani et al. 2007; Sinkus et al. 2005; Zhang et al. 2008b), but much remains to be studied. It is remarkable that clinical trials and technology development have been so successful given the paucity of understanding and data on the developmental aspects and mechanisms of altered biomechanics in specific disease states.

Another important direction for the field is the development of advanced estimators that go beyond relative stiffness and Young's modulus. Specifically, viscosity, anisotropy, nonlinearity, dispersion, and their changes with disease have only recently been estimated and there is enormous opportunity for devising the accurate estimation techniques and applying them to developmental models of major diseases. The opportunity is great; however, the challenges are also great, particularly since the study of elastography imaging and the biomechanics of normal and diseased tissue is highly multidisciplinary, covering classical mechanics, wave propagation, biomechanics, imaging, digital signal processing, radiology and pathology plus other specialties. Progress will require highly interactive teams covering a broad range of expertise.

Acknowledgments

I am deeply grateful to Dr. D. J. Rubens and Professor M. M. Doyley for their contributions and advice, and to the great colleagues who have produced insightful results. This work is supported by NIH grant R01AG02980-04.

References

Adler, R., J. M. Rubin, P. Bland, and P. Carson. 1989. Characterization of transmitted motion in fetal lung: Quantitative analysis. *Med Phys* 16:333–337.

Adler, R. S., J. M. Rubin, P. H. Bland, and P. L. Carson. 1990. Quantitative tissue motion analysis of digitized M-mode images: Gestational differences of fetal lung. *Ultrasound Med Biol* 16:561–569.

Alam, S. K., D. W. Richards, and K. J. Parker. 1994. Detection of intraocular pressure change in the eye using sonoelastic Doppler ultrasound. *Ultrasound Med Biol* 20:751–758.

An, L., B. Mills, Z. Hah, S. Mao, J. Yao, J. Joseph, D. J. Rubens, J. Strang, and K. J. Parker. 2011. Crawling wave detection of prostate cancer: Preliminary in vitro results. *Med Phys* 38:2563–2572.

An, L., D. J. Rubens, and J. Strang. 2010. Evaluation of crawling wave estimator bias on elastic contrast quantification. In American Institute of Ultrasound in Medicine Annual Convention, San Diego, CA. S1-14.

Axel, L., and L. Dougherty. 1989. Heart wall motion: Improved method of spatial modulation of magnetization for MR imaging. *Radiology* 172:349–350.

Bamber, J. C., P. E. Barbone, N. L. Bush, D. O. Cosgrove, M. M. Doyely, F. G. Fuechsel, P. M. Meaney, N. R. Miller, T. Shiina, and F. Tranquart. 2002. Progress in freehand elastography of the breast. *Ieice T Inf Syst* E85d:5–14.

Bamber, J. C., and N. L. Bush. 1996. Freehand elasticity imaging using speckle decorrelation rate. *Acoust Imag* 22:285–292.

Barbone, P. E., and N. H. Gokhale. 2004. Elastic modulus imaging: On the uniqueness and nonuniqueness of the elastography inverse problem in two dimensions. *Inverse Probl* 20:283–296.

Barnes, S. L., P. P. Young, and M. I. Miga. 2009. A novel model-gel-tissue assay analysis for comparing tumor elastic properties to collagen content. *Biomech Model Mechanobiol* 8:337–343.

Barr, R. G. 2006. Clinical applications of real-time elastography technique in breast imaging. In 5th International Conference on the Ultrasonic Measurement and Imaging of Tissue Elasticity, Snowbird, UT, p. 51.

Berry, G. P., J. C. Bamber, C. G. Armstrong, N. R. Miller, and P. E. Barbone. 2006. Towards an acoustic model-based poroelastic imaging method: I. Theoretical foundation. *Ultrasound Med Biol* 32:547–567.

Birnholz, J. C., and E. E. Farrell. 1985. Fetal lung development: Compressibility as a measure of maturity. *Radiology* 157:495–498.

Bishop, J., G. Poole, M. Leitch, and D. B. Plewes. 1998. Magnetic resonance imaging of shear wave propagation in excised tissue. *J Magn Reson Imaging* 8:1257–1265.

Bishop, J., A. Samani, J. Sciarretta, and D. B. Plewes. 2000. Two-dimensional MR elastography with linear inversion reconstruction: Methodology and noise analysis. *Phys Med Biol* 45:2081–2091.

Biswas, R., P. Patel, D. W. Park, T. J. Cichonski, M. S. Richards, J. M. Rubin, J. Hamilton, and W. F. Weitzel. 2010. Venous elastography: Validation of a novel high-resolution ultrasound method for measuring vein compliance using finite element analysis. *Semin Dial* 23:105–109.

Boursier, J., J. Vergniol, A. Sawadogo, T. Dakka, S. Michalak, Y. Gallois, V. Le Tallec, et al. 2009. The combination of a blood test and Fibroscan improves the non-invasive diagnosis of liver fibrosis. *Liver Int* 29:1507–1515.

Burnside, E. S., T. J. Hall, A. M. Sommer, G. K. Hesley, G. A. Sisney, W. E. Svensson, J. P. Fine, J. Jiang, and N. J. Hangiandreou. 2007. Differentiating benign from malignant solid breast masses with US strain imaging. *Radiology* 245:401–410.

Carstensen, E. L., K. J. Parker, and R. M. Lerner. 2008. Elastography in the management of liver disease. *Ultrasound Med Biol* 34:1535–1546.

Castaneda, B., L. An, S. Wu, L. L. Baxter, J. L. Yao, J. V. Joseph, K. Hoyt, J. Strang, D. J. Rubens, and K. J. Parker. 2009. Prostate cancer detection using crawling wave sonoelastography. S. A. McAleavey, and J. D'Hooge, eds. SPIE, Lake Buena Vista, FL, USA. 726513-726511-726513-726510.

Castaneda, B., K. Hoyt, M. Zhang, D. Pasternack, L. Baxter, P. Nigwekar, A. di Sant'Agnese, et al. 2007. P1C-9 Prostate Cancer Detection based on three dimensional sonoelastography. In Ultrasonics Symposium, New York, NY, 2007. IEEE. 1353–1356.

Castera, L., X. Forns, and A. Alberti. 2008. Non-invasive evaluation of liver fibrosis using transient elastography. *J Hepatol* 48:835–847.

Castera, L., J. Foucher, P. H. Bernard, F. Carvalho, D. Allaix, W. Merrouche, P. Couzigou, and V. de Ledinghen. 2010. Pitfalls of liver stiffness measurement: A 5-year prospective study of 13,369 examinations. *Hepatology* 51:828–835.

Catheline, S., F. Wu, and M. Fink. 1999. A solution to diffraction biases in sonoelasticity: The acoustic impulse technique. *J Acoust Soc Am* 105:2941–2950.

Cho, Y. T., Z. Hah, and L. An. 2010. Theoretical investigation of strategies for generating crawling waves using focused beams. In American Institue for Ultrasound in Medicine Annual Convention, San Diego, CA. S1-14.

Cox, M., and P. H. Rogers. 1987. Automated noninvasive motion measurement of auditory organs in fish using ultrasound. *J Vib Acoust* 109:55–59.

Curiel, L., R. Souchon, O. Rouviere, A. Gelet, and J. Y. Chapelon. 2005. Elastography for the follow-up of high-intensity focused ultrasound prostate cancer treatment: Initial comparison with MRI. *Ultrasound Med Biol* 31:1461–1468.

De Jong, P. G. M., T. Arts, A. P. G. Hoeks, and R. S. Reneman. 1990. Determination of tissue motion velocity by correlation interpolation of pulsed ultrasonic echo signals. *Ultrasonic Imaging* 12:84–98.

de Korte, C. L., S. G. Carlier, F. Mastik, M. M. Doyley, A. F. van der Steen, P. W. Serruys, and N. Bom. 2002. Morphological and mechanical information of coronary arteries obtained with intravascular elastography; feasibility study *in vivo*. *Eur Heart J* 23:405–413.

de Korte, C. L., G. Pasterkamp, A. F. van der Steen, H. A. Woutman, and N. Bom. 2000. Characterization of plaque components with intravascular ultrasound elastography in human femoral and coronary arteries in vitro. *Circulation* 102:617–623.

Dickinson, R. J., and C. R. Hill. 1982. Measurement of soft tissue motion using correlation between A-scans. *Ultrasound Med Biol* 8:263–271.

Dighe, M., J. Kim, S. Luo, and Y. Kim. 2010. Utility of the ultrasound elastographic systolic thyroid stiffness index in reducing fine-needle aspirations. *J Ultrasound Med* 29:565–574.

Doyley, M. M. 2012. Model-based elastography: A survey of approaches to the inverse elasticity problem. *Phys Med Biol* 57:R35–R73.

Doyley, M. M., E. E. Van Houten, J. B. Weaver, S. Poplack, L. Duncan, F. Kennedy, and K. D. Paulsen. 2004. Shear modulus estimation using parallelized partial volumetric reconstruction. *IEEE Trans Med Imaging* 23:1404–1416.

Dresner, M. A., G. H. Rose, P. J. Rossman, R. Muthupillai, and R. L. Ehman. 1998. Magnetic resonance elastography of the prostate. *Radiology* 209P:181–181.

Dresner, M. A., G. H. Rose, P. J. Rossman, R. Muthupillai, A. Manduca, and R. L. Ehman. 2001. Magnetic resonance elastography of skeletal muscle. *J Magn Reson Imaging* 13:269–276.

Eisensher, A., E. Schweg-Toffer, G. Pelletier, and G. Jacquemard. 1983. La palpation echographique rythmee-echosismographie. *J Radiol* 64:225–261.

Eskandari, H., S. E. Salcudean, and R. Rohling. 2008. Viscoelastic parameter estimation based on spectral analysis. *IEEE Transactions on Ultrasonics, Ferroelectrics, and Frequency Control* 55:1611–1625.

Fehrenbach, J. 2007. Influence of Poisson's ratio on elastographic direct and inverse problems. *Phys Med Biol* 52:707–716.

Fowlkes, J. B., S. Y. Emelianov, J. G. Pipe, A. R. Skovoroda, P. L. Carson, R. S. Adler, and A. P. Sarvazyan. 1995. Magnetic-resonance imaging techniques for detection of elasticity variation. *Med Phys* 22:1771–1778.

Garra, B. S., E. I. Cespedes, J. Ophir, S. R. Spratt, R. A. Zuurbier, C. M. Magnant, and M. F. Pennanen. 1997. Elastography of breast lesions: Initial clinical results. *Radiology* 202:79–86.

Ginat, D. T., S. V. Destounis, R. G. Barr, B. Castaneda, J. G. Strang, and D. J. Rubens. 2009. US elastography of breast and prostate lesions. *Radiographics* 29:2007–2016.

Guo, J. Y., Y. P. Zheng, Q. H. Huang, and X. Chen. 2008. Dynamic monitoring of forearm muscles using one-dimensional sonomyography system. *J Rehab Res Dev* 45:187–195.

Hah, Z., C. Hazard, Y. T. Cho, D. Rubens, and K. Parker. 2010. Crawling waves from radiation force excitation. *Ultrason Imaging* 32:177–189.

Hall, T. J., Y. Zhu, and C. S. Spalding. 2003. *In vivo* real-time free-hand palpation imaging. *Ultrasound Med Biol* 29:427–435.

Hardy, P. A., A. C. Ridler, C. B. Chiarot, D. B. Plewes, and R. M. Henkelman. 2005. Imaging articular cartilage under compression—Cartilage elastography. *Magn Reson Med* 53:1065–1073.

Holen, J., R. C. Waag, and R. Gramiak. 1985. Representations of rapidly oscillating structures on the doppler display. *Ultrasound Med Biol* 11:267–272.

Hoyt, K., B. Castaneda, and K. J. Parker. 2007a. Muscle tissue characterization using quantitative sonoelastography: Preliminary results. In Ultrasonics Symposium, New York, NY, 2007. IEEE. 365–368.

Hoyt, K., B. Castaneda, and K. J. Parker. 2007b. P4F-4 feasibility of two-dimensional quantitative sonoelastographic imaging. In Ultrasonics Symposium, New York, NY, 2007. IEEE. 2032–2035.

Hoyt, K., B. Castaneda, and K. J. Parker. 2008a. Two-dimensional sonoelastographic shear velocity imaging. *Ultrasound Med Biol* 34:276–288.

Hoyt, K., B. Castaneda, M. Zhang, P. Nigwekar, P. A. di Sant'agnese, J. V. Joseph, J. Strang, D. J. Rubens, and K. J. Parker. 2008b. Tissue elasticity properties as biomarkers for prostate cancer. *Cancer Biomark* 4:213–225.

Hoyt, K., T. Kneezel, B. Castaneda, and K. J. Parker. 2008c. Quantitative sonoelastography for the *in vivo* assessment of skeletal muscle viscoelasticity. *Phys Med Biol* 53:4063–4080.

Hoyt, K., K. J. Parker, and D. J. Rubens. 2006. Sonoelastographic shear velocity imaging: Experiments on tissue phantom and prostate. In Ultrasonics Symposium, Vancouver, BC, 2006. IEEE. 1686–1689.

Hoyt, K., K. J. Parker, and D. J. Rubens. 2007c. Real-time shear velocity imaging using sonoelastographic techniques. *Ultrasound Med Biol* 33:1086–1097.

Huang, S. R., R. M. Lerner, and K. J. Parker. 1990. On estimating the amplitude of harmonic vibration from the doppler spectrum of reflected signals. *J Acoust Soc America* 88:2702–2712.

Huang, S. R., R. M. Lerner, and K. J. Parker. 1992. Time domain doppler estimators of the amplitude of vibrating targets. *J Acoust Soc America* 91:965–974.

Huang, Y. P., Y. P. Zheng, S. F. Leung, and A. P. Choi. 2007. High frequency ultrasound assessment of skin fibrosis: Clinical results. *Ultrasound Med Biol* 33:1191–1198.

Insana, M. F., C. Pellot-Barakat, M. Sridhar, and K. K. Lindfors. 2004. Viscoelastic imaging of breast tumor microenvironment with ultrasound. *J Mammary Gland Biol Neoplasia* 9:393–404.

Itoh, A., E. Ueno, E. Tohno, H. Kamma, H. Takahashi, T. Shiina, M. Yamakawa, and T. Matsumura. 2006. Breast disease: Clinical application of US elastography for diagnosis. *Radiology* 239:341–350.

Jun, S., Y.-P. Zheng, Q.-H. Huang, and C. Xin. 2008. Continuous monitoring of sonomyography, electromyography and torque generated by normal upper arm muscles during isometric contraction: Sonomyography assessment for arm muscles. *IEEE Trans Biomed Eng* 55:1191–1198.

Kallel, F., R. E. Price, E. Konofagou, and J. Ophir. 1999. Elastographic imaging of the normal canine prostate in vitro. *Ultrason Imaging* 21:201–215.

Kemper, J., R. Sinkus, J. Lorenzen, C. Nolte-Ernsting, A. Stork, and G. Adam. 2004. MR elastography of the prostate: Initial *in vivo* application. *Rofo* 176:1094–1099.

Kiss, M. Z., M. A. Hobson, T. Varghese, J. Harter, M. A. Kliewer, E. M. Hartenbach, and J. A. Zagzebski. 2006. Frequency-dependent complex modulus of the uterus: Preliminary results. *Phys Med Biol* 51:3683–3695.

Kiss, M. Z., T. Varghese, and T. J. Hall. 2004. Viscoelastic characterization of in vitro canine tissue. *Phys Med Biol* 49:4207–4218.

Krouskop, T. A., D. R. Dougherty, and F. S. Vinson. 1987. A pulsed Doppler ultrasonic system for making noninvasive measurements of the mechanical properties of soft tissue. *J Rehabil Res Dev* 24:1–8.

Krouskop, T. A., T. M. Wheeler, F. Kallel, B. S. Garra, and T. Hall. 1998. Elastic moduli of breast and prostate tissues under compression. *Ultrason Imaging* 20:260–274.

Kruse, S. A., G. H. Rose, K. J. Glaser, A. Manduca, J. P. Felmlee, C. R. Jack, Jr., and R. L. Ehman. 2008. Magnetic resonance elastography of the brain. *Neuroimage* 39:231–237.

Kruse, S. A., J. A. Smith, A. J. Lawrence, M. A. Dresner, A. Manduca, J. F. Greenleaf, and R. L. Ehman. 2000. Tissue characterization using magnetic resonance elastography: Preliminary results. *Phys Med Biol* 45:1579–1590.

Lau, J. C., C. W. Li-Tsang, and Y. P. Zheng. 2005. Application of tissue ultrasound palpation system (TUPS) in objective scar evaluation. *Burns* 31:445–452.

Lee, F., Jr., J. P. Bronson, R. M. Lerner, K. J. Parker, S. R. Huang, and D. J. Roach. 1991. Sonoelasticity imaging: Results in in vitro tissue specimens. *Radiology* 181:237–239.

Lerner, R. M., S. R. Huang, and K. J. Parker. 1990. "Sonoelasticity" images derived from ultrasound signals in mechanically vibrated tissues. *Ultrasound Med Biol* 16:231–239.

Lerner, R. M., and K. J. Parker. 1987. Sonoelasticity images derived from ultrasound signals in mechanically vibrated targets. In Seventh European Communities Workshop. J. Tjissen, editor, Nijmegen, The Netherlands.

Lerner, R. M., K. J. Parker, J. Holen, R. Gramiak, and R. C. Waag. 1988. Sonoelasticity: Medical elasticity images derived from ultrasound signals in mechanically vibrated targets. *Acoust Imag* 16:317–327.

Levinson, S. F. 1987. Ultrasound propagation in anisotropic soft-tissues—The application of linear elastic theory. *J Biomech* 20:251–260.

Levinson, S. F., M. Shinagawa, and T. Sato. 1995. Sonoelastic determination of human skeletal-muscle elasticity. *J Biomech* 28:1145–1154.

Ling, H. Y., Y. P. Zheng, and S. G. Patil. 2007. Strain dependence of ultrasound speed in bovine articular cartilage under compression in vitro. *Ultrasound Med Biol* 33:1599–1608.

Lorenzen, J., R. Sinkus, D. Schrader, M. Lorenzen, C. Leussler, M. Dargatz, and P. Roschmann. 2001. Imaging of breast tumors using MR elastography. *Rofo* 173:12–17.

Mahdavi, S. S., M. Moradi, X. Wen, W. J. Morris, and S. E. Salcudean. 2011. Evaluation of visualization of the prostate gland in vibro-elastography images. *Medical Image Analysis* 15:589–600.

Manduca, A., V. Dutt, D. T. Dorup, R. Muthupillai, J. F. Greenleaf, and R. L. Ehman. 1998. An inverse approach to the calculation of elasticity maps for magnetic resonance elastography. In *SPIE*. 426–436.

Manduca, A., R. Muthupillai, P. J. Rossman, J. F. Greenleaf, and R. L. Ehman. 1996. Local wavelength estimation for magnetic resonance elastography. In Image Processing, 1996. Proceedings, International Conference, Lausanne, Switzerland, 1996. 527–530 vol.523.

McCracken, P. J., A. Manduca, J. Felmlee, and R. L. Ehman. 2005. Mechanical transient-based magnetic resonance elastography. *Magn Reson Med* 53:628–639.

McKnight, A. L., J. L. Kugel, P. J. Rossman, A. Manduca, L. C. Hartmann, and R. L. Ehman. 2002. MR elastography of breast cancer: Preliminary results. *AJR Am J Roentgenol* 178:1411–1417.

McLaughlin, J., D. Renzi, K. Parker, and Z. Wu. 2007. Shear wave speed recovery using moving interference patterns obtained in sonoelastography experiments. *J Acoust Soc Am* 121:2438–2446.

Muthupillai, R., V. Dutt, A. Manduca, J. A. Smith, J. F. Greenleaf, and R. L. Ehman. 1997. Resolution limits of MR elastography (MRE). *Radiology* 205:1386–1386.

Muthupillai, R., D. J. Lomas, P. J. Rossman, J. F. Greenleaf, A. Manduca, and R. L. Ehman. 1995. Magnetic-resonance elastography by direct visualization of propagating acoustic strain waves. *Science* 269:1854–1857.

Nava, A., E. Mazza, M. Furrer, P. Villiger, and W. H. Reinhart. 2008. *In vivo* mechanical characterization of human liver. *Med Imag Anal* 12:203–216.

O'Donnell, M., A. R. Skovoroda, B. M. Shapo, and S. Y. Emelianov. 1994. Internal displacement and strain imaging using ultrasonic speckle tracking. *IEEE T Ultrason Ferr* 41:314–325.

Oestreicher, H. L. 1951. Field and impedance of an oscillating sphere in a viscoelastic medium with an application to biophysics. *J Acoust Soc Am* 23:704–714.

Ophir, J., I. Cespedes, H. Ponnekanti, Y. Yazdi, and X. Li. 1991. Elastography: A quantitative method for imaging the elasticity of biological tissues. *Ultrason Imaging* 13:111–134.

Ophir, J., F. Kallel, T. Varghese, M. Bertrand, E. I. Cespedes, and H. Ponnekanti. 1997. Elastography: A systems approach. *Int J Imaging Syst Technol* 8:89–103.

Parker, K. J., S. R. Huang, R. M. Lerner, F. Lee, Jr., D. Rubens, and D. Roach. 1993. Elastic and ultrasonic properties of the prostate. In Ultrasonics Symposium, Baltimore, MD, 1993. Proceedings, IEEE 1993. 1035–1038 vol. 1032.

Parker, K. J., S. R. Huang, R. A. Musulin, and R. M. Lerner. 1990. Tissue response to mechanical vibrations for "sonoelasticity imaging". *Ultrasound Med Biol* 16:241–246.

Parker, K. J., and R. M. Lerner. 1992. Sonoelasticity of organs: Shear waves ring a bell. *J Ultrasound Med* 11:387–392.

Parker, K. J., L. S. Taylor, S. Gracewski, and D. J. Rubens. 2005. A unified view of imaging the elastic properties of tissue. *J Acoust Soc Am* 117:2705–2712.

Plewes, D. B., I. Betty, S. N. Urchuk, and I. Soutar. 1995. Visualizing tissue compliance with MR imaging. *J Magn Reson Imaging* 5:733–738.

Plewes, D. B., J. Bishop, A. Samani, and J. Sciarretta. 2000. Visualization and quantification of breast cancer biomechanical properties with magnetic resonance elastography. *Phys Med Biol* 45:1591–1610.

Porter, B. C., D. J. Rubens, J. G. Strang, J. Smith, S. Totterman, and K. J. Parker. 2001. Three-dimensional registration and fusion of ultrasound and MRI using major vessels as fiducial markers. *Medical Imaging, IEEE Transactions on* 20:354–359.

Righetti, R., F. Kallel, R. J. Stafford, R. E. Price, T. A. Krouskop, J. D. Hazle, and J. Ophir. 1999. Elastographic characterization of HIFU-induced lesions in canine livers. *Ultrasound Med Biol* 25:1099–1113.

Rubens, D. J., M. A. Hadley, S. K. Alam, L. Gao, R. D. Mayer, and K. J. Parker. 1995. Sonoelasticity imaging of prostate cancer: In vitro results. *Radiology* 195:379–383.

Sack, I., E. Gedat, J. Bernarding, G. Buntkowsky, and J. Braun. 2004. Magnetic resonance elastography and diffusion-weighted imaging of the sol/gel phase transition in agarose. *J Magn Reson* 166:252–261.

Salcudean, S. E., D. French, S. Bachmann, R. Zahiri-Azar, X. Wen, and W. J. Morris. 2006. Viscoelasticity modeling of the prostate region using vibro-elastography. *Med Image Comput Comput Assist Interv* 9:389–396.

Samani, A., J. Bishop, C. Luginbuhl, and D. B. Plewes. 2003. Measuring the elastic modulus of *ex vivo* small tissue samples. *Phys Med Biol* 48:2183–2198.

Samani, A., J. Zubovits, and D. Plewes. 2007. Elastic moduli of normal and pathological human breast tissues: An inversion-technique-based investigation of 169 samples. *Phys Med Biol* 52:1565–1576.

Shiina, T., N. Nitta, E. Ueno, and J. C. Bamber. 2002. Real time elasticity imaging using the combined autocorrelation method. *J Med Ultrasonics* 29:119–128.

Sinkus, R., J. Lorenzen, D. Schrader, M. Lorenzen, M. Dargatz, and D. Holz. 2000. High-resolution tensor MR elastography for breast tumour detection. *Phys Med Biol* 45:1649–1664.

Sinkus, R., K. Siegmann, T. Xydeas, M. Tanter, C. Claussen, and M. Fink. 2007. MR elastography of breast lesions: Understanding the solid/liquid duality can improve the specificity of contrast-enhanced MR mammography. *Magn Reson Med* 58:1135–1144.

Sinkus, R., M. Tanter, T. Xydeas, S. Catheline, J. Bercoff, and M. Fink. 2005. Viscoelastic shear properties of *in vivo* breast lesions measured by MR elastography. *Magn Reson Imaging* 23:159–165.

Sumi, C. 2008. Displacement vector measurement using instantaneous ultrasound signal phase—Multidimensional autocorrelation and Doppler methods. *IEEE T Ultrason Ferr* 55:24–43.

Svensson, W. E., and D. Amiras. 2006. *Ultrasound Elasticity Imaging*. Cambridge University Press. 7. http://journals.cambridge.

org/action/displayFulltext?type=1&fid=446403&jid=BCO&
volumeId=9&issueId=06&aid=446402&bodyId=&member
shipNumber=&societyETOCSession=

Taylor, K. J. 1976. Absolute measurement of acoustic particle velocity. *J Acoust Soc Am* 59:691–694.

Taylor, L. S., B. C. Porter, G. Nadasdy, P. A. di Sant'Agnese, D. Pasternack, Z. Wu, R. B. Baggs, D. J. Rubens, and K. J. Parker. 2004. Three-dimensional registration of prostate images from histology and ultrasound. *Ultrasound Med Biol* 30:161–168.

Taylor, L. S., B. C. Porter, D. J. Rubens, and K. J. Parker. 2000. Three-dimensional sonoelastography: Principles and practices. *Phys Med Biol* 45:1477–1494.

Taylor, L. S., M. S. Richards, and A. J. Moskowitz. 2001. Viscoelastic effects in sonoelastography: Impact on tumor detectability. In Ultrasonics Symposium, Atlanta, GA, 2001 IEEE. 1639–1642 vol. 1632.

Taylor, L. S., D. J. Rubens, B. C. Porter, Z. Wu, R. B. Baggs, P. A. di Sant'Agnese, G. Nadasdy, et al. 2005. Prostate cancer: Three-dimensional sonoelastography for in vitro detection. *Radiology* 237:981–985.

Taylor, L. S., and M. Zhang. 2002. In-vitro imaging of thermal lesions using three-dimensional vibration sonoelatography. *In Second International Symposium on Therapeutic Ultrasound*, Seattle, WA, 2002. 176–184.

Thitaikumar, A., T. A. Krouskop, B. S. Garra, and J. Ophir. 2007. Visualization of bonding at an inclusion boundary using axial-shear strain elastography: A feasibility study. *Phys Med Biol* 52:2615–2633.

Thomas, A., M. Warm, M. Hoopmann, F. Diekmann, and T. Fischer. 2007. Tissue Doppler and strain imaging for evaluating tissue elasticity of breast lesions. *Acad Radiol* 14:522–529.

Tikkakoski, T. 2007. Magnetic resonance elastography of brain tumors. *Acta Radiol* 48:249.

Tristam, M., D. C. Barbosa, D. O. Cosgrove, J. C. Bamber, and C. R. Hill. 1988. Application of Fourier analysis to clinical study of patterns of tissue movement. *Ultrasound Med Biol* 14:695–707.

Tristam, M., D. C. Barbosa, D. O. Cosgrove, D. K. Nassiri, J. C. Bamber, and C. R. Hill. 1986. Ultrasonic study of *in vivo* kinetic characteristics of human tissues. *Ultrasound Med Biol* 12:927–937.

Valtorta, D., and E. Mazza. 2005. Dynamic measurement of soft tissue viscoelastic properties with a torsional resonator device. *Med Imag Anal* 9:481–490.

Van Houten, E. E., M. M. Doyley, F. E. Kennedy, J. B. Weaver, and K. D. Paulsen. 2003. Initial *in vivo* experience with steady-state subzone-based MR elastography of the human breast. *J Magn Res Imag: JMRI* 17:72–85.

Van Houten, E. E., M. I. Miga, J. B. Weaver, F. E. Kennedy, and K. D. Paulsen. 2001. Three-dimensional subzone-based reconstruction algorithm for MR elastography. *Magn Reson Med* 45:827–837.

Van Houten, E. E., K. D. Paulsen, M. I. Miga, F. E. Kennedy, and J. B. Weaver. 1999. An overlapping subzone technique for MR-based elastic property reconstruction. *Magn Reson Med* 42:779–786.

Varghese, T., and J. Ophir. 1997. A theoretical framework for performance characterization of elastography: The strain filter. *IEEE Trans Ultrason Ferroelectr Freq Control* 44:164–172.

Varghese, T., J. Ophir, and I. Cespedes. 1996. Noise reduction in elastograms using temporal stretching with multicompression averaging. *Ultrasound Med Biol* 22:1043–1052.

Varghese, T., U. Techavipoo, W. Liu, J. A. Zagzebski, Q. Chen, G. Frank, and F. T. Lee, Jr. 2003. Elastographic measurement of the area and volume of thermal lesions resulting from radiofrequency ablation: Pathologic correlation. *AJR Am J Roentgenol* 181:701–707.

Von Gierke, H. E., H. L. Oestreicher, E. K. Franke, H. O. Parrack, and W. W. Wittern. 1952. Physics of vibrations in living tissues. *J Appl Physiol* 4:886–900.

Wang, J. H., C. S. Changchien, C. H. Hung, H. L. Eng, W. C. Tung, K. M. Kee, C. H. Chen, T. H. Hu, C. M. Lee, and S. N. Lu. 2009. FibroScan and ultrasonography in the prediction of hepatic fibrosis in patients with chronic viral hepatitis. *J Gastroenterol* 44:439–446.

Wang, Q., Y. P. Zheng, G. Leung, W. L. Lam, X. Guo, H. B. Lu, L. Qin, and A. F. Mak. 2008. Altered osmotic swelling behavior of proteoglycan-depleted bovine articular cartilage using high frequency ultrasound. *Phys Med Biol* 53:2537–2552.

Wang, Q., Y. P. Zheng, H. J. Niu, and A. F. Mak. 2007. Extraction of mechanical properties of articular cartilage from osmotic swelling behavior monitored using high frequency ultrasound. *J Biomech Eng* 129:413–422.

Weaver, J. B., M. Doyley, Y. Cheung, F. Kennedy, E. L. Madsen, E. E. W. Van Houten, and K. Paulsen. 2005. Imaging the shear modulus of the heel fat pads. *Clin Biomech* 20:312–319.

Weaver, J. B., E. Van Houten, M. I. Miga, F. E. Kennedy, and K. D. Paulsen. 2000. Three-dimensional steady state acquisition and reconstruction in MR elastography. *Radiology* 217:206–206.

Weaver, J. B., E. E. Van Houten, M. I. Miga, F. E. Kennedy, and K. D. Paulsen. 2001. Magnetic resonance elastography using 3D gradient echo measurements of steady-state motion. *Med Phys* 28:1620–1628.

Weitzel, W. F., K. Kim, P. K. Henke, and J. M. Rubin. 2009. High-resolution ultrasound speckle tracking may detect vascular mechanical wall changes in peripheral artery bypass vein grafts. *Ann Vasc Surg* 23:201–206.

Wilson, L. S., and D. E. Robinson. 1982. Ultrasonic measurement of small displacements and deformations of tissue. *Ultrason Imaging* 4:71–82.

Wojcinski, S., A. Farrokh, S. Weber, A. Thomas, T. Fischer, T. Slowinski, W. Schmidt, and F. Degenhardt. 2010. Multicenter study of ultrasound real-time tissue elastography in 779 cases for the assessment of breast lesions: Improved diagnostic performance by combining the BI-RADS(R)-US classification system with sonoelastography. *Ultraschall Med* 31:484–491.

Wu, T., J. P. Felmlee, J. F. Greenleaf, S. J. Riederer, and R. L. Ehman. 2000. MR imaging of shear waves generated by focused ultrasound. *Magn Reson Med* 43:111–115.

Wu, Z. 2005. Shear wave interferometry and holography, an application of sonoelastography. In *Electrical & Computer Engineering*. University of Rochester, Rochester, NY. 104.

Wu, Z., K. Hoyt, D. J. Rubens, and K. J. Parker. 2006. Sonoelastographic imaging of interference patterns for estimation of shear velocity distribution in biomaterials. *J Acoust Soc Am* 120:535–545.

Wu, Z., L. S. Taylor, D. J. Rubens, and K. J. Parker. 2002. Shear wave focusing for three-dimensional sonoelastography. *J Acoust Soc Am* 111:439–446.

Wu, Z., L. S. Taylor, D. J. Rubens, and K. J. Parker. 2004. Sonoelastographic imaging of interference patterns for estimation of the shear velocity of homogeneous biomaterials. *Phys Med Biol* 49:911–922.

Xie, H., K. Kim, S. R. Aglyamov, S. Y. Emelianov, X. Chen, M. O'Donnell, W. F. Weitzel, S. K. Wrobleski, D. D. Myers, T. W. Wakefield, and J. M. Rubin. 2004. Staging deep venous thrombosis using ultrasound elasticity imaging: Animal model. *Ultrasound Med Biol* 30:1385–1396.

Yamakoshi, Y., J. Sato, and T. Sato. 1990. Ultrasonic imaging of internal vibration of soft tissue under forced vibration. *IEEE Trans Ultrason Ferroelectr Freq Control* 37:45–53.

Zerhouni, E. A., D. M. Parish, W. J. Rogers, A. Yang, and E. P. Shapiro. 1988. Human heart: Tagging with MR imaging—A method for noninvasive assessment of myocardial motion. *Radiology* 169:59–63.

Zhang, M., B. Castaneda, J. Christensen, W. E. Saad, K. Bylund, K. Hoyt, J. G. Strang, D. J. Rubens, and K. J. Parker. 2008a. Real-time sonoelastography of hepatic thermal lesions in a swine model. *Med Phys* 35:4132–4141.

Zhang, M., B. Castaneda, Z. Wu, P. Nigwekar, J. V. Joseph, D. J. Rubens, and K. J. Parker. 2007. Congruence of imaging estimators and mechanical measurements of viscoelastic properties of soft tissues. *Ultrasound Med Biol* 33:1617–1631.

Zhang, M., P. Nigwekar, B. Castaneda, K. Hoyt, J. V. Joseph, A. di Sant'Agnese, E. M. Messing, J. G. Strang, D. J. Rubens, and K. J. Parker. 2008b. Quantitative characterization of viscoelastic properties of human prostate correlated with histology. *Ultrasound Med Biol* 34:1033–1042.

Zheng, Y. P., S. F. Leung, and A. F. Mak. 2000. Assessment of neck tissue fibrosis using an ultrasound palpation system: A feasibility study. *Med Biol Eng Comput* 38:497–502.

Zhi, H., X. Y. Xiao, H. Y. Yang, B. Ou, Y. L. Wen, and B. M. Luo. 2010. Ultrasonic elastography in breast cancer diagnosis: Strain ratio vs 5-point scale. *Acad Radiol* 17:1227–1233.

13

Acoustic Radiation Force Imaging

Jeremy J. Dahl
Duke University

13.1 Introduction

Manual palpation of tissues for the purpose of the identification of disease and diagnosis has been a staple of practicing clinicians for hundreds of years. The utility of manual palpation is based on the ability to identify the diseased tissue by its difference in stiffness from the surrounding tissue. Manual palpation, however, is limited to more superficial and larger nodules or structures because deeper and smaller structures become obscured by overlying tissues. For example, breast cancers may often be detected by manual palpation; however, liver cancers often go undetected until pain occurs or liver function is impaired because they lie too deep within the body for manual palpation.

The ability to noninvasively image the mechanical properties of tissues is therefore of great benefit. Although many lesions can be visualized with conventional ultrasound, a significant portion of them are isoechoic, meaning they return the same echo brightness as the tissue surrounding the lesion, making them difficult to visualize or invisible to the radiologist. In some cases, the exact borders of the lesion may not be apparent even though the lesion is easily observed.

Methods for imaging the mechanical properties of tissues involve some form of mechanical excitation of the tissue and observation of the tissue response. Mechanical excitation can be static or dynamic; however, the majority of techniques used often use some form of dynamic mechanical excitation of the tissue. In dynamic excitation techniques, such as those derived from acoustic radiation force, dynamic forces are used to elicit a response from tissue and the mechanical properties are then derived from the dynamic response of the tissue. These methods attempt to differentiate the tissues based on their stiffness, Young's modulus, shear modulus, or acoustic resonance.

This chapter will cover the physical basis for the generation of radiation force from acoustic waves, and explore several imaging methods based on acoustic radiation force. These methods include acoustic radiation force impulse (ARFI) imaging, which is based on the mechanical response of tissue to an acoustic impulse, shear wave elasticity imaging (SWEI), which derives the mechanical properties of tissue through shear waves induced by radiation force, and vibro-acoustography and harmonic motion imaging (HMI), which derive the mechanical properties from the acoustic resonance of tissue by applying a forcing function. We also present nonimaging techniques that are similar to these approaches and also estimate the mechanical properties of tissue. We review the current clinical uses of each of these radiation force imaging techniques as well as the most promising clinical applications. The potential bioeffects associated with radiation force-based imaging methods are also examined.

13.2 Physical Basis for Acoustic Radiation Force

Acoustic radiation force is a unidirectional force that arises when absorbing or reflecting targets are in the propagation path of an acoustic wave. The force results from a transfer of momentum from the acoustic wave to the propagation medium. The basis for acoustic radiation force can be derived directly from the basic equations; however, there are many approaches to this derivation. Here, we take the approach in which we assume that acoustic radiation force is entirely a product of the absorption of the acoustical wave. We have taken liberties with the derivation here so that readers of all backgrounds may benefit from this introduction to radiation force; however, knowledge of vector calculus is required. In this derivation, we combine the classic derivations of radiation force from Eckart (1948) and Nyborg (1965).

13.2.1 Acoustic Radiation Force

Consider a homogeneous isotropic fluid (i.e., a fluid with the same acoustic and mechanical properties in all directions) in which the density, pressure, and particle velocity of a small element of the fluid are described by $\rho(x,y,z,t)$, $p(x,y,z,t)$, and $\mathbf{u}$, respectively, where $\mathbf{u}$ is a vector containing components in x, y, z, and t. The particle velocity describes the velocity of this small element, which is much smaller than the wavelength of the acoustic wave, yet much larger than the interatomic spacing of the fluid. The particle velocity is a vector with dependencies on time and space, and should not be confused with the speed of sound, c_0.

Nearly all derivations of the acoustic radiation force start with Newton's second law, $\mathbf{F} = m\mathbf{a}$; the well-known relation that says a force is equal to a mass times an acceleration. Because acceleration is the time derivative of velocity and density is just a mass per unit volume, this formula can be written as $\mathbf{F} = \rho d\mathbf{u}/dt$. Recall, however, that $\mathbf{u}$ is a function of x, y, and z as well, and the correct acceleration obtained from the derivative of the particle velocity requires differentiation with respect to the spatial dimensions as well. Thus, we must use something called the *material derivative* in order to obtain the correct expression. Applying the material derivative to Newton's second law expresses the law in a form called the momentum balance equation:

$$\mathbf{F} = \rho \frac{\partial \mathbf{u}}{\partial t} + \rho(\mathbf{u} \cdot \nabla)\mathbf{u}, \tag{13.1}$$

where ∇ is the vector operator defined as

$$\nabla = \frac{\partial}{\partial t}\hat{\mathbf{x}} + \frac{\partial}{\partial t}\hat{\mathbf{y}} + \frac{\partial}{\partial t}\hat{\mathbf{z}}.$$

In the simplest case of a plane acoustic wave propagating in the z-direction (i.e., one spatial dimension), Equation 13.1 would be expressed as

$$\mathbf{F} = \left[\rho \frac{\partial u}{\partial t} + \rho \frac{\partial u}{\partial z}\frac{\partial z}{\partial t}\hat{\mathbf{z}} \right]. \tag{13.2}$$

In this case, $u = u(z, t)\hat{\mathbf{z}}$ and velocity $\mathbf{u}$ is replaced by the classic definition $\partial z/\partial t$, which is simply the derivative of the position of the particle. We now introduce the continuity equation, which is an equation that says that the rate at which mass enters a system is equal to the rate at which mass exits the system:

$$\frac{\partial \rho}{\partial t} + \nabla \cdot (\rho \mathbf{u}) = 0. \tag{13.3}$$

In our case, the system is just our small element of fluid defined above. Multiplying Equation 13.3 by the vector $\mathbf{u}$, and taking the advantage of the chain rule in the partial derivative $(\partial(\rho\mathbf{u})/\partial t) = \rho(\partial\mathbf{u}/\partial t) + \mathbf{u}(\partial\rho/\partial t)$, we can rearrange Equation 13.3 to obtain

$$\rho \frac{\partial \mathbf{u}}{\partial t} = \frac{\partial(\rho\mathbf{u})}{\partial t} + \mathbf{u}\nabla \cdot (\rho\mathbf{u}). \tag{13.4}$$

Substituting Equation 13.4 into Equation 13.1 gives

$$\mathbf{F} = \frac{\partial(\rho\mathbf{u})}{\partial t} + \mathbf{u}\nabla \cdot (\rho\mathbf{u}) + \rho(\mathbf{u} \cdot \nabla)\mathbf{u}. \tag{13.5}$$

The $\mathbf{F}$ in Equation 13.5 is often replaced by a sum of forces to achieve the Navier–Stokes equation for fluid dynamics. This would be useful, for example, in determining the streaming velocity of a fluid resulting from acoustic radiation force. However, we are merely concerned with determining the force induced by a propagating acoustic wave, and thus it is only necessary to carry our derivation through Equation 13.5. Unfortunately, the true velocities, pressures, and densities of a fluid (or tissue) are complicated functions, making it difficult to solve for the exact solutions of Equation 13.5. We therefore approximate the density, pressure, and velocity by

$$\rho = \rho_0 + \rho_1 + \rho_2 + \cdots$$
$$p = p_0 + p_1 + p_2 + \cdots$$
$$\mathbf{u} = \mathbf{u}_0 + \mathbf{u}_1 + \mathbf{u}_2 + \cdots.$$

In this approximation, the density, pressure, and particle velocity are broken up into increasingly higher-order terms. Each successive higher-order term is smaller than the preceding term and has less effect on the overall solution. By breaking up our parameters in such a manner, we can employ a perturbation method in order to solve an approximation to Equation 13.5. To determine which terms are most important in the solution, we create nondimensionalized versions of our equations. For example, we let U be some value in units of meters per second, T be a value in units of seconds, and Z be a value in units of meters, such that $\mathbf{v} = u/U$, $\tau = t/T$, and $\zeta = z/Z$. We then create a dimensionless parameter, called the perturbation parameter, $N = UT/Z$. By replacing all u's with $\mathbf{v}U$, t's with τT, and z's with ζZ, we can rewrite Equations 13.3 and 13.5 as

$$\frac{\partial \rho}{\partial \tau} + N\nabla \cdot (\rho\mathbf{v}) = 0 \tag{13.6}$$

$$\mathbf{F} = \frac{\partial(\rho\mathbf{v})}{\partial \tau} + N\mathbf{v}\nabla \cdot (\rho\mathbf{v}) + N\rho(\mathbf{v} \cdot \nabla)\mathbf{v}, \tag{13.7}$$

where the density term, ρ, has been left as a dimensionalized parameter and the x and y components of ∇ have been nondimensionalized with Z. The pressure, density, and velocity approximations are then written to include the perturbation parameter

$$\rho = \rho_0 + N\rho_1 + N^2\rho_2 + \cdots$$
$$p = p_0 + Np_1 + N^2 p_2 + \cdots$$
$$\mathbf{v} = \mathbf{v}_0 + N\mathbf{v}_1 + N^2\mathbf{v}_2 + \cdots.$$

To perform our perturbation of Equations 13.6 and 13.7, we insert the above approximations up to the first-order terms and collect terms based on the power of the perturbation parameter N. For example, in the nondimensionalized continuity equation (Equation 13.6), we insert $\rho_0 + N\rho_1$ for ρ and $\mathbf{v}_0 + N\mathbf{v}_1$ for $\mathbf{v}$ to obtain

$$\frac{\partial \rho_0}{\partial \tau} + N\frac{\partial \rho_1}{\partial \tau} + N\nabla \cdot (\rho_0 + N\rho_1)(\mathbf{v}_0 + N\mathbf{v}_1) = 0$$

$$\frac{\partial \rho_0}{\partial \tau} + N\frac{\partial \rho_1}{\partial \tau} + N\nabla \cdot (\rho_0 \mathbf{v}_1) + N^2\nabla \cdot (\rho_0 \mathbf{v}_1)$$
$$+ N^2\nabla \cdot (\rho_1 \mathbf{v}_0) + N^3\nabla \cdot (\rho_1 \mathbf{v}_1) = 0.$$

We then collect terms according to the power of the perturbation parameter and replace the nondimensionalized parameters with their original dimensionalized parameters. The zeroth-order continuity equation (N^0) is, therefore

$$\frac{\partial \rho_0}{\partial t} = 0, \tag{13.8}$$

which states that ρ_0 is a constant (i.e., it is the steady-state density of the propagation medium). The first-order solution (N^1) to the continuity equation is

$$\frac{\partial \rho_1}{\partial t} + \rho_0 \nabla \cdot \mathbf{u}_0 = 0. \tag{13.9}$$

Using Equation 13.9 and performing the same perturbation analysis with Equation 13.7, we arrive at the first-order solution for the instantaneous force:

$$\mathbf{F} = \rho_0 \frac{\partial \mathbf{u}_1}{\partial t} + \frac{\partial(\rho_1 \mathbf{u}_0)}{\partial t} + \rho_0(\mathbf{u}_0 \nabla \cdot \mathbf{u}_0 + \mathbf{u}_0 \cdot \nabla \mathbf{u}_0). \tag{13.10}$$

To obtain the acoustic radiation force, we must compute the time average of Equation 13.10. The time average of the force is given by

$$\mathbf{F}_{\mathbf{rad}} = \langle \mathbf{F} \rangle = \frac{1}{T} \int_{-T/2}^{T/2} (F_x\hat{\mathbf{x}} + F_y\hat{\mathbf{y}} + F_z\hat{\mathbf{z}}) \, dt. \tag{13.11}$$

In this step, it will help simplify the problem to one spatial dimension by modeling the wave as an exponentially decaying plane acoustic wave. That is, $\mathbf{u}_0 = U_0 \exp(-\alpha z)\cos(2\pi ft - kz)\hat{\mathbf{z}}$, where f is the frequency of the wave, $k = 2\pi/\lambda$ is the wavenumber, λ is the wavelength, α is the absorption coefficient, which is a function of the bulk and shear viscosities of the fluid, and T is selected such that it is much longer than the period of the wave. Because of the sinusoidal nature of this wave, the time-dependent derivatives in Equation 13.10 drop out in the time-averaged force. Therefore, the acoustic radiation force for a plane wave is given by

$$\mathbf{F}_{\mathbf{rad}} = 2\rho_0 \left\langle u_0 \frac{\partial u_0}{\partial z} \right\rangle \hat{\mathbf{z}}. \tag{13.12}$$

Carrying out the partial derivatives, we are left with

$$\mathbf{F}_{\mathbf{rad}} = -4\rho_0\alpha U_0^2 \exp(-2\alpha z)\big\langle \cos^2(2\pi ft - kz) + k\sin(4\pi ft - 2kz)\big\rangle \hat{\mathbf{z}}. \tag{13.13}$$

In the time average of Equation 13.13, the sinusoidal term on the right drops out, and the $\cos^2(\cdot)$ term equals 1/2. The acoustic radiation force is then given by

$$\mathbf{F}_{\mathbf{rad}} = -2\alpha\rho_0 U_0^2 \exp(-2\alpha z) = -\frac{2\alpha I}{c_0}\hat{\mathbf{z}}, \tag{13.14}$$

where I is the time-averaged intensity of the acoustic wave given by $I = \rho_0 c_0 U_0^2 \exp(-2\alpha z)$, and c_0 is the steady-state speed of sound of the propagating medium. In practice, the attenuation coefficient (i.e., the decrease in the wave's amplitude due to both scattering and absorption) is typically used for α rather than the absorption coefficient because the attenuation due to scattering and absorption are difficult to separate empirically. In addition, this derivation neglects the radiation force associated with scattering (see Westervelt (1951) for a derivation including scattering). In most applications to human tissue, however, scattering is weak and absorption dominates the attenuation coefficient, so the attenuation coefficient is a good approximation for the absorption coefficient and Equation 13.14 is a good approximation to the radiation force. Figure 13.1 demonstrates the radiation force field created in a medium with an attenuation coefficient of 0.7 dB/cm/MHz by a linear ultrasound transducer array.

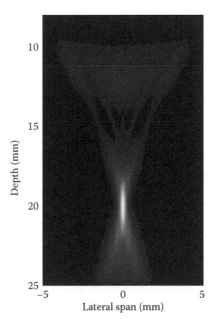

FIGURE 13.1 Image of the *xz* plane of a radiation force field created by a linear transducer array. The medium has an attenuation coefficient of 0.7 dB/cm/MHz. The linear array focuses the ultrasonic waves at a depth of 20 mm, so the intensity, and thus the radiation force, is maximal at that location, indicated by the brighter white pixels.

13.2.2 Practical Generation of Acoustic Radiation Force

Although plane waves provide a simple solution for the acoustic radiation force, they are not particularly useful in practical applications. First, a plane wave cannot provide a fine-enough interrogation of a medium for a radiation force-based imaging system to provide detailed resolution of that medium. Second, greater radiation force can be achieved with less transmit power by focusing the ultrasonic wave. Even though the radiation force derived in Equation 13.14 results from a plane wave, the formula is still applicable to focused ultrasound beams. In this case, the radiation force field is a function of the spatial distribution of the transmitted wave's intensity. The spatial distribution of the intensity is dictated by both the tissue properties, such as speed of sound and attenuation, and the transmitted pulse characteristics, such as aperture size, pulse length, and frequency. Typically, the intensity is relatively weak in the region just beneath the surface of the transducer, and higher intensities are observed in the focal region of the acoustic pulse.

The magnitude of acoustic radiation force resulting from conventional diagnostic ultrasound in soft tissue is relatively small, so its effects are not generally apparent. In order to achieve the displacements that are detectable by ultrasonic equipment (i.e., displacements greater than a micron), the intensity of the acoustic wave must be increased well above that used for diagnostic imaging. Intensities greater than those used in diagnostic ultrasound can be achieved by increasing the length or pressure of the acoustical pulse. Generally, increasing the pulse length is preferred because increasing the pressure may lead to detrimental or dangerous effects, such as cavitation.

13.3 Radiation Force-Based Imaging Methods

Radiation force-based imaging methods have several potential advantages over mechanical force-based imaging techniques, such as elastography or strain imaging. The application of radiation force is not operator dependent, provided sufficient contact between transducer and patient is applied. This means that radiation force-based images tend to be more similar regardless of who is operating the ultrasound scanner. Second, radiation force-based techniques provide better resolution than mechanical force-based techniques because the radiation force is applied to localized regions of tissue rather than over a broad swath of tissue. Radiation force methods can be applied to selected remote locations and with considerably smaller forces. Generally, global mechanical compression requires large strains near the transducer in order to generate small strains at depth, which introduce artifacts into the image.

In the last decade, several novel imaging methods based on acoustic radiation force have been introduced, with some methods finding niches in clinical applications and others just a few years from clinical practice. Although each of these methods employs radiation force, the mechanical properties that they image are different and each method demonstrates its own advantages.

13.3.1 Acoustic Radiation Force Impulse Imaging

ARFI imaging is a radiation force-based imaging method that uses short-duration (tens to hundreds of microseconds) acoustical pulses to generate a localized region of radiation force in tissue (Nightingale et al., 2002). Although the pulses are an order of magnitude longer than conventional B-mode imaging pulses, the pulses are relatively short compared to the dynamic response of the tissue to the radiation force, and hence the word "impulse" in the name of the imaging method.

The response of the tissue to this localized radiation force is much slower than the propagation speed of acoustic waves. Thus, conventional pulse-echo ultrasound can be used to monitor the displacement of the localized region of tissue (typically called the excitation region) as it moves away and recovers back to its original position. To produce an image, the "push-and-track" is applied many times over a wide field-of-view, and the achieved displacements at each pixel location are displayed.

Because the response of the tissue is dynamic, there are several possible choices for showing displacements, such as maximum displacement over time or the displacements occurring an arbitrary number of milliseconds after radiation force application. Furthermore, parametric images can be created from the displacements, such as the time to reach the peak displacement, or the time it takes for the tissue to recover to some percentage of its peak displacement.

ARFI imaging is advantageous because it can be implemented on currently available commercial ultrasound scanners and transducers without additional equipment. Thus, for scanner manufacturers, this amounts to a change in software. The same transducers that are used to create the conventional B-mode images can be used to generate the sufficient radiation force to induce the micron-sized displacements in tissue and track those displacements.

The radiation force in ARFI imaging depends on the local acoustic intensity (see Equation 13.14), which can be affected by the spatial distribution of tissue types and their associated acoustic parameters (e.g., α and c_0). In addition, the intensity tends to be greater at the focal depth of the acoustic pulses. Therefore, the images of displacement or other parametric properties are *relative* measures of the mechanical properties of the tissue. This means that the displacements observed at one location can only be compared with the displacements observed at a close or neighboring position. For example, if equal displacements are observed at two locations that are far apart in ARFI imaging, they do not necessarily have the same amount of stiffness. Similarly, the same type of tissue at two locations may have different overlying tissues with different attenuation, or they may be at different depths receiving the different acoustic intensities, thereby displaying different displacements. When the two points are close to each other, it is assumed that the overlying tissues and intensities are approximately equivalent, giving rise to the same amount of radiation force and allowing for absolute comparisons in displacement.

Figure 13.2 compares a conventional B-mode ultrasound image with an ARFI image of a stiff 3 mm lesion surrounded by a much softer background medium in a tissue-mimicking phantom. The lesion is 3.7 times stiffer (according to its Young's modulus) than the background region. The ARFI image shows the displacements that occur 0.47 milliseconds after application of the radiation force impulse. Figure 13.2c shows the response of the tissue inside the lesion and within the background medium. The softer background medium displaces much further than the stiff lesion, and the time-to-peak (TTP) displacement is much quicker for the background medium than it is for the stiff lesion. The recovery time (RT), which in this case is the time to return to 63% of the peak displacement, of the soft background is much longer than the RT of the stiff lesion. Parametric images of TTP and RT can be constructed from the response of tissue as well as the displacement image shown in Figure 13.2b.

13.3.1.1 Clinical Applications of ARFI Imaging

Current applications of ARFI imaging include breast and liver imaging. Clinical applications of ARFI imaging to breast tumors have only been recently introduced (Meng et al., 2011); however, research into breast imaging with ARFI has shown considerable promise (Nightingale et al., 2002; Sharma et al., 2004). Figure 13.3 shows an example of ARFI imaging applied to an invasive ductal adenocarcinoma. The figure demonstrates features of the lesion that are not apparent using the conventional ultrasound imaging. For example, the nodule visible in the B-mode image shows a small lesion with well-defined and smooth borders. These characteristics are typically consistent with a benign, complex cyst. However, the ARFI image demonstrates that the lesion is much larger, based on the stiffness of the tissue, than indicated in the B-mode image. In addition, the lesion is much stiffer than the surrounding breast parenchyma, which is more indicative of a carcinoma than a cyst.

ARFI imaging is commercially available in Europe and Asia, and has had considerable success in imaging liver nodules (Cho et al., 2010; Shuang-Ming et al., 2011). ARFI imaging

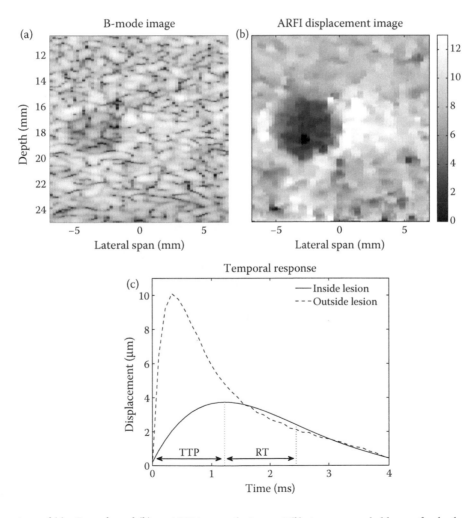

FIGURE 13.2 Comparison of (a) a B-mode and (b) an ARFI image of a 3 mm stiff lesion surrounded by a softer background. (c) The dynamic response of a region inside the lesion and in the background due to radiation force.

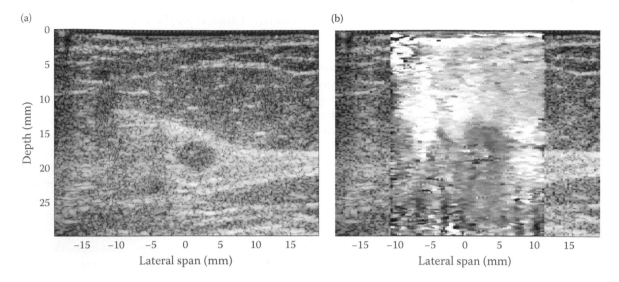

(a)

(b)

FIGURE 13.3 (a) B-mode and (b) ARFI images of a breast lesion. In the B-mode image, a nodule is visible in the center of the image. In the ARFI image, the nodule is much stiffer than the surrounding breast parenchyma, and the size of the lesion is much larger than that indicated in the B-mode image. Biopsy of the lesion determined that it was an invasive ductal adenocarcinoma. (Images courtesy of Dr. Kathryn Nightingale.)

has demonstrated the capability of identifying or clarifying 18% more cancers than conventional ultrasound (Cho et al., 2010). ARFI imaging has also shown to differentiate hepatocellular carcinoma from metastatic disease in the liver (Fahey et al., 2008a). Figure 13.4 demonstrates ARFI imaging applied to metastatic disease in the liver. In this figure, three metastatic nodules are apparent in the liver, one in the center and two on the right side of the box outlined in the B-mode image. There is a blood vessel visible on the left side of this box. In the ARFI image, the displacement image is shown only for the region outlined by the box in the B-mode image. The nodules in the ARFI

image are much larger, and indicate much stiffer tissue than the surrounding healthy liver tissue. The displacements at the location of the blood vessel show extreme dark and bright values, which are commonly observed in blood vessels because of the motion of the blood.

Applications of ARFI imaging in the near future include imaging of the mechanical makeup of carotid plaques (Dahl et al., 2009; Allen et al., 2011) and prostate tumors (Zhai et al., 2010, 2012), the monitoring of ablation nodules in cardiac and other tissues (Fahey et al., 2005b, 2008; Hsu et al., 2007b), and observing natural and pathological-based changes in stiffness in

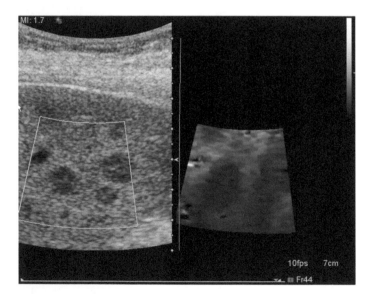

FIGURE 13.4 B-mode (left) and ARFI (right) images of three metastatic lesions in the liver, taken on a commercial scanner. In the B-mode image, the three lesions are visible in the center and right of the box (there is a blood vessel on the left side of the image). Upon ARFI imaging, the lesions appear much larger than that indicated in the B-mode image. In addition, the blood vessel appears to show either randomly large or small displacements due to the movement of blood in the vessel. (Image courtesy of David Bradway.)

myocardial tissue (Hsu et al., 2007a). Carotid plaques have the potential to rupture, resulting in stroke and potentially death. The ability to determine which plaques may be vulnerable to rupture is therefore important in preventing the stroke. ARFI imaging has demonstrated the ability to detect regions of plaque containing soft (potentially lipid core) tissue as well as to demonstrate plaques that are hard throughout (Dahl et al., 2009; Allen et al., 2011). The current technology for imaging prostate cancer is lacking due to its poor ability to differentiate the prostate cancer from the healthy tissue. Although needle biopsy is the gold standard for the prostate cancer diagnosis, the method by which the biopsy is performed is suspect because it relies on a grid of biopsy cores rather than specifically targeting suspicious regions. ARFI imaging has demonstrated, in both *in vivo* (Zhai et al., 2012) and *ex vivo* (Zhai et al., 2010) studies, the ability to identify the various prostate pathologies that correlated well with the histological findings.

ARFI imaging is also useful in monitoring radiofrequency ablation (RFA) or cryoablation lesions in liver tissues (Fahey et al., 2006, 2008b). One complication with ablation procedures is that for many patients the cancerous lesion is inadequately visualized to guide the electrode placement. In addition, the amount of time necessary to ablate a region of tissue and obtain the appropriate ablation margin may vary with patient or protocol. Active monitoring of the ablation lesion size can result in more accurate and potentially safer procedures for individuals. Because the resulting ablation lesion is much stiffer than either the healthy or cancerous tissue, ablation lesions are well defined in ARFI images.

In cardiac RFA, typically a line of RFA lesions are used to isolate or destroy offensive circuit pathways of common supraventricular tachyarrhythmias. Although these procedures are effective, gaps in lesion lines will facilitate conduction of residual arrhythmias. Therefore, lesion placement and size are the primary determinants of the efficacy of the treatment. ARFI imaging has been shown to be able to visualize these gaps using intracardiac echocardiography probes (Fahey et al., 2005; Hsu et al., 2007b; Eyerly et al., 2010). In addition to cardiac ablation, ARFI imaging has demonstrated its ability to detect the normal physiological changes in cardiac tissue stiffness during normal contraction of the heart (Hsu et al., 2007c, 2009, 2010). In one study, ARFI was able to show the propagation of the contraction of the canine heart (via stiffness) in a canine heart as the heart was stimulated artificially with pacing electrodes (Hsu et al., 2007c). It is hypothesized that ARFI imaging would be able to detect the regions of dysfunction in the heart, such as an infarcted region, based on the inability of that region to contract. Thus, while normal healthy tissue would display a dynamically varying stiffness throughout the cardiac cycle, an infarcted region would demonstrate little or no change in stiffness.

13.3.2 Shear Wave Elasticity Imaging

The radiation force field that generates displacement of tissue also creates shear stresses at the boundaries of the region of excitation.

These stresses induce low-frequency, transverse oscillations (i.e., a shear wave) in the tissue that propagate in a direction normal to the direction of applied radiation force (see Figure 13.5). These oscillations can be described by (Sarvazyan et al., 1998)

$$F_z = \frac{\partial^2 s_z}{\partial t^2} - c_t^2 \Delta_\perp s_z + \frac{\eta}{\rho} \frac{\partial}{\partial t} \Delta_\perp s_z, \tag{13.15}$$

where F_z is the applied radiation force in the z direction, η is the shear viscosity, s_z is the z component of the displacement vector, and $\Delta_\perp$ is the Laplacian operator for the x and y components. c_t is the shear wave velocity, defined as

$$c_t = \sqrt{\mu/\rho}, \tag{13.16}$$

where μ is the shear modulus, and the tissue is assumed to be an isotropic, elastic, incompressible medium. The first two terms on the right-hand side of Equation 13.15 describe the linear wave equation with forcing function F_z, and the last term on the right is a dissipative term. Equation 13.15 essentially describes a low-frequency shear wave that propagates, with speed c_t, away from the axis of the acoustic beam. The shape of the shear wave in the xy plane is much like an enlarging ring, similar to what one would observe by dropping a small stone into a pool of water.

SWEI is a radiation force-based method that derives the mechanical properties of tissue from the shear waves (Sarvazyan et al., 1998). In SWEI, the tissue is assumed to be an isotropic, elastic, and incompressible medium such that the shear modulus of the wave can be estimated using Equation 13.16. Thus, in SWEI, the aim is to induce a shear wave in tissue and measure its propagation speed, which can then be converted to the shear modulus.

The are two primary, shear wave-inducing methods that are currently implemented: the radiation force impulse (ARFI) (Nightingale et al., 2003) and the supersonic shear source (Bercoff et al., 2004b). Shear wave generation with ARFI uses a radiation force impulse identical to that used in ARFI imaging, where the

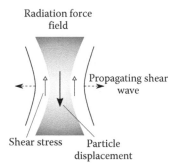

FIGURE 13.5 Shear waves are generated when radiation force is exerted on a local region of tissue. The displacement of the tissue induces shear forces at the edges of the radiation force field, or region of excitation. The shear stresses create outward propagating shear waves. The direction of tissue movement and shear wave propagation are indicated by the black-tipped arrows.

radiation force is applied for a relatively short duration of tens to hundreds of microseconds. The supersonic shear source is a shear wave with a large axial (*z*-axis) component, created by applying many radiation force impulses at increasingly greater depths. The "supersonic" name is given to this shear wave source because the compressional waves used for the radiation force impulses travel at velocities an order of magnitude greater than the shear waves, allowing multiple radiation force impulses to be applied over a large range before the shear waves propagate a substantial distance. Nominally, the compressional waves travel at speeds around 1500 m/s in tissue, whereas shear waves travel at speeds of around 1–5 m/s. The difference between the impulsive and supersonic sources is primarily in its application; the impulse is used for quantitative measures at a specific location in the tissue, whereas the supersonic source is used to create the 2D images.

SWEI differs from ARFI imaging primarily in the tracking methods used to obtain the estimates of tissue displacement. While ARFI imaging tracks displacement of tissue at the center of the region of excitation, shear wave imaging tracks displacements lateral to the region of excitation (see Figure 13.6). Because the displacements are only tracked in a single plane, only two points on the enlarging ring are observed, thus giving Figure 13.6 the appearance of two different shear waves traveling out from the region of excitation. Parallel tracking methods (Bercoff et al., 2004b; Dahl et al., 2007) are required to monitor the propagation of shear waves traveling perpendicular to the direction of radiation force. In addition, the primary objective in SWEI is to create an image of the shear modulus, which is a direct measure of the material properties of the tissue. In ARFI imaging, the primary objective is to create an image of displacement, which is a relative measure of the mechanical properties of the tissue.

There are two primary methods for computing the shear modulus from the propagation of shear waves. In the lateral TTP algorithm (Palmeri et al., 2008), the shear wave is observed by ultrasonically monitoring the displacements induced by the shear wave at locations lateral to the region of excitation. The time for each lateral location to reach its peak in displacement is estimated, and the shear wave velocity is computed from the inverse of the slope of the line between the time to reach the peak displacement and lateral location. Using the shear wave velocity, the shear modulus can be estimated from Equation 13.16.

The second method, described by Bercoff et al. (2004b), estimates the shear modulus from the measured displacements (*u*) by inversion of the Helmholtz equation:

$$\rho \frac{\partial^2 u_i(x)}{\partial t^2} = \mu \nabla^2 u_i(x), \tag{13.17}$$

where *x* is the lateral direction, which is perpendicular to the beam axis. Because the measured displacements are a function of time and two dimensions of space (*x* and *z*), the shear modulus can be estimated as a function of *x* and *z* using

$$\mu(x,z) = \frac{\rho}{N} \sum_{\omega} \frac{\mathcal{FT}\left\{\dfrac{\partial^2 u(x,z)}{\partial t^2}\right\}}{\mathcal{FT}\left\{\dfrac{\partial^2 u(x,z)}{\partial x^2} + \dfrac{\partial^2 u(x,z)}{\partial z^2}\right\}}. \tag{13.18}$$

In this solution for μ, the problem is broken up into several solutions at *N* different frequencies, ω. The inverse problem is then solved *N* times in the frequency domain, where $\mathcal{FT}\{\cdot\}$ implies the Fourier transform. The resulting solutions are then summed together to complete the full solution.

Like most inverse problems, however, the inversion process is very sensitive to noise, which comes in the form of jitter errors from displacement estimation. However, this method obtains significantly better resolution than the TTP algorithm because Equation 13.18 can be solved for every point of measured

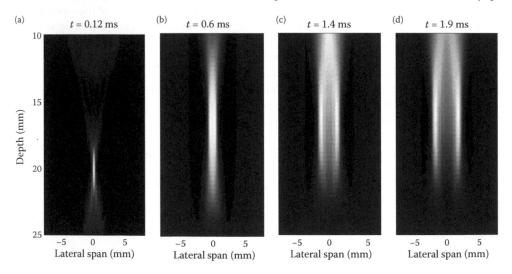

FIGURE 13.6 Time sequence of a finite element model of the tissue displacement generated by radiation force. (a) Initially, the displacements resemble the radiation force field. (b) Because of the shear stresses, a shear wave is generated that propagates out from the excitation region. (c, d) In these planes, the shear wave appears as two waves propagating in opposite directions. Shear wave elasticity imaging and quantitative measures are derived from ultrasonically measuring the speed at which this wave propagates.

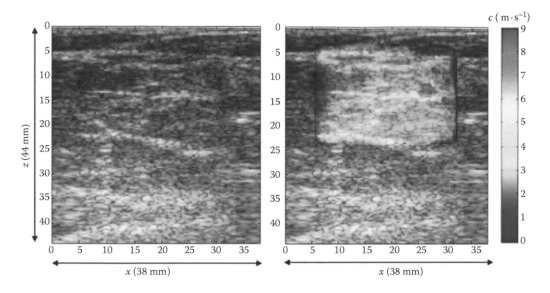

FIGURE 13.7 (**See color insert.**) Invasive ductal carcinoma, grade III, evaluated with shear wave imaging. The lesion is undetectable in the B-mode image but is shown to be detectable in the shear wave image. This cancerous lesion is showing to be significantly stiffer than the surrounding parenchymal tissue. (Reprinted from *Ultrasound Med Biol 34*, Tanter, M. et al., Quantitative assessment of breast lesion viscoelasticity: Initial clinical results using supersonic shear imaging, 1373–1386, Figure 7, Copyright 2008, with permission from Elsevier.)

displacement. The TTP algorithm requires a lateral spatial extent of a few millimeters in order to determine the velocity, and therefore suffers a loss of resolution. The TTP algorithm, however, is much less susceptible to jitter errors.

13.3.2.1 Clinical Applications of SWEI

The primary application of SWEI thus far has been limited to imaging the mechanical properties of breast tissue (Tanter et al., 2008; Athanasiou et al., 2009; Evans et al., 2010). Figure 13.7 demonstrates an example of imaging mechanical properties in breast tissue using the SWEI method. The B-mode image in this figure demonstrates no significant findings. In the SWEI image, however, a small lesion with high contrast in mechanical properties is well visualized. The lesion shows a stiffness twice as hard as the surrounding breast tissue. On biopsy, this lesion was identified as a grade III, invasive ductal carcinoma.

Future applications of SWEI imaging are similar to that of ARFI imaging. Potential applications include differentiating benign and malignant thyroid nodules (Sebag et al., 2010), broad mapping of liver stiffness for fibrosis staging (Muller et al., 2009), observation of the contractility of myocardial tissue under normal and diseased states (Couade et al., 2011; Pernot et al., 2011), and the monitoring of thermally induced lesions with the ability to monitor the temperature changes (Bercoff et al., 2004a; Arnal et al., 2011a,b). SWEI imaging, however, appears to be more restricted to shallower applications due to the difficulty in generating the large-enough displacements at depth. Typically, the amplitude of the shear wave decreases as it propagates, due to the cylindrical spreading of the shear wave. For deep applications, smaller initial displacements (less amplitude) are generated by the supersonic shear source, so the shear wave dies off quicker, and the field-of-view becomes severely restricted. Most applications of SWEI imaging have demonstrated good images down to 25 mm, but are limited beyond that.

While most applications of SWEI and ARFI imaging overlap, the temperature monitoring of thermally induced lesions (Arnal et al., 2011a) presents an interesting application for SWEI. Because the speed of a shear wave in tissue is temperature dependent, the relative temperature change of the tissue can be calculated from the change in estimated shear modulus. Arnal et al. (2011a) showed that the temperature dependence of the shear modulus in tissue was approximately linear and repeatable in the range of 30–45°C, which is the temperature range where proteins are not denatured and the shear modulus is not permanently affected by ablation of the tissue. Each tissue type has its own temperature dependence for shear modulus, and thus the tissue type needs to be known in order to compute the temperature. The advantage of this method over ultrasound- and magnetic resonance-based thermometry techniques is that it is not as susceptible to motion artifacts, and can therefore be applicable to *in vivo* imaging.

13.3.3 Vibro-Acoustography

In ARFI and SWEI techniques, an acoustic radiation force impulse was used to elicit a mechanical response of the tissue. We now consider the radiation force applied in a continuous manner. Consider our plane-wave model of the particle velocity used in Equation 13.12. We now apply an amplitude modulation to the particle velocity, and write our plane wave as

$$\mathbf{u}_0 = U_0 \exp(-\alpha z)\cos(2\pi ft - kz)\cos\left(\frac{2\pi\Delta f_0 t}{2}\right)\hat{\mathbf{z}}, \qquad (13.19)$$

where $\Delta f_0/2$ is the modulating frequency and $\Delta f_0 \ll f$. We next assume that the period T in the time average of the radiation force is much longer than the period of the acoustic frequency f, but much shorter than the period of the modulating frequency Δf_0, or $1/f \ll T \ll 1/\Delta f_0$. Then, inserting Equation 13.19 into Equation 13.13, we obtain

$$
\begin{aligned}
\mathbf{F_{rad}} &= -4\rho_0\alpha U_0^2\exp(-2\alpha z)\cos^2\!\left(\frac{2\pi\Delta f_0 t}{2}\right) \\
&\quad \times \left\langle \cos^2(2\pi ft - kz) + k\sin(4\pi ft - 2kz)\right\rangle\hat{\mathbf{z}} \\
&= -2\rho_0\alpha U_0^2\cos^2\!\left(\frac{2\pi\Delta f_0 t}{2}\right)\hat{\mathbf{z}}.
\end{aligned}
\tag{13.20}
$$

Using the trigonometric identity $\cos^2(\theta) = (1 + \cos 2\theta)/2$, and substituting in the definition of time-averaged intensity from Section 13.2.1, Equation 13.20 can be rewritten as

$$
\mathbf{F_{rad}} = -\left(\frac{\alpha I}{c_0} + \frac{\alpha I}{c_0}\cos\Delta f_0 t\right)\hat{\mathbf{z}}.
\tag{13.21}
$$

The left-hand term of Equation 13.21 is a steady-state term resulting from the time-averaged intensity of the acoustic wave, and is identical to Equation 13.14 except for a factor of two. The right-hand term of Equation 13.21 is a modulating term that varies the total radiation force between 0 and $-2\alpha I/c_0$.

Ultrasound-stimulated vibro-acoustography (USVA) is one such technique that applies modulating radiation forces to image tissue (Fatemi and Greenleaf, 1998, 1999). The technique works by scanning a focused, modulating radiation force over a plane and recording the amplitude and phase of the vibrations of the tissue in response to the modulating radiation force. Images are then constructed of the magnitude of the vibrations.

A modulating radiation force can be created by an amplitude-modulated ultrasound beam; however, this strategy creates a modulating radiation force field that extends over a very large range in the medium. A better strategy involves the use of two *unmodulated* ultrasound beams that differ in frequency by Δf_0, where Δf_0 is on the order of tens of kHz. In this method, the two beams are crossed such that they interfere for only a short range. In the region where the two beams cross, the interference pattern is an amplitude-modulating acoustic beam. The smaller the region over which these two beams interfere, the better the resolution of the imaging system.

There are several strategies for creating an amplitude-modulated beam by crossing two unmodulated beams (Chen et al., 2004b). The simplest, but still effective, method is to use two piston transducers to form the crossing beams at a specified focal depth into the medium. A similar strategy involves the use of a confocal arrangement, in which a ring-shaped transducer surrounds a piston transducer. This method allows two beams to be crossed at a specified focal depth while minimizing the

positioning errors and expensive setups for two separate transducers. Conventional diagnostic transducers can also be used to generate the two unmodulated beams (Urban et al., 2011). This configuration may be the most ideal, particularly for clinical use, because the linear array of elements in a diagnostic transducer allows beams to be steered electronically. This would allow the scanning of the intersected beam over a line without a mechanical steering system as is required with the dual transducer or confocal setups. In addition, the implementation on a diagnostic transducer allows concurrent formation of conventional B-mode images and does not incur additional expense for the mechanical steering.

With a modulating force, the tissue within the region of excitation vibrates at the modulation frequency. The vibration of this tissue creates an acoustic field in the surrounding medium, which is dependent upon the viscoelastic properties of the vibrating tissue as well as the modulated beam's cross-section. The acoustic field can be detected by a hydrophone at the surface of the medium. The hydrophone detects the amplitude of the acoustic field, and the phase is determined relative to the transmitted acoustic waves. Given the phase and amplitude, an image of a plane in the medium is created by computing the magnitude of the acoustic field from the vibration of every point in that plane. An example of a vibro-acoustography system is detailed in Figure 13.8.

Vibro-acoustography makes very detailed images and lacks the speckle signature that accompanies conventional diagnostic ultrasound imaging. In addition, the resolutions of the system is isotropic, as the images are often created from the C-plane (the plane perpendicular to the transducer axis). Conventional B-mode ultrasound has different resolutions in the depth and lateral directions. The disadvantages of vibro-acoustography systems are that scans currently take several minutes to complete.

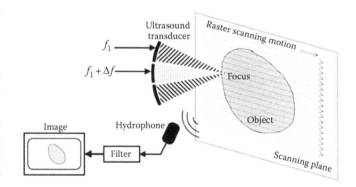

FIGURE 13.8 Demonstration of a vibro-acoustography setup. The confocal transducer supplies two focused and unmodulated beams with frequencies that differ by Δf. This creates an amplitude-modulated force at the focal point in the object. The vibration at the focal point creates an acoustic emission, which is recorded by the hydrophone. An image is created by scanning the focus of the two beams across a plane normal to the direction of the applied force, and displaying the response at each point. (Reprinted from *Phys Med Biol 45*, Fatemi, M. and J. F. Greenleaf, Probing the dynamics of tissue at low frequencies with the radiation force of ultrasound, 1449–1464, Figure 2, Copyright 2000, with permission from Elsevier.)

This makes the vibro-acoustography susceptible to patient and physiological motion, and prevents it from being used on targets such as the heart. In addition, to obtain the isotropic resolution desired with USVA, a conventional diagnostic transducer would still require a mechanical steering stage to create a C-scan (Urban et al., 2011).

13.3.3.1 Clinical Applications of Vibro-Acoustography

Vibro-acoustography has potential applications in breast imaging (Fatemi et al., 2002; Alizad et al., 2004, 2006), visualizing brachytherapy seeds and cryotherapy lesions in prostate imaging (Mitri et al., 2008, 2009), and observing calcifications in carotid or peripheral arterial stenosis (Pislaru et al., 2008, 2011). In breast imaging, the major advantage of vibro-acoustography is its ability to detect the microcalcifications (Fatemi et al., 2002). The small, hard crystals of a microcalcification are ideal targets for vibro-acoustography because they resonate with ease under the modulating radiation force. Thus, microcalcifications show up as bright targets in a vibro-acoustography image, much like they do in x-ray images. In fact, the images produced by vibro-acoustography of the breast are very similar to conventional x-ray mammography, making the images easily recognizable by radiologists (see Figures 13.9 and 13.10). In addition, vibro-acoustography can potentially detect the boundaries of breast lesions (Alizad et al., 2006).

Like microcalcifications, brachytherapy seeds are very hard compared to normal human tissue, and thus make excellent candidates for mechanical vibration. Brachytherapy seeds are small (typically 800 μm in diameter by 4.5 mm in length) rod-shaped structures made most commonly of iodine 125 or palladium 103 and used for radiation therapy in prostate cancer. They are often placed in a grid within the prostate to deliver the small doses of radiation to the neighboring tissue in order to kill the cancer cells. Placement of the seeds is often performed under the guidance of transrectal ultrasound; however, no current diagnostic imaging modalities are ideal for confirming seed placement. Seed orientation is problematic for conventional B-mode ultrasound because a seed positioned with its long axis in line with the ultrasonic beam appears invisible, due to the small reflecting surface and the speckle of the surrounding tissue. Mitri et al. (2009) proposed vibro-acoustography as a method to detect the brachytherapy seeds, independent of seed orientation. This method could potentially be complementary to B-mode ultrasound and allow for immediate and quick determination of seed placement. In addition, Mitri et al. (2008) propose the use of vibro-acoustography for the observation of cryotherapy lesions in the prostate. The concepts here are similar to the use of SWEI and ARFI imaging for ablation lesions.

Another potential target for vibro-acoustography is the observation of carotid and peripheral arterial plaque morphology (Pislaru et al., 2008, 2011). Calcification is often present in these plaques, which, like microcalcifications, are hard structures that vibrate well with vibro-acoustography. Pislaru et al. (2008) demonstrated the visualization of *in vivo* and *in situ* porcine femoral arteries demonstrating the ability to detect the calcification in plaques.

Although vibro-acoustography has primarily been demonstrated in excised tissues, Urban et al. (2011) have demonstrated the feasibility of vibro-acoustography on clinical scanners. The addition of vibro-acoustography to clinical scanners allows current diagnostic utilities to be complemented with the information provided by vibro-acoustography. The current major hurdle in vibro-acoustography on clinical systems, besides its susceptibility to motion, is obtaining images in the same plane as diagnostic B-mode imaging. Currently, B-mode imaging provides a tomographic view (i.e., a plane parallel to the transducer axis), whereas vibro-acoustography provides a view perpendicular to B-mode imaging. With localized focusing on a linear array, vibro-acoustography could be implemented in the same plane as B-mode imaging, but the current intersection of the two unmodulated beams requires an axial extent much longer than the resolution of B-mode imaging, making it currently unsuitable to complement B-mode imaging. In addition, creating the C-scan for the vibro-acoustic image requires a mechanical translation stage. These issues could potentially be resolved with the recent introduction of matrix arrays, which have the capability of creating the conventional ultrasound scans in the same plane as the vibro-acoustic scans.

13.3.3.2 Harmonic Motion Imaging

Konofagou and Hynynen (2003) proposed a method which uses a modulating radiation force field like vibro-acoustography, but instead of measuring the acoustic field resulting from the

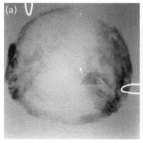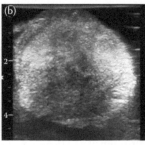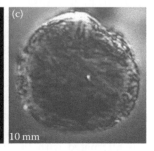

FIGURE 13.9 Comparison of (a) x-ray, (b) conventional B-mode ultrasound, and (c) ultrasound-stimulated vibro-acoustography imaging of an excised prostate sample. A calcification is visible in the center of the x-ray image. Under conventional ultrasound, the calcification is not visible. In the USVA image, the calcification is shown with good contrast. (Reprinted from *Ultrasonics 48*, Alizad A. et al., Image features in medical vibro-acoustography: *In vitro* and *in vivo* results, 559–562, Figure 2, Copyright 2008, with permission from Elsevier.)

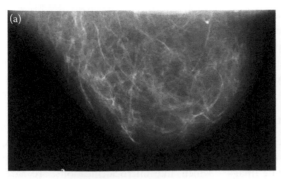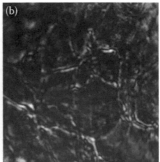

FIGURE 13.10 Comparison of (a) x-ray and (b) ultrasound-stimulated vibro-acoustography images of *in vivo* breast. The connective tissue and breast parynchema appear quite similar in appearance to the x-ray image. The x-ray image presents a two-dimensional projection of the entire breast, whereas the USVA image presents a single slice through the breast at 2 cm depth. (Reprinted from *Ultrasonics 48*, Alizad A. et al., Image features in medical vibro-acoustography: *In vitro* and *in vivo* results, 559–562, Figure 4, Copyright 2008, with permission from Elsevier.)

vibrating tissue, the vibration of the tissue itself is measured directly with conventional ultrasound, giving rise to the name, harmonic motion imaging. Tracking the vibration of tissue in HMI is similar to the displacement tracking used in ARFI imaging, except here the tracking occurs *during* the application of the radiation force rather than postexcitation. In HMI, the amplitude of the modulating tissue displacements are measured, and an image of the displacement amplitude is created. A key difference between HMI and vibro-acoustography is the Δ*f* used in vibrating the tissue. In order to monitor the amplitude of the vibration with conventional ultrasound, the vibrations in HMI must be on the order of hundreds of hertz, while vibro-acoustic emissions are observed from vibrations of tens of kilohertz.

In order to effectively measure the vibrating tissue displacement with conventional ultrasound, an imaging transducer must be placed on the same axis in which the modulating radiation force is applied. This can be accomplished by crossing the beams of two piston transducers to generate the radiation force and placing the imaging transducer between the two piston transducers (Konofagou and Hynynen, 2003). While this configuration creates viable HMI images, the setup is difficult for practical uses in clinics.

HMI, however, has demonstrated potential use in the area of high-intensity focused ultrasound (HIFU) (Maleke and Konofagou, 2008, 2010). In HIFU, a large, concave piston transducer is used to apply the high-intensity ultrasound to a targeted region in tissue. The high intensity of the ultrasound generates heating of the tissue, enough to ablate a small region of tissue. Focused ultrasound surgery (FUS) is the application of HIFU to clinical uses, such as the ablation of cancerous lesions in the liver. In order to monitor these ablations with HMI, the HIFU transducer is created with a hole in the middle for an imaging transducer. The HIFU transducer then uses an amplitude-modulated beam to create an oscillating radiation force. This beam is alternated with the high-intensity beam used to create the ablations. The imaging transducer is activated during the amplitude-modulated beam in order to monitor the tissue displacements during HMI (Maleke and Konofagou, 2008). HMI has successfully demonstrated the visualization of ablation lesions in mammary tumors of mice (Maleke and Konofagou, 2010).

13.3.4 Nonimaging Uses of Radiation Force

Some of the radiation force-based techniques used for imaging have found nonimaging applications, primarily in the form of "virtual biopsy" techniques in which a sole location in the tissue is interrogated. Although these methods are not imaging methods in and of themselves, they are primarily guided by or exclusively used with conventional ultrasonic imaging.

13.3.4.1 Lateral Time-to-Peak

Although SWEI and ARFI imaging provide nice images of the mechanical properties of tissue, sometimes quantitative measures are desired. The ARFI-based lateral TTP method provides a robust method for quantitative measures of shear wave velocity, particularly when resolution is not of interest. The lateral TTP method applies a radiation force impulse to a location, and then monitors the locations surrounding the region of excitation to observe how long it takes for the shear wave to pass through those locations, giving a simple estimate of the shear wave velocity.

The lateral TTP method has demonstrated good-quality measures for measuring the shear wave velocity and shear modulus in the liver (Palmeri et al., 2008). This has led to commercial implementation of the lateral TTP technique as a method for staging liver fibrosis. These studies have demonstrated that liver fibrosis can be well differentiated, particularly between fibrosis stages 2, 3, and 4 (Friedrich-Rust et al., 2009; Lupsor et al., 2009; Takahashi et al., 2009; Haque et al., 2010). In addition, this ARFI-based technique also showed capabilities in detecting the different stiffnesses associated with chronic liver diseases (Kim et al., 2010).

Zhai et al. (2010) correlated findings between regional measurements by SWEI (using the lateral TTP algorithm) and ARFI imaging of excised prostate tissue. This method created a grid measurement of shear moduli overlayed on an ARFI image. Higher shear moduli corresponded to darker, or stiffer, regions in the ARFI image, and smaller shear moduli corresponded to brighter, or softer, regions of tissue.

13.3.4.2 Spatially Modulated Ultrasonic Radiation Force

One limitation of the lateral TTP method is that significant variance in the shear wave speed estimate exists because of measurement bias of the arrival time of the shear wave at the observation point. The measurement bias is a product of ultrasonic speckle (McAleavey et al., 2009), which is different at every observation point and therefore gives rise to a different bias at every observation point. Significant spatial averaging, therefore, is required to reduce the variance.

A similar technique to the lateral TTP method, proposed by McAleavey et al. (2007), eliminates the arrival time bias. The relatively new method, called spatially modulated ultrasonic radiation force (SMURF) (McAleavey et al., 2007, 2009; Elegbe et al., 2011), excites multiple locations in a region of interest with a radiation force impulse, and monitors a single location near that region to observe when the resulting shear waves pass through that point. The speed of the shear waves are then estimated from the difference in the arrival time between the shear waves. Because the shear waves are all observed at the same point, they all experience the same bias in displacement, which can then be ignored. Not only does this method reduce variance in the shear wave speeds, but it can also be applied over a much smaller area than the lateral TTP method, and is therefore capable of creating images of the shear modulus, much like SWEI (McAleavey et al., 2009).

13.3.4.3 Shear Wave Dispersion Ultrasound Vibrometry

Shear wave dispersion ultrasound vibrometry (SDUV) is a method based on vibro-acoustography in which the vibration technique described in Section 13.3.3 is used to oscillate the tissue at a desired low frequency, in the range of a few hundred Hz (Chen et al., 2004a, 2009). This low frequency range permits the formation of a shear wave rather than the acoustic emission described in Section 13.3.3. Like the SWEI methods, detection of the shear wave is performed with two ultrasonic transducers positioned laterally to the excitation point. These two transducers detect both the magnitude and phase of the resulting shear wave. Because the frequency of the oscillations is known, the shear wave speed can be calculated from both the phase of the shear wave and the positions of the two sensors. In addition, the frequency of the oscillation is varied over a range of frequencies (e.g., 100–500 Hz). Observing the response of the shear wave over a range of frequencies allows one to characterize the dispersion of the medium and compute the shear viscosity. The shear wave speed and viscosity are related using the Voigt model by

$$c_s(f_s) = \sqrt{\frac{2(\mu^2 + 2\pi f_s^2 \eta^2)}{\rho_0\left(\mu + \sqrt{\mu^2 + 2\pi f_s^2 \eta^2}\right)}}, \qquad (13.22)$$

where f_s is the frequency of the shear wave. SDUV has been used to estimate the shear modulus and viscosity in excised human prostate samples (Mitri et al., 2011). SDUV demonstrated estimates of shear modulus and viscosity consistent with the values of prostate shear modulus and viscosity measured using other means.

13.3.5 Safety Considerations for Radiation Force Imaging Methods

13.3.5.1 Nonthermal

Radiation force-based imaging techniques often require acoustic intensities beyond what is required for conventional B-mode imaging in order to elicit a sufficient response from the tissue. However, the U.S. Food and Drug Administration (FDA) and the International Electrotechnical Commission (IEC) have placed limitations on the amount of acoustic exposure that is introduced into the body. The main limitations of acoustic exposure are applied to the spatial-peak, temporal average intensity (I_{spta}) and the spatial-peak, and spatial average intensity (I_{sppa}). These two measures estimate the maximum acoustic intensity applied over the course of an exam (I_{spta}) and the maximum intensity applied for any given ultrasonic beam (I_{spta}). These acoustic intensities are measured in water with a hydrophone, and are then "derated" by assuming that the acoustic beam is propagating in a tissue with an attenuation coefficient of 0.3 dB/cm/MHz. Currently, the FDA and IEC limit the I_{spta} to 720 mW/cm² and the I_{sppa} to 190 mW/cm² (Food and Drug Administration—Center for Devices and Radiological Health, 1997; International Electrotechnical Commission, 2001).

In addition, limitations are also applied to the mechanical index (MI) of the transmitted pulses. The MI is a measure of the potential mechanical bioeffects, most importantly cavitation. Cavitation is the formation of empty cavities in a fluid, or tissue, by the rapid changes in pressure. These empty cavities, or bubbles, then collapse and are forceful enough that they can potentially cause damage to the surrounding tissue. Cavitation is mainly impacted by the rarefactional pressure, where the negative pressure effectively spreads the particles in the medium. The MI is calculated as

$$\mathrm{MI} = \frac{p_-}{\sqrt{f}}, \qquad (13.23)$$

where p_- is the peak negative pressure and f is the transmitted frequency of the wave. The MI is mainly a threshold effect, meaning that the potential for cavitation does not occur until a threshold is met. The maximum MI allowed by the FDA and IEC is 1.9 (Food and Drug Administration—Center for Devices and Radiological Health, 1997; International Electrotechnical Commission, 2001). In addition, the FDA and IEC allow the I_{sppa} to exceed 190 mW/cm² if the MI is kept below 1.9.

This exception is what permits acoustic radiation force techniques to be implemented in diagnostic settings. Radiation force is achieved by increasing the acoustic pulse duration, effectively increasing the I_{sppa} to approximately five times the FDA limit. However, by maintaining a reasonable output pressure, the I_{spta}

and MI can be kept well below their respective limits, and thus comply with FDA and IEC requirements and maintain a safe acoustical exposure for the patient.

13.3.5.2 Thermal

The increases in acoustic intensity associated with radiation force techniques deposit significantly more energy into tissue than conventional B-mode imaging. While some of the energy absorbed by the tissue is converted into a mechanical force, the vast majority of the energy is converted to heat. The FDA and IEC limit the amount of heating that an ultrasonic system can produce. This limit is determined by the thermal index (TI), which is effectively an estimation of the amount of temperature rise the tissue is experiencing. The FDA and IEC limit this temperature rise to be 6°C (Food and Drug Administration—Center for Devices and Radiological Health, 1997; International Electrotechnical Commission, 2001).

The thermal increases from a single radiation force pulse used in ARFI imaging are typically <0.1°C (Palmeri and Nightingale, 2004; Dahl et al., 2007), but the cumulative effects of real-time, two-dimensional imaging with ARFI, however, can have a significant impact on heating. Palmeri and Nightingale (2004) used finite element models to show that temperature increases in tissue can reach up to 4.5°C with an ARFI frame rate of just three frames per second (fps), 50 image lines, and an acoustic absorption of 0.5 dB/cm/MHz. In vibro-acoustography, the peak heating of the tissue after a full scan was estimated to be approximately 0.035°C (Chen et al., 2010).

Transducer heating is also of concern in acoustic radiation force imaging techniques, perhaps more so than heating of the tissue by absorption of the wave, because the heating is much more rapid. At relatively low frame rates, the transducer heating for ARFI imaging was shown to increase the surface temperature by approximately 2°C after approximately 2 seconds of imaging (Dahl et al., 2007; Bouchard et al., 2009). In vibro-acoustography, the heating at the surface of the confocal transducer was measured to be 0.27°C for a single scan (Chen et al., 2010).

While SWEI and vibro-acoustography techniques require fewer application or less intense radiation force than ARFI imaging, ARFI imaging can be used in real time and at significant imaging depth. Techniques for reducing the heating and increasing the frame rates in ARFI imaging include parallel (Dahl et al., 2007), multiplexed, and multitime (Bouchard et al., 2009) tracking techniques. Parallel tracking increases the ratio of spatial locations observed to applications of radiation force. Multiplex tracking observes two separate locations within the time often used to track the response at a single location, and multitime tracking observes many spatial locations within the time used to track the response at a single location by combining the parallel and multiplex tracking. Because heating is linked to the rate of application of the radiation force, either an increase in frame rate or a decrease in heating can be obtained with these techniques, but not both (Dahl et al., 2007; Bouchard et al., 2009).

A special communication from the Technical Standards Committee of the American Institute of Ultrasound in Medicine (AIUM) recently proposed changes to the way TI is measured (Bigelow et al., 2011). As a part of this special communication, particular emphasis was placed on the new imaging techniques based on the radiation force of acoustic waves, as the current methods for assessing the TI do not properly estimate the appropriate TI for radiation force techniques. Although current measures of heating from acoustic radiation force imaging methods have thus far been shown to create relatively small amounts of thermal increases, it will be important for manufacturers of systems, including radiation force methods, to properly address the measurements of TI so that ultrasonic imaging modalities remain a safe imaging modality for everyone.

References

Alizad, A., M. Fatemi, L. E. Wold, and J. F. Greenleaf. 2004. Performance of vibro-acoustography in detecting microcalcifications in excised human breast tissue: A study of 74 tissue samples. *IEEE Trans Med Imaging* 23(3), 307–312.

Alizad, A., D. H. Whaley, J. F. Greenleaf, and M. Fatemi. 2006. Critical issues in breast imaging by vibro-acoustography. *Ultrasonics* 44(Suppl 1), e217–e220.

Allen, J. D., D. M. Dumont, B. Sileshi, R. Mitchell, G. E. Trahey, and J. J. Dahl. 2011. The development and potential of acoustic radiation force impulse (ARFI) imaging for carotid artery plaque characterization. *Vasc Med* 16(4), 302–311.

Arnal, B., M. Pernot, and M. Tanter. 2011a. Monitoring of thermal therapy based on shear modulus changes: I. Shear wave thermometry. *IEEE Trans Ultrason Ferroelect Freq Contr* 58(2), 369–378.

Arnal, B., M. Pernot, and M. Tanter. 2011b. Monitoring of thermal therapy based on shear modulus changes: II. Shear wave imaging of thermal lesions. *IEEE Trans Ultrason Ferroelect Freq Contr* 58(8), 1603–1611.

Athanasiou, A., A. Tardivon, L. Ollivier, F. Thibault, C. E. Khoury, and S. Neuenschwander. 2009. How to optimize breast ultrasound. *Eur J Radiol* 69(1), 6–13.

Bercoff, J., M. Pernot, M. Tanter, and M. Fink. 2004a. Monitoring thermally-induced lesions with supersonic shear imaging. *Ultrason Imaging* 26(2), 71–84.

Bercoff, J., M. Tanter, and M. Fink. 2004b. Supersonic shear imaging: A new technique for soft tissue elasticity mapping. *IEEE Trans Ultrason Ferroelect Freq Contr* 51(4), 396–409.

Bigelow, T. A., C. C. Church, K. Sandstrom, J. G. Abott, M. C. Ziskin, P. D. Edmonds, B. Herman, and K. E. Thomenius. 2011. The thermal index: Its strengths, weaknesses, and proposed improvements. *J Ultrasound Med* 30(5), 714–734.

Bouchard, R. R., J. J. Dahl, S. J. Hsu, M. L. Palmeri, and G. E. Trahey. 2009. Image quality, tissue heating, and frame-rate trade-offs in acoustic radiation force impulse imaging. *IEEE Trans Ultrason Ferroelect Freq Contr* 56(1), 63–76.

Chen, S., W. Aquino, A. Alizad, M. Urban, R. Kinnick, J. F. Greenleaf, and M. Fatemi. 2010. Thermal safety of vibro-acoustography using a confocal transducer. *Ultrasound Med Biol 36*(2), 343–349.

Chen, S., M. Fatemi, and J. F. Greenleaf. 2004a. Quantifying elasticity and viscosity from measurement of shear wave speed dispersion. *J Acoust Soc Am 115*(6), 2781–2785.

Chen, S., M. Fatemi, R. Kinnick, and J. F. Greenleaf. 2004b. Comparison of stress field forming methods for vibro-acoustography. *IEEE Trans Ultrason Ferroelect Freq Contr 51*(3), 313–321.

Chen, S., M. Urban, C. Pislaru, R. Kinnick, Y. Zheng, A. Yao, and J. F. Greenleaf. 2009. Shearwave dispersion ultrasound vibrometry (SDUV) for measuring tissue elasticity and viscosity. *IEEE Trans Ultrason Ferroelect Freq Contr 56*(1), 55–62.

Cho, S. H., J. Y. Lee, J. K. Han, and B. I. Choi. 2010. Acoustic radiation force impulse elastography for the evaluation of focal solid hepatic lesions: Preliminary findings. *Ultrasound Med Biol 36*(2), 202–208.

Couade, M., M. Pernot, E. Messas, A. Bel, M. Ba, A. Hagége, M. Fink, and M. Tanter. 2011. *In vivo* quantitative mapping of myocardial stiffening and transmural anisotropy during the cardiac cycle. *IEEE Trans Med Imaging 30*(2), 295–305.

Dahl, J. J., D. M. Dumont, E. M. Miller, J. D. Allen, and G. E. Trahey. 2009. Acoustic radiation force impulse imaging for noninvasive characterization of carotid artery atherosclerotic plaques: A feasibility study. *Ultrasound Med Biol 35*(5), 707–716.

Dahl, J. J., G. F. Pinton, M. L. Palmeri, V. Agrawal, K. R. Nightingale, and G. E. Trahey. 2007. A parallel tracking method for acoustic radiation force impulse imaging. *IEEE Trans Ultrason Ferroelect Freq Contr 54*(2), 301–312.

Eckart, C. 1948. Vortices and streams caused by sound waves. *Phys Rev 73*(1), 68–76.

Elegbe, E. C., M. G. Menon, and S. A. McAleavey. 2011. Comparison of two methods for the generation of spatially modulated ultrasound radiation force. *IEEE Trans Ultrason Ferroelect Freq Contr 58*(7), 1344–1354.

Evans, A., P. Whelehan, K. Thomson, D. McLean, K. Brauer, C. Purdie, L. Jordan, L. Baker, and A. Thompson. 2010. Quantitative shear wave ultrasound elastography: Initial experience in solid breast masses. *Breast Cancer Res 12*(6), R104.

Eyerly, S. A., S. J. Hsu, S. H. Agashe, G. E. Trahey, Y. Li, and P. D. Wolf. 2010. An *in vitro* assessment of acoustic radiation force impulse imaging for visualizing cardiac radiofrequency ablation lesions. *J Cardiovasc Electrophysiol 21*(5), 557–563.

Fahey, B., R. C. Nelson, D. P. Bradway, S. J. Hsu, D. M. Dumont, and G. E. Trahey. 2008a. *In vivo* visualization of abdominal malignancies with acoustic radiation force elastography. *Physics Med Biol 53*(1), 279–293.

Fahey, B., R. C. Nelson, S. J. Hsu, D. P. Bradway, D. M. Dumont, and G. E. Trahey. 2008b. *In vivo* guidance and assessment of liver radio-frequency ablation with acoustic radiation force elastography. *Ultrasound Med Biol 34*(10), 1590–1603.

Fahey, B. J., K. R. Nightingale, S. A. McAleavey, M. L. Palmeri, P. D. Wolf, and G. E. Trahey. 2005. Acoustic radiation force impulse imaging of myocardial radiofrequency ablation: Initial *in vivo* results. *IEEE Trans Ultrason Ferroelect Freq Contr 52*(4), 631–641.

Fahey, B. J., M. L. Palmeri, and G. E. Trahey. 2006. Frame rate considerations for real-time abdominal acoustic radiation force impulse imaging. *Ultrason Imaging 28*(4), 193–210.

Fatemi, M. and J. F. Greenleaf. 1998. Ultrasound-stimulated vibro-acoustic spectrography. *Science 280*(3), 82–85.

Fatemi, M. and J. F. Greenleaf. 1999. Vibro-acoustography: An imaging modality based on ultrasound-stimulated acoustic emission. *Proc Natl Acad Sci 96*(12), 6603–6608.

Fatemi, M., L. E. Wold, A. Alizad, and J. F. Greenleaf. 2002. Vibro-acoustic tissue mammography. *IEEE Trans Med Imaging 21*(1), 1–8.

Food and Drug Administration—Center for Devices and Radiological Health (1997, September). Information for manufacturers seeking marketing clearance of diagnostic ultrasound systems and transducers. Technical report, US Department of Health and Human Services. Retrieved from http://www.fda.gov/cdrh/ode/ulstran.pdf.

Friedrich-Rust, M., K. Wunder, S. Kriener, F. Sotoudeh, S. Richter, J. Bojunga, E. Herrmann et al. 2009. Liver fibrosis in viral hepatitis: Noninvasive assessment with acoustic radiation force impulse imaging versus transient elastography. *Radiology 252*(2), 595–604.

Haque, M., C. Robinson, D. Owen, E. M. Yoshida, and A. Harris. 2010. Comparison of acoustic radiation force impulse imaging (ARFI) to liver biopsy histologic scores in the evaluation of chronic liver disease: A pilot study. *Ann Hepatol 9*(3), 289–293.

Hsu, S. J., R. R. Bouchard, D. M. Dumont, P. D. Wolf, and G. E. Trahey. 2007a. *In vivo* assessment of myocardial stiffness with acoustic radiation force impulse imaging. *Ultrasound Med Biol 33*(11), 1706–1719.

Hsu, S. J., R. R. Bouchard, D. M. Dumont, P. D. Wolf, and G. E. Trahey. 2009. On the characterization of left ventricular function with acoustic radiation force impulse imaging. In *Proc IEEE Ultrason Symp*, pp. 1942–1945.

Hsu, S. J., D. P. Bradway, R. R. Bouchard, P. J. Hollender, P. D. Wolf, and G. E. Trahey. 2010. Parametric pressure-volume analysis and acoustic radiation force impulse imaging of left ventricular function. In *Proc IEEE Ultrason Symp*, pp. 698–701.

Hsu, S. J., B. J. Fahey, D. M. Dumont, P. D. Wolf, and G. E. Trahey. 2007b. Challenges and implementation of radiation-force imaging with an intracardiac ultrasound transducer. *IEEE Trans Ultrason Ferroelect Freq Contr 54*(5), 996–1009.

Hsu, S. J., J. L. Hubert, P. D. Wolf, and G. E. Trahey. 2007c. Acoustic radiation force impulse imaging of mechanical stiffness propagation within myocardial tissue. In *Proc IEEE Ultrason Symp*, pp. 864–867.

International Electrotechnical Commission. 2001. Medical electrical equipment—Particular requirements for the safety of ultrasonic medical diagnostic and monitoring equipment. Technical report, International Electrotechnical Commission, Geneva, Switzerland. Report No. 60601-2-37.

Kim, J. E., J. Y. Lee, Y. J. Kim, J. H. Yoon, S. H. Kim, J. M. Lee, J. K. Han, and B. I. Choi. 2010. Acoustic radiation force impulse elastography for chronic liver disease: Comparison with ultrasound-based scores of experienced radiologists, Child-Pugh scores and liver function tests. *Ultrasound Med Biol* 36(10), 1637–1643.

Konofagou, E. E. and K. Hynynen. 2003. Localized harmonic motion imaging: Theory, simulations, and experiments. *Ultrasound Med Biol* 29(10), 1405–1413.

Lupsor, M., R. Badea, H. Stefanecu, Z. Sparchez, H. Branda, A. Serban, and A. Maniu. 2009. Performance of a new elastographic method (ARFI technology) compared to unidimensional transient elastography in the noninvasive assessment of chronic hepatitis C. Preliminary results. *J Gastrointestin Liver Dis* 18(3), 303–310.

Maleke, C. and E. E. Konofagou. 2008. Harmonic motion imaging for focused ultrasound (HMIFU): A fully integrated technique for sonication and monitoring of thermal ablation in tissues. *Physics Med Biol* 53(6), 1773–1793.

Maleke, C. and E. E. Konofagou. 2010. *In vivo* feasibility of real-time monitoring of focused ultrasound surgery (FUS) using harmonic motion imaging (HMI). *IEEE Trans Biomed Eng* 57(1), 7–11.

McAleavey, S. A., M. Menon, and E. Elegbe. 2009. Shear modulus imaging with spatially modulated ultrasound radiation force. *Ultrason Imaging* 31(4), 217–234.

McAleavey, S. A., M. Menon, and J. Orszulak. 2007. Shear-modulus estimation by application of spatially-modulated impulsive acoustic radiation force. *Ultrason Imaging* 29(2), 87–104.

Meng, W., G. Zhang, C. Wu, G. Wu, Y. Song, and Z. Lu. 2011. Preliminary result of acoustic radiation force impulse (ARFI) ultrasound imaging of breast lesions. *Ultrasound Med Biol* 37(9), 1436–1443.

Mitri, F. G., B. J. Davis, A. Alizad, J. F. Greenleaf, T. M. Wilson, L. A. Mynderse, and M. Fatemi. 2008. Prostate cryotherapy monitoring using vibroacoustography: Preliminary results of an *ex vivo* study and technical feasibility. *IEEE Trans Biomed Eng* 55(11), 2584–2592.

Mitri, F. G., B. J. Davis, M. W. Urban, A. Alizad, J. F. Greenleaf, G. H. Lischer, T. M. Wilson, and M. Fatemi. 2009. Vibro-acoustography imaging of permanent prostate brachytherapy seeds in an excised human prostate—Preliminary results and technical feasibility. *Ultrasonics* 49(3), 389–394.

Mitri, F. G., M. W. Urban, M. Fatemi, and J. F. Greenleaf. 2011. Shear wave dispersion ultrasonic vibrometry for measuring prostate shear stiffness and viscosity: An *in vitro* pilot study. *IEEE Trans Biomed Eng* 58(2), 235–242.

Muller, M., J.-L. Gennisson, T. Deffieux, M. Tanter, and M. Fink. 2009. Quantitative viscoelasticity mapping of human liver using supersonic shear imaging: Preliminary *in vivo* feasibility study. *Ultrasound Med Biol* 35(2), 219–229.

Nightingale, K. R., S. A. McAleavey, and G. E. Trahey. 2003. Shear-wave generation using acoustic radiation force: *In vivo* and *ex vivo* results. *Ultrasound Med Biol* 29(12), 1715–1723.

Nightingale, K. R., M. S. Soo, R. W. Nightingale, and G. E. Trahey. 2002. Acoustic radiation force impulse imaging: *In vivo* demonstration of clinical feasibility. *Ultrasound Med Biol* 28(2), 227–235.

Nyborg, W. L. M. 1965. Acoustic streaming. In W. P. Mason (Ed.), *Physical Acoustics*, Volume IIB, Chapter 11, pp. 265–331. New York: Academic Press, Inc.

Palmeri, M. L. and K. R. Nightingale. 2004. On the thermal effects associated with radiation force imaging of soft tissue. *IEEE Trans Ultrason Ferroelect Freq Contr* 51(5), 551–565.

Palmeri, M. L., M. H. Wang, J. J. Dahl, K. D. Frinkley, and K. R. Nightingale. 2008. Quantifying hepatic shear modulus *in vivo* using acoustic radiation force. *Ultrasound Med Biol* 34(4), 546–558.

Pernot, M., M. Couade, P. Mateo, B. Crozatier, R. Fischmeister, and M. Tanter. 2011. Real-time assessment of myocardial contractility using shear wave imaging. *J Am Coll Cardiol* 58(1), 65–72.

Pislaru, C., J. F. Greenleaf, B. Kantor, and M. Fatemi. 2011. *Atherosclerosis Disease Management*, Chapter 21: Vibro-Acoustography of Arteries, pp. 675–698. New York: Springer.

Pislaru, C., B. Kantor, R. R. Kinnick, J. L. Anderson, M.-C. Aubry, M. W. Urban, M. Fatemi, and J. F. Greenleaf. 2008. *In vivo* vibroacoustography of large peripheral arteries. *Invest Radiol* 43(4), 243–252.

Sarvazyan, A. P., O. V. Rudenko, S. D. Swanson, J. B. Fowlkes, and S. Y. Emelianov. 1998. Shear wave elasticity imaging: A new ultrasonic technology of medical diagnostics. *Ultrasound Med Biol* 24(9), 1419–1435.

Sebag, F., J. Vaillant-Lombard, J. Berbis, V. Griset, J. F. Henry, P. Petit, and C. Oliver. 2010. Shear wave elastography: A new ultrasound imaging mode for the differential diagnosis of benign and malignant thyroid nodules. *J Clin Endocrinol Metab* 95(12), 5281–5288.

Sharma, A. C., M. S. Soo, G. E. Trahey, and K. R. Nightingale. 2004. Acoustic radiation force impulse imaging of *in vivo* breast masses. In *Proc IEEE Ultrason Symp*, Volume 1, pp. 728–731.

Shuang-Ming, T., Z. Ping, Q. Ying, C. Li-Rong, Z. Ping, and L. Rui-Zhen. 2011. Usefulness of acoustic radiation force impulse imaging in the differential diagnosis of benign and malignant liver lesions. *Acad Radiol* 18(7), 810–815.

Takahashi, H., N. Ono, Y. Eguchi, T. Eguchi, Y. Kitajima, Y. Kawaguchi, S. Nakashita et al. 2009. Evaluation of acoustic radiation force impulse elastography for fibrosis staging of chronic liver disease: A pilot study. *Liver Int* 30(4), 538–545.

Tanter, M., J. Bercoff, A. Athanasiou, T. Deffieux, J.-L. Gennisson, G. Montaldo, M. Muller, A. Tardivon, and M. Fink. 2008. Quantitative assessment of breast lesion viscoelasticity: Initial clinical results using supersonic shear imaging. *Ultrasound Med Biol* 34(9), 1373–1386.

Urban, M. W., C. Chalek, R. R. Kinnick, T. M. Kinter, B. Haider, J. F. Greenleaf, K. E. Thomenius, and M. Fatemi. 2011. Implementation of vibro-acoustography on a clinical ultrasound system. *IEEE Trans Ultrason Ferroelect Freq Contr* 58(6), 1169–1181.

Westervelt, P. J. 1951. The theory of steady forces caused by sound waves. *J Acoust Soc Am* 23(4), 312–315.

Zhai, L., J. Madden, W.-C. Foo, M. L. Palmeri, V. Mouraviev, T. J. Polascik, and K. R. Nightingale. 2010. Acoustic radiation force impulse imaging of human prostates *ex vivo*. *Ultrasound Med Biol* 36(4), 576–588.

Zhai, L., T. J. Polascik, W.-C. Foo, S. Rosenzweig, M. L. Palmeri, J. Madden, and K. R. Nightingale. 2012. Acoustic radiation force impulse imaging of human prostates: Initial *in vivo* demonstration. *Ultrasound Med Biol* 38(1), 50–61.

VI

Multimodality Imaging

Multimodality: Positron Emission Tomography-Magnetic Resonance Imaging

Jinsong Ouyang
Harvard Medical School

Quanzheng Li
Harvard Medical School

Georges El Fakhri
Harvard Medical School

14.1 Introduction

Positron emission tomography-magnetic resonance imaging (PET-MR) is a novel and promising imaging modality that combines the strengths of both positron emission tomography (PET) and magnetic resonance (MR). It paves the way for a wealth of new clinical imaging applications [1]. PET-MR has comparable performance to PET-computed tomography (PET-CT) in the detection, diagnosis, and monitoring of suspected tumors [2]. In addition, it has the potential to become an important tool for the development of antitumor drugs, such as inhibitors of angiogenesis or modulators of the immune system. For brain imaging, it is obvious that simultaneous acquisition of various metabolic and functional parameters is a natural application of PET-MR that may yield new insights into the organization of the brain and its associated changes in disease [3]. PET imaging of the thorax and abdomen is limited by poor spatial resolution as well as by cardiac and respiratory motion. Simultaneous PET-MR provides a promising approach to continuously compensate for motion during PET acquisition, particularly in cardiology but also in the accurate detection of lesions in the abdomen or thorax [4].

The combination of late-enhancement MRI and fluorodeoxyglucose (FDG) uptake in a single imaging session can potentially push cardiac imaging to a new level. Combining cardiac MR, MR angiography, and cardiac PET may enable detection and differentiation of vulnerable plaques [5]. Furthermore, simultaneous whole-body PET-MR systems allow the measurement of the heart's 3D nonrigid motion field, including wall motion, by using tagged MR without the additional radiation associated with CT. Accurate estimation of motion using MR will significantly improve the resolution and signal-to-noise ratio (SNR) of cardiac PET imaging, and consequently result in accurate quantification of myocardial blood flow (MBF) and precise assessment of peri-infarct ischemia in patients with nontransmural infarcts. More accurate and robust quantification of MBF reserve will enhance the prognostic value of cardiac PET [6,7] in coronary artery disease (CAD), which represents an enormous public health burden in the United States [8], particularly among the patients with diffuse atherosclerosis (e.g., diabetes; HIV; s/p heart transplantation) in whom the amount of ischemia is likely to be underestimated. Better characterization of peri-infarct ischemia would allow for better identification of patients with prior CAD who may benefit from coronary or surgical revascularization. Given that current techniques (MRI perfusion, stress echo, SPECT, and PET), all have inherent limitations in evaluating the amount of ischemia in areas of prior nontransmural infarction, PET-MR with improved resolution and SNR will play an important role in the identification of ischemia in such patients.

PET-MR can also play synergistic roles in abdominal cancer imaging. For example, combined PET-MR can aid in identifying the liver metastases, especially in challenging cases of diffuse hepatic metastases, where PET can discriminate among numerous treated lesions that obscure identification of potentially new, recurrent, or untreated lesions while MR is crucial in detecting the small hepatic lesions. Additionally, accurate response to treatment is often difficult to judge on liver MR alone and may be significantly aided by combined PET-MR. PET-MR

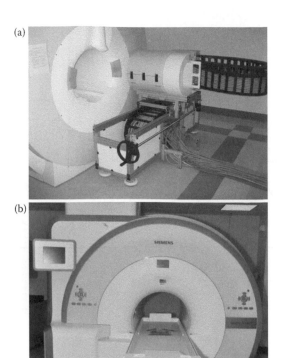

(a)

(b)

FIGURE 14.1 Siemens PET-MR scanners installed in the Athinoula A. Martinos Center for Biomedical Imaging, Radiology Department, MGH. (a) Brain PET-MR scanner. The PET insert is to the right of the 3T MR scanner. (b) Whole-body Biograph mMR PET-MR scanner. (Courtesy of C. Catana and B. Rosen, MGH.)

also has the added potential to identify the extrahepatic disease, which is often less conspicuous and easily overlooked on MR alone. In addition, PET-MR overcomes several limitations of PET-CT in patients undergoing cancer restaging examinations, including (1) allergies to CT contrast agents, (2) renal dysfunction (GFR < 30), and (3) greater detection of small lesions by MR.

Presently, all three major imaging manufacturers are working on finalizing the designs of integrated or tandem back-to-back PET-MR systems. A large number of scanners are scheduled for delivery by the same manufacturers in the next 2 years in the United States and worldwide, underlining the increasing acceptance of the potential for PET-MR by the imaging community. Figure 14.1 shows the Siemens brain and whole-body PET-MR scanners installed at Athinoula A. Martinos Center for Biomedical Imaging, Massachusetts General Hospital (MGH). Figure 14.2 shows a coronal slice of a simultaneous ^{18}FDG-PET-MR acquisition using the whole-body PET-MR scanner. About 5MCi of FDG were injected to the subject. The PET list-mode data were acquired 30 min postinjection. Five bed positions (3 min/bed) were used to scan from the head to the upper thighs. The attenuation correction for each bed position was derived from the MR data acquired with a 3D volumetric interpolated breath-hold examination (VIBE) Dixon sequence (TR/TE1/TE2 = 3.6/1.23/2.46 ms, flip angle = 10°, left-right field of view (FOV) = 500 mm, anterior-posterior FOV = 500/333 mm, slices per slab = 128, voxel size $4.1 \times 2.6 \times 2.6$ mm^3). The PET images were reconstructed using an iterative algorithm (OP-OSEM, 1 iteration, 21 subsets). The MR data for each bed position were acquired using a T1 turbo spin echo sequence with the following parameters: TR/TE = 500/9.5 ms, base resolution = 384, FOV along readout direction = 450 mm, FOV along phase encoding direction = 68.8%; voxel size $1.6 \times 1.2 \times 5$ mm^3. PET-compatible MR coils were used for this study (head and neck, spine + body arrays).

14.2 PET-MR Instrumentation

The integration of PET and MR was first proposed many years ago. However, such integration is challenging due to the interference between the PET and MR and due to the lack of space

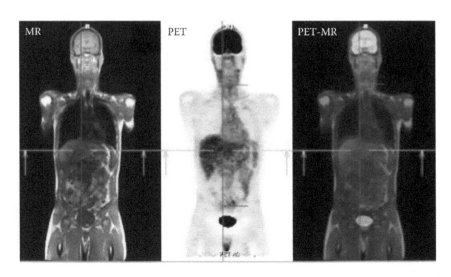

MR   PET   PET-MR

FIGURE 14.2 (**See color insert.**) Coronal slice of a simultaneous ^{18}FDG-PET-MR acquisition. (Courtesy of C. Catana and B. Rosen, MGH.)

inside the MR bore. After significant research and development effort, a fully integrated simultaneous PET-MR system has only recently become a reality.

For the integration of PET and MR, the primary technical challenge is the effect of strong static magnetic field, gradient, and radiofrequency (RF) signals on the PET detectors. Conventional photomultiplier tubes (PMT) are very sensitive to magnetic field and can also pick up easily the RF signals from MR. Furthermore, the homogeneity of the main magnetic field in MR is a very important factor for MR imaging quality and must not be significantly degraded by the PET detectors. Finally, the PET detectors must be compact so that they can fit into the bore of the MR scanner.

One method to build an MR-compatible PET system is to use long optical fibers to transport the scintillation light outside the magnetic field region [9–12]. However, the light loss by the transmission via long optical fibers is significant. Another method is to use the avalanche photodiodes (APDs), which are insensitive to magnetic fields and have good compactness. APDs, which were developed in the 1980s [13], have become an established technology in photo-detection. APDs measure the intensity of an optical signal and take advantage of the internal gain provided by ionization. The feasibility of using APDs for PET in a high magnetic field was first studied in 1998 [14], and the first preclinical PET-MR scanner using lutetium oxyorthosilicate (LSO) and APD detectors was reported in 2006 [15].

One major disadvantage of presently available APDs is that they have two to three orders of magnitude lower gain than conventional PMTs, and thus require sophisticated preamplifiers. APDs also suffer from limited timing resolution and cannot be used for time-of-flight (TOF) PET. The silicon photomultiplier (SiPM) is a densely packed matrix of small Geiger-mode APD cells, in which the APD is biased above the avalanche breakdown voltage, with individual quenching resisters [16]. SiPMs are insensitive to magnetic fields and have high gains $\sim 10^5$–10^6. Furthermore, SiPMs have intrinsic timing resolution of about 60 ps and can thus be used for TOF PET. The feasibility of using SiPMs in an MR system was first demonstrated in 2007 [17,18]. A PET-MR prototype system for small-animal imaging using LSO and SiPM arrays has been recently reported [19].

Siemens Biograph mMR shown in Figure 14.1b is the first commercially available clinical simultaneous PET-MR scanner. It uses LSO/APD PET detectors that are placed between MR body coil and gradient coils. It has been shown that the performance of the PET system in the mMR scanner is equivalent to that of the Siemens Biograph PET/CT. The MR system in mMR performs essentially like a stand-alone system [20].

Sequential PET-MR systems have also been developed and proposed in the past. Unlike the simultaneous PET-MR system, a sequential system does not need a redesign of the PET system. For example, Philips Ingenuity TF is a clinical whole-body sequential PET-MR scanner with a TOF PET and 3T MR system [21]. The PET system is almost the same as the one used in the Philips PET/CT systems except the PMTs are shielded from the magnetic fields. This design does not allow simultaneous PET and MR acquisition; it allows for acquisition of coregistered PET and MR images acquired sequentially.

14.3 MR-Based PET Attenuation Correction

After the successful integration of MR and PET hardware, PET attenuation correction remains a key unsolved challenge facing PET-MR. For PET, the photon attenuation coefficients must be measured accurately in order to compute a faithful reconstruction of the tracer concentration maps. In standard PET imaging, the photon attenuation coefficients are measured separately using an external source of positron emitters. In PET-CT scanning, the photon attenuation coefficients are measured with x-ray transmission and the attenuation coefficients for the x-rays are remapped to estimate the values for 511 keV photons. For PET-MR imaging, the PET attenuation map generally needs to be derived from the MR image. In most cases, MR signals are related to hydrogen density and mobility. On the other hand, the attenuation experienced by photons in a PET scan is entirely related to electron density. Therefore, photon attenuation coefficient is not directly related to the measured MRI signal. Mapping from an MRI image to a PET attenuation image is challenging.

There have been many reports in the literature on MRI-based PET attenuation correction (AC) in the past. One method is to use atlas registration [22–26]. In order to obtain the attenuation map for an acquired MR image, a reference CT image is registered to the MR image using nonrigid intermodality registration. Alternatively, a reference MR image, which is already registered to a reference CT image, can be first registered to the MR image using nonrigid intramodality registration followed by the same transformation of the reference CT image. Atlas registration solves the problem if a perfect deformable registration between different subjects can be achieved. However, this is rarely the case especially for whole-body applications due to the anatomical difference between subjects and the limitations of the nonrigid deformable registration method itself (e.g., local minima due to nonconvexity of the similarity measure). The potential for the atlas registration approach to be used in clinical settings is therefore limited. Another approach, which may have the potential for clinical use, is to segment an MRI image into different tissue types and then assign the corresponding photon attenuation coefficients to them. This approach has been applied to both brain [27–31] and whole-body imaging [24,32–35] in the past.

In the past, MR-based PET AC for brain imaging has been studied extensively using the tissue segmentation; however, such research for torso or whole-body imaging is still in its early stage. For whole-body studies, Martinez-Moller et al. developed an MR-based four-class segmentation approach (air, lungs, fat, and other tissue) and reported that the bias of standardized uptake value (SUV) due to the segmentation, which is CT-based (i.e., applying segmentation on CT images), was <10% for all the lesions except one in the pelvis using their 35 patient PET-CT samples [36]. They also studied lesion detection and found there were no obvious differences in the clinical interpretations of the lesions due to the segmentation. Hu et al. reported SUV bias <10% using a three-class segmentation method (air, lungs, and other tissues) for all the lesions in their patient samples except the ones in pelvis [37]. Schulz et al. showed that the SUV difference between CT- and MR-based three-class segmentation methods is <5% for the bone lesions found in their 15 PET-CT/MR patient scans [34]. Eiber et al. also showed that no significant SUV difference could be found for 81 positive lesions in low-dose CT compared to Dixon-based MR (CT images were actually used for segmentation to avoid the registration errors between CT and MR images) [35]. Recently, Hofmann et al. also found that SUV bias is <10% using both segmentation and atlas-based/pattern recognition methods from 11 patient scans [26]. All these whole-body studies showed that the SUV bias was generally <10% using either three- (air, lungs, and other tissues) or four-class (air, lungs, fat, and other tissues) MR-based segmentation methods. However, most of these conclusions were based on limited bias studies on SUV values for some lesions already identified at some particular locations in their patient samples. The use of the SUV as an endpoint to evaluate bias is too restrictive. It is the bias in the local PET concentration values that is important for evaluating the absolute quantitative potential of PET-MR. Another limitation is that these studies provided a limited assessment of the performance of these segmentation methods because the locations of the lesions studied did not cover the entire body. Moreover, a region of interest (ROI) was first drawn around an existing lesion previously identified on the PET images reconstructed using the original CT for AC, then the SUV bias for the lesion was calculated for the segmentation method used. The SUV bias can be affected by how the ROI is drawn. Furthermore, it is likely that an incorrect attenuation map causes artifacts in the reconstructed PET images. This is different from other degrading effects, such as partial volume effects, which cause blurring on the images. Unlike blurring effects, artifacts caused by biased AC can either increase or decrease lesion detectability. Therefore, the performance of AC methods should be based on an activity estimation task across the entire ROI for the desired clinical application, that is, a bias study, instead of a lesion detection task. Furthermore, PET-MR applications are not limited to lesion detection or staging. Many clinical PET-MR applications will require absolute quantitation and therefore small SUV bias within the entire volume of interest.

Ouyang et al. applied multiple thresholds directly to the CT images of 23 patients to classify tissue classes [38]. All the 23 FDG whole-body PET-CT scans were acquired on a Siemens Biograph-64 PET-CT scanner. Segmentation methods to classify air, lung, other tissues (3-C); air, lung, fat, other tissues (4-C); and air, lung, fat, nonfat soft tissues, bones (5-C) were considered. Four thresholds, T1(−950), T2(−350), T3(−20), and T4(140), were applied on the CT Hounsfield (HU) numbers to segment tissues into air (CT-HU-Number < T1), lungs (T1 ≤ CT-HU-Number < T2), fat (T2 ≤ CT-HU-Number < T3), nonfat soft tissues (T3 ≤ CT-HU-Number < T4), and bones

(T4 ≤ CT-HU-Number). All PET images were reconstructed with the Siemens clinical OSEM protocol with eight subsets and two iterations. PET bias image volume for each patient was computed using the PET reconstructed with the original CT as the gold standard. A reference patient with normal body mass index was chosen. The CT images of all the other 22 patients were registered to that of the reference patient. The transformation obtained from the registration was used to transform the corresponding bias image volumes. A population-based bias atlas for each segmentation method was then obtained by averaging over all the patients. For the reference patient, Figure 14.3a shows the original CT and the pseudo-CT for each segmentation method. Figure 14.3b shows the bias slice for each segmentation method. Figure 14.3c shows the PET bias atlas. A patient-independent bias atlas provides a complete view of the bias across the body and can be used to determine the minimum segmentation level that is required for imaging a particular region of interest. Conventional MR images can be used to identify body contour and lungs to perform 3-C. This CT-based study, which was based on the 23 PET-CT patient scans, found that the voxelwise bias for 3-C is 5.8 ± 14.6%, 14.2 ± 9.3%, −0.8 ± 8.0%, and −15.4 ± 10.5% within lungs, fat, nonfat soft tissues, and bones, respectively. Current PET-MR scanners use the Dixon sequence, which makes it possible to perform 4-C, for PET attenuation correction. It has been found in this study that fat/in-phase ratio images provide more accurate fat identification as compared to fat images. The voxelwise bias for 4-C is 4.4 ± 14.4%, 2.5 ± 3.2%, 4.0 ± 5.3%, and −9.0 ± 9.2% within lungs, fat, nonfat soft tissues, and bones, respectively. If an MR bone sequence can be used in addition to the Dixon sequence, 5-C can be performed. The voxelwise bias for 5-C is 4.5 ± 14.2%, 1.2 ± 2.7%, 0.2 ± 3.0%, and −0.1 ± 7.9% within lungs, fat, nonfat soft tissues, and bones, respectively. It is important to realize that even though the average bias is small, the probability of having a large bias in a voxel can still be high if the standard deviation is large.

For whole-body PET-MR systems, patient anatomy can extend beyond the MR FOV in a transverse plane. The attenuation map derived from MR images without accounting for the truncated anatomy can lead to significant bias. Attempts have been made to estimate the attenuation map outside the MR FOV using phantom data or uncorrected PET images [39]. However, such approaches still need to be validated.

14.4 MR-Based PET Motion Correction

The intrinsic spatial resolution of modern whole-body PET scanners is in the range of 4–5 mm full-width-half-maximum (FWHM) for stationary objects (2–4 mm for brain scanners) [40–43]. However, due to inevitable respiratory and cardiac motion of the subject, this resolution cannot be achieved in clinical PET imaging. There have been many attempts to develop the practical and effective methodology to remove the effects of respiratory and cardiac motion (e.g., [44–54]). Cardiac and/or respiratory gating strategies that "freeze" motion are popular approaches in static PET, but have been neither effective nor

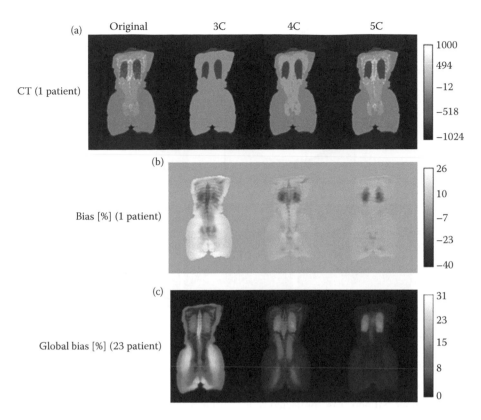

FIGURE 14.3 One coronal slice through the heart for the CT-based study. (a) The CTs for the reference patient. (b) The bias images for the reference patient. (c) The bias atlas calculated using all the 23 patient data sets.

successful in dynamic PET imaging of rapid dynamic functions, such as MBF, due to the substantial noise associated with rejecting a large number of detected events in a short dynamic frame. Furthermore, it is important to highlight that it is not "just" the emission data that is corrupted by motion; there are often large reconstruction errors due to the elastic deformation of the attenuating medium. In order to use the emission data from all phases of the cardiac and respiratory cycles, one must know the corresponding attenuation correction at all times. Due to concerns about radiation exposure, this data cannot be obtained routinely by conventional PET-CT.

For myocardial PET imaging, the myocardial motion field can be measured using tagged MR during subject breath-holds. Tagging is a method of encoding MR-visible markers by imposing a magnetization pattern (M_z) prior to the acquisition of the image in order to assess the motion between tagging preparation and image acquisition. The ECG signal is recorded during the entire PET acquisition, and used to assign a cardiac phase to each PET event. A separate tagged cine MR scan of the whole heart over the complete cardiac cycle can be acquired in only a few breath-holds. The MR acquisition sequence can be either conventionally tagged segmented gradient echo (GRE) or nontagged balanced steady-state free precession (TrueFISP) [55]. TrueFISP has the advantage of being acquired as quickly as a single breath-hold, but does not provide as much base-to-apex rotation information as tagged GRE. The myocardial motion

field can also be measured during subject free breathing. This requires navigator scans to track the respiratory phase through diaphragmatic motion. Bellows can also be used for this purpose if chest wall and diaphragmatic motions have a constant relationship. Because the respiratory cycle is much longer than the cardiac cycle, the navigator scan can be performed right after the ECG trigger. Tagging and GRE (or TrueFISP) can be performed afterwards. The data is then combined into a single 3D time-varying motion field. Each tag unit is evaluated as a series of displacement steps, unevenly spaced over the cardiac and respiratory cycles. Finding the displacement steps is a direct application of harmonic phase (HARP) or nonrigid registration methods as previously demonstrated [56]. Tagging and navigator scans can be also used to measure the motion field for liver imaging. Given the measured motion field, a PET reconstruction, which uses the measured motion field within the system matrix, can be used to yield a unique reference frame for all the events in all the cardiac and respiratory cycles, while preserving Poisson's statistics.

To assess the performance of this motion-correction approach in a challenging acquisition, we constructed and tested a PET-MR cardiac phantom that mimics cardiac motion (schematic design shown in Figure 14.4). Two different size balloons were suspended in a hot background gel. "Hot" gel between the two balloons were used to mimic a myocardium (myocardium/background = 3). A small "cold" gel within the myocardium was

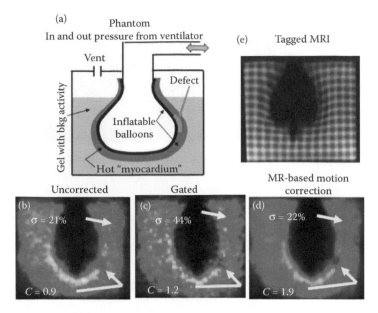

FIGURE 14.4 (a) Deformable respiratory phantom. Coronal slice (6 min) with no motion correction (b), respiratory gating (c), and simultaneous tagged-MR motion correction (d). Note the dramatic improvement in ischemic lesion contrast "C" in (d) compared to (b) and (c) (arrow) without an increase in noise as in (c) (arrow). Tagged MR is shown in (e). Noise σ was computed over 10 noise realizations.

used to mimic a defect. The inner balloon was connected to a Harvard 613 Ventilator. This deformable phantom was scanned in the simultaneous PET-MR brain scanner at MGH. We have successfully collected tagged MR and ^{18}FDG PET list-mode data simultaneously using a GRE for the MR acquisition, with TE = 2.41 ms, TR = 100 ms, a flip angle = 25°, and tagging distance = 8 mm as described in [57]. The motion field was computed using B-spline-based image registration with mutual information [58,59] and incorporated into the system matrix of OSEM along with the motion-dependent attenuation map [60]. Figure 14.5 shows that both gating and MR motion correction improve the contrast of the defect, but only MR motion correction can do so without increasing the background noise or acquisition time. A non-prewhitening (NPW) observer and channelized

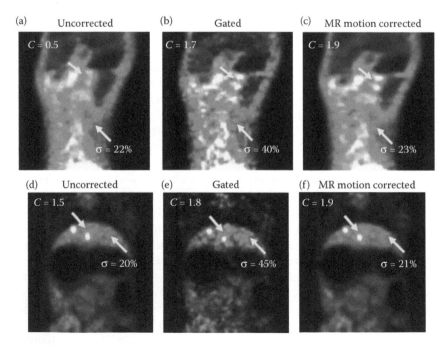

FIGURE 14.5 Respiratory motion deteriorates contrast (C) in monkey (top) and rabbit (bottom) studies. Unlike gating, MR motion correction improves contrast without amplifying the noise.

hotelling observer (CHO) were applied to 20 sets of simultaneous PET-MR acquisition (6 min each) to assess the improvement in lesion detection SNR with the MR motion corrections [58,59,61]. Figure 14.5 shows that both gating and MR motion correction improve the contrast of lesions ($C = 1.4$ and 1.6 vs. 0.7), but only MR motion correction can do so without increasing the background noise (similar σ in b and d) or acquisition time. The NPW and CHO studies confirm the quantitative improvement in SNR, which ranges from 129% to 266%, when comparing MR motion-correction methods to the gating method ($p < 0.001$). This study also shows improvement between 49% and 58% ($p < 0.001$) in lesion detection SNR for large motion as compared to the uncorrected method [56]. The greater the motion, the greater is the improvement in lesion detection SNR.

Simultaneous PET-MR *in vivo* animal studies were also performed using nonhuman primates and rabbits after implanting beads to mimic hot lesions under an approved IACUC animal protocol. Air-pressure sensing respiratory bellows were used for gating for all *in vivo* studies. The anesthetized monkey was scanned for 32 min. In the rabbit studies, six FDG beads were implanted in the liver and the diaphragm of two rabbits. Two FDG doses were injected into each rabbit to mimic low and high lesion to background contrast. Simultaneous PET-MR acquisitions were performed for 1 h. Figure 14.5 demonstrates that MR-based motion correction improves contrast (between 27% and 280%) and removes the blurring associated with breathing without increasing the noise (as is the case with gating) [56]. The CHO detection SNR study in rabbits demonstrates that performance in detection and estimation tasks can be improved in the *in vivo* setting by as much as 60% and 30%, respectively, even in lesions with small motion range.

14.5 PET Reconstruction

The image reconstruction algorithms used in PET-MR scanners could, in principle, be similar to those employed in traditional PET scanners once an attenuation correction map is available. However, the simultaneously measured MR image can provide further information to improve the PET reconstruction. The motion field estimated from MR images could be used to correct the motion artifacts in the emission data and in the attenuation map, and the high-resolution anatomical information in the MR image could be applied to facilitate the PET image reconstruction.

14.5.1 System Modeling

Commercial PET-CT systems usually have a measured point spread function (PSF), which is generally acquired by placing a point source at different locations using a computer-controlled robot, for the reconstruction. Such a measured PSF is not yet available for simultaneous PET-MR scanners because it is difficult to use a robot due to MR compatibility issues. However, there is a PSF model that has been validated by many applications [62–69]. The model effectively tracks the progression from

positron emission to photon pair production, to passage of the photons through the body, and finally to coincidence detection. The PSF model includes two main effects: First, a geometric projection matrix that maps from the image space to the surface of the detectors; while this matrix typically assumes that the detector crystals are arranged uniformly on the surface of a cylinder [64,69], we later adapted the model to specifically account for the piecewise planar arrangement of crystals in a block-detector-based scanner and for the gaps between blocks [70]. Second, a detector response model that accounts for the interaction of the photon pair with the detectors and includes the effects such as block structure, intercrystal scatter, and intercrystal penetration. The sinogram blurring matrix is at the core of the ability to optimize the resolution recovery.

14.5.2 Static Reconstruction

The MR-measured motion field can be integrated into a unified maximum *a posteriori* (MAP; or penalized maximum likelihood) framework for static reconstruction. The time-dependent attenuation map will be incorporated into the system matrix $\tilde{a}_{ni}(t)$, and combined into the log-likelihood function of a moving subject (Equation 14.1) along with an image prior, yielding the following penalized log-likelihood function:

$$\hat{f} = \underset{p \geq 0}{\mathrm{argmax}}\left(L + \mu D(X)\right)$$

$$= \sum_{n=1}^{N} \log\left[\sum_{i=1}^{M} \tilde{a}_{ni}(t_n)\rho_i + Sc_n(t_n) + R_n(t_n)\right] - \sum_{i=1}^{M} \rho_i \tilde{s}_i + \mu D(X),$$

$$(14.1)$$

where Sc_n represents scatter coincidences, R_n represents random coincidences, $\tilde{s}_i$ is the sensitivity of voxel i, μ is the hyperparameter to control the trade-off between data fidelity and penalty, the function $L(.)$ is the Poisson log-likelihood, and $D(.)$ is the penalty function. The motion-free PET image is denoted as $\rho = [\rho_1, \rho_2, ..., \rho_N]^T$, where ρ_i represents the activity at voxel index i. The prior can be the MR-based anatomical information or a simple quadratic smoothing function. The time-dependent system matrix including MR-based attenuation and motion can then be computed. The reconstruction is performed using a preconditioned conjugate gradient (PCG) as previously reported in Refs. [71,72].

14.5.3 Dynamic Reconstruction

The above static reconstruction method can be extended to motion-corrected, frame-by-frame dynamic reconstruction, as demonstrated in Figure 14.6. The list-mode data is divided into many frames according to the requirements of kinetic analysis. For example, when performing MBF imaging, the initial frames will be chosen to be short (5 s) followed by longer frames (15 s, 30 s, 1 min). In each frame, motion correction is performed for the whole list-mode data using the motion field estimated for

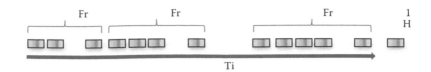

FIGURE 14.6 Schematic illustration of the motion-corrected, frame-by-frame dynamic reconstruction framework.

each motion phase so that we can obtain one reference sinogram that is statically reconstructed. The reconstructed volume at that time then represents the reference activity in that particular frame. By accurately correcting the motion in a dynamic study, many previously impossible kinetic analyses of the moving subject become feasible, especially for liver, lung, and heart, opening new avenues for clinical applications.

14.5.4 Incorporating MR Anatomical Priors in PET Reconstruction

Previous studies have demonstrated the use of a Bayesian framework to incorporate anatomical information into the reconstruction of functional images [73–76] through priors based on information-theoretic similarity measures. The images reconstructed using the information priors show improved resolution recovery, as anatomical prior-based reconstruction (Figure 14.7) can identify the peri-aqueductal gray structures while traditional reconstruction without anatomical prior cannot (Figure 14.7a). It also demonstrates that the recovery coefficient versus

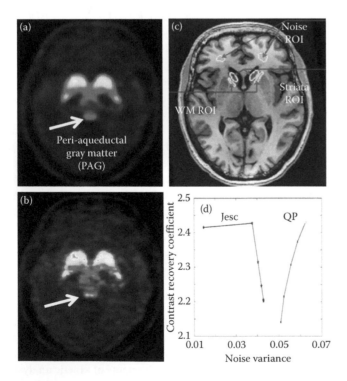

FIGURE 14.7 PET without anatomical prior (a) and JE-scale with priors (b). (c) MR template for ROIs computation of CRC and noise variance. (d) Contrast recovery coefficient versus noise variance plots as the hyperparameter is varied.

noise variance trade-off (Figure 14.7d) of the joint entropy prior reconstructed image calculated from ROI in Figure 14.7c is substantially better than that obtained using a quadratic prior.

However, the information in these studies is obtained from coregistered anatomical images, and therefore registration errors can propagate into the reconstruction. Since simultaneous acquisition of PET and MR data can freeze motion and provide near-perfect image registration, incorporation of MR-based anatomical prior information in the PET reconstruction is more meaningful and has thus become an important area of research. In addition, MR and dynamic contrast-enhanced MR (DCE-MR) provide excellent soft tissue contrast and consequently rich anatomic information about the liver when compared with CT, which makes MR more suitable as a source of anatomical priors for PET reconstruction. Consequently, the resulting PET images will have better quantitation and lesion detectability. In this section, we will demonstrate how to develop and optimize a Bayesian framework for simultaneous PET-MR reconstruction using information-theory-based priors that do not enforce the overlap of boundaries. Scale-space theory will be applied to define the feature vectors used to compute priors, and an efficient fast Fourier transform (FFT)-based method will be developed to compute these priors and their derivatives.

14.5.4.1 A Bayesian PET Reconstruction Method Using Simultaneous MR as Prior

To use anatomic priors derived from simultaneous MR, we replace the penalty function in Equation 14.2 with a new function computed from both the PET image and the MR image:

$$\hat{f} = \arg\max_{p \geq 0}(L + \mu D(X,Y))$$

$$= \sum_{n=1}^{N} \log\left[\sum_{i=1}^{M} \tilde{a}_{ni}(t_n)\rho_i + Sc_n(t_n) + R_n(t_n)\right] - \sum_{i=1}^{M} \rho_i \tilde{s}_i + \mu D(X,Y),$$

$$(14.2)$$

where $D(.)$ is the information-theoretic similarity metric that is defined between two random feature vectors X and Y that are computed using motion-free MR and PET reference images. The Ns feature vectors extracted from the PET and anatomical images will be represented as x_i and y_i, respectively, for $i = 1,2,...,Ns$. These can be considered as independent realizations of the random vectors X and Y. We use mutual information (MI) and joint entropy (JE) as measures of similarity between the anatomical and functional images. The mutual information between two continuous random variables X and Y with marginal distributions $p(x)$ and $p(y)$ and joint distribution $p(x,y)$ is defined as

$$
\begin{aligned}
I(X,Y) &= H(X) + H(Y) - H(X,Y) \\
&= -\int p(x)\log p(x)\mathrm{d}x - \int p(y)\log p(y)\mathrm{d}y \\
&\quad + \iint p(x,y)\log p(x,y)\mathrm{d}x\mathrm{d}y,
\end{aligned}
\tag{14.3}
$$

where $H(X)$ and $H(Y)$ are the entropy, and $H(X,Y)$ is the joint entropy. Mutual information can be interpreted as the reduction in uncertainty (entropy) of X by the knowledge of Y or vice versa [77].

14.5.4.2 A Scale-Space Framework to Extract Feature Vectors PET-MR Reconstruction

Since the information-theoretic priors are global measurements that are naturally insensitive to differences in intensity between the two images, spatial information is introduced into the information-theoretic priors by defining the feature vectors that capture local morphology. The combination of local measures of image morphology and information-theoretic measures can improve PET reconstruction of structures similar to those of the simultaneously acquired MR. Scale-space theory provides a framework to extract useful information from an image using a multiscale representation of the image [78]. In this approach, a family of images is generated from an image by blurring it with Gaussians of increasing width (scale parameter), thus generating images at decreasing levels of detail. Relevant features can then be extracted by analyzing this family of images rather than just the original image. The feature vectors in this approach can be defined as (1) the original image, (2) the image blurred by a Gaussian kernel of width σ_1, and (3) the Laplacian of the blurred image. By analyzing the image at two different scales, the original and image at scale σ_1, we are giving more emphasis to those boundaries that remain in the image at the higher scale, and attach less importance to those that are blurred out at the higher scale. Though the scale-space images are correlated, one could choose to make an independence assumption to reduce the computational cost of the prior term, and define it as

$$
D(X,Y) = \sum_{i=1}^{n} \xi_i D(X_i, Y_i),
\tag{14.4}
$$

where X_i and Y_i are the random variables corresponding to the ith scale-space images, and ξ_i are the weights on each of the scale-space features to provide more emphasis to some features over others, depending on the application. More features could be added by using different kernel sizes. The effect of weighting of the scale-space priors and the optimal number of scale-space images could be studied and optimized for specific applications.

The anatomical priors that impose piecewise smoothness can yield a mosaic-like appearance and information-theoretic priors that only use intensity information can lead to isolated hot and cold spots in the image [76]. Instead, the scale-space approach to incorporate the spatial correlation into the information-based priors can avoid these artifacts [76]. The proper selection of parameters (weights and number of scale-space images) of the scale-space method for certain applications is important to minimize the potential artifacts.

14.5.4.3 An Efficient FFT-Based Method to Compute Entropy and Its Gradient

In order to estimate the PET image, there are many quantities that must be computed, including the density function of the feature vector, the joint density of the feature vector, and the gradients of marginal entropy and joint entropy. To speed up the computation, a nonparametric approach is used to estimate the joint probability density functions (pdfs) using the Parzen window method [79]. It has been proven that when $Ns \to \infty$, the Parzen estimate converges to the true pdf in the mean-squared sense. For 3D PET reconstruction Ns is large, so there is a sufficient number of samples to give an accurate estimate of the pdf. The gradients of marginal entropy and joint entropy have been derived in Ref. [76]. The computation of the Parzen window estimate $p(x)$ at M locations using Ns samples is $O(MNs)$. In addition, the gradient of joint entropy is also of complexity $O(MNs)$, because for each f_k, a summation of M multiplications need to be computed. This can be expensive for the application such as PET reconstruction. Therefore, an approach similar to Ref. [80] is usually used to compute the entropy measures and their gradients efficiently using FFTs.

14.6 Conclusion

PET-MR combines two powerful imaging modalities and has great potential for many important clinical applications. For example, PET-MR can aid in identifying the liver metastases, especially in challenging cases of diffuse hepatic metastases, where PET can discriminate among numerous treated lesions that obscure identification of potentially new, recurrent, or untreated lesions while MR is crucial in detecting the small hepatic lesions. Additionally, accurate response to treatment is often difficult to judge on liver MR alone and may be significantly aided by combined PET-MR. PET-MR also has added potential to identify the extrahepatic disease, often less conspicuous and easily overlooked on MR alone. Identification of extrahepatic metastatic disease may have profound clinical impact, often altering patient management. Likewise, simultaneous cardiac PET-MR can measure the 3D nonrigid respiratory and ventricular wall motion without additional radiation dose as with CT. There are many promising clinical applications of PET-MR in pediatric imaging, where simultaneous PET-MR allows to reduce the radiation dose as well as the number of sedations required. PET-MR is also a promising new modality for imaging muculoskeletal tumors, neuroblastomas, and OB-GYN metastatic staging.

References

1. Judenhofer, M.S. et al., Simultaneous PET-MRI: A new approach for functional and morphological imaging. *Nat Med*, 2008. 14: 459–465.

2. Wehrl, H. et al., Combined PET/MR imaging-technology and applications. *Technol Cancer Res Treat*, 2010. 9(1): 5–20.

3. Albert, M. et al., The use of MRI and PET for clinical diagnosis of dementia and investigation of cognitive impairment: A comsensus report, in *Alzheimer's Association Neuroimaging Work Group Consensus Report*, Alzheimer's Association. p. 1–15.

4. Chun, S. et al., 4D tagged MR-based motion correction in simultaneous PET-MR. *J Nucl Med*, 2010. 51(Supplement 2): 80.

5. Nekolla, S., A. Martinez-Moller, and A. Saraste, PET and MRI in cardiac imaging: From validation studies to integrated applications. *Eur J Nucl Mol Imaging*, 2009. 36(suppl 1): S121–S130.

6. Tonino, P.A. et al., Fractional flow reserve versus angiography for guiding percutaneous coronary intervention. *N Engl J Med*, 2009. 360(3): 213–24.

7. Hachamovitch, R. et al., Comparison of the short-term survival benefit associated with revascularization compared with medical therapy in patients with no prior coronary artery disease undergoing stress myocardial perfusion single photon emission computed tomography. *Circulation*, 2003. 107(23): 2900–7.

8. AHA, *Heart Disease and Stroke Statistics—2009 Update*, Dallas, TX: American Heart Association, 2009.

9. Christensen, N.L. et al., Positron emission tomography within a magnetic field using photomultiplier tubes and lightguides. *Phys Med Biol*, 1995. 40(4): 691–7, ISSN: 0031-9155.

10. Shao, Y. et al., Simultaneous PET and MR imaging. *Phys Med Biol*, 1997. 42(10): 1965–70, ISSN: 0031-9155.

11. Marsden, P. et al., Simultaneous PET and MR. *Br J Radiol*, 2002. 75: S53–S59.

12. Lucas, A. et al., Development of a combined microPET-MR system. *Technol Cancer Res Treat*, 2006. 5(4): 337–341.

13. Lightstone, A. et al., A bismuth germanate-avalanche photodiode module designed for use in high resolution positron emission tomograph. *IEEE Trans Nucl Sci*, 1986. NS(33): 456–459.

14. Pichler, B. et al., Performance test of a LSO-APD PET module in a 9.4 Tesla magnet. In *Nuclear Science Symposium and Medical Imaging Conference*. 1997. Albuquerque, NM.

15. Canata, C. et al., Simultaneous acquisition of multislice PET and MR images: Inital results with a MR-compatible PET scanner. *J Nucl Med*, 2006. 47: 1968–1976.

16. Renker, D., Geiger-mode avalache photodiodes, history, properties and problems. *Nucl Instrum Methods Phys Res A*, 2006. 567: 48–56.

17. Hawkes, R. et al., Silicon photomultiplier performance tests in magnetic resonance pulsed fields. In *Nuclear Science Symposium (NSS) and Medical Imaging Conference (MIC) Record*, Honolulu, Hawaii, 2007.

18. Espana, S. et al., Performance evaluation of SiPM detectors for PET imaging in the presence of magnetic fields. In *Nuclear Science Symposium (NSS) and Medical Imaging Conference (MIC) Record*, Dresden, Germany, 2007.

19. Schulz, V. et al., A preclinical PET/MR insert for a human 3T MR scanner. In *Nuclear Science Symposium (NSS) and Medical Imaging Conference (MIC) Record*, Orlando, Florida, 2009.

20. Delso, G. et al., Performance measurement of the Siemens mMR integrated whole-body PET/MR scanner. *J Nucl Med*, 2001. 52(12): 1914–22.

21. Zaidi, H. et al., Design and performance evaluation of a whole-body Ingenuity TF PET-MRI system. *Phys Med Biol*, 2011. 56: 3091–3106.

22. Rota Kops, E. and H. Herzog, Alternative methods for attenuation correction for PET images in MR-PET scanners. *IEEE Nucl Sci Symp Conf Rec*, 2007. 6: 4327–4330.

23. Rota Kops, E. et al., MRI based attenuation correction for brain PET images. *Springer Proc In Phys*, 2007. 114: 93–97.

24. Beyer, T. et al., MR-based attenuation correction for torso-PET/MR imaging: Pitfalls in mapping MR to CT data. *Eur J Nucl Mol Imaging*, 2008. 35: 1142–1146.

25. Hofmann, M. et al., MRI-based attenuation correction for PET/MRI: A novel approach combining pattern recognition and atlas registration. *J Nucl Med*, 2008. 49: 1875–1883.

26. Hofmann, M. et al., MRI-based attenuation correction for whole-body PET/MRI: Quantitative evaluation of segmentation- and atlas-based methods. *J Nucl Med*, 2011. 52(9): 1392–1399.

27. Le Goff-Rougetet, R. et al., Segmented MR images for brain attenuation correction in PET. *SPIE*, 1994. 2167: 725–736.

28. El Fakhri, G., M.F. Kijewski, and S.C. Moore, Absolute activity quantitation from projections using an analytical approach: Comparison with iterative methods in Tc-99 m and I-123 brain SPECT. *IEEE Trans Nucl Sci*, 2001. 48: 768–773.

29. El Fakhri, G. et al., MRI-guided SPECT perfusion measures and volumetric MRI in prodromal Alzheimer's disease. *Arch Neurol*, 2003. 60: 1066–1072.

30. Zaidi, H., M.L. Montandon, and S.R. Meikle, Strategies for attenuation compensation in neurological PET studies. *NeuroImage*, 2007. 34: 518–541.

31. Zaidi, H., M.L. Montandon, and D.O. Slosman, Magnetic resonance imaging-guided attenuation and scatter corrections in three-dimensional brain positron emission tomography. *Med Phys*, 2003. 30: 937–948.

32. Steinberg, J. et al., Three-region MR-based whole-body attenuation corrcetion for automated PET reconstruction. *Nucl Med Biol*, 2010. 37(2): 227–235.

33. Martinez-Moller, A. et al., Tissue classification as a potential approach for attenuation correction in whole-body PET/MRI: Evaluation with PET/CT data. *J Nucl Med*, 2009. 50: 520–526.

34. Schulz, V. et al., Automatic, three-segment, MR-based attenuation correction for whole-body PET/MR data. *Eur J Nucl Mol Imaging*, 2011. 38: 138–152.

35. Eiber, M. et al., Value of a Dixon-based MR/PET attenuation correction sequence for the localization and evaluation of PET-positive lesions. *Eur J Nucl Mol Imaging*, 2011. 38: 1691–1701.

36. Martinez-Moller, A. et al., MR-based attenuation correction for whole-body MR/PET. *J Nucl Med*, 2008. 49: 65P.

37. Hu, Z. et al. MR-based attenuation correction for a whole-body sequential PET/MR system. In *IEEE Trans Nucl Sci Symposium Conf.* 2009. Orlando, FL.

38. Ouyang, J., Chun, SY., Alpert, N., El Fakhri, G., Evaluation of MR-based PET attenuation correction using PET-CT and MR. *Med. Phys.*, 2012. Submitted.

39. Delso, G. et al., The effect of limited MR field of view in MR/PET attenuation correction. *Med Phys*, 2010. 37(6): 2804–2812.

40. Stickel, J.R. and S.R. Cherry, High-resolution PET detector design: Modelling components of intrinsic spatial resolution. *Phys Med Biol*, 2005. 50: 179–188.

41. Sureau, F.C. et al., Impact of image-space resolution modeling for studies with the high-resolution research tomograph. *J Nucl Med*, 2008. 49: 1000–1008.

42. Wiant, D. et al., Evaluation of the spatial dependence of the point spread function in 2D PET image reconstruction using LOR-OSEM. *Med Phys*, 2010. 37: 1169–1182.

43. Alessio, A.M. et al., Application and evaluation of a measured spatially variant system model for PET image reconstruction. *IEEE Trans Med Imaging*, 2010. 29: 938–949.

44. Boucher, L. et al., Respiratory gating for 3-dimensional PET of the thorax: Feasibility and initial results. *J Nucl Med*, 2004. 45: 214–219.

45. Nehmeh, S.A. et al., Effect of respiratory gating on quantifying PET images of lung cancer. *J Nucl Med*, 2002. 43: 876–881.

46. Martinez-Moller, A. et al., Dual cardiac-respiratory gated PET: Implementation and results from a feasibility study. *Eur J Nucl Med Mol Imaging*, 2007. 34(9): 1447–1454.

47. Buther, F. et al., List mode-driven cardiac and respiratory gating in PET. *J Nucl Med*, 2009. 50(5): 674–681.

48. Montgomery, A.J. et al., Correction of head movement on PET studies: Comparison of methods. *J Nucl Med*, 2006. 47(12): 1936–1944.

49. Klein, G.J., R. Reutter, and R.H. Huesman, Four-dimensional affine registration models for respiratory-gated PET. *IEEE Trans Nucl Sci*, 2002. 48(3): 756–760.

50. Liveratos, L., Stegger, L., Bloomfield, P.M., Schafers K., Bailey D.L., and Camici P.G., Rigid-body transformation of list-mode projection data for respiratory motion correction in cardiac PET. *Phys Med Biol*, 2005. 50(14): 3313–22.

51. Lamare, F. et al., List-mode-based reconstruction for respiratory motion correction in PET using non-rigid body transformations. *Phys Med Biol*, 2007. 52(17): 5187–5204.

52. Li, T. et al., Model-based image reconstruction for four-dimensional PET. *Med Phys*, 2006. 33(5): 1288–1298.

53. Qiao, F. et al., A motion-incorporated reconstruction method for gated PET studies. *Phys Med Biol*, 2006. 51: 3769–3783.

54. Reyes, M. et al., Model-based respiratory motion compensation for emission tomography image reconstruction. *Phys Med Biol*, 2007. 52(12): 3579–3600.

55. Bundy, J. et al. Segmented TrueFISP cine imaging of the heart. In *ISMRM*. 1999. Philadelphia.

56. Chun, SY., Reese, T., Ouyang, J., Guerin, B., Catana, C., Zhu, X., Alpert, NM., El Fakhri, G., MRI-based nonrigid motion correction in simultaneous PET/MR. *J Nucl Med*, 2012 Aug. 53(8): 1284–1291.

57. Axel, L. and L. Dougherty, MR imaging of motion with spatial modulation of magnetization. *Radiology*, 1989. 171(3): 841–845.

58. Chun, S. et al. Compensation for nonrigid motion using B-spline image registration in simultaneous MR-PET. In *International Society for Magnetic Resonance in Medicine*, Stockholm, Sweden, 2010.

59. Reese, T. et al., Respiratory motion correction of PET using simultaneously acquired tagged MRI. In *International Society for Magnetic Resonance in Medicine*, Stockholm, Sweden, 2010.

60. Lamare, F. et al., List-mode-based reconstruction for respiratory motion correction in PET using non-rigid body transformations. *Phys Med Biol*, 2007. 52(17): 5187–5204.

61. Guerin, B. et al., Non-rigid PET motion compensation using tagged-MRI in simultaneous PET-MR imaging. *Med Phys*, 2011. 38: 3025–3038.

62. Yang, Y. et al., Optimization and performance evaluation of the micro-PET II scanner for in vivo small-animal imaging. *Phys Med Biol*, 2004. 49: 2527–2546.

63. Bai, B. et al., Positron range modeling for statistical PET image reconstruction. In *Nuclear Science Symposium (NSS) and Medical Imaging Conference (MIC) Record,* Portland, Oregon, 2003.

64. Qi, J. et al., High-resolution 3D Bayesian image reconstruction using the microPET small-animal scanner. *Phys Med Biol*, 1998. 43(4): 1001–13, ISSN: 0031-9155.

65. Qi, J. et al., Fully 3D Bayesian image reconstruction for the ECAT EXACT HR+. *IEEE Trans Nucl Sci*, 1998. 45: 1096–1103.

66. Mumcuoglu, E.U. et al., Accurate geometric and physical response modelling for statistical image reconstruction in high resolution PET. In *Nuclear Science Symposium (NSS) and Medical Imaging Conference (MIC) Record,* Anaheim, California, 1996.

67. Mumcuoglu, E.U. et al., A phantom study of the quantitative behavior of Bayesian PET reconstruction methods. In *Nuclear Science Symposium (NSS) and Medical Imaging Conference (MIC) Record,* San Francisco, California, 1995.

68. Leahy, R. and X.H. Yan. Statistical models and methods for PET image reconstruction. In *Proc. of Stat. Comp. Sect. of Amer. Stat. Assoc,* 1991.

69. Ruangma, A. et al., Three-dimensional maximum a posteriori (MAP) imaging with radiopharmaceuticals labeled with three Cu radionuclides. *Nucl Med Biol*, 2006. 33(2): 217–226.

70. Yang, Y. et al., Optimization and performance evaluation of the micro-PET II scanner for in vivo small-animal imaging. *Phys Med Biol*, 2004. 49: 2527–2546.

71. Qi, J. et al., High-resolution 3D Bayesian image reconstruction using the micro-PET small-animal scanner. *Phys Med Biol*, 1998. 43(4): 1001–1013, ISSN: 0031-9155.

72. Qi, J. et al., Fully 3D Bayesian image reconstruction for the ECAT EXACT HR+. *IEEE Trans Nucl Sci*, 1998. 45: 1096–1103.

73. Somayajula, S., E. Asma, and R.M. Leahy, PET image reconstruction using anatomical information through mutual information based priors. In *Nuclear Science Symposium (NSS) and Medical Imaging Conference (MIC) Record*, Wyndham El Conquistador, Puerto Rico, 1995.

74. Somayajula, S., A. Rangarajan, and R.M. Leahy, PET image reconstruction using anatomical information through mutual information based priors: A scale space approach. In *4th IEEE International Symposium on Biomedical Imaging*, Arlington, VA, 2007: p. 165–168.

75. Panagiotou, C. et al., Information theoretic regularization in diffuse optical tomography. *J Opt Soc Am A*, 2009. 26(5): 1277–1290.

76. Somayajula, S. et al., PET image reconstruction using information theoretic anatomical priors. *IEEE Trans Med Imaging*, 2010. 30: 537–549.

77. Cover, T. and J. Thomas, *Elements of Information Theory* (*Wiley Series in Telecommunications*, Wiley-Interscience, New York, 1991).

78. Lindeberg, T., *Scale-Space Theory in Computer Vision*. 1994: Netherlands: Kluwer Academic.

79. Duda, R.O., P.E. Hart, and D.G. Stork, *Pattern Classification*. Wiley-Interscience, New York, 2001.

80. Shwartz, S., M. Zibulevsky, and Y.Y. Schechner, Fast kernel entropy estimation and optimization. *Signal Process*, 2005. 85(5): 1045–1058.

15

Multimodal Photoacoustic and Ultrasound Imaging

Peter van Es
University of Twente

Ton G. van Leeuwen
University of Twente
University of Amsterdam

Wiendelt Steenbergen
University of Twente

Srirang Manohar
University of Twente

15.1 Multimodal Imaging

Imaging with two or more imaging modalities is widely becoming standard practice because of their needed additional information (Esenaliev et al. 1993). Much research is being performed to develop multimodal systems such as MRI/PET and CT/PET and to optimize them to acquire complementary information of disease in the form of both functional and anatomical details. Anatomical imaging visualizes internal structures of the body, and functional imaging visualizes and often quantifies physiological processes such as blood flow, oxygenation, and metabolism (Moseley and Donnan 2004).

Optical imaging in the visible and near-infrared (NIR) wavelength range is gaining great interest because of the possibility to exploit the wavelength-dependent absorption of oxy- (HbO_2) and deoxygenated hemoglobin (Hb) in tissue, thereby providing functional information. Techniques that are most likely to be available in the clinic in near future, such as photoacoustic tomography (PAT) and diffuse optical tomography (DOT), are now capable of imaging Hb and HbO_2 at tissue depths of several centimeters (Heijblom et al. 2012; Beard 2011). These techniques will most possibly be combined with high-resolution magnetic resonance imaging (MRI) (Lin et al. 2011), x-ray computed tomography (CT) (Yang et al. 2010), or ultrasound (US) (Zhu et al. 1999) scanners to form multimodal systems that can be used for imaging breast cancer, small extremities, and small animals.

In this review, we will concentrate on various combinations of the fast-growing PAT technique with ultrasound imaging which was first reported by Emelianov et al. (2004). We begin with a discussion on the motivation for optical absorption imaging using PAT. Linear and planar multimodal PAT/US systems that are currently being investigated and tested will be discussed in the second section. The third section will show how US (tomography) can also be used to improve the image reconstruction for PAT.

15.1.1 Optical Absorption Contrast as a Marker for Disease

In optical imaging modalities, the differences between malignant and healthy tissue can be perceived in a range of interactions that can occur between light and the investigated material. The main interactions encompass absorption, emission (i.e., fluorescence or phosphorescence), or scattering processes (i.e., Rayleigh or Raman) (Mishchenko et al. 2002).

The most common chromophore (containing) molecules are oxyhemoglobin (HbO_2), deoxyhemoglobin (Hb), melanin, water, and lipid (Nachabe et al. 2010), each with a signature optical absorption spectrum (Figure 15.1). In many situations, chromophores such as Hb, HbO_2, and melanin are associated with pathological conditions and can be used as intrinsic markers for disease. For example, the elevated presence of hemoglobin can be a sign of angiogenesis, the growth of new blood vessels. While angiogenesis is an essential process in the body associated with, for example, the repair of wounded tissue such as ulcers and burn wounds (Mishchenko et al. 2002), it is also encountered in several disorders such as cancer (Baeriswyl and Christofori 2009) and a wide variety of inflammatory diseases (Carmeliet 2005),

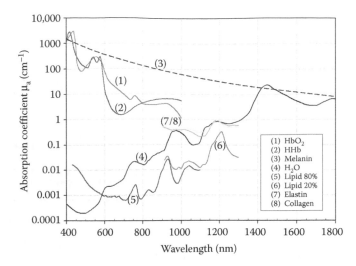

FIGURE 15.1 Absorption coefficients of oxyhemoglobin ((1), 150 g/L), deoxyhemoglobin ((2), 150 g/L), melanin in skin (3), water (4), lipid ((5) 80% in tissue, (6) 20% in tissue), elastin (7), and collagen (8). (Reprinted from Beard, P. 2011. *Interface Focus* 1(4):602–31. With permission from Royal Society Publishing.)

including rheumatoid arthritis (Wang et al. 2007). The possibility to detect the presence of Hb and HbO$_2$ and their concentrations can be used as an early marker for such disorders (Van Veen et al. 2005). Melanin associated with melanoma, the cancer of melanocytes, is a strong absorber of light and is used as an optical marker of the disease (Jose et al. 2011a; Zhang et al. 2010).

Optical imaging modalities such as diffuse optical imaging (DOI) and diffuse optical tomography (DOT) can detect the spectroscopic variations in the visible and NIR light absorption of hemoglobin in blood and of melanin in melanomas shedding light on the metabolic state of the diseased region (Lee 2011). The method has been applied considerably for breast cancer imaging using NIR light, providing detection based on functional abnormalities. Approximately 2000 women have been imaged using optical mammography to evaluate the breast, and results show that ~85% of breast lesions are detectable with optical mammography. Methods are being investigated to improve the performance of DOT such as better spectral coverage and improved modeling of photon propagation in tissues. However, the technique faces a fundamental challenge in that it has poor spatial resolution, beyond ~1 mm penetration depth, due to high optical scattering in tissue (Wang 2008).

Photoacoustic imaging does not suffer from the photon scattering-induced poor resolution of purely optical imaging. This hybrid technique combines the high contrast of optical imaging with the high spatial resolution of ultrasound imaging. Pulsed laser light applied to tissue is absorbed by chromophores causing a local rise in temperature, which results in the generation of acoustic waves by thermoelastic expansion. The acoustic waves reach the surface with different time delays depending on the depth of their origin. These waves are typically a broadband emission up to tens of MHz and are detected by an array of ultrasound receivers at the tissue surface.

15.1.2 Underlying Physics of Photoacoustic Imaging

The physics of PA imaging can be divided into two main processes: (1) the interaction of light with biological tissue and (2) the interaction of acoustic waves with tissue. The first step in PA imaging is the use of the far-red and NIR part of the optical spectrum, referred to as the optical imaging window. In this wavelength range, approximately between 650 and 1100 nm, light penetration in soft tissue is high. Further, all the important chromophores such as Hb, HbO$_2$, melanin, and water have distinctive signatures in this wavelength region, which allows them to be distinguished from each other by spectral analysis.

Two conditions require to be met to guarantee spatial resolution and signal strength of the generated acoustic pulses, namely thermal and stress confinement. Under the thermal confinement condition, the duration of the absorbed laser pulse (τ_p) should be shorter than the thermal relaxation time (τ_{th}) for the tissue volume. The second condition, stress confinement, calls for τ_p to be shorter than the transition time of the acoustic stress wave (τ_s) from the absorber into the surrounding medium. Adherence to these two regimes ensures efficient thermal heating and strong acoustic pulse generation in a confined region.

The spatially and time-dependent heating of an object can be represented by the heating function $H(r,t)$. This function represents the volume and time-dependent amount of generated heat. The resulting temperature rise can be modeled by the heat conduction equation, which relates the temperature distribution $T(r,t)$ [K] to the heating function $H(r,t)$ [J/(m^3 s)]:

$$\rho C_p \frac{\partial T(r,t)}{\partial t} = \lambda \nabla^2 T(r,t) + H(r,t) \qquad (15.1)$$

in which the density (ρ [kg/m^3]), specific heat capacity (C_p [J/(K kg)]) at constant pressure and the thermal conductivity (λ [J/K m s] are properties of the material. Neglecting thermal diffusion under thermal confinement

$$\rho C_p \frac{\partial T(r,t)}{\partial t} \approx H(r,t) \qquad (15.2)$$

The change in temperature is related to the acoustic pressure $p(r,t)$ at location r by the general photoacoustic equation:

$$\nabla^2 p(r,t) - \frac{1}{v^2} \frac{\partial^2}{\partial t^2} p(r,t) = -\beta \rho \frac{\partial^2}{\partial t^2} T(r,t) \qquad (15.3)$$

in which v [m/s] is the speed of sound and β [K^{-1}] is the volume thermal expansion coefficient. Using Equation 15.2 in Equation 15.3, one can relate the heating function with the generated acoustic pressure:

$$\nabla^2 p(r,t) - \frac{1}{v^2} \frac{\partial^2}{\partial t^2} p(r,t) = -\frac{\beta}{C_p} \frac{\partial}{\partial t} H(r,t) \qquad (15.4)$$

From this equation, it can be concluded that time-invariant heating will not produce a pressure wave, because the right-hand side of Equation 15.4 takes the first derivative of H. The heating function can be described as the product between the spatial distribution ($A(r)$) and the time-dependent illumination of the laser pulse ($I(t)$).

$$H(r,t) = A(r)I(t) \tag{15.5}$$

A solution to the generated pressure in terms of the absorption distribution $A(r)$ can be found by the use of Green's function (see Wang 2008), which results in

$$p(\boldsymbol{r},t) = -\frac{\beta}{4\pi C_p} \frac{\partial}{\partial t}\left(\frac{1}{t} \iint\limits_{r'-r=vt} A(r')dr'\right) \tag{15.6}$$

This means that the generated pressure is related to the time derivative of the integrated absorption distribution over the surface of sphere, where the radius of the sphere is proportional to the time that has elapsed since firing the laser source. The initial pressure at $t = 0$ is then given by

$$p_0(r) = \frac{\beta v_s^2}{C_p} A(r) = \Gamma A(r) \tag{15.7}$$

where Γ is the Grüneisen parameter, which represents the photoacoustic efficiency and A is the energy density locally deposited.

Before the generated acoustic waves reach the surface of the tissue where they can be detected by an array of ultrasound transducers, they undergo attenuation and refraction within the tissue. Both scattering and absorption are low for the low frequency range, but are not negligible for higher frequencies. The detection of ultrasound is often achieved using an array of piezoelectric sensors, but capacitive micromachined ultrasound transducers (CMUT) (Vaithilingam et al. 2009) and purely optical methods (Zhang et al. 2008) are gaining in popularity.

15.1.3 Imaging and Applications Explored and Foreseen

Several photoacoustic imaging configurations exist, but most fall under either PAT or photoacoustic microscopy (PAM).

15.1.3.1 Photoacoustic Tomography

Imaging can be performed with ultrasound detection following a linear or planar view or following a circular or spherical view. In the former, the acoustic signals measured by the unfocused transducers can be reconstructed by back-projection (Xu and Wang 2005) or using a phased-array approach (Hoelen et al. 1998) along a line or a plane. In this case, the source of the acoustic waves can then be triangulated when at least three measurements are taken from each different locations. The

recorded acoustic signals are back-projected over a spherical surface from which the radius (r) is obtained by the product between the time delay and the speed of sound.

Linear arrays are often used for imaging of blood vessels beneath the skin. Some examples of photoacoustic images made by linear arrays are discussed in Section 15.2. Planar arrays have been used to produce three-dimensional (3D) images of organs such as the breast (Piras et al. 2010) using piezoelectric elements. Detection in a planar view can also be achieved using optical interferometric techniques such as in the UCL system, which employs a Fabry–Perot polymer film sensor to detect the pressure signals (Zhang et al. 2011).

The axial resolution of linear and planar systems depends on the pulse duration of the light (thermal and stress confinements) and, as in US imagers, on the bandwidth and center frequency of the acoustic transducer. The lateral resolution of planar arrays is limited by the detection aperture of the elements (Beard 2011). The imaging speed of these systems is limited by the repetition frequency and energy of the laser pulses. In theory, shorter imaging times can also be obtained by higher pulse energy, which would increase the signal-to-noise ratio (SNR) and reduce the amount of averaging needed. However, safety regulations limit the energy levels that can be used by the maximum permissible exposure (MPE).

Circular- or spherical-view systems represent more complex geometries for PA detection. Tomography setups with circular geometries often consist of a curvilinear or linear transducer array that records the acoustic signals while rotating along a circular path around the sample. Applications of circular-view systems include imaging of brain vasculature (Gamelin et al. 2008), brain lesions (Wang et al. 2003), and tumors in small-animal imaging (Ku et al. 2005), as well as imaging of small joints for investigation of inflammatory diseases such as rheumatoid arthritis (Wang et al. 2008a). Spherical-view systems usually represent hemispherical geometries that have the advantage of recording the acoustic signals from more angles around the sample, which improves the spatial resolution compared to other imaging geometries. Some recent examples in breast imaging (Kruger et al. 2010) and small-animal imaging (Brecht et al. 2009; Lam et al. 2010) have used spherical arrays. Figure 15.2a shows an example of PA image imaging system with a spherical geometry used on a sacrificed mouse. The right image (Figure 15.2b) is made by rotation of the arc-shaped linear array around the mouse in the axial direction.

15.1.3.2 Photoacoustic Microscopy

PAM images are obtained by raster scanning A-lines obtained from mechanically scanning a focused ultrasound detector (acoustic resolution (AR)) or a combination of unfocused detector and focused laser-beam (optical resolution (OR)) over a surface area. PA microscopy has the advantage that it can use relatively large detection elements compared to PAT. Small piezoelectric elements have the disadvantage of being less sensitive and difficult to produce. Figure 15.3 shows an example of results obtained by AR-PAM. The image shows multispectral imaging

(a)

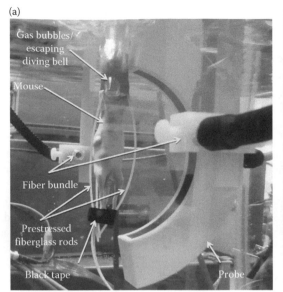

(b)

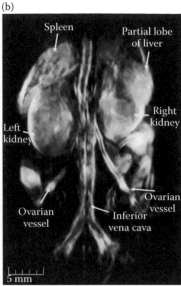

FIGURE 15.2 Photoacoustic mouse imaging setup with a spherical detection geometry. (a) Overview of the setup with an arc-shaped detector array and side illumination. (b) Photoacoustic scan of organs of a nude mouse measured at 755 nm. (Reprinted from Brecht, H.-P. et al. 2009. *Journal of Biomedical Optics* 14(6):064007. With permission from SPIE.)

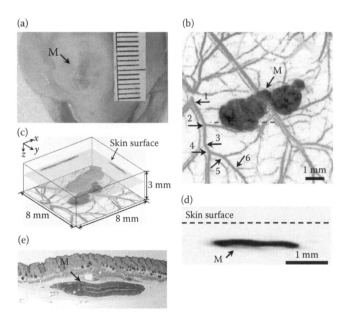

FIGURE 15.3 Microvasculature and melanoma detection by photo acoustic microscopy. (a) Photograph of the melanoma. (b) Result of multispectral PA microscopy of a melanoma (at 584 nm) and blood vessels (at 764 nm) that were pseudocolored brown and red after maximum-amplitude projection. The image shows six orders of branching in the blood vessels (1–6). (c) Three-dimensional rendering of the melanoma at 764 nm. (d) A B-scan of the melanoma. (e) Hematoxylin and eosin (HE) staining of the melanoma. (Reprinted by permission from Macmillan Publishers Ltd. [*Nature Biotechnology*] (Zhang, H.F. et al. Functional photoacoustic microscopy for high-resolution and noninvasive *in vivo* imaging. 24(7):848–51, Copyright 2006.)

of melanomas and blood vessels in 3D, as a means to staging the malignancy. For this system, the authors reported an axial resolution of 15 μm and a lateral resolution of 45 μm by which they were able to measure the objects down to 6 μm in diameter. In AR-PAM, the imaging depth can be increased by changing the focal distance of the transducer and lowering the central frequency. Axial scanning can be applied to increase the resolution at increasing depths by placing the focus of the transducer deeper into the tissue since the acoustic signal degrades when measured outside the transducer focus.

OR-PAM that utilizes focused laser beams and unfocused US transducers is capable of higher resolutions in the order of a few micrometers. OR systems do not enable high-resolution imaging beyond 1 mm penetration depth because of loss of focus due to optical scattering. OR-PAM has been used not only to image the individual capillaries and red blood cells in superficial structures such as the skin, but also to visualize the contrast-enhanced vasculature in mouse brains (Hu et al. 2011).

Many other imaging modalities that exploit the PA effect in a similar way as those mentioned before can be thought of. Minimally invasive imaging techniques such as PA endoscopy (Yang et al. 2009) and intravascular photoacoustic imaging (Jansen et al. 2011; Sethuraman et al. 2007) have the ability to image at locations beyond the optical transport mean free path. The developments in PA sensing techniques are also far from finished. New techniques such as PA Doppler flowmetry can in the future be combined with spectral imaging with the aim to acquire the metabolic rate (MRO_2) of tissue (Yao et al. 2011). PA thermometry can enable temperature monitoring of tissue during photothermal therapy or high-intensity focused ultrasound (HIFU) (Larina et al. 2005; Schüle et al. 2004).

15.1.4 Summary

Photoacoustic imaging is growing rapidly since its relative rediscovery in 1993 (Esenaliev et al. 1993; Kruger 1994). Photoacoustic imaging combines light and sound to produce high-resolution, high-contrast three-dimensional (3D) images to indicate the presence of various disease states, including cancer. It is based on imaging functional parameters instead of anatomical features while being cost-effective and patient-friendly, and could represent a breakthrough in imaging technology in the near future. While research and investigations into technology and application continue, there is an increasing trend to augment these systems to develop multimodal photoacoustic ultrasound systems. Multimodal imaging can provide additional information that will support the medical practitioners in decision making.

15.2 Engineering Aspects and Applications of Combined Photoacoustic Ultrasound Imaging Systems in Linear and Planar Geometries

This section focuses on the wide variety of multimodal PAT/ultrasound imaging devices introduced by various groups. A distinction has been made between linear geometries, based on the well-known B-mode ultrasonography setups, and planar geometries that have recently been developed.

15.2.1 Linear Geometries

Since there are many conventional ultrasound imaging systems widely available, and photoacoustics is based on detecting ultrasound, researchers in photoacoustics have made efforts to enhance these ultrasound systems with the addition of ns-lasers to develop multimodal photoacoustic ultrasound systems. Systems, mentioned in this section, were developed by the research groups of Wang (Kim et al. 2011), Frenz (Niederhauser et al. 2005), Oraevsky (Fronheiser et al. 2010), Wang (Wang et al. 2008b), Haisch (Haisch et al. 2010),

Emelianov (Park et al. 2006), Khuri-Yakub (Vaithilingam et al. 2009), and van Leeuwen/Steenbergen (Kolkman et al. 2008). Some of these groups have chosen to work with modified versions of preexisting ultrasound setups and others have chosen to design new data acquisition setups. Currently, two different types of data acquisition techniques are used: those which collect data sequentially (conventional) and those in parallel (ultrahigh speed).

15.2.1.1 Conventional Ultrasound Probe: Sequential Processing

Kolkman et al. (2008) developed a multimodal imaging system by attaching an optical module to the ultrasound probe of a conventional ultrasound system. The optical system was composed of a beam expander and a stationary mirror that illuminated an area of 5×20 mm^2 as shown in Figure 15.4a. Light pulses (8 ns) from a 1 kHz Nd:YAG laser (1064 nm) were provided to the optical system by means of a 600 μm optical fiber (Figure 15.4). The US system's hardware and software were modified to switch between US and PA imaging by switching off the US emission and synchronizing the imager using the laser's trigger pulse. The 128 elements of the linear probe were sequentially read out at the pulse frequency (1 kHz) of the laser which resulted in an imaging speed of 8 frames per second (fps). The vasculature of a human hand *in vivo* could be imaged to an imaging depth up to 12.5 mm (Figure 15.4b). The authors were limited by the decreased sensitivity of the transducer for blood vessels larger than ~100 μm due to the transducer's choice of central frequency (7.5 MHz) and bandwidth (75%–6 dB).

Haisch et al. (2010) also developed a multimodal system from a preexisting ultrasound system. A 100 Hz pulsed Nd:YAG laser pumping an OPO was synchronized with the US system that could switch between US and photoacoustic imaging (Figure 15.5). The optical fiber (1 mm) from the laser was bifurcated and both new cable ends (0.6 mm) were connected side by side to the ultrasound probe by a clip which allowed for changes in the illumination angle. The ultrasound probe had 192 elements with a central frequency of 8.8 MHz and a bandwidth of 4.1 MHz (−6 dB). Half of the 192 elements are used for photoacoustics and the other half for ultrasound imaging. The fact that each of the

	Acquisition	Number of Probe Elements	Center Frequency	−6 dB Bandwidth	Wavelength	F_Laser	fps
Linear Arrays from							
Kolkman et al. (2008)	Sequential	128	7.5 MHz	75%	1064 nm	1000 Hz	8 Hz
Haisch et al. (2010)	Sequential	96 (196)	8.8 MHz	50%	OPO	100 Hz	1.1 Hz
Niederhauser et al. (2005)	Parallel	64	7.5 MHz	76%	760 nm	7.5 Hz	7.5 Hz
Erpelding et al. (2010)	Parallel	128	6 MHz	67%	Tuneable dye	10 Hz	1 Hz
Wang et al. (2008b)	Parallel	64 or 128	≈2.7 MHz	100%	OPO	10 Hz	5 Hz (128) or 10 Hz (64)
Fronheiser et al. (2010)	Parallel	128	5 MHz	75%	1064 nm	10 Hz	10 Hz
Planar Arrays from							
Vaithilingam et al. (2009)	Sequential	$16 \times 16 = 256$	3.48 MHz	93.40%	OPO	10 Hz	NA (offline)
Erpelding et al. (2011)	Semiparallel	$50 \times 50 = 2500$	≈4.5 MHz	111%	650 nm	10 Hz	NA (offline)

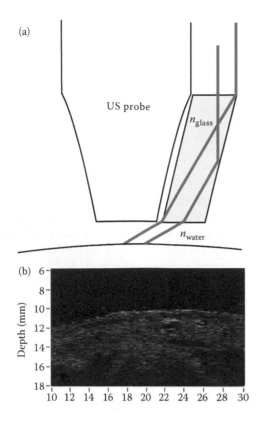

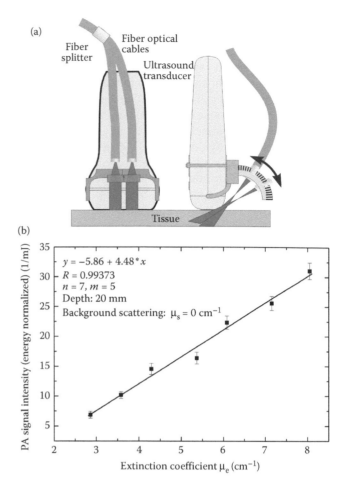

FIGURE 15.4 (See color insert.) (a) Schematic drawing of the US probe with the attached optical system of the sequentially recording setup of Kolkman et al. (2009). (b) Combined US (gray scale) and photoacoustic (red) image showing absorption in blood vessels and the skin. Acquisition time was 128 ms at a fluence of 1.9 mJ/cm². (Reprinted from Kolkman, R.G.M. et al. 2008. *Journal of Biomedical Optics* 13(5):050510. With permission from SPIE.)

96 elements needs to be read out subsequently at 100 Hz pulse frequency results in imaging at 1.1 fps, which makes the laser repetition frequency a bottleneck for real-time *in vivo* measurements. The image reconstruction was performed by a fast 50 ms Fourier-based reconstruction algorithm that does not compromise the imaging speed. Basic measurements of the quantitative oxygenation levels proved possible by averaging the data five times and subsequently taking the integral intensity over all pixels in the area of interest. The precision (SD in %) in determining the oxygenation level of these measurements, performed in nonscattering (8%) and scattering phantoms (9.6–14%), was however not close to the 3% precision of clinically used devices for determination of the oxygenation level. Better results could have been obtained if the laser pulse-to-pulse intensity did not vary over 32% and if the coupling of light into the optical fiber was less influenced by the OPO (Haisch et al. 2010).

15.2.1.2 High-Speed Ultrasound Probes: Parallel Processing

Niederhauser et al. (2005) used a custom-made high-speed ultrasound system from the Fraunhofer Institute for Biomedical Engineering (St. Ingbert, Germany). This system

FIGURE 15.5 (a) Schematic drawing of the US probe with the attached optical system of the sequentially recording setup of Haisch et al. (2010). (b) Corresponding fit for the correlation between the absorption and PA signal intensity in a flow channel filled with absorbing liquid placed in nonscattering phantom material at a depth of 20 mm. (With kind permission from Springer Science + Business Media: Combined optoacoustic/ultrasound system for tomographic absorption measurements: Possibilities and limitations, *Analytical and Bioanalytical Chemistry*, 397(4), 2010, 1503-10-10, Haisch, C., et al.)

was connected to a computer that allowed for real-time reconstruction at 7.5 fps due to the use of a modified Fourier reconstruction algorithm (Kostli and Beard 2003). The illumination from an optical fiber was provided from two sides around the probe by an optical system consisting of a lens, a prism, and two mirrors (Figure 15.6a). All transducer elements were read out simultaneously after one laser pulse and then reconstructed. At first sight, the images that the system produced showed higher SNR than the previously mentioned sequential recording setups while using a lower amount of elements in their probe (64 vs. 96 and 128). The difference in quality might be caused by the higher fluence of light (15 mJ) and more appropriate wavelength (760 nm) that was used. Because of that, only one pulse was required for each frame. This was

(a)

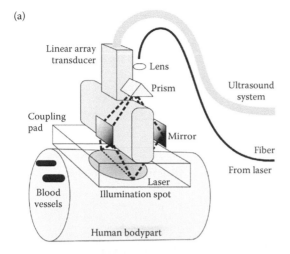

(b)

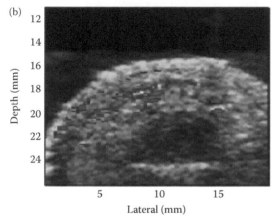

FIGURE 15.6 (a) A schematic overview of a US probe that is controlled by a parallel data acquisition system. (b) A combined US (gray scale) and photoacoustic (red) image made on a more recent version of the setup showing absorption by blood vessels in a cross section of the middle finger. The US image clearly shows the outlines of the skin and bone while the PA image made with 70 mJ/cm² at 1064 nm shows the blood vessels. (Reprinted from Niederhauser, J.J. et al. 2005. *IEEE Transactions on Medical Imaging* 24(4):436–40. With permission from IEEE; Courtesy of M. Jaeger, IAP, University of Bern.)

possible, without exceeding the MPE, because of the lower laser pulse repetition rate. The combination of high contrast and the possibility for real-time imaging makes this parallel acquisition system a more suitable candidate for clinical use than the sequential data acquisition systems. The fact that this group worked with a custom-made data acquisition system will most likely result in higher costs compared to the systems modified from a preexisting ultrasound systems. Figure 15.6b shows a more recent example of a multimodal cross-section image of the middle finger measured by the latest version of the setup. The gray-scale US image clearly shows the outlines of the skin and bone. The red PA image shows absorbing structures such as blood vessels in red.

Erpelding et al. (2010) also developed a multimodal imaging device based on a parallel acquisition US system comprising a 128-element probe. The channel board was modified to send both US and PA signals to a data acquisition computer without changing the US capabilities of the device itself. Light pulses (6.5 ns) from a tunable dye laser, pumped by a Q-switched Nd:YAG laser, were coupled into a bifurcated optical fiber whose free ends were attached to either side of the US probe with a bandwidth ranging from 4 to 8 MHz. A Fourier-based reconstruction algorithm was used to calculate and display the PA images at ~1 fps, while the data acquisition was limited by the 10 Hz pulse repetition frequency of the laser. With this system, rat sentinel lymphatic nodes (SLN) stained by methylene blue as contrast agent were visualized after being covered by 2.5 cm thick chicken breast. Shortly after, Kim et al. (2010) used the same system to show 7 mm tube filled with methylene blue (30 mM) covered by 5.2 cm chicken breast. They also showed that it was possible to image the blood vessels without contrast agent and guide a needle to a methylene blue-labeled SLN (Figure 15.7) with high contrast (Kim et al. 2011).

Fronheiser et al. (2010) reported a system with the capability to image the larger blood vessels in the forearm with sufficient sharpness of the vessel boundaries and real-time (10 Hz) visualization. The laser light was generated by a Q-switched Nd:YAG laser that produced 6 ns pulses of 1064 nm with a repetition frequency of 10 Hz. The light was coupled into a bifurcated fiber and illuminated the surface from both sides of the probe with a maximum fluence of 15 mJ/cm². The PA probe consisted of 128 elements with a central frequency of 5 MHz and a fractional bandwidth of 75% that were focused at 20 mm in elevation by an acoustic lens. The PA data in Figure 15.8 was sampled at 25 MHz and reconstructed in real time with 10 fps (Figure 15.9(a)). The US images that were made at the same location as the PA images are shown in Figure 15.9(b).

The sharpness of the vessel boundaries was ascribed to a five-scaled high-frequency wavelet transform implemented during

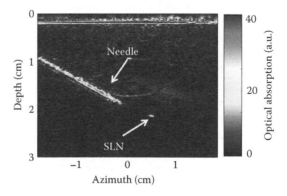

FIGURE 15.7 Photoacoustic guidance of a metal needle (18-gauge) for SLN biopsies. SLN = sentinel lymph node. (Reprinted from Kim, C. et al. 2010. Deeply penetrating *in vivo* photoacoustic imaging using a clinical ultrasound array system. *Biomedical Optics Express* 1(1):278–84. With permission from Optical Society of America.)

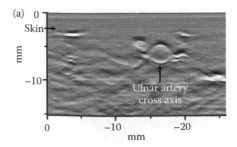
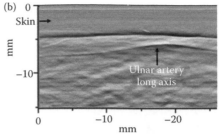

FIGURE 15.8   (a) Cross-sectional and (b) longitudinal PA image of larger blood vessels in the wrist. (Reprinted from Fronheiser, M.P. et al. 2010. *Journal of Biomedical Optics* 15(2):021305. With permission from SPIE.)

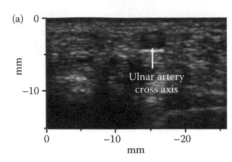
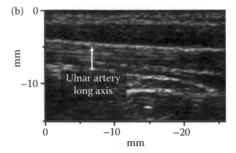

FIGURE 15.9   (a) Cross-sectional and (b) longitudinal US image of larger blood vessels in the wrist. (Reprinted from Fronheiser, M.P. et al. 2010. *Journal of Biomedical Optics* 15(2):021305. With permission from SPIE.)

signal processing (Patrickeyev and Oraevsky 2003). Due to the 10 Hz imaging speed and the sharp edges of the larger blood vessels that were imaged, it proved possible to measure the heart rate from the change in cross section.

Wang et al. (2008b) designed a multimodal system for measuring the small *ex vivo* samples with a handheld probe. A 128-element parallel US probe was connected to a 64-channel 200 MHz data acquisition system. The probe, mentioned 1.5–4.2 MHz, measured the acoustic signals generated by the 5 ns pulses from the 10 Hz Nd:YAG-pumped OPO system that directed the illumination orthogonally to the acoustic beam path. The acquisition and visualization frame rate was 5 fps for 128 elements and 10 fps for 64 elements which was limited by the pulse rate of the laser and by the 64-channel data acquisition system. The axial and lateral resolutions of the system were 0.65 and 0.88 mm, respectively. Combined ultrasound and photoacoustic measurements were performed on a fresh piece of canine prostate that was embedded in porcine gel (Figure 15.10a and b). A subsurface lesion, to simulate a prostate cancer tumor, was generated by a subsurface injection of a mixture of fresh canine blood and porcine blood. The optical absorption coefficient of this mixture was 50% compared to that of whole blood. The lesion was detected by the PAT system at a depth of 1.5 cm while it was not visible in US alone. Spectral analysis at 720 and 868 nm showed that the optical absorption was matching with deoxygenated blood.

In general, all these sequential or parallel systems are capable of performing PAT. The downside of using the available US probes is that these are designed for both transmission and reception of US signals. The chosen piezoelectric crystals are therefore, as a compromise, not optimized for the most sensitive and broadband reception of PAT signals. Technical developments in the field of CMUT, with high sensitivities and pulse transmission power, are expected to improve the imaging (Butrus and Ömer 2011). Imaging in three dimensions can be obtained by mechanical scanning or by use of planar array geometries. The latter is discussed in the next section.

The resolution and penetration depth of all linear US probes is limited by the central frequency and bandwidth. The aforementioned systems were capable of visualizing the absorbing structures at several centimeter imaging depth but at a limited resolution because all worked with US probes with low frequencies in the range of 1–12 MHz. These systems do not yet have the capability to image at axial resolution below 100–300 µm. As an answer to limited resolution, Song et al. (2008) developed a multimodal system that was based on a linear US array with a central frequency of 30 MHz and bandwidth up to 70%. The axial, lateral, and elevational resolution were 25, 70, and 200 µm, respectively. An imaging depth greater than 3 mm for scattering biological tissues was reported. This PA system is fast with a scan rate of 50 B-scans per seconds. However it is not capable of multimodal imaging (Song et al.

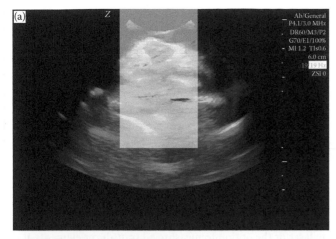

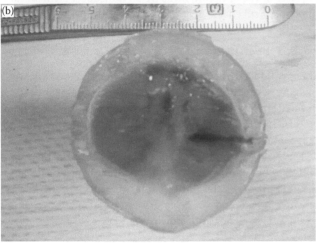

FIGURE 15.10 (a) US and PAT dual-modality imaging of a canine prostate with a generated subsurface lesion mimicking prostate tumor. (b) Photo of the canine prostate with prostate tumor mimicking lesion. Measurement performed at 720 nm. (Reprinted from Wang, X. et al. 2008b. Experimental evaluation of a high-speed photoacoustic tomography system based on a commercial ultrasound unit. Paper presented at the 2008 IEEE International Ultrasonics Symposium (IUS), November 2–5, 2008. With permission from IEEE.)

2008). The first commercial high-resolution multimodal system was launched by VisualSonics Inc. (Toronto, Canada). These systems utilize high-frequency ultrasound probes with the capability to measure up to 80 MHz. The resolution of these systems can be up to 30 microns at imaging depths of up to 25 mm (Petrov et al. 2011).

Reflection mode setups are well suited to image parts of the body that cannot be imaged in transmission mode due to obstruction of acoustic waves by, for instance, bone. Imaging in reflection mode does however have the disadvantage of systematic image background that originates from subsurface transients generated closely beneath the irradiated skin. These transients travel into the tissue and are scattered back, obscuring PA signals from deeper parts of tissue, resulting in limited imaging depths. Attempts to reduce this image background

were investigated by Jaeger et al. (2009) using displacement-compensated averaging (DCA) of series of images during deformation of the tissue (Jaeger et al. 2009, 2011).

15.2.2 Planar Geometries

Two dimensional (2D) arrays are the ideal candidates for photoacoustic imaging at high frame rates. Probes with 2D arrays and parallel data acquisition are in theory capable of making a whole 3D image of an object using only one laser pulse. The fact that these are currently not widely available can be attributed to several technical and financial obstacles that still need to be surmounted. Multielement 2D arrays require as much connections, signal amplifiers, and data acquisition channels as there are elements, which, while being technically possible, are exorbitantly expensive. More and smaller transducers are also desired for 2D arrays in order to accomplish the higher resolutions. Unfortunately, small piezoelectric elements need more complex electrical interconnects, are difficult to fabricate, and, additionally, also suffer from lower SNR. Current studies that should lead to more sensitive and smaller US transducers are now focusing on CMUT transducers (Te-Jen et al. 2010; Vaithilingam et al. 2009). Arrays consisting of CMUT transducers are preferred over piezoelectric transducers because they are fabricated by a silicon micromachining process that allows a broad range in frequencies, element sizes, and array geometries. Another advantage is that the manufacturing process allows integrated circuits (IC) to be placed close to the CMUT transducers. The signal amplifiers and multiplexing hardware in the IC help in maximizing the signal integrity and allow a reduction in the amount of cables toward the signal-processing system.

Vaithilingam et al. (2009) showed preliminary results of a 2D array consisting of 16×16 CMUTs (Figure 15.11). The 2D array was used to image four polyethylene tubes filled with water, 5 µM indocyanine green (ICG), blood, or with ICG and blood. These tubes, with an inner diameter of 1.19 mm and an outer diameter of 1.70 mm, were embedded in a phantom at a depth of 1.8 cm (0.8 cm chicken breast and 1.0 cm agarose gel). Instead of using water or water-containing gels, vegetable oil was used for acoustic coupling. This electrically nonconducting medium allowed the not electrically isolated array to be used. A simulation of a 64×64 elements array was performed by scanning the 16×16 array over a 4×4 grid with a step resolution equal to the length and width of the CMUT array. Side illumination from two sides was performed using free space optics. The fluence of the two beams from the 775 nm Nd:YAG laser was 9 mJ/cm^2 and 14 mJ/cm^2, respectively. Because no acoustic lenses were added to the transducer, a synthetic aperture technique and coherence factor weighting were needed to reduce the distortions in focusing (Hollman et al. 1999). The dataset was reconstructed in three dimensions by a delay and sum algorithm. An example of a three-dimensional image is shown in Figure 15.11b. This figure shows an overlay of a pulse-echo image (gray) and a PA image (orange) showing the four tubes inside the phantom. All four tubes are visible in

(a)

16 × 16
CMUT

Integrated
circuit

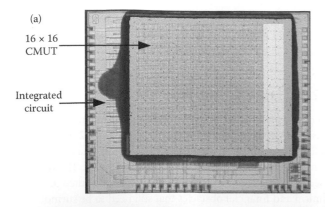

(b)

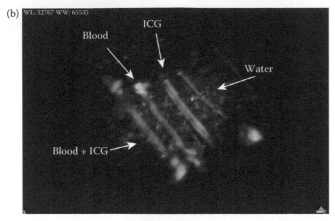

FIGURE 15.11 (**See color insert.**) (a) Photograph of a 16 × 16 CMUT array flip-chip bonded to custom-designed integrated circuit. (b) 3D-rendered photoacoustic (hot-metal color scale) image overlaid on a pulse-echo image (gray scale) displayed at 70 dB dynamic range. The image shows four tubes filled with water, ICG, blood, blood + ICG. (Reprinted from Vaithilingam, S. et al. 2009. *IEEE Transactions on Ultrasonics, Ferroelectrics and Frequency Control* 56(11):2411–19. With permission from IEEE, Courtesy of S. Vaithilingam, Stanford University, Stanford.)

the ultrasound image. The tubes containing absorbing chromophores such as ICG and blood are visible in the PA image that has been overlaid on the ultrasound image. The SNR of the PA image was found to be 87 dB between the phantom background and the tube filled with 5 µM ICG. This was obtained after coherence factor weighting.

While the future plan is to enable the parallel detection of all elements at the same time, the IC of the probe now extracts one channel at a time. This means that 256 light pulses were required for one 3D scan.

PAT with a commercial US system using a 2500-element planar array was shown by Erpelding et al. (2011). The raw PA and US data could be acquired by modifications to the channel board architecture. Thirty-six laser shots were required to measure a PA volume image because microbeamforming was used to reduce the amount of data acquisition channels. The acoustic signals were generated by a 10 Hz tunable dye laser pumped by a Q-switched Nd:YAG laser with 6.5 ns pulse duration.

Illumination was supplied by a bifurcated optical fiber with the distal ends connected to both sides of the probe. All data for an image of $4 \times 4 \times 5$ cm^3 could be collected in 20 s and subsequently reconstructed offline. The resolution along the azimuthal and elevational direction was 0.77 and 0.96 mm, respectively, and the axial resolution was 1.02 mm. The system was used for *in vivo* 3D mapping of methylene blue accumulation in SLNs. The SLN was mapped at 0, 5, and 30 min after dye injection which resulted in the PA images in Figure 15.12. The PA signal generated at the location of the SLN was shown to increase in time.

The developers suggest that this system might be used to image the multiple SLNs and to guide the percutaneous needle biopsies of SLNs more efficiently. Clinical use of this setup is currently limited by the data collection speed. A faster pulsating laser or the option to measure all transducer elements at the same time will enable 3D imaging at real-time frame rates.

15.2.3 Summary

The mentioned hybrid systems show that combined PA and US imaging have the potential to be implemented into clinical trials in the near future. These systems are capable of visualizing the blood vessels, lymph nodes, tumors, and many more tissue types that contain either natural or added chromophores.

The setups that are capable of parallel data acquisition are preferred over the sequentially acquisitioning setups because of the high imaging rates. Future improvements will include more sensitive broadband US detectors and tunable lasers with high repetition frequency and sufficient power.

15.3 Simultaneous Imaging of Optical and Acoustic Parameters in Computed Tomography Geometries

The rationale for extracting the speed of sound distribution of the object under photoacoustic investigation was to improve the photoacoustic reconstruction. Variations in the acoustic speed (AS) can result in radial and tangential displacement of the photoacoustic signals during reconstruction (Jin and Wang 2006). Radial displacement occurs when the assumed AS is incorrect along the linear projection lines toward the detector element. Tangential displacement occurs on the boundary of two tissues that have a variation in the AS. The acoustic waves refract from the otherwise linear path and are detected at another detector position. Back-projection of the acoustic signals without compensation for AS variations results in blurring and displacement of the reconstructed optical absorption distribution.

15.3.1 Acoustic Speed Correction

Jin and Wang (2006) first implemented such an approach in a thermoacoustic tomography (TAT) setup expanded with the capability of separately performing ultrasonic transmission

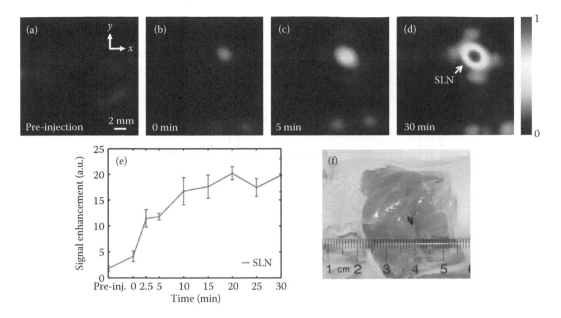

FIGURE 15.12 Noninvasive PA images of rat axillary region. (a) Control PA maximum amplitude projection (MAP) image before dye injection into the rat forepaw. (b–d) PA MAP images collected 0, 5, and 30 min after showing an increase in signal enhancement at the SLN in time. (e) Dynamic changes in PA signal amplitude in time from the SLN. (f) Postmortem photograph of the rat acquired after photoacoustic imaging and skin removal. (Reprinted from Erpelding, T.N. et al. 2011. Three-dimensional photoacoustic imaging with a clinical two-dimensional matrix ultrasound transducer. Paper presented at the Proceedings of SPIE. With permission from SPIE.)

tomography (UTT) (Figure 15.13). Two single-element unfocused ultrasonic transducers (2.25 MHz) and the sample were immersed in mineral oil during the experiments. One element was used to transmit ultrasonic pulses and the other element was used to receive the pulses after passing through the sample. Scanning was performed by fan-beam scanning and circular scanning. During fan-beam scanning, the transmitter and receiver were moved over 120 steps that covered the whole object over 67.5°. The circular scanning was then followed by rotating the transmitter and receiver along the center axis of the fan beam in steps of 2.25°.

A quantitative distribution map of the AS inside tissue samples was calculated from the time-of-flight (TOF) of acoustic waves that travel between the US transmitter and receiver. A travel-time perturbation δT, where

$$\delta T = T - T_0 \tag{15.8}$$

was calculated, with T and T_0 are the TOFs along ray paths ($l(r_0)$) for unknown sample and known homogeneous reference sample. The calculation of the travel times T and T_0 was given as the integral over the inverse of the AS ($c(r)$) along the ray path:

$$\delta T = \int_{l(r_0)} \left(\frac{1}{c(r)} - \frac{1}{c_0} \right) dl \tag{15.9}$$

Equation 15.9 can be interpreted as a Radon transform, which was used to reconstruct the AS distribution by a filtered acoustic back-projection.

Two pieces of porcine muscle were embedded into a piece of 61×39 mm porcine fat of 12 mm thickness (Figure 15.14). The small piece of muscle had a diameter of ~2.5 mm. The dimensions of the larger piece of muscle were 17×21 mm. The AS difference between these absorbers and background was around 10%. Figure 15.8b shows the measured AS distribution. The TAT image without compensation for AS variations in Figure 15.8c shows that the small piece of muscle is blurred while the TAT image with compensation in Figure 15.8d is sharper and therefore shows less blurring.

Another example of AS correction was shown by Deán-Ben et al. (2011a). With *a priori* knowledge of the speed of sound distribution, it proved possible to reconstruct the PAT maps by use of a modified version of the filtered back-projection algorithm. Three phantoms with one, two, or three absorbers were placed inside water (1482 m/s). The AS (1670 m/s) was homogeneous throughout the whole phantom. Figure 15.15a through c shows the reconstructions that assume a similar AS for both water and the phantom. Figure 15.15d through f shows the reconstruction of the same dataset but then performed with the modified-filtered back-projection algorithm that takes the actual AS of the phantom into account. The images Figure 15.15d through f show a more representative reconstruction of the absorber shape and distribution.

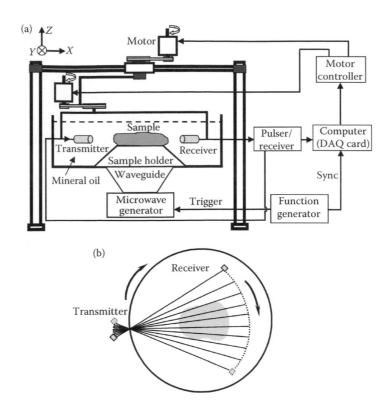

FIGURE 15.13 Experimental setup and scanning geometry: (a) Experimental setup for the combined TAT/UTT imaging setup and (b) schematic of the scanning geometry in top view. In UTT, the transmitter sent pulsed ultrasonic signals, and the receiver on the opposite side of the transmitter received the ultrasonic pulses. In TAT, the transmitter used in UTT was used as the receiver, which circularly scanned the tissue sample. (Reprinted from Jin, X. and L.V. Wang. 2006. Thermoacoustic tomography with correction for acoustic speed variations. *Physics in Medicine and Biology* 51(24):6437–48. With permission from IOP Publishing.)

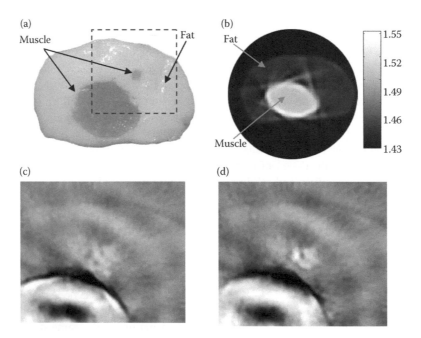

FIGURE 15.14 Phantom experiment: (a) Photograph of the phantom sample in top view. (b) The speed-of-sound image of the phantom sample. To illustrate the blurring more clearly, we only showed the close-up TAT image of the small absorber as marked by the black dashed square in (a). (c) Close-up TAT image obtained by adjusting the average AS (boundary is denoted by arrows). (d) Close-up TAT image obtained by AS compensation using the AS distribution (boundary is denoted by arrows). (Reprinted from Jin, X. and L.V. Wang. 2006. Thermoacoustic tomography with correction for acoustic speed variations. *Physics in Medicine and Biology* 51(24):6437–48. With permission from IOP Publishing.)

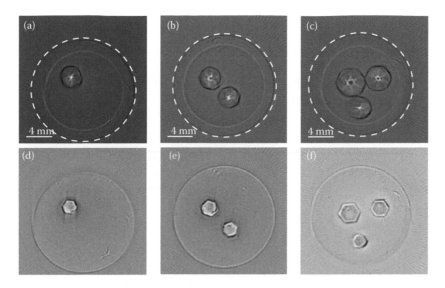

FIGURE 15.15 (a–c) Tomographic reconstructions while considering a uniform AS for both water and the phantom (dashed circumferences). (d–f) Tomographic reconstructions of the same datasets but this time reconstructed by the modified filtered back-projection algorithm that also considers the different AS of the phantom. (Reprinted from Dean-Ben, X.L. et al. 2011a. Time-shifting correction in optoacoustic tomographic imaging for media with nonuniform speed of sound. Paper presented at the Proceedings of SPIE. With permission from SPIE.)

15.3.2 Passive Element-Enriched Tomography: Simultaneous Imaging of Acoustic and Optical Properties

Manohar et al. (2007) also estimated the AS distribution by TOF measurements, but then without the need for a separate ultrasound transmitter or extra measurements. A photoacoustic imager in a computed tomography geometry was modified with the addition of a highly absorbing passive element with a small cross section that was placed into the path of the illuminating laser (Figure 15.16).

The acoustic waves that fan out from the passive element are distorted by the interaction with the object and detected by ultrasound array that was located at the opposing side. The wavefront either advances or retards at object regions of higher or lower ASs, respectively. The measured TOF of the acoustic waves through the sample were cross-correlated with the TOF of the acoustic waves through the reference medium (water). The AA was derived from the decrease in signal intensity through the sample.

The advantage of this technique is that no additional measurements are necessary since both the PA signals from the sample and passive element can be measured in the same projection. The addition of passive element which usually takes the form of a carbon fiber, colored nylon line, or a horse-tail hair also allowed the SOS imaging to be inexpensive because it does not require ultrasound transmission hardware.

15.3.2.1 Estimation of the Acoustic Transmission Parameters: Acoustic Speed and Acoustic Attenuation

The estimated time delay and integrated attenuation along a ray path between the passive element and the detector element were

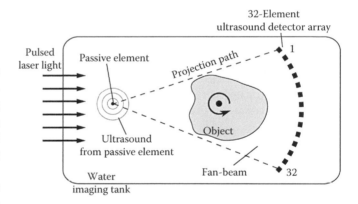

FIGURE 15.16 A schematic overview of the passive element-enriched photoacoustic computed tomography setup (PER-PACT) that allows to measure the AS and acoustic attenuation (AA) cross-sectional images of a sample. Ultrasound transients are generated in the passive element (carbon fiber) placed in the path of light illuminating the sample. The time-of-flight and AA of the ultrasound transient are measured at each element of the ultrasound detector array. The photoacoustic signals and passive element signals are generated and measured at the same time. No extra measurements were required. (Reprinted from Jose, J. et al. 2011b. *Optics Express* 19(3):2093–104. With permission from Optical Society of America.)

used for the reconstruction of the AS and AA distribution maps (Dean-Ben et al. 2011a; Jose et al. 2011b).

15.3.2.1.1 Estimating Acoustic Speed

The signal from the passive element through water as reference measurement is correlated to the time domain signals through the sample by a matched filter approach (Jose et al. 2011b). This resulted in an estimation of the time delays expressed in an integer multiple of the time spacing $\Delta t = (1/f_s)$. The accuracy of the estimation was however not limited by the height of the sampling frequency.

The signal from the passive element $h(t)$ is shifted over time shift τ and scaled by a factor A before the signal from the measurement $f(t)$ has reached the detector element.

$$f(t) = Ah(t + \tau) \tag{15.10}$$

The same can be written for a time-windowed and sampled signal vector z as

$$z = h_z(x) + n_z = Ah(t_i + \tau) + n_z \tag{15.11}$$

where parameter vector $x = [\tau, A]^T$ and n_z is the added Gaussian white noise with a variance σ_z^2. A maximum-likelihood estimate of x is obtained by solving

$$\hat{x}_{ML} = \arg_x \min z - h_z(x)^2 \tag{15.12}$$

A nonlinear measurement function of x combined with the initial estimate $x^{(1)} = \left[\tau^{(1)}, A^{(1)}\right]^T$ on the matched filter results in a solution by the use of the Gauss–Newton optimization method (Fletcher 1987) that is based on quadratic cost function that iteratively linearizes the measurement function around the current estimated parameter value. The minimum of the cost function is used as the next estimate for the subsequent iteration.

15.3.2.1.2 Estimating Acoustic Attenuation

The integrated AA of the object and the reference medium is expressed as

$$Y_o(\omega) = H(\omega)Y_w(\omega) \tag{15.13}$$

Here, Y_O and Y_w are the transfer functions for the measured object signal and reference signal, respectively, and $H(\omega)$ is the transfer function of the object that contains the individual transfer functions of the attenuation, reflection, and time delay (τ). The attenuation and sound dispersion of the reference signal are assumed as negligible. Equation 15.13 is rewritten as

$$Y_o(\omega) = Y_w(\omega)\exp\left[h_{att}(x) + jh_{phase}(x)\right] \tag{15.14}$$

with

$$x = \begin{bmatrix} \alpha_0 \\ \alpha_r \\ \tau_0 \end{bmatrix}$$

where the attenuation and phase functions are described as

$$h_{att,i}(x) = -\alpha_0 |\omega|_i^y d - \alpha_r \tag{15.15}$$

$$\hbar_{phase,i}(x) = -\omega_i\left(\tau_0 + \alpha_0 \tan\left(\frac{\pi}{2} y\right)\left(|\omega_i|^{y-1} - |\omega_0|^{y-1}\right)\right)d \tag{15.16}$$

The attenuation is described in terms of frequency-dependent attenuation caused by reflections (α_r) and the phase function as a function of distance (d) and time delay (τ_0).

$$\tau_0 = \frac{1}{c(\omega_{0)}} - \frac{1}{c_w} \tag{15.17}$$

The object signal as a function in the time domain can be expressed as

$$h_t(x) = Re\left\{2\tilde{V}\begin{pmatrix} Y_w(\omega_2)\exp[h_{att,2}(x) + jh_{phase,2}(x)] \\ \vdots \\ Y_w(\omega_m)\exp[h_{att,m}(x) + jh_{phase,m}(x)] \end{pmatrix}\right. \tag{15.18}$$

where $2V$ is two times the truncated inverse discrete Fourier transform that would otherwise have resulted in redundant information since we only take the real part of the signals. The DC component of the signal (ω_1) was neglected and all frequencies above the Nyquist frequency (ω_n) were not included. This defines subscript m as

$$m = \frac{n}{2}$$

The measured time domain signal by the system is also corrupted by additive Gaussian white noise (n_{zt})

$$z_t = h_t(x) + n_{z_t} \tag{15.19}$$

15.3.2.2 Reconstruction of Acoustic Properties Distributions

Two uniformly spaced grids are produced in order to discretize the two acoustic properties distributions that have to be obtained. These grids contain the sample points that each represents the value of either the As or AA at that location of the sample. Bilinear interpolation was applied in order to adjust for off-grid projections from the reconstruction. Vector x_a contains the values for the attenuation distribution (α_0 and α_r) and vector x_c contains the object slowness values that were normalized with the background (Willemink et al. 2008).

$$x_{c,k} = \frac{1}{c(r_k)} - \frac{1}{c_w} \qquad (15.20)$$

The position of pixel k was defined by r_k. All projection measurements were denoted by vector z_t and all ray paths were described in the linear relation via matrix operation H_t. This relates back to the initial acoustic sound distribution:

$$z_t = H_t(x_c)x_c \qquad (15.21)$$

$H_t(x_c)$ is the projection matrix that was created from the given slowness distribution x_c. This problem was solved by calculating the solution to the regularized cost function iteratively with an initial guess of vector $x_c^{(i)}$.

$$\hat{x}_c^{(i+1)} = \arg_{x_c}\max(z_t - H_t(\hat{x}_c^{(i)}x_c)^2 + \lambda H_{G_x}x_c^2 + \lambda H_{G_y}x_c^2) \qquad (15.22)$$

This was solved with the LSQR method (Paige and Saunders 1982).

15.3.2.3 Sample Distribution Maps

Two examples of the results that were obtained by this imaging technique were given by Jose et al. (2011b) in which the AS and AA distribution were reconstructed: One phantom that contained acoustic inhomogeneities and a second one that was made of biological specimens consisting of combinations of muscle and adipose tissue.

The setup, called PER-PACT (short for passive element-enriched photoacoustic tomography), used a Q-switched Nd:YAG laser that delivered 6 ns pulses at a wavelength of 532 nm. The acoustic signals were detected by a curvilinear array that consisted of 32 elements from which the elements had a central frequency of 6.25 MHz and a bandwidth of over 80%. The interelement spacing of the 10×0.25 mm elements was 1.85 mm. At a distance of 48 mm from the detector surface, the elements produced an elevational plane focus of 1 mm. The detector signals were amplified by 60 dB at a sampling rate of 80 MHz with a 32-channel pulse receiver system (Lecoeur, Paris). A black horse-tail hair of 250 μm was used as passive element and placed at a distance of 90 mm from the detector surface. The whole sample was always measured within the fan-beam-shaped line of sight between the passive element and the detector elements.

The first example showed a 3% agar phantom with an outer diameter of 26 mm that contained our inhomogeneities of 4% agar mixed with 80% milk (Figure 15.17). The reconstructed distribution maps for the AS and attenuation show all four objects. The measured AS in the 4.6 mm squares was in good agreement with values from the literature. In the 2.6 mm spheres, the AS was slightly underestimated. This was attributed to the resolution limit of the system, which resulted in an overestimation of the size of the spherical inhomogeneity (3.2 mm instead of 2.6 mm). The AA distribution map showed more faithful reconstruction of the shape and dimensions of the inhomogeneities. Also, the values for the AA of both small and large inhomogeneities were according to the values from the literature. Some artifacts, in the form of concentric rings, did remain though. These were attributed to multipath propagation.

Another example that was given to demonstrate the value of acoustic property imaging is shown in Figure 15.18. The AS and attenuation of four layers of pork tissue were mapped simultaneously. The four layers were made from muscle tissue, adipose tissue (fat), muscle tissue, and a mixture of adipose and muscle tissue subsequently. The photoacoustic absorption distribution only shows the edges of the tissues, since green light was used. The AS distribution in subset (b) shows that the AS in muscle tissue (1572 ms^{-1}) is higher than in fat (1525 ms^{-1}), which shows that this map can be used to distinguish two different tissues from each other. The AA map also resulted in additional contrast-rich

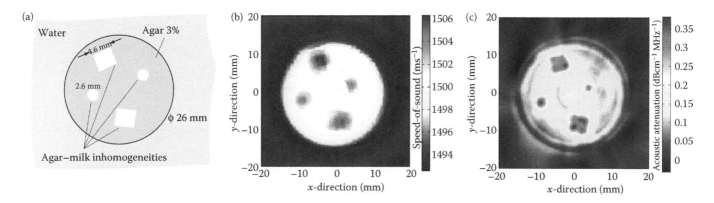

FIGURE 15.17 Results of extraction of acoustic property projections and the reconstruction of acoustic property distributions on phantom 3. (Reprinted from Jose, J. et al. 2011b. Passive element-enriched photoacoustic computed tomography (PER PACT) for simultaneous imaging of acoustic propagation properties and light absorption. *Optics Express* 19(3):2093–104. With permission from Optical Society of America.)

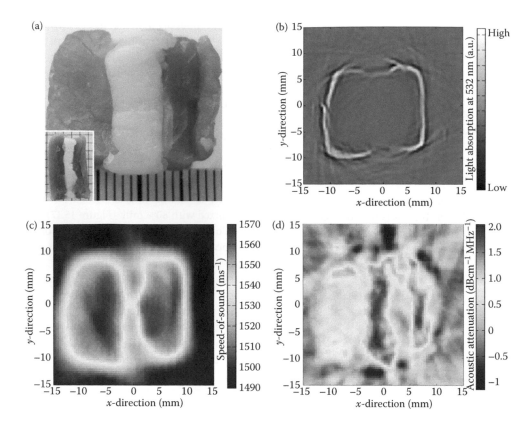

FIGURE 15.18 (**See color insert.**) (a) Photograph of four layers of pork tissue showing from left to right: muscle, fat, muscle and a mixture of fat and muscle. (b) The photoacoustic image shows the boundaries of the phantom. (c) The AS distribution shows the presence of the muscle tissue with high AS (red) and the fat with low AS (blue). (d) The AA image shows the presence of the highest attenuators in red which are the second layer (fat) and the fourth layer (mixture of fat and muscle). (Reprinted from Jose, J. et al. 2011b. Passive element enriched photoacoustic computed tomography (PER PACT) for simultaneous imaging of acoustic propagation properties and light absorption. *Optics Express* 19(3):2093–104. With permission from Optical Society of America.)

images. The second layer (fat) showed considerately higher AA than the muscle tissue. The fourth layer that was made of both fat and muscle also showed increased attenuation due to the high amount of fat. Both acoustic property distributions proved to pay off in sensitivity and will be valuable for diagnostic purposes.

A drawback of the PER-PACT system is that it relies on the measurement of a large number of views or projection angles. The accompanying long measurement times were required to avoid the aliasing problems. A method to shorten the measurement time without compromising the resolution and contrast was introduced shortly after (Resink et al. 2011). The proposed solution involved the placement of a higher number of passive elements (*N*) that effectively increase the number of views per projection with *N* × 32 (Figure 15.19).

The concept was proven by performing experiments on the previously mentioned agar phantom (3%) with four agar–milk (4% vs. 80%) inhomogeneities. The same PA setup was used with a varying amount of passive elements (1–9 elements). One experiment was performed with one passive element while taking 80 projections of 4.5° over 32 detector elements. The other experiments used 1, 3, 6, or 9 passive elements

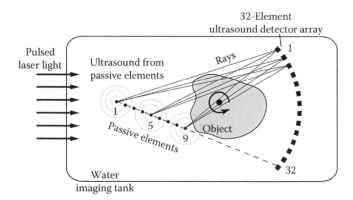

FIGURE 15.19 Schematic of the multi-PER-PACT which allows short measurement times while maintaining the image quality. Nine passive elements are shown, and for simplicity only a few projections rays of ultrasound propagation are shown. (Reprinted from Resink, S. et al. 2011. Multiple passive element enriched photoacoustic computed tomography. *Optics Letters* 36(15):2809–11. With permission from Optical Society of America.)

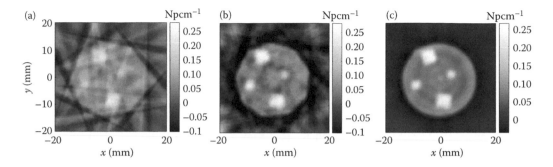

FIGURE 15.20 Reconstruction of the AA in nine projections by (a) one passive element and (b) nine passive elements. (c) An example of an AA map measured with one passive element in 80 projections. (Reprinted from Resink, S. et al. 2011. Multiple passive element enriched photoacoustic computed tomography. *Optics Letters* 36(15):2809–11. With permission from Optical Society of America.)

while the number of projections, each over 40°, was limited to 9. The reconstruction of the inhomogeneities was not precise for the case during which only one passive element was used for nine projections (Figure 15.20a). An increase in the number of passive elements to contribute to the reconstruction improved the quality of the images considerably while working with a limited number of actual physical views (Figure 15.20b). A comparison between the experiment where 80 projections with one passive element (Figure 15.20c) were used and the experiment where nine projections with nine passive elements were used shows that the number of projections can be reduced without compromising the resolution and contrast significantly. The figure shows the example with reconstructions of the AA.

15.3.3 Summary

Imaging of acoustic properties in a PA setup has the profound advantage that a more representative reconstruction of the absorber distribution can be obtained. This additional information does not only improve the imaging resolution, but it will also allow one to obtain more quantitative data in the future. It should be noted that the previously shown examples are only applicable for inhomogeneities from which the difference in AS compared to the background is not too large. Inhomogeneities in soft biological tissue vary at most 10% in AS, but structures such as bone cannot be imaged accurately by these methods. For strongly scattering structures, more reliable results can be obtained by a method that also works with a passive element but set up in reflection mode (Dean-Ben et al. 2011b). The location of the scattering inhomogeneities can be used to reduce the artifacts that are generated by scattering.

Acknowledgment

P. van Es and S. Manohar are funded by the Netherlands Organization for health research and development (ZonMw) under the program New Medical Devices for Affordable Health.

References

Baeriswyl, V. and G. Christofori. 2009. The angiogenic switch in carcinogenesis. *Seminars in Cancer Biology* 19(5):329–37.

Beard, P. 2011. Biomedical photoacoustic imaging. *Interface Focus* 1(4):602–31.

Brecht, H.-P., R. Su, M. Fronheiser, S.A. Ermilov, A. Conjusteau, and A.A. Oraevsky. 2009. Whole-body three-dimensional optoacoustic tomography system for small animals. *Journal of Biomedical Optics* 14(6):064007.

Butrus, T.K.-Y. and O. Ömer. 2011. Capacitive micromachined ultrasonic transducers for medical imaging and therapy. *Journal of Micromechanics and Microengineering* 21(5):054004.

Carmeliet, P. 2005. Angiogenesis in life, disease and medicine. *Nature* 438(7070):932–36.

Deán-Ben, X.L., R. Daniel, and N. Vasilis. 2011. The effects of acoustic attenuation in optoacoustic signals. *Physics in Medicine and Biology* 56(18):6129.

Dean-Ben, X.L., V. Ntziachristos, and D. Razansky. 2011a. Time-shifting correction in optoacoustic tomographic imaging for media with non-uniform speed of sound. Paper presented at the Proceedings of SPIE.

Dean-Ben, X.L., D. Razansky, and V. Ntziachristos. 2011b. Measurement of the acoustic scatterers distribution within the imaged sample in an optoacoustic tomographic setup. Paper presented at the Proceedings of SPIE.

Emelianov, S.Y., S.R. Aglyamov, J. Shah, S. Sethuraman, W.G. Scott, R. Schmitt, M. Motamedi, A. Karpiouk, and A.A. Oraevsky. 2004. Combined ultrasound, optoacoustic, and elasticity imaging. Paper presented at the Proceedings of SPIE.

Erpelding, T.N., C. Kim, M. Pramanik, L. Jankovic, K. Maslov, Z. Guo, J.A. Margenthaler, M.D. Pashley, and L.V. Wang. 2010. Sentinel lymph nodes in the rat: Noninvasive photoacoustic and us imaging with a clinical us system1. *Radiology* 256(1):102–10.

Erpelding, T.N., Y. Wang, L. Jankovic, Z. Guo, J.-L. Robert, G. David, C. Kim, and L.V. Wang. 2011. Three-dimensional photoacoustic imaging with a clinical two-dimensional matrix ultrasound transducer. Paper presented at the Proceedings of SPIE.

Esenaliev, R.O., A.A. Oraevsky, V.S. Letokhov, A.A. Karabutov, and T.V. Malinsky. 1993. Studies of acoustical and shock waves in the pulsed laser ablation of biotissue. *Lasers in Surgery and Medicine* 13(4):470–84.

Fletcher, R. 1987. *Practical Methods of Optimization*. 2nd ed. Chichester: Wiley.

Fronheiser, M.P., S.A. Ermilov, H.-P. Brecht, A. Conjusteau, R. Su, K. Mehta, and A.A. Oraevsky. 2010. Real-time optoacoustic monitoring and three-dimensional mapping of a human arm vasculature. *Journal of Biomedical Optics* 15(2):021305.

Gamelin, J., A. Aguirre, A. Maurudis, F. Huang, D. Castillo, L.V. Wang, and Q. Zhu. 2008. Curved array photoacoustic tomographic system for small animal imaging. *Journal of Biomedical Optics* 13(2):024007.

Haisch, C., K. Eilert-Zell, M. Vogel, P. Menzenbach, and R. Niessner. 2010. Combined optoacoustic/ultrasound system for tomographic absorption measurements: Possibilities and limitations. *Analytical and Bioanalytical Chemistry* 397(4):1503-10-10.

Heijblom, M., D. Piras, W. Xia, J.C.G. van Hespen, J.M. Klaase, F.M. van den Engh, T.G. van Leeuwen, W. Steenbergen, and S. Manohar. 2012. Visualizing breast cancer using the Twente photoacoustic mammoscope: What do we learn from twelve new patient measurements? *Optics Express* 20(11): 11582–11597.

Hoelen, C.G.A., F.F.M. De Mul, R. Pongers, and A. Dekker. 1998. Three-dimensional photoacoustic imaging of blood vessels in tissue. *Optics Letters* 23(8):648–50.

Hollman, K.W., K.W. Rigby, and M. O'donnell. 1999. Coherence factor of speckle from a multi-row probe. Paper presented at the 1999 IEEE Ultrasonics Symposium Proceedings, 1999.

Hu, S., K. Maslov, and L.V. Wang. 2011. Second-generation optical-resolution photoacoustic microscopy with improved sensitivity and speed. *Optics Letters* 36(7):1134–36.

Jaeger, M., S. Preisser, M. Kitz, D. Ferrara, S. Senegas, D. Schweizer, and M. Frenz. 2011. Improved contrast deep optoacoustic imaging using displacement-compensated averaging: Breast tumour phantom studies. *Physics in Medicine and Biology* 56(18):5889.

Jaeger, M., L. Siegenthaler, M. Kitz, and M. Frenz. 2009. Reduction of background in optoacoustic image sequences obtained under tissue deformation. *Journal of Biomedical Optics* 14(5):054011.

Jansen, K., A.F.W. Van Der Steen, H.M.M. Van Beusekom, J.W. Oosterhuis, and G. Van Soest. 2011. Intravascular photoacoustic imaging of human coronary atherosclerosis. *Optics Letters* 36(5):597–99.

Jin, X. and L.V. Wang. 2006. Thermoacoustic tomography with correction for acoustic speed variations. *Physics in Medicine and Biology* 51(24):6437–48.

Jose, J., D.J. Grootendorst, T.W. Vijn, M. Wouters, H.V. Boven, T.G.V. Leeuwen, W. Steenbergen, T.J.M. Ruers, and S. Manohar. 2011a. Initial results of imaging melanoma metastasis in resected human lymph nodes using photoacoustic computed tomography. *Journal of Biomedical Optics* 16(9):096021.

Jose, J., R.G.H. Willemink, S. Resink, D. Piras, J.C.G. Hespen Van, C.H. Slump, W. Steenbergen, T.G. Leeuwen Van, and S.

Manohar. 2011b. Passive element enriched photoacoustic computed tomography (PER PACT) for simultaneous imaging of acoustic propagation properties and light absorption. *Optics Express* 19(3):2093–104.

Kim, C., T.N. Erpelding, L. Jankovic, M.D. Pashley, and L.V. Wang. 2010. Deeply penetrating *in vivo* photoacoustic imaging using a clinical ultrasound array system. *Biomedical Optics Express* 1(1):278–84.

Kim, C., T.N. Erpelding, L. Jankovic, and L.V. Wang. 2011. Combined ultrasonic and photoacoustic system for deep tissue imaging. Proceedings of SPIE 7899, no. 1:789935.

Kolkman, R.G.M., P.J. Brands, W. Steenbergen, and T.G.V. Leeuwen. 2008. Real-time *in vivo* photoacoustic and ultrasound imaging. *Journal of Biomedical Optics* 13(5):050510.

Kolkman, R.G.M., P.J. Brands, W. Steenbergen, and T.G.V. Leeuwen. 2009. Real-time photoacoustic and ultrasound imaging of human vasculature. *Journal of Biomedical Optics* 13(5):050510.

Kostli, K.P., and P.C. Beard. 2003. Two-dimensional photoacoustic imaging by use of Fourier-transform image reconstruction and a detector with an anisotropic response. *Applied Optics* 42(10):1899–1908.

Kruger, R.A., R.B. Lam, D.R. Reinecke, S.P.D. Rio, and R.P. Doyle. 2010. Photoacoustic angiography of the breast. *Medical Physics* 37(11):6096–100.

Ku, G., X. Wang, X. Xie, G. Stoica, and L.V. Wang. 2005. Imaging of tumor angiogenesis in rat brains *in vivo* by photoacoustic tomography. *Applied Optics* 44(5):770–75.

Lam, R.B., R.A. Kruger, D.R. Reinecke, S.P. Delrio, M.M. Thornton, P.A. Picot, and T.G. Morgan. 2010. Dynamic optical angiography of mouse anatomy using radial projections. Proceedings of SPIE 7564, no. 1:756405.

Larina, I.V., K.V. Larin, and R.O. Esenaliev. 2005. Real-time optoacoustic monitoring of temperature in tissues. *Journal of Physics D: Applied Physics* 38(15):2633.

Lee, K. 2011. Optical mammography: Diffuse optical imaging of breast cancer. *World Journal of Clinical Oncology* 2(1): 64–72.

Lin, Y., D. Thayer, O. Nalcioglu, and G. Gulsen. 2011. Tumor characterization in small animals using magnetic resonance -guided dynamic contrast enhanced diffuse optical tomography. *Journal of Biomedical Optics* 16(10):106015.

Manohar, S., R.G.H. Willemink, F. Van Der Heijden, C.H. Slump, and T.G. Van Leeuwen. 2007. Concomitant speed-of-sound tomography in photoacoustic imaging. *Applied physics letters* 91(13):131911–132100.

Mishchenko, M.I., L.D. Travis, and A.A. Lacis. 2002. *Scattering, Absorption, and Emission of Light by Small Particles*: Cambridge University Press, Cambridge.

Moseley, M. and G. Donnan. 2004. Multimodality imaging. *Stroke* 35(11, Suppl 1):2632–34.

Nachabe, R., B.H. Hendriks, M. Van Der Voort, A.E. Desjardins, and H.J. Sterenborg. 2010. Estimation of biological chromophores using diffuse optical spectroscopy: Benefit of extending the uv-vis wavelength range to include 1000 to 1600 nm. *Biomedical Optics Express* 1(5):1432–42.

Niederhauser, J.J., M. Jaeger, R. Lemor, P. Weber, and M. Frenz. 2005. Combined ultrasound and optoacoustic system for real-time high-contrast vascular imaging *in vivo*. *IEEE Transactions on Medical Imaging* 24(4):436–40.

Paige, C.C. and M.A. Saunders. 1982. Algorithm 583: Lsqr: Sparse linear equations and least squares problems. *ACM Transactions on Mathematical Software* 8(2):195–209.

Park, S., J. Shah, S.R. Aglyamov, A.B. Karpiouk, S. Mallidi, A. Gopal, H. Moon, X.J. Zhang, W.G. Scott, and S.Y. Emelianov. 2006. Integrated system for ultrasonic, photoacoustic and elasticity imaging. Proceedings of SPIE 6147, no. 1:61470H.

Patrickeyev, I. and A.A. Oraevsky. 2003. Multiresolution reconstruction method to optoacoustic imaging. Paper presented at the Proceedings of SPIE.

Petrov, I.Y., Y. Petrov, D.S. Prough, and R.O. Esenaliev. 2011. High-resolution ultrasound imaging and noninvasive optoacoustic monitoring of blood variables in peripheral blood vessels. Paper presented at the Proceedings of SPIE.

Piras, D., X. Wenfeng, W. Steenbergen, T.G. Van Leeuwen, and S.G. Manohar. 2010. Photoacoustic imaging of the breast using the twente photoacoustic mammoscope: Present status and future perspectives. *IEEE Journal of Selected Topics in Quantum Electronics* 16(4):730–39.

Resink, S., J. Jose, R.G.H. Willemink, C.H. Slump, W. Steenbergen, T.G. Van Leeuwen, and S. Manohar. 2011. Multiple passive element enriched photoacoustic computed tomography. *Optics Letters* 36(15):2809–11.

Schüle, G., G. Hüttmann, C. Framme, J. Roider, and R. Brinkmann. 2004. Noninvasive optoacoustic temperature determination at the fundus of the eye during laser irradiation. *Journal of Biomedical Optics* 9(1):173–79.

Sethuraman, S., J.H. Amirian, S.H. Litovsky, R.W. Smalling, and S.Y. Emelianov. 2007. *Ex vivo* characterization of atherosclerosis using intravascular photoacoustic imaging. *Optics Express* 15(25):16657–66.

Song, L., K. Maslov, R. Bitton, K.K. Shung, and L.V. Wang. 2008. Fast 3-D dark-field reflection-mode photoacoustic microscopy *in vivo* with a 30-mhz ultrasound linear array. *Journal of Biomedical Optics* 13(5):054028.

Te-Jen, M., S.R. Kothapalli, S. Vaithilingam, O. Oralkan, A. Kamaya, I.O. Wygant, Z. Xuefeng, S.S. Gambhir, R.B. Jeffrey, and B.T. Khuri-Yakub. 2010. 3-D deep penetration photoacoustic imaging with a 2-D CMUT array. Paper presented at the Ultrasonics Symposium (IUS), 2010 IEEE, 11–14 Oct. 2010.

Vaithilingam, S., T.J. Ma, Y. Furukawa, I.O. Wygant, Z. Xuefeng, A. De La Zerda, O. Oralkan, A. Kamaya, S.S. Gambhir, R.B. Jeffrey, and B.T. Khuri-Yakub. 2009. Three-dimensional photoacoustic imaging using a two-dimensional CMUT array. *IEEE Transactions on Ultrasonics, Ferroelectrics and Frequency Control* 56(11):2411–19.

Van Veen, R.L.P., A. Amelink, M. Menke-Pluymers, C. Van Der Pol, and H.J.C.M. Sterenborg. 2005. Optical biopsy of breast tissue using differential path-length spectroscopy. *Physics in Medicine and Biology* 50(11):2573.

Wang, L.V. 2008. Tutorial on photoacoustic microscopy and computed tomography. *IEEE Journal of Selected Topics in Quantum Electronics* 14(1):171–79.

Wang, X., D.L. Chamberland, J.B. Fowlkes, P.L. Carson, and D.A. Jamadar. 2008a. Photoacoustic tomography of small-animal and human peripheral joints. Proceedings of SPIE 6856, no. 1:685604.

Wang, X., D.L. Chamberland, and D.A. Jamadar. 2007. Noninvasive photoacoustic tomography of human peripheral joints toward diagnosis of inflammatory arthritis. *Optics Letters* 32(20):3002–04.

Wang, X., J.B. Fowlkes, P.L. Carson, and L. Mo. 2008b. Experimental evaluation of a high-speed photoacoustic tomography system based on a commercial ultrasound unit. Paper presented at the 2008 IEEE International Ultrasonics Symposium (IUS), November 2–5, 2008.

Wang, X., Y. Pang, G. Ku, G. Stoica, and L.V. Wang. 2003. Three-dimensional laser-induced photoacoustic tomography of mouse brain with the skin and skull intact. *Optics Letters* 28(19):1739–41.

Willemink, R.G.H., S. Manohar, Y. Purwar, C.H. Slump, F.V.D. Heijden, and T.G.V. Leeuwen. 2008. Imaging of acoustic attenuation and speed of sound maps using photoacoustic measurements. Paper presented at the Proceedings of SPIE.

Xu, M. and L.V. Wang. 2005. Universal back-projection algorithm for photoacoustic computed tomography. *Physical Review E* 71(1):016706.

Yang, J.-M., K. Maslov, H.-C. Yang, Q. Zhou, K.K. Shung, and L.V. Wang. 2009. Photoacoustic endoscopy. *Optics Letters* 34(10):1591–93.

Yang, X., H. Gong, G. Quan, Y. Deng, and Q. Luo. 2010. Combined system of fluorescence diffuse optical tomography and microcomputed tomography for small animal imaging. *Review of Scientific Instruments* 81(5):054304–8.

Yao, J., K.I. Maslov, and L.V. Wang. 2011. Noninvasive quantification of metabolic rate of oxygen (mro2) by photoacoustic microscopy. Proceedings of SPIE 7899, no. 1:78990N.

Zhang, C., K. Maslov, and L.V. Wang. 2010. Subwavelength-resolution label-free photoacoustic microscopy of optical absorption *in vivo*. *Optics Letters* 35(19):3195–97.

Zhang, E.Z., J. Laufer, and P. Beard. 2008. Backward-mode multiwavelength photoacoustic scanner using a planar fabry-perot polymer film ultrasound sensor for high-resolution three-dimensional imaging of biological tissues. *Applied Optics* 47(4):561–77.

Zhang, E.Z., B. Povazay, J. Laufer, A. Alex, B. Hofer, B. Pedley, C. Glittenberg, B. Treeby, B. Cox, P. Beard, and W. Drexler. 2011. Multimodal photoacoustic and optical coherence tomography scanner using an all optical detection scheme for 3d morphological skin imaging. *Biomedical Optics Express* 2(8):2202–15.

Zhang, H.F., K. Maslov, G. Stoica, and L.V. Wang. 2006. Functional photoacoustic microscopy for high-resolution and noninvasive *in vivo* imaging. *Nature Biotechnology* 24(7):848–51.

Zhu, Q., T. Durduran, V. Ntziachristos, M. Holboke, and A.G. Yodh. 1999. Imager that combines near-infrared diffusive light and ultrasound. *Optics Letters* 24(15):1050–52.

Multimodality Imaging: Photoacoustic Tomography/Diffuse Optical Tomography

Adam Q. Bauer
Washington University School of Medicine

Ralph E. Nothdurft
Washington University School of Medicine

Todd N. Erpelding
Philips Research North America

Lihong V. Wang
Washington University

Joseph P. Culver
Washington University School of Medicine

16.1 Introduction

Photoacoustic imaging (PAI) is a relatively new imaging modality that blends the strong optical absorption contrast of biological tissue with the high spatial resolution capabilities of ultrasound. The high contrast provided by endogenous chromophores alone has allowed PAI to image anatomy (Maslov et al. 2008), brain structure (Wang et al. 2003), functional organization of the cerebral cortex (Wang et al. 2003; Zhang et al. 2006; Liao, Li et al. 2010), and to detect the breast cancer in humans (Oraevsky et al. 2001; Manohar et al. 2007) and melanoma cells in rats (Zhang et al. 2006). With the advent of bioconjugated, tunable optical contrast agents (e.g., gold nanoparticles or carbon nanotubes), molecular PAI is also possible (Copland et al. 2004; Portnov et al. 2005; De La Zerda et al. 2008; Li et al. 2008; Kim et al. 2009; Pan et al. 2009; Xiang et al. 2009; Kim et al. 2010; Ntziachristos and Razansky 2010). However, the ability to interpret the molecular or functional contrast (the photoacoustic images) depends on the reliability of the PAI absorption spectroscopy (Laufer et al. 2005). If the absorption spectrum of a chromophore is known, then it is possible to calculate the concentration. For example, if the optical absorption is dominated by hemoglobin, then the absorption coefficient can be written as

$$\mu_\alpha(x,\lambda) = \varepsilon_{HbO_2}(\lambda)[HbO_2] + \varepsilon_{HbR}(\lambda)[HbR]. \quad (16.1)$$

Traditional spectroscopy requires the measurements at multiple wavelengths to invert the above equation to arrive at the concentrations of oxy- and deoxyhemoglobin. Current photoacoustic spectroscopic methods assume that the recovered initial pressure distributions (the PA data) reflect accurately (at least to within a multiplicative constant) the absorption properties of the medium under study, but his may not always be a safe assumption. Without a light transport model, the fluence is unknown, and the fluence depends on the chromophore concentrations that one is attempting to recover in addition to depending on the absorption and scattering spectra of everything that the light has passed through to get to the point of interest (Cox et al. 2008, 2009). Because the relative photoacoustic (PA) signal is not an intrinsic property of a medium, but is instead a product of the optical absorption coefficient (the quantity of interest) and the local light fluence, spatial and spectral inhomogeneities in the fluence may undermine the spectral interpretation of PA images.

A noninvasive solution to this problem is to combine PAI with diffuse optical tomography (DOT). DOT is a clinically relevant imaging technology enabling researchers to study physiological processes (e.g., metabolism (Culver, Durduran et al. 2003) and hemodynamics (McBride et al. 1999; Culver, Durduran et al. 2003; Culver, Siegel et al. 2003; Zhou et al. 2006)), and is capable of reconstructing quantitative maps of optical properties (Culver, Siegel et al. 2003; Patwardhan and Culver 2008), albeit at lower resolution compared with PAI. DOT reconstructions of scattering and absorption can be used in conjunction with diffuse light modeling to generate the fluence information required to improve the PA absorption spectroscopy.

Although there has been some success in extracting quantitative information from PA images, previous studies have largely been done in simulation (Ripoll and Ntziachristos 2005; Cox et al. 2006; Banerjee et al. 2008; Cox et al. 2009; Yuan and Jiang 2009; Bal and Uhlmann 2010; Zemp 2010), or applied iterative approaches using the PA image in conjunction with a light transport model to arrive at a least squares solution of the absorption coefficient by assuming uniform bulk optical properties (Cox et al. 2006; Yuan and Jiang 2006; Yin et al. 2007; Jetzfellner et al. 2009; Yao et al. 2009). In this chapter, we describe the use of a noninvasive hybrid imaging modality that combines PAI with DOT to circumvent sources of artifact in PAI. With our hybrid technique, DOT is used to recover the low-resolution absorption and reduced scattering coefficient maps of a tissue-mimicking phantom that was initially imaged in a PAI system. The optical properties and the nonuniform surface fluence pattern of the PAI system are input parameters to a light-tissue model that calculates the fluence throughout the phantom. This fluence distribution is then used to correct the PA image of the phantom, resulting in an accurate quantitative image of the absorption coefficient.

16.2 Materials and Methods

16.2.1 Tissue-Mimicking Phantoms

In order to evaluate the sources of artifact in PA images and the quantitative accuracy of our compensation algorithms, tissue-mimicking phantoms were designed with heterogeneous optical properties (Figure 16.1a). The agarose (Sigma-Aldrich, Saint Louis, MO), intralipid (20% fat emulsion, Fresenius Kabi, Germany), and India ink (Speedball, Statesville, NC) reference mixture (absorption coefficient $\mu_a = 0.06$ cm^{-1} at 780 nm and 0.07 cm^{-1} at 650 nm, and reduced scattering coefficient $\mu_s' = 5$ cm^{-1} at 780 nm and 7.2 cm^{-1} at 650 nm) is poured into a mold and allowed to solidify at room temperature. Within this reference mixture, two types of inclusions were studied: 500 μm inner-diameter capillary tubes (BD Intramedic PE 50, Ontario, Canada) filled with 2% India ink ($\mu_a = 30$ cm^{-1} at 780 nm, 36 cm^{-1} at 650 nm) placed at a depth of 12 mm, and larger (length × width × height, 10 cm × 1 cm × 0.2 cm) rectangular scattering (4 × background) and absorption (5 × background) perturbations placed 4 mm deep in the phantom, 8 mm above the tubes. The larger agarose inclusions were designed to aberrate the fluence profile while maintaining the physiological values for absorption and scattering contrasts.

16.2.2 Photoacoustic Imaging

The PAI system used in this experiment (Figure 16.1b) was modified from a clinical ultrasound (US) array system (iU22, Philips Healthcare, Andover, MA) and is described in a previous publication (Erpelding et al. 2010). Briefly, the original channel board architecture of the US imaging system was modified to acquire raw, per-channel PA and US data. Raw radiofrequency data were transferred to a data acquisition computer where postprocessing was performed. The data acquisition system controlled the

laser firing and optical-wavelength tuning. PA images were processed using Fourier beam-forming reconstruction (Kostli et al. 2001), and displayed at ~1 fps. The maximum data acquisition rate is 10 fps, limited by the current laser repetition rate. A linear 128-element US probe with a nominal bandwidth of 4–8 MHz (L8-4, Philips Healthcare) was physically integrated with a bifurcated optical fiber bundle (Light Guide Optics, Los Angeles, CA), forming a hand-held probe. Laser pulses with a 6.5 ns pulse duration and 10 Hz repetition rate were generated from a tunable dye laser (Precision Scan-P, Sirah, Germany) tuned to 650 nm and pumped by a Q-switched Nd:YAG laser (Quanta-Ray Pro-350, Newport, Irvine, CA). The US probe and incident fluence were coupled to the phantom via water bath. Twenty-one sites along the phantom were investigated in the orientation shown in Figure 16.1b by translating the integrated US probe/fiber bundle in 3 mm increments along the targets. At each location, 100 frames were collected and averaged to improve the signal-to-noise ratio in the PA images. The US signals collected by the probe extend through 5 cm in depth, but the PA images reported have been cropped to a depth of 1.8 cm to only display the relevant phantom information.

16.2.3 Diffuse Optical Tomography

The details of the experimental DOT system (Figure 16.1c) can be found in a previous publication (Patwardhan and Culver 2008). Briefly, a mode-locked Ti:Sapphire laser (pulse width <100 fs, pulse repetition rate = 80 MHz, MTS, Kapteyn-Murnane Labs, Boulder, CO) operating at 780 nm peak wavelength is pumped by a 6 W, 532 nm DPSS laser (Verdi, Coherent, Santa Clara, CA) and illuminates the phantom in a transmission geometry. The beam is steered by an x–y galvanometer pair (AO, Model 6230, Cambridge Technology, Lexington, MS) to illuminate the imaging chamber at 26 locations separated by 2 mm. An ultrafast image intensifier (PicoStarHR-12, LaVision, Inc., Ypsilanti, MI) relays time-gated images of the transmitted light levels on the detection plane to an EMCCD camera (iXon 877 f, Andor Technologies, South Windsor, CT). In this configuration, 48 time gates (400 ps wide, separated by 50 ps) sampled the transmitted light pulse at each of the 26 sites. Fourier transformation of the time-domain data provides frequency-domain information from 156 MHz to 1.6 GHz. For the reconstructions reported here, we used 313 MHz; this modulation frequency was found to provide an adequate balance of contrast between the absorption and scattering inclusions (Patwardhan and Culver 2008).

Differential image reconstructions were obtained using a linear Rytov approximation approach. In this scheme, the total diffuse fluence, φ is written as

$$\varphi = \varphi_0 e^{\varphi_{scat}}. \qquad (16.2)$$

The total fluence φ consists of a background field, φ_0, which depends on background optical properties, and a perturbed field, φ_{scat}, which is linearly related to a set of spatial variations in the optical properties μ_α and D, the diffusion coefficient. D

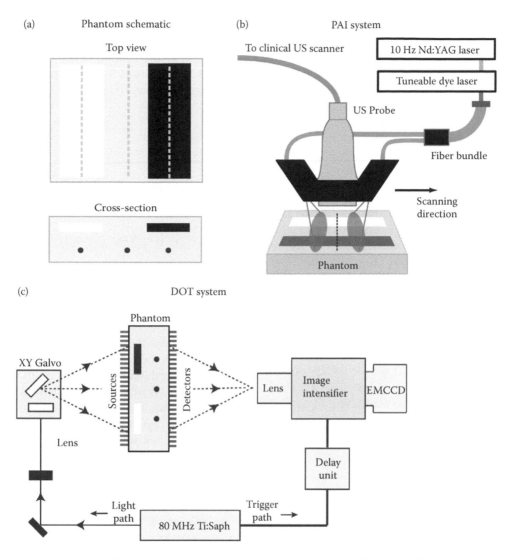

FIGURE 16.1 Experimental layout: (a) Schematic of phantom showing the cross-sectional and overhead distribution of imbedded targets. Two large rectangular targets provide absorption and scattering contrast (5× background and 4× background, respectively) and produce inhomogeneous fluence profiles for the three identical deeper capillary tubes. (b) Subset of PAI system: fiber-coupled light from a pumped dye laser irradiates the surface of the phantom in a dark-field illumination configuration. PA signals are acquired by a clinical ultrasound linear array. Twenty-one PA images were acquired by translating the probe in the direction shown; vertical dotted line under ultrasonic probe marks approximate location of PA image with respect to the probe. (c) Time-domain DOT system: a pulsed source beam is steered by a pair of galvanometer scanning mirrors to the source side of an imaging cassette. Light emitted from the detector plane of the cassette is collected by a lens and temporally gated by an ultrafast-gated image intensifier and detected by an EMCCD camera.

and reduced scattering coefficient, $3\mu_s'$, are related by $D = v/3\mu_s'$, where v is the speed of light in the medium. Experimental measurements of φ from the sample surface lead to images of spatially varying absorption and scattering via solution of a least-squares problem. In the work described in this chapter, absorption and scattering perturbations were reconstructed from intensity data. The scattered field was modeled using a linear Rytov approximation, $y = Ax$. In this formulation, y is the scattered field:

$$y_i = \ln\left[\frac{\varphi(\mathbf{r}_{s(i)}, \mathbf{r}_{d(i)}, \omega)}{\varphi_0(\mathbf{r}_{s(i)}, \mathbf{r}_{d(i)}, \omega)}\right]. \quad (16.3)$$

$A = [W^a W^D]$, where W^a and W^D represent the sensitivity functions for the absorption and diffusion coefficients:

$$W_{ij}^a = -\frac{vh^3}{D_0}\frac{G(\mathbf{r}_{s(i)}, \mathbf{r}_j, \omega_i)G(\mathbf{r}_j, \mathbf{r}_{d(i)}, \omega_i)}{G(\mathbf{r}_{s(i)}, \mathbf{r}_{d(i)}, \omega_i)}.$$

$$\quad (16.4)$$

$$W_{ij}^D = \frac{h^3}{D_0}\frac{\nabla G(\mathbf{r}_{s(i)}, \mathbf{r}_j, \omega_i)\nabla G(\mathbf{r}_j, \mathbf{r}_{d(i)}, \omega_i)}{G(\mathbf{r}_{s(i)}, \mathbf{r}_{d(i)}, \omega_i)}.$$

h^3 is the voxel volume, $\mathbf{r}_{s(i)}$ and $\mathbf{r}_{d(i)}$ are the positions of the ith source and jth detector, respectively, and G is Green's function solution to the photon diffusion equation, and $x = [\Delta\mu_a(r)\ \Delta D(r)]^T$ is the optical property map. Diffuse photons were modeled with frequency-domain diffuse photon density waves using an extrapolated zero-boundary, semi-infinite media solution for source–detector plane separations of 1.8 cm. Separating the real (R) and imaginary (I) components, the forward problem takes the form:

$$\begin{bmatrix} Y_R \\ Y_I \end{bmatrix} = \begin{bmatrix} W_{aR} & W_{aI} \\ W_{DR} & W_{DI} \end{bmatrix} \begin{bmatrix} x_a \\ x_D \end{bmatrix}. \tag{16.5}$$

Because the absorption and diffusion variables differ by orders of magnitude, the Jacobian matrix needs to be preconditioned to reduce the crosstalk between $\Delta\mu_a(r)$ and $\Delta D(r)$. The absorption and diffusion variables are normalized prior to inversion using the substitution $\tilde{x} = Lx$, where

$$L = \begin{bmatrix} \| W_{aR}W_{aI} \| & 0 \\ 0 & \| W_{DR}W_{DI}\| \end{bmatrix}. \tag{16.6}$$

Measurement noise is also incorporated into the forward model by the substitution of variables, $\tilde{y} = Cy$ and $\tilde{A} = CAL^{-1}$, where the diagonal of C is a shot noise model. With these modifications, we invert $\tilde{y} = \tilde{A}\tilde{x}$ using Tikhonov regularization:

$$\min\left\{\| \tilde{y} - \tilde{A}\tilde{x} \|_2^2 + \alpha \| Q\tilde{x} \|_2^2 \right\}. \tag{16.7}$$

The penalty term for image variance, $\alpha Q\tilde{x}_2^2$ is a pseudonorm, and depth-dependent regularization was used where the diagonal of $Q = \mathrm{sqrt}\left[\mathrm{diag}(\tilde{A}^T\tilde{A})\right] + \alpha$. A solution, $\tilde{x} = A_\alpha^\# \tilde{y}$ was obtained using a Moore–Penrose generalized inverse with $A_\alpha^\# = Q^{-1}\tilde{A}^T(\tilde{A}^T\tilde{A} + \alpha I)^{-1}\tilde{y}$, where $\tilde{A}^T = \tilde{A}L^{-1}$. The value of α was optimized to provide even imaging across the field of view. Maps of the diffusion coefficient were converted to those of the reduced scattering coefficient after reconstruction.

16.2.4 Hybrid DOT-PAI

A photoacoustic image directly reports the acoustic pressure distribution, $p_0(x,\lambda)$, arising from localized optical absorption. The absorbed optical energy density, $h(x,\lambda)$, and $p_0(x,\lambda)$, are related by

$$p_0(x,\lambda) = \Gamma(x)h(x,\lambda), \tag{16.8}$$

where the Grüneisen parameter, $\Gamma(x)$, is a thermodynamic property of the tissue (Duck 1990). The absorbed optical energy density, $h(x,\lambda)$, is equal to the optical fluence distribution, $\varphi(x,\lambda)$, multiplied by the optical absorption coefficient, $\mu_a(x,\lambda)$, within the irradiated medium, that is

$$h(x,\lambda) = \varphi(x,\lambda)\mu_a(x,\lambda). \tag{16.9}$$

PAI directly reconstructs the pressure field, $p_0(x,\lambda)$, which is proportional to both the optical absorption coefficient and optical fluence distribution within the irradiated medium. Rearranged, Equation 16.9 can be expressed as a formula for the absorption coefficient:

$$\mu_a(x,\lambda) = \frac{1}{\Gamma(x)} \frac{p_0(x,\lambda)}{\varphi(x,\lambda)}. \tag{16.10}$$

In our approach, the continuous-wave fluence in the phantom was calculated by solving the diffuse photon density wave equation

$$\nabla \cdot (D(r)\nabla\varphi(r)) - \nu\mu_a(r)\varphi(r) = -\nu S(r) \tag{16.11}$$

using a finite difference (FD) method (Holboke et al. 2000). Inputs into the calculation were the spatially varying absorption and scattering maps of the DOT reconstructions, and the surface illumination pattern of the PAI system for the source term, $S(r)$. Because the DOT and PAI systems operate at different wavelengths, the absorption and scattering coefficients were spectrally mapped from 780 to 650 nm. The absorption properties of the India ink were fully characterized from 400 to 900 nm with a spectrophotometer (DU640, Beckman Coulter, Inc., Brea, CA). For the background concentrations used in the phantom, μ_a of the India ink was measured to be 0.07 cm^{-1} at 650 nm and 0.06 cm^{-1} at 780 nm, while μ_s' of the intralipid is 5 cm^{-1} at 780 nm and calculated to be 7.2 cm^{-1} at 650 nm using Mie theory (Michels et al. 2008). The scattering inclusion in the phantom was mixed to have a reduced scattering coefficient 4× that of the background, and the large absorption inclusion and was designed to be 5× the background absorption. After reconstructing the optical property maps at 780 nm with the DOT system, the absorption image was scaled by the ratio of the absorption coefficient at 650 and 780 nm and the reduced scattering images were scaled by the ratio of the reduced scattering coefficient at 650 and 780 nm. Although it is unfortunate that the time-domain DOT system did not operate over the same wavelength range as the PAI system, this difference in wavelength should not pose a significant problem to the methodology as both intralipid and India ink are well characterized and relatively spectrally flat (Canpolat and Mourant 2000; Michels et al. 2008) (i.e., the absorption and scattering properties monotonically decrease with increasing wavelength, with no peaks or troughs over the spectral region investigated).

16.3 Results

The PAI system reconstructs a cross-sectional image (depth vs. azimuth) of the phantom at the dashed line between the two elliptical illumination patterns of the fiber bundles in Figure 16.1b). The PA reconstruction (Figure 16.2a) shows with high resolution all three capillary tubes and the large absorbing target. However, PAI cannot reconstruct the scattering

(a)

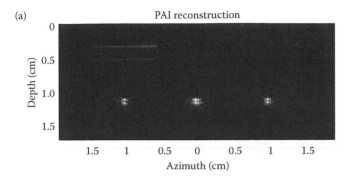

(b)

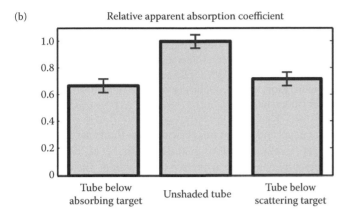

FIGURE 16.2 Raw photoacoustic reconstruction: (a) PA image showing a cross section of the phantom. Note the complete absence of the scattering target in addition to the different signal magnitudes of the three 1.2 mm deep, optically identical capillary tubes. (b) Volume-integrated PA signal magnitude of the three capillary tubes, normalized to the middle tube. Due to the inhomogeneous fluence distribution within the bulk of the phantom, the three capillary tubes produce PA signals of different magnitudes. The mean PA signal of the two outer, shaded tubes differs in magnitude with the unshaded middle tube by ~33%. Errors bars represent the standard deviation of the mean PA signal magnitude of each tube measured at the 21 sites (standard deviation for each tube is between 7% and 8% of the mean).

perturbations. PA signals are not generated from optical scatterers, so the rectangular scattering target does not provide any source of photoacoustic contrast in the PA image. Two artifacts of PAI can also be seen in Figure 16.2a: (1) Pressure fields generated outside the effective bandwidth of the transducer are not detected. This results in the spatial-derivative-like appearance of the objects detected by this system (most notably the larger rectangular absorber near the surface). The top and bottom surfaces of the large absorbing target and the smaller capillary tubes are the only portions that produce US frequency components within the nominal 4–8 MHz bandwidth of the US probe. (2) The PA signal from the three optically identical capillary tubes is significantly different in magnitude, presumably due to the uneven fluence profile created by the presence of the two shallower absorbing and scattering targets above, as well as the incident shape of the surface fluence pattern. Artifact (2) is characterized in Figure 16.2b, where the signals from each

target are volume integrated (FWHM, with an equal number of pixels included in the integral) and normalized to the middle tube. Over the 21 sites investigated, the mean PA signals from the two shaded outer targets are as much as 33% less than the mean PA signal of the middle target. In addition, the PA signal emerging from the bottom surface of the rectangular absorber (~5 mm deep in Figure 16.2a) is less than the signal from the top surface (~3 mm deep). Even over this 2 mm length scale, fluence-related artifacts are apparent in a single target. By combining PAI with DOT, one can recover scattering contrasts (to which PAI is not sensitive) and address the fluence-related Artifact 2. Artifact 1, while not addressed in this study, can be reduced through wide-band ultrasound detection schemes (Karabutov et al. 2000; Zhang et al. 2008; Chen et al. 2009).

DOT reconstructions of the absorption and scattering targets (Figure 16.3a) depict very little crosstalk between the two contrasts. Although the resolution of the DOT reconstructions is markedly poorer than the PA image, the volume-integrated signals from the objects are quantitative with respect to background property values (Culver, Siegel et al. 2003; Patwardhan and Culver 2008). The surface fluence distribution of the PAI system (Figure 16.3b) was obtained by photographing from below a piece of white paper placed at the same distance from the transducer as the surface of the phantom imaged with this system. The lobed pattern of the source field in the PAI system is designed to minimize the surface PA signals from saturating the US probe (situated between the two lobes, not shown). This technique is effective at generating better images near the surface, but this pattern will also produce inhomogeneous fluence throughout the imaging domain.

The calculated fluence (Figure 16.3c) within the phantom using the optical property maps from the DOT reconstructions at the location of the PA image in Figure 16.2a clearly illustrates the inhomogeneities caused by the inclusions and the surface illumination pattern. It can be seen from a normalized line profile of the fluence at the depth of the capillary tubes (Figure 16.3d) that the reduction of the fluence at the location of the two shaded outer targets (located at −1 cm and 1 cm azimuthally) is approximately the same magnitude as the percent difference in the PA signals between the outer shaded tubes and the middle unshaded tube (Figure 16.2b).

The fluence-related artifacts present in the original PA image can be corrected by dividing the raw PA image in Figure 16.2a by the fluence image in Figure 16.3c pixel-by-pixel. From this corrected image (Figure 16.4a), the capillary tubes are all of equal brightness. It can also be seen that the top and bottom surfaces of the larger, shallower absorber are also of comparable brightness, though the overall brightness of this object is reduced from the original image because it is markedly lower in absorption compared with the capillary tubes. The fluence-compensated results are compared to the initial images using the volume-integrated signals over each capillary tube (Figure 16.4b). The average error between the outer shaded tubes is now within ~6% of the middle unshaded tube, that is, the error in the original PA image has been reduced by ~6×.

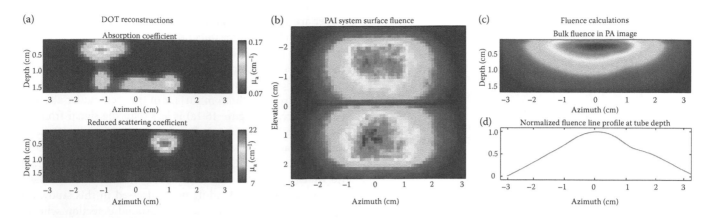

FIGURE 16.3 (**See color insert.**) DOT reconstruction data and fluence calculations: (a) DOT reconstructions of the absorption and scattering properties of the phantom. (b) Fluence pattern on the surface of the phantom during PA data collection. This source distribution and DOT reconstructions are used as the input to the finite difference solution of Equation 16.11. (c) Cross-sectional slice of the bulk fluence distribution calculated using the finite-difference method. (d) Fluence line profile at the depth of the capillary tubes (1.2 cm) showing azimuthal inhomogeneity. Note the reduction in fluence magnitude of ~33% at the location of the capillary tubes (±1 cm azimuth).

16.4 Discussion

Absorption spectroscopy is a powerful aspect of PAI, yet it is highly susceptible to the effects of fluence heterogeneity. In a small-animal imaging scenario, it is not uncommon to have orders-of-magnitude changes in light levels as one irradiates different sections of the animal (Patwardhan and Culver 2008). In the study described in this chapter, where the fluence varies over an order of magnitude within the phantom, we have shown that a reduction in the light level by only a third can produce appreciable misrepresentations of observed optical absorption. Because estimations of the absorption coefficient are linearly related to chromophore concentration, in this scenario, the concentrations would be underestimated by 33%. PAI without fluence correction is susceptible to artifacts caused by (1) structure in the illumination pattern, (2) attenuation of the light fluence due to bulk optical properties and determined by the tissue geometry, and (3) internal absorption and scattering heterogeneities. For accurate spectral PAI, these factors need to be addressed and appropriately corrected. This work demonstrates the feasibility of correcting PAI for all three types of fluence inhomogeneities using experimental data from a time-resolved DOT system and diffuse light modeling.

Previous experimental efforts to quantify the photoacoustic images with diffusing-light measurements (Yin et al. 2007) used the measured light intensity along the surface of a phantom as a constraint on an iterative solution to a fluence calculation; however, homogeneous absorption and scattering were assumed in estimating the distribution of optical fluence in the phantom studied. In addition, the absorption perturbations imaged with the PA system were relatively small, low-contrast targets that minimally affected the bulk fluence. Although it does not apply to iterative methods in general, some experimental investigations toward quantifying PA images using this technique to converge on optical property data have shown that too many iterations can produce deleterious effects (Jetzfellner et al. 2009), while

more novel and robust quantification methods using sparse signal representation of PA signals (Rosenthal et al. 2009) may still be beset by nonuniform surface illumination. The advantage of our method is that it properly accounts for both scattering and absorption targets and collectively solves for all three types of fluence inhomogeneities.

The compensation performed in this study assumed that the Grüneisen parameter, Γ, in Equations 16.8 and 16.10 was constant and part of the calibration. This assumption is not unreasonable in this particular study as the targets in the phantom are all made of the same material or materials having very similar mechanical and thermodynamic properties. However, applying this methodology to another medium (e.g., mouse brain) may require a different calibration factor, and provided the tissue types are not drastically different (such as the mechanical and thermodynamic properties of fat and blood (Duck 1990; Toubal et al. 1999)), the assumption that Γ remains approximately constant may still be valid.

Ideally, the images acquired from the DOT and PAI systems should be obtained using the same wavelength. This restriction was relaxed here because the spectral characteristics of the absorption and scattering contrasts (India ink and intralipid) are slowly varying over a 400–900 nm range so that the optical properties at the PAI wavelength could be predicted from the DOT data at a different wavelength (Canpolat and Mourant 2000; Michels et al. 2008). The DOT and PAI data sets were matched by mapping the optical properties acquired with the DOT system to those of the PAI system using the Mie theory for the intralipid, and quantitative broadband experimental spectroscopy for the India ink. Using the data from the spectrophotometer at 650 nm, this PAI system can be calibrated to the middle unshaded tube (which presumably represents the true absorption properties of all three tubes) to provide quantitative images of optical absorption. In moving this method forward to an *in vivo* setting, where the contrasts are more spectrally rich, it will be preferable to match the wavelengths of the DOT and PAI systems.

(a)

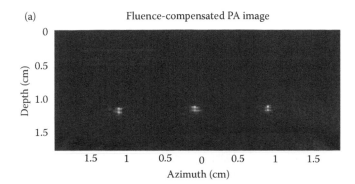

(b)

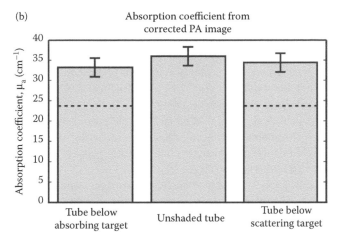

FIGURE 16.4 PA data from Figure 16.2 compensated by the DOT and surface fluence data shown in Figure 16.3 using Equation 16.10. (a) After DOT-assisted fluence correction, the three capillary tubes are all approximately of equal magnitude, and the top and bottom surfaces of the larger agarose inclusion are also of equal magnitude. (b) Volume-integrated absorption coefficient of the three capillary tubes. The discrepancy in the absorption coefficient in Figure 16.2 can be reduced to 6% between the tubes with this method. Horizontal dotted lines indicate what would be estimated as the magnitude of the absorption coefficient of the outer tubes from original uncompensated PA image of Figure 16.2a. Errors bars represent the standard deviation of the mean PA signal magnitude of each tube measured at the 21 sites (standard deviation for each tube is between 7% and 8% of the mean).

With DOT being used to compensate for PAI illumination, one strategy, that might be to have the DOT system, uses the same illumination structure as PAI. However that is not the case here. In either the DOT or PAI system, the type of illumination is chosen to maximize the sensitivity of the modality, with the structure of the interrogating light being best suited for the method used. Uniform, planar illumination is desired for many PAI applications, while discrete optodes are ideal for DOT measurements. For the compensation method presented here to work, it is not required that the DOT source pattern (discrete point sources) and the PAI source pattern (a lobed, dark-field surface pattern) have the same source illumination pattern. Provided the objects in the phantom have been sufficiently sampled by the DOT source grid, the optical property maps recovered from DOT measurements should be nearly unaffected by

the density of measurements (Culver et al. 2001; Graves et al. 2004). The illumination pattern of the PAI system is taken into account during the final diffuse light modeling of the fluence throughout the PA imaging domain.

The typical resolving capability of a DOT system, and therefore the resolution limit of the fluence estimate using DOT, is ~2–3 mm. However, the distribution of light fluence is, in general, comprised of low spatial frequency components—even high spatial frequency changes in μ_a (e.g., the transition from the bulk of the phantom to the rectangular absorbing target or capillary tubes) will not be observed in analysis of fluence profiles. Although the imaging resolution of DOT is relatively poorer than PA imaging, an advantage of using DOT to estimate the fluence profiles is that targets falling below the imaging resolution of the DOT system are still recovered, that is, the optical property maps become volume-averaged quantities (e.g., cortical activations occurring on the capillary bed), which can still be used to accurately estimate the light fluence. In a PA imaging system, objects falling below the imaging resolution of the US transducer (which depends on the bandwidth and center frequency of the probe) are not seen at all—as in the case of the bulk of the larger rectangular targets in the phantom, and the bulk of the phantom itself. In the case of a spatially rapid absorption change (e.g., a thrombus within a vessel during an ischemic event), the rapid change in blood flow at the occlusion will be spread over a relatively larger area in the DOT image, and more localized in a PA image. In this instance, a higher-resolution PAM image of smaller vessels (~10 micron), or a lower-resolution PAT image of larger vessels (100s of microns to mm), would each provide complementary data sets to the DOT image.

Here, we have demonstrated a method to correct PA images for fluence heterogeneities at a depth significantly deeper than 1 mm using DOT where each pixel in the 2D source pattern from the PA imaging system is modeled as an exponentially decaying collimated source within the diffusion approximation (Holboke et al. 2000). This method has been shown to represent more accurately the remitted flux of photons on the object's surface, as well as the flux of photons propagating forward in the object (Haskell et al. 1994; Boas et al. 2002). Because photoacoustic microscopy operates at distances near to and shallower than 1 mm, in this situation, it may be beneficial to utilize the light modeling methods based on the more accurate radiative transport equation (e.g., using Monte Carlo techniques). Although PA signals will always depend on the fluence profile of the illumination, PA signals originating from objects within tens to hundreds of microns of the surface may not be subject to the other "shading" types of fluence-related artifacts also addressed in this study.

For some clinical applications, it will likely be necessary to also have the DOT data set acquired with reflection-mode instrumentation. For example, in human neuroimaging, fluence distributions over many orders of magnitude have been detected and modeled with sufficient signal-to-noise ratio to be capable of reconstructing the endogenous chromophore concentrations in the adult (Boas et al. 2004; Zeff et al. 2007) and infant brain

(Liao, Gregg et al. 2010), as well as in brain-mimicking phantoms (Pogue et al. 2000; Dehghani et al. 2009). So, in principle, compensating PA images with DOT measurements should work equally well in an imaging scenario involving larger variations in light fluence. While the relative error in the data presented here was reduced by a factor of ~6 (33–6%), the remaining uncertainty in the optical absorption coefficient of the capillary tubes is most likely due to inhomogeneities in the optical properties of the ingredients in the phantom (the India Ink is a suspension of carbon powder, which is prone to aggregation), the lateral sensitivity of the US probe, and aberrations in the detected pressure field causing the phase cancellation artifacts at the US receiver surface (Bauer et al. 2008).

16.5 Conclusions

We have shown that quantitative PAI is possible when coupled with DOT. Traditional PAI may contain fluence-related errors that can render photoacoustic spectroscopy both quantitatively and qualitatively inaccurate. To compensate for PA images acquired in this study, low-resolution DOT reconstructions of a phantom's optical properties were used in conjunction with the surface fluence profile of the PAI system to numerically calculate the fluence throughout the phantom. This fluence distribution was then used to correct PA images, yielding quantitative information about targets 1.2 cm deep in the phantom. Before compensation, three optically identical PA targets were found to differ in PA signal magnitude by 33%. This considerable error was reduced to 6% with the methods described herein. These results motivate development of an integrated PAI-DOT system for concurrent *in vivo* imaging.

Acknowledgments

This work was supported in part by NIH grants R01-EB008085 and U54 CA136398.

References

Bal, G. and G. Uhlmann 2010. Inverse diffusion theory of photoacoustics. *Inverse Problems* 26: 085010.

Banerjee, B., S. Bagchi et al. 2008. Quantitative photoacoustic tomography from boundary pressure measurements: Noniterative recovery of optical absorption coefficient from the reconstructed absorbed energy map. *Journal of the Optical Society of America* 25: 2347–2356.

Bauer, A. Q., K. R. Marutyan et al. 2008. Negative dispersion in bone: The role of interference in measurements of the apparent phase velocity of two temporally overlapping signals. *The Journal of the Acoustical Society of America* 123: 2407–2414.

Boas, D., J. Culver et al. 2002. Three dimensional Monte Carlo code for photon migration through complex heterogeneous media including the adult human head. *Optics Express* 10: 159–170.

Boas, D. A., K. Chen et al. 2004. Improving the diffuse optical imaging spatial resolution of the cerebral hemodynamic response to brain activation in humans. *Optics Letters* 29: 1506–1508.

Canpolat, M. and J. R. Mourant 2000. Monitoring photosensitizer concentration by use of a fiber-optic probe with a small source-detector separation. *Applied Optics* 39: 6508–6514.

Chen, S. L., S. W. Huang et al. 2009. Polymer microring resonators for high-sensitivity and wideband photoacoustic imaging. *IEEE Transactions on Ultrasonics Ferroelectrics and Frequency Control* 56: 2482–2491.

Copland, J. A., M. Eghtedari et al. 2004. Bioconjugated gold nanoparticles as a molecular based contrast agent: Implications for imaging of deep tumors using optoacoustic tomography. *Molecular Imaging and Biology* 6: 341–349.

Cox, B. T., S. R. Arridge et al. 2009. Estimating chromophore distributions from multiwavelength photoacoustic images. *Journal of the Optical Society of America. A, Optics, Image Science, and Vision* 26: 443–455.

Cox, B. T., S. R. Arridge et al. 2008. Simultaneous estimation of chromophore concentration and scattering distributions from multiwavelength photoacoustic images. *SPIE* 6856: 68560Y-1–68560Y-12.

Cox, B. T., S. R. Arridge et al. 2006. Two-dimensional quantitative photoacoustic image reconstruction of absorption distributions in scattering media by use of a simple iterative method. *Applied Optics* 45: 1866–1875.

Culver, J. P., T. Durduran et al. 2003. Diffuse optical tomography of cerebral blood flow, oxygenation, and metabolism in rat during focal ischemia. *Journal of Cerebral Blood Flow and Metabolism* 23: 911–924.

Culver, J. P., V. Ntziachristos et al. 2001. Optimization of optode arrangements for diffuse optical tomography: A singular-value analysis. *Optics Letters* 26: 701–703.

Culver, J. P., A. M. Siegel et al. 2003. Volumetric diffuse optical tomography of brain activity. *Optics Letters* 28: 2061–2063.

De La Zerda, A., C. Zavaleta et al. 2008. Carbon nanotubes as photoacoustic molecular imaging agents in living mice. *Nature Nanotechnology* 3: 557–562.

Dehghani, H., B. R. White et al. 2009. Depth sensitivity and image reconstruction analysis of dense imaging arrays for mapping brain function with diffuse optical tomography. *Applied Optics* 48: D137–143.

Duck, F. 1990. *Physical Properties of Tissue: A Comprehensive Reference Book*. London: Academic Press.

Erpelding, T. N., C. Kim et al. 2010. Sentinel lymph nodes in the rat: Noninvasive photoacoustic and US imaging with a clinical US system. *Radiology* 256: 102–110.

Graves, E. E., J. P. Culver et al. 2004. Singular-value analysis and optimization of experimental parameters in fluorescence molecular tomography. *Journal of the Optical Society of America. A, Optics, Image Science, and Vision* 21: 231–241.

Haskell, R. C., L. O. Svaasand et al. 1994. Boundary conditions for the diffusion equation in radiative transfer. *Journal of the Optical Society of America. A, Optics, Image Science, and Vision* 11: 2727–2741.

Holboke, M. J., B. J. Tromberg et al. 2000. Three-dimensional diffuse optical mammography with ultrasound localization in a human subject. *Journal of Biomedical Optics* 5: 237–247.

Jetzfellner, T., D. Razansky et al. 2009. Performance of iterative optoacoustic tomography with experimental data. *Applied Physics Letters* 95: 013703.

Karabutov, A. A., E. V. Savateeva et al. 2000. Backward mode detection of laser-induced wide-band ultrasonic transients with optoacoustic transducer. *Journal of Applied Physics* 87: 2003–2014.

Kim, C., E. C. Cho et al. 2010. *In vivo* molecular photoacoustic tomography of melanomas targeted by bioconjugated gold nanocages. *ACS Nano* 4: 4559–4564.

Kim, J. W., E. I. Galanzha et al. 2009. Golden carbon nanotubes as multimodal photoacoustic and photothermal high-contrast molecular agents. *Nature Nanotechnology* 4: 688–694.

Kostli, K. P., M. Frenz et al. 2001. Temporal backward projection of optoacoustic pressure transients using Fourier transform methods. *Physics in Medicine and Biology* 46: 1863–1872.

Laufer, J., C. Elwell et al. 2005. *In vitro* measurements of absolute blood oxygen saturation using pulsed near-infrared photoacoustic spectroscopy: Accuracy and resolution. *Physics in Medicine and Biology* 50: 4409–4428.

Li, P. C., C. R. C. Wang et al. 2008. *In vivo* photoacoustic molecular imaging with simultaneous multiple selective targeting using antibody-conjugated gold nanorods. *Optics Express* 16: 18605–18615.

Liao, L. D., M. L. Li et al. 2010. Imaging brain hemodynamic changes during rat forepaw electrical stimulation using functional photoacoustic microscopy. *NeuroImage* 52: 562–570.

Liao, S. M., N. M. Gregg et al. 2010. Neonatal hemodynamic response to visual cortex activity: high-density near-infrared spectroscopy study. *Journal of Biomedical Optics* 15: 026010.

Manohar, S., S. E. Vaartjes et al. 2007. Initial results of *in vivo* non-invasive cancer imaging in the human breast using near-infrared photoacoustics. *Optics Express* 15: 12277–12285.

Maslov, K., H. F. Zhang et al. 2008. Optical-resolution photoacoustic microscopy for *in vivo* imaging of single capillaries. *Optics Letters* 33: 929–931.

McBride, T. O., B. W. Pogue et al. 1999. Spectroscopic diffuse optical tomography for the quantitative assessment of hemoglobin concentration and oxygen saturation in breast tissue. *Applied Optics* 38: 5480–5490.

Michels, R., F. Foschum et al. 2008. Optical properties of fat emulsions. *Optics Express* 16: 5907–5925.

Ntziachristos, V. and D. Razansky 2010. Molecular imaging by means of multispectral optoacoustic tomography (MSOT). *Chemical Reviews* 110: 2783–2794.

Oraevsky, A. A., A. A. Karabutov et al. 2001. Laser optoacoustic imaging of breast cancer in vivo. *Proceedings of SPIE* 4256: 6–15.

Pan, D. P., M. Pramanik et al. 2009. Molecular photoacoustic tomography with colloidal nanobeacons. *Angewandte Chemie—International Edition* 48: 4170–4173.

Patwardhan, S. V. and J. P. Culver 2008. Quantitative diffuse optical tomography for small animals using an ultrafast gated image intensifier. *Journal of Biomedical Optics* 13: 011009.

Pogue, B. W., K. D. Paulsen et al. 2000. Calibration of near-infrared frequency-domain tissue spectroscopy for absolute absorption coefficient quantitation in neonatal head-simulating phantoms. *Journal of Biomedical Optics* 5: 185–193.

Portnov, A., Y. Ganot et al. 2005. Probing molecular dynamics using action, Doppler and photoacoustic spectroscopy. *Journal of Molecular Structure* 744: 107–115.

Ripoll, J. and V. Ntziachristos 2005. Quantitative point source photoacoustic inversion formulas for scattering and absorbing media. *Physical Review E* 71: 031912.

Rosenthal, A., D. Razansky et al. 2009. Quantitative optoacoustic signal extraction using sparse signal representation. *IEEE Transactions on Medical Imaging* 28: 1997–2006.

Toubal, M., M. Asmani et al. 1999. Acoustic measurement of compressibility and thermal expansion coefficient of erythrocytes. *Physics in Medicine and Biology* 44: 1277–1287.

Wang, X. D., Y. J. Pang et al. 2003. Noninvasive laser-induced photoacoustic tomography for structural and functional *in vivo* imaging of the brain. *Nature Biotechnology* 21: 803–806.

Xiang, L. Z., Y. Yuan et al. 2009. Photoacoustic molecular imaging with antibody-functionalized single-walled carbon nanotubes for early diagnosis of tumor. *Journal of Biomedical Optics* 14: 021008.

Yao, L., Y. Sun et al. 2009. Quantitative photoacoustic tomography based on the radiative transfer equation. *Optics Letters* 34: 1765–1767.

Yin, L., Q. Wang et al. 2007. Tomographic imaging of absolute optical absorption coefficient in turbid media using combined photoacoustic and diffusing light measurements. *Optics Letters* 32: 2556–2558.

Yuan, Z. and H. B. Jiang 2006. Quantitative photoacoustic tomography: Recovery of optical absorption coefficient maps of heterogeneous media. *Applied Physics Letters* 88: 231101.

Yuan, Z. and H. B. Jiang 2009. Simultaneous recovery of tissue physiological and acoustic properties and the criteria for wavelength selection in multispectral photoacoustic tomography. *Optics Letters* 34: 1714–1716.

Zeff, B. W., B. R. White et al. 2007. Retinotopic mapping of adult human visual cortex with high-density diffuse optical tomography. *Proceedings of the National Academy of Sciences of the United States of America* 104: 12169–12174.

Zemp, R. J. 2010. Quantitative photoacoustic tomography with multiple optical sources. *Applied Optics* 49: 3566–3572.

Zhang, E., J. Laufer et al. 2008. Backward-mode multiwavelength photoacoustic scanner using a planar Fabry-Perot polymer film ultrasound sensor for high-resolution three-dimensional imaging of biological tissues. *Applied Optics* 47: 561–577.

Zhang, H. F., K. Maslov et al. 2006. Functional photoacoustic microscopy for high-resolution and noninvasive *in vivo* imaging. *Nature Biotechnology* 24: 848–851.

Zhou, C., G. Q. Yu et al. 2006. Diffuse optical correlation tomography of cerebral blood flow during cortical spreading depression in rat brain. *Optics Express* 14: 1125–1144.

17

Combined Magnetic Resonance Imaging and Near-Infrared Spectral Imaging

Michael A. Mastanduno
Dartmouth College

Kelly E. Michaelsen
Dartmouth College

Scott C. Davis
Dartmouth College

Shudong Jiang
Dartmouth College

Brian W. Pogue
Dartmouth College

Keith D. Paulsen
Dartmouth College

Multimodality imaging systems such as PET-CT and PET-MRI provide valuable anatomic as well as metabolic information about tissue and have become standards-of-care in certain clinical settings. The synergistic combination of high spatial resolution with rich functional information offers improved clinical decision-making and disease management because of increased understanding of the characteristics and progression of disease. An emerging and potentially promising multimodality technology combines near-infrared spectroscopic (NIRS) imaging with MRI. NIRS provides data on the functional status of tissue, including its hemoglobin, oxygen saturation, water, and lipid concentrations. Differences in these parameters are often signatures of malignant tissue; thus, including NIRS with MRI can augment identification of the boundaries of disease with indicators of its aggressiveness. Combined NIRS and MR imaging is a technique in which MRI significantly improves the accuracy of the NIRS results and NIRS significantly improves the specificity of MRI for diagnosis of disease such as breast cancer.

17.1 Introduction

Near-infrared (NIR) light is a nonionizing form of radiation in the 650–950 nm wavelength range. A detailed description of the physics of light–tissue interactions is given in Chapter 8, and applications to breast and brain imaging are described in Chapters 9 and 10, respectively. Briefly, in tissue, absorption and scattering dominate light interactions and they determine the path and final destination of the signal. At a given wavelength, light transport in tissue can be described by the absorption and scattering coefficients, μ_a and μ_s'. Scattering is the more common event, occurring about 100 times more frequently than absorption. Scattering results when light photons meet an interface where a difference in index of refraction exists, such as within mitochondria or in collagen fibrils. Due to the high likelihood of scattering, many photons will travel far from their source and the probability that any propagate through tissue without scattering is extremely small. To obtain a reasonable signal-to-noise ratio (SNR), measurement of scattered light is essential, but the exact path of each photon is unknown; thus, scatter is modeled stochastically and the spatial resolution that can be achieved in an image formed from these highly scattered photons is low. The disadvantage of low spatial resolution can be overcome by combining the optical data with higher-resolution structural information obtained from another modality such as MRI. Co-registration of NIR with MRI produces optical property maps that are much easier to interpret than NIR images acquired from stand-alone systems.

In addition to scattering, NIR light photons are absorbed as they pass through the tissue. The level of absorption is relatively low in the NIR window (650–950 nm) and allows detectable light signals to travel 10 cm or more. Hemoglobin

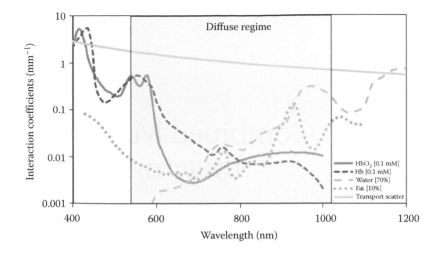

FIGURE 17.1 Absorption spectra of chromophores oxy- and deoxyhemoglobin, water, and fat vary as a function of wavelength. If tissue is probed with multiple wavelengths of near-infrared light, these spectra and the measured absorption coefficients can be used to quantify their concentrations.

as well as water and lipids preferentially absorb light in this spectral band. Each of these tissue chromophores has a well-defined absorption spectrum, reviewed in Figure 17.1, in which the amount of absorption varies as a function of wavelength. Using the known absorption spectra and the measured absorption coefficients, concentrations of these chromophores can be determined if tissue is probed with several different wavelengths of NIR light. Methods based on this technique are known as diffuse optical spectroscopy or near-infrared spectroscopy.

Creating maps of hemoglobin, oxygen saturation, water, and lipid levels generate contrast because changes in these parameters can indicate the presence or absence of disease. In breast cancer, cell-signaling pathways are altered, which can lead to overexpression of growth factors involved in angiogenesis—the formation of new blood vessels (Schneider and Miller, 2005). In addition, these blood vessels are often disorganized and permeable; thus, vessels formed in response to malignancy promote higher hemoglobin concentrations within the associated tissue. Cancer cells also have higher rates of metabolism leading to lower oxygen saturation (Vander Heiden, 2011). Within tumors, inflammation and edema can be present, resulting in higher water content, lower lipid fraction, and higher scattering coefficients. These intrinsic differences in physiology represent potential cancer biomarkers that are measurable with NIRS imaging.

Hundreds of patients have undergone NIRS breast studies at multiple academic centers across the United States and in Europe. In addition, several commercial systems have been developed over the years (Intes, 2005, Poellinger et al., 2008). The sensitivity and specificity of breast cancer detection with NIRS varies depending on the system geometry but has been reported to be in the range of 91–96% and 93–95%, respectively (Chance et al., 2005, Kukreti et al., 2010). High specificity is particularly advantageous for a combined MRI and NIRS system because breast MR is very sensitive but less specific—the latter

being a factor that has limited its acceptance in the diagnostic setting (Kriege et al., 2004).

17.1.1 Benefits of Combining NIRS with MRI

NIRS is an ideal adjunct to MRI for characterizing the tissue because it provides unique, rich, and complementary functional data. Additionally, safe and low-cost technology is involved and the technique does not require the injection of contrast agents (as in PET imaging) or expose patients to ionizing radiation (as in x-ray imaging). The optical data may also be obtained concurrently with the MR scans; hence, no additional exam time is necessary to acquire the NIRS results.

Since scattering is the principal interaction of light in tissue, stand-alone NIRS imaging systems suffer from low spatial resolution (~1 cm). Spatial information from MR can be used to guide the reconstruction of the optical data thereby mitigating this weakness. MRI is well matched to NIRS because it provides structural information on the breast itself. With MRI, suspicious lesions can be identified and separated from normal adipose and fibroglandular tissues. Knowledge of the internal structure of the breast has been shown to improve the spatial resolution of NIRS imaging to 1 mm (Pogue et al., 2006). In addition, each type of tissue possesses different average chromophore concentrations (Brooksby et al., 2006) and the incorporation of MR spatial information has been found to increase the accuracy with which these quantities can be imaged with NIRS (Brooksby et al., 2005b).

Beyond improving the localization and accuracy of NIRS imaging, separation of benign from malignant lesions is possible (Ntziachristos et al., 2002b). The addition of spatial priors from MRI improves the area under the curve (AUC) of the receiver-operating characteristics (ROC) for identifying the breast cancers using NIRS alone and illustrates the ultimate benefit of the coupled technology (Pogue et al., 2011). The sensitivity and specificity of combined MRI/NIRS imaging is anticipated to be greater than that of either modality alone, although larger patient studies are still needed to quantify the diagnostic performance of this multimodality breast imaging platform.

17.1.2 Potential Clinical Application of MRI/NIRS

Current clinical care includes breast MRI for surgical staging and screening of high-risk populations. NIRS imaging can readily be integrated in these settings and may provide additional benefits. Dynamic-contrast breast MR imaging is recommended for the screening of women at high risk for the development of breast cancer because it has greater sensitivity than standard mammography. However, false-positive findings are prevalent and often lead to unnecessary biopsies. Conventional breast imaging modalities are compared in Figure 17.2 in terms of sensitivity, specificity, spatial resolution, and cost. NIRS imaging is a technique with high sensitivity, specificity, and low cost, but low spatial resolution. Simultaneously obtaining optical data with MRI may decrease the number of false-positives by providing information on the metabolic properties of the regions of interest (ROIs). More generally, a combined MRI/NIRS scan may be useful prior to biopsy in women who have had a suspicious mammogram. Unlike other common forms of clinical

breast imaging, neither MRI nor NIRS expose the tissue to ionizing radiation. Therefore, the combined technology is well suited to situations requiring the repeated imaging of the same individual over a relatively short period. For example, monitoring response to neoadjuvant chemotherapy requires longitudinal imaging to ensure that the chemotherapy is working (or permit modification of treatment if not). Alternatively, the method may be attractive for evaluation of macroscopic tumor response to a novel drug or therapeutic regimen.

17.1.3 Challenges for Combined MRI/NIRS Imaging

One drawback to any multimodality imaging system is the cost of the additional information that is generated—more images are collected and need to be interpreted. In the case of MRI/NIRS, the high cost stems primarily from the expense of maintaining and operating the MRI and is not greatly influenced by the addition of NIRS imaging. Nonetheless, overall costs will limit the use of this technology in some cases and require significantly greater diagnostic benefits when compared to lower-cost options. In addition, substantial engineering challenges are involved in the creation of such systems. The breast interface for the optical imaging must be maintained within the bore of the MR scanner, and, thus, cannot contain paramagnetic

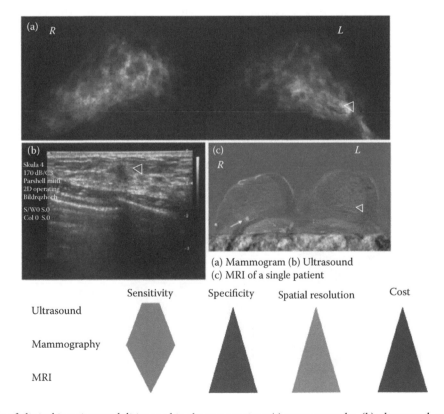

(a) Mammogram (b) Ultrasound
(c) MRI of a single patient

FIGURE 17.2 Example of clinical imaging modalities used in the same patient: (a) mammography, (b) ultrasound, and (c) MRI. Their relative performances in terms of sensitivity, specificity, spatial resolution, and cost are indicated. Comparatively, NIRS imaging is a technique with high sensitivity and specificity, and low cost, but low spatial resolution. (From Kuhl, C. K. et al. 2005. *J Clin Oncol*, 23, 8469–8476.)

materials—a constraint that has been met in different ways by research groups from Dartmouth and the University of Pennsylvania (UPenn).

Despite the drawbacks, combining the metabolic data obtained from NIRS imaging with the soft-tissue contrast and anatomic resolution available with MRI has high potential for clinical impact in the breast care and health management of women. More research is needed to define explicitly how such multimodality systems can be incorporated in the healthcare setting and what overall benefit is actually derived. The remaining part of this chapter will provide an overview of the technical considerations involved in creating an integrated MRI/NIRS imaging system, present details on reconstructing optical data when spatial priors from MR images are available, and summarize results from the multimodality MRI/NIRS breast imaging platforms that have been reported in the literature. Comments on the future of combined MRI/NIRS imaging and discussion of its clinical potential will also be offered.

17.2 Combining MRI and NIRS Instrumentation

Perhaps the biggest challenge in realizing an MRI-guided NIRS breast imaging system is the restriction on using metals with ferrous content (and nonferrous metals in general which can cause eddy-current-related artifacts). Additionally, the space inside the bore of an MR scanner is very limited especially when accounting for the presence of the MR breast coils. These requirements can be met in a number of ways through the design and implementation of MR-specific hardware.

17.2.1 MR Hardware

The primary function of the MR portion of the MRI/NIRS system is acquisition of diagnostic quality images of the breast. In

this case, NIRS is added to complement MRI, and, thus, must not interfere with MR image quality. From the optical imaging perspective, the MR images guide optical image reconstruction (discussed in the next section). Chance et al. found the sensitivity and specificity of optical imaging to be excellent in breast (96% and 93%, respectively), although its spatial resolution is not sufficient to reliably detect the tumors smaller than 1.0 cm (Chance et al., 2005, Brooksby et al., 2006, Pogue et al., 2006). Pogue et al. and Tromberg et al. have suggested that the spatial resolution of NIRS can be increased to nearly 1 mm, if the technique is used synergistically with other imaging modalities (Pogue et al., 2006, Tromberg et al., 2008).

Because NIRS is added to clinical MRI, the particulars of the MRI scanner (e.g., vendor) are not generally critical. Considerations do arise when coupling the optical fibers into a specific MR breast coil, but the NIRS technology is otherwise widely adaptable to almost any type of scanner. Patients are typically placed in the same prone position used in breast MR during the multimodality MRI/NIRS breast exam. Both breasts are pendant within the MR coil platform and one side is lightly compressed to accommodate the NIRS fiber positioning. Representative breast coils and scanners with integrated NIRS imaging arrays are shown in Figure 17.3. Breast coils best suited to MRI/NIRS integration have large open spaces beneath the patient to allow fiber positioning close to the chest wall. Padding for patient comfort and the thickness of the breast coil platform make positioning of the optical fibers against the chest wall very challenging but is important because many breast lesions are located close to the chest wall or even adjacent to the pectoral muscle.

17.2.2 Optical Hardware

Adapting NIRS to operate in an MR scanner also involves modifications to the optical hardware. A clinical MR scanner

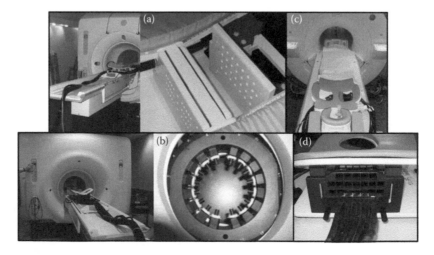

FIGURE 17.3 Examples of MRI/NIRS systems described in the literature: (a) at UPenn using a parallel-plate geometry, (b) at Dartmouth in a circular geometry, (c) at Stanford in a parallel-plate geometry at 1.5T, and (d) at Dartmouth in a parallel-plate geometry at 3T. (From Tromberg, B. J. et al. 2008. *Med Phys*, 35, 2443–2451.)

is sited within a magnetically shielded room and the hardware components of an optical imaging system cannot be placed next to the exam table as they would be for a stand-alone system. For fiber-based delivery and collection of light, extensions of the fibers with low attenuation losses can be used, although they must be lengthened to 10–15 m (from the 2–3 m usually employed) in order to deliver and receive light when the optical instrumentation resides outside the scanner bay. The cost of these fibers scale with material use and the large fiber bundles needed for breast imaging (6 mm diameter) can quickly become the most expensive component of an MRI/NIRS imaging system. For example, the MRI-guided NIRS imaging systems at Dartmouth use 0.4 and 4 mm silica fiber bundles to deliver and collect the light with a unit cost of approximately $700 and $4000 per channel, respectively (Brooksby et al., 2004, Davis et al., 2008).

Attenuation over the increased fiber length is a concern but is typically much less than the light losses occurring at the fiber junctions with the lasers (light sources) or the detectors, which can be 50%. Transmission losses are maintained below a few percent if high-grade materials such as silica are used, as in the fiber-optic communications industry. Multiple fiber bundles of this size can be bulky and become cumbersome to manipulate, but many MRI suites are equipped with conduits for passing cables through the walls. Fibers are easily concealed below the padding on conventional MRI tables and an MRI/NIRS-specific patient platform could be incorporated into a dedicated breast scanner.

17.2.3 Optical-MRI Breast Coupling

As with stand-alone NIRS tomography, the most important function of the breast interface is maintenance of fiber contact with the patient's tissue in order to satisfy the assumptions of diffusion theory. Thus, adjustability and degrees of freedom in positioning the optical fibers in contact with the breast are paramount and an effective interface must not only be reasonably comfortable for the patient, but also enable robust and repeatable fiber positioning relative to the ROI within the breast. Approaches to optical breast imaging have appeared in the literature with fiber arrangements sufficient for volumetric imaging and flexible tissue sampling through arrays in planar or circular geometries.

One planar array produced at UPenn consisted of a 9 × 5 grid of 45 fibers attached to a plate placed on the breast during an exam (Culver et al., 2003, Choe et al., 2005). Schmitz et al. created a circular array of up to 25 fibers which were optimized for fast data collection (Schmitz et al., 2002, 2005) at SUNY Downstate Medical Center. A combined optical imaging and x-ray tomosynthesis system was developed at Massachusetts General Hospital with a denser coverage of 40 source fibers arranged in a fixed grid and nine avalanche photodiode detectors located on the opposite side of the breast (Li et al., 2003). Each of these designs exploits NIR's ability to cover large tissue volumes with arrays of measurement probes connected to the dedicated channels of data-acquisition hardware.

More recently, measurement geometries have been adapted specifically to each patient through conformal fiber arrays. The SUNY Downstate group designed a dual breast optical tomography system that acquired data through as many as 31 fibers that completely covered each breast within two semispheres having adjustable radii. The associated fiber array could be altered for size, tilt, lift, and pitch during an individual breast exam (Schmitz et al., 2005). Similarly, efforts at Dartmouth have used circular rings of fibers to sample three separate anatomically coronal planes through the breast with adjustable radial and coronal positions (Jiang et al., 2003, Pogue et al., 2004). An interface evaluated at University College London (UCL) consisted of interconnected fiber bundle rings mounted to an adjustable conical frame that allowed construction of shapes customized for each breast to be imaged.

Translating these types of fiber interfaces into an MRI-guided breast exam presents a number of challenges. Since MRI/NIRS systems are coupled into a clinical scanner, the patient is placed in an MR breast coil with limited space for deploying elaborate fiber configurations. Cost is also a factor because every sampling channel requires a long (and expensive) optical fiber. Large numbers of fibers quickly increase costs. The group at UPenn produced the first MRI-guided NIRS system, which consisted of an 8 × 3 source fiber grid and a 4 × 2 detector fiber panel arranged in a parallel-plate geometry (Ntziachristos et al., 2002b). Brooksby et al. developed an MRI/NIRS platform with 16 fibers arranged in a circular array inserted in a clinical breast coil (Brooksby et al., 2004). These early fiber interfaces could accommodate different breast sizes but their coverage of breast volume was limited relative to stand-alone NIRS systems.

While MR produces a fully three-dimensional (3D) image volume, the optical instrumentation developed by Brooksby was limited to data collection from a single plane of fibers. The imaging plane would then need to intersect the region of interest/suspicion in the breast (Carpenter et al., 2008). Mastanduno et al. addressed the single fixed plane limitation by designing a system to reposition the optical fibers in order to move a single or multiple planes of data acquisition to the areas of suspicion while the breast was positioned in the coil and the patient was in the bore of the MR scanner. The approach holds promise for increasing the optical access to the full breast volume and leads to more accurate quantification of the relevant ROIs in the breast (Mastanduno et al., 2011). The fiber interface, shown in Figure 17.4, is based on a pneumatic control system that raises and/or lowers 16 fibers arranged in a single-plane parallel-plate geometry to a predetermined (user selected) height. A custom MR breast coil enables a large range of motion in the (anatomically) coronal direction, and full breast coverage is achieved with only 16 fiber-optic cables.

While combining NIRS instrumentation with a clinical MRI scanner is challenging, the opportunity to exploit the structural/functional strengths of MRI and NIRS in a coregistered, simultaneous breast exam is very appealing. The ideal MR scanner is one with a large bore to accommodate the patients and a breast coil with sufficient space for optical fiber

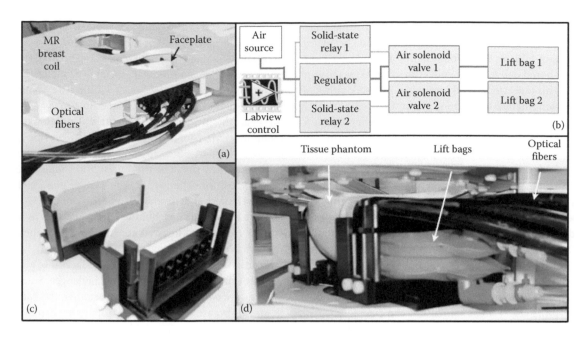

FIGURE 17.4 (a) MR breast coil with integrated optical breast interface. (b) Block diagram of control system used to remotely move fibers. (c) Parallel-plate interface showing location of the optical fibers (black circles). (d) Side view of MR breast coil with lift bags, optical fibers, and tissue-simulating phantom in place demonstrating a remotely actuated fiber array. (From Mastanduno, M. A. et al. 2011. *J Biomed Opt*, 16, 066001.)

placement. High-quality diagnostic images (especially in the contralateral breast) can still be obtained and patient comfort is still maintained. The optical instrumentation, itself, only requires the longer fiber-optic cables. The biggest challenge is maintaining breast–fiber coupling. Multiple groups have addressed the issue and MRI/NIRS systems are available that provide a comfortable and adjustable coupling between the breast and the optical fibers.

17.3 Image Reconstruction

Once data is acquired from an NIRS imaging system, iterative image reconstruction is commonly used to generate maps of tissue functional parameters, as discussed in detail in Chapter 8. While the underlying diffusion equation modeling is essentially the same, the formulation of the inversion needed for image recovery is different in the MRI-guided NIRS case. Because the propagation of light through tissue is dominated by scattering, the overall photon transport is usually modeled by diffusive processes governed by the radiative transfer equation (RTE).

17.3.1 Light Transport in Tissue

The intrinsic optical properties of tissue affect how light propagates and changes in these properties can provide information on tissue metabolic status. Absorption and scattering are two properties that are described by the absorption and scattering coefficients (μ_a and μ_s'), respectively. These coefficients represent the likelihood of a photon undergoing an absorption or scattering event per unit length in tissue. The reciprocal of the

absorption or scattering coefficient is the mean free path or average distance that a photon will travel without experiencing an absorptive or scattering event. Although photons can scatter in any direction in tissue, most are transported in the forward direction. The reduced scattering coefficient (μ_s') incorporates information on photon directionality after a collision, and is written as $\mu_s' = \mu_s(1 - g)$, where g is the average cosine of the scattering angle (~0.9 in tissue).

The RTE can be simplified under two assumptions: (1) optical radiance is only linearly anisotropic, all higher-order terms are insignificant, and (2) the rate of change of photon flux is much lower than the photon collision frequency. The first assumption holds for regions far (>1 mm) from light sources and tissue boundaries, and is readily satisfied in breast imaging because the tissue volume is relatively large. The second assumption is true because scattering occurs much more frequently than absorption in breast tissue in the NIR regime. With these assumptions met, a simplification in the RTE known as the diffusion approximation is possible and is represented mathematically by

$$\frac{1}{c}\frac{\partial \Phi(\mathbf{r},t)}{\partial t} - \nabla \cdot D\nabla\Phi(\mathbf{r},t) = -\mu_a\Phi(\mathbf{r},t) + S(\mathbf{r},t)$$

where Φ is the photon fluence, $D = (1/3(\mu_a + \mu_s'))$, is the diffusion coefficient, μ_a is the absorption coefficient, c is the speed of light and S is a light source. This equation describes the fluence of light at a certain position, $\mathbf{r}$, at time t. The four terms represent changes in flux, diffusion, loss due to absorption, and gain from a light source and are balanced in the form of a conservation law.

A more rigorous derivation of the diffusion equation (from the RTE) can be found in Pogue et al. (2001). In addition to modeling light as it travels through tissue, whether it is reflected back inside or exits at a tissue surface is accommodated through incorporation of appropriate boundary conditions.

17.3.2 Forward Problem

Accurate modeling of light propagation (in this case diffusion) in tissues of known optical properties is essential. Computing the light fluence at external boundaries for a tissue having specified optical properties is known as the *forward problem*, and several approaches are available to attain its solution. For example, analytical solutions of the RTE exist (Arridge, 1995). They are relatively easy to implement and are computationally efficient. However, they are only accurate for simple geometries such as spheres and semi-infinite domains. Hence, analytical solutions are impractical for most situations that arise in clinical imaging, especially of the breast. Monte Carlo methods represent another class of solutions to the forward problem. In Monte Carlo computations, the propagation of individual photons are simulated stochastically until they are absorbed in tissue or exit an external boundary (Patterson et al., 1989, Jacques, 1995). The technique is amenable to any tissue geometry; however, it is computationally intensive and often not practical in the imaging setting (Jacques and Pogue, 2008). Finite element modeling (FEM) is another alternative. It can be applied to complicated tissue shapes and is less computationally demanding than Monte Carlo methods. As a result, FEM is currently the method of choice for modeling light propagation in NIRS systems (Dehghani et al., 2008). Figure 17.5 provides an overview of the image reconstruction process and illustrates how an MR image volume is segmented into regions based on tissue type, meshed, and used to generate an optical image.

FEM solves the diffusion approximation on a discretized tissue domain. The individual points within the computational mesh are known as nodes and are interconnected to form elements that are concatenated to form volumes and associated surfaces representative of the whole breast or other internal tissues of interest as shown in Figure 17.5. This solution, calculated at each node, corresponds to the photon fluence at that location but is expressed in terms of a piecewise linear basis function that provides a smooth solution at any point in the mesh (i.e., within the breast). The number of nodes contained in a mesh can vary, but generally, more nodes lead to a more accurate solution at the cost of larger computational requirements. A detailed discussion of FEM is beyond the scope of this chapter but can be found in Lynch (2005) and Arridge and Hebden (1997).

17.3.3 Inverse Problem

In NIRS imaging, measurements are recorded at light detectors external to the tissue volume of interest, and μ_a and μ'_s can be estimated from this data at every node in the mesh by solving the *inverse problem*. The inverse problem is nonlinear, ill posed, and underdetermined, and, thus, challenging to solve. Underdetermined refers to the fact that many more unknowns (the tissue optical properties at each node) exist than the number of measurements. In an ill-posed problem, small changes in the signal or noise can result in large changes in the solution. The general inverse solution requires an initial estimate of the optical properties of the tissue of interest. Using these values, the forward problem is solved and compared with the collected measurement data. Differences between the model and the data

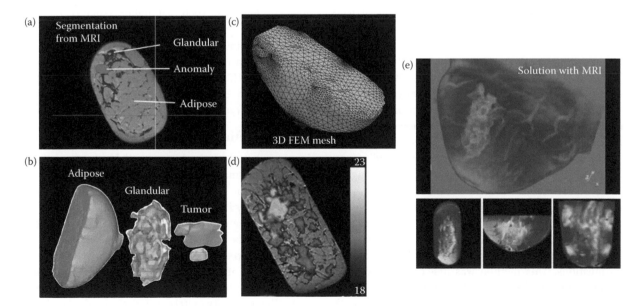

FIGURE 17.5 (**See color insert.**) (a) Optical reconstruction is guided by MRI information generated from segmented DICOM images. (b) 3D regions of glandular, adipose, and tumor regions are identified and assigned. (c) A finite element mesh is generated. (d) Optical images are generated and projected onto slices of coregistered MR. (e) Optical solutions are presented in volumetric form within the coregistered MR image volume.

are used to update the approximation of the optical properties. This process is repeated until a stopping criterion is reached. Methods are available for solving the inverse problems, which have advantages and disadvantages in NIRS image reconstruction, and a number of research groups are working on optimizing these techniques (Yalavarthy et al., 2007b, Dehghani et al., 2008, Fang et al., 2009, Pogue et al., 2011).

17.3.4 MRI-Guided Reconstructions

Obtaining MRI data simultaneously with NIRS image acquisition can assist in the optical image reconstruction. The breast is a heterogeneous organ comprised largely of adipose and fibroglandular tissue, which have very different optical properties (Brooksby et al., 2005a). Stand-alone NIRS cannot spatially resolve these two tissue types throughout the breast volume. In addition, wide variations in the ratio of adipose to fibroglandular tissue occur in women. MRI can readily discriminate between these two tissue types, and thereby provides important information for guiding the NIRS image reconstruction.

17.3.4.1 Creating Meshes from MRI Images

MRI image volumes of the breast can be used as the basis for finite element mesh generation through either commercial medical imaging software such as Materialise's MIMICS© program or freeware developed at academic institutions such as Dartmouth's NIRFAST platform. In either case, DICOM image volumes are uploaded and a series of user-guided transformations are applied to assign each pixel to a region that may include (but is not limited to) adipose tissue, fibroglandular tissue, possible malignancy, skin, chest wall muscle, other organs, and air. These tissues are typically separated based on location as well as pixel intensity through tools provided by the software package. Once all pixels have been labeled, the optical sources and detectors are placed around the external surface of the breast and a mesh is created.

One component of mesh creation unique to MRI/NIRS imaging is tumor (or suspicious tissue) definition. The boundaries of suspected lesions are obtained from the coregistered MRI image volume before optical image reconstruction occurs, and many methods are available to define these regional boundaries. The most common approach is based on DCE-MRI data where multiple (identical) MR acquisitions occur before and after an injection of gadolinium (Gd) contrast agent. By subtracting precontrast from postcontrast image volumes, areas of Gd uptake are accentuated. A clinically accepted practice segments the tumor region based on the fitting of each pixel's contrast uptake to an exponential function describing the pharmacokinetics of the vasculature using all (typically 5) of the postcontrast image acquisitions. This method was first described by Tofts and Kermond and is applied in commercially available software such as Cadstream© (Tofts et al., 1995). Figure 17.6 illustrates the results from different methods for segmentation based on the DCE-MRI data. Clearly, these differences will affect the final optical image reconstruction, but their influences are much

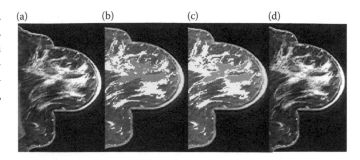

FIGURE 17.6 (**See color insert.**) (a) Methods can be used to segment suspicious regions from breast MRI. (b) Segmentation based on subtraction of pre- and postcontrast scans (suspicion for tumor in red). (c) Segmentation based on curve fitting to contrast washout kinetics from several postcontrast scans (suspicion for tumor in pink). (d) Differences in region classifications between (b) and (c) with suspicion of tumor differences shown in green.

smaller than the differences in the tissue optical properties between the segmented regions.

The MIMICS software package and the Dartmouth NIRFAST platform offer certain advantages when developing an MRI-guided NIRS mesh. MIMICS provides many useful functions for efficient and accurate segmentation of tissue structures. The software is used by medical professionals and radiologists who are comfortable applying it to identify ROIs. An alternative platform, NIRFAST, was recently released by Dartmouth as software designed to create meshes for NIRS imaging utilizing DICOMs from conventional imaging modalities. The software has fewer tools for image segmentation (relative to commercial packages such as MIMICS) but is free and includes the most important and basic functionalities needed for DICOM image to NIRS mesh generation. Nonuniform meshes can also be created with NIRFAST such that higher resolution can exist in the tumor region, preserving high computational accuracy in the ROI while maintaining the lower computational costs for image reconstruction in the surrounding tissues. The final outcome of the MRI/NIRS mesh generation process is a computational domain, where each node is assigned as belonging to a specific tissue region (typically adipose, fibroglandular, or ROI).

17.3.4.2 Spatial Priors in Reconstruction

In general, two methods are available for incorporating the regional information encoded in the mesh during the reconstruction of the NIRS data and are known as *hard* and *soft* priors. In the *hard priors* approach, all nodes assigned to a single tissue type are constrained to return the same optical property values. The method substantially decreases the number of unknowns and transforms the inverse problem from an underdetermined to an overdetermined estimation, which also significantly reduces the problem size and its associated computational costs. The sensitivity matrix, or Jacobian, calculated during each iteration of the image reconstruction, is decreased in one dimension from the number of nodes (typically tens of thousands) to the number of regions (typically <10). This contraction decreases the computation time significantly and can make previously intractable problems solvable on a laptop. Hard prior information improves the ROC curve in identifying the simulated cancerous lesions compared to NIRS data reconstructed without prior information (Pogue et al., 2011). However, incorrect or inaccurate region assignment can produce spatial bias during the inversion process and false values (i.e., false-positives or false-negatives) may be generated during image reconstruction (Yalavarthy et al., 2007a).

The false-value issue can be mitigated through the alternative approach known as *soft priors*. In this method, variation within a region is minimized but not eliminated and is often achieved through a Baysian or Laplacian regularization within each region. *Soft priors* have shown up to 30% improvement in localization and chromophore quantification in simulated and phantom measurements versus reconstructions without priors (Intes et al., 2004, Brooksby et al., 2005a). They have been successfully applied to patient imaging as well (Carpenter et al., 2008). Although soft prior approaches return more accurate values when tissue region assignments are incorrect, some deviations will still exist; hence, careful segmentation is important. Additionally, soft prior reconstructions do not have the same computational savings as hard priors since the size of the Jacobian is unchanged relative to the no priors approach.

17.3.5 Software Packages for NIR Image Reconstruction

Clearly, nuances exist when reconstructing NIRS data, but several software packages have been developed to assist the users through the process including the Virtual Photonics Initiative, TOAST, and NIRFAST (Tromberg, 2011) (Schweiger and Arridge 2011), (Dehghani and Pogue, 2011). These software packages are generally open source, offer simple graphical user interfaces, and provide tutorials and support documentation. Versatile and robust software is important for advancing the technology toward clinical adoption and research groups are committed to designing and supporting free software.

17.4 NIRS/MRI Imaging Results

17.4.1 Contrast Recovery and Spatial Resolution

The spatial resolution of diffuse optical imaging is not sufficient to define the detailed anatomical structure/tissue—the literature usually quotes a spatial resolution of 1 cm (Tromberg et al., 2008). Photon paths spread with distance between source and detector, typically originating from a small fiber, spreading out (diffusing), and being collected by a small fiber (Moon et al., 1993, Moon and Reintjes, 1994). Thus, NIRS spatial resolution is better near a source or detector where spread in the photon path is smallest, and detection of deeply embedded optical heterogeneity requires higher contrast than when near the boundary. With inclusion of MRI for spatial guidance, the spatial resolution of NIRS imaging, especially at depth, is much more

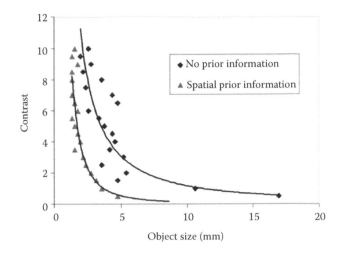

FIGURE 17.7 At representative contrasts (<200%), stand-alone NIRS can be used to reliably detect the objects approximately 1 cm in size. With the incorporation of spatial priors, the minimum detectable size can be improved by almost an order of magnitude. (From Pogue, B. W. et al. 2006. *J Biomed Opt*, 11, 33001.)

favorable and the combined technique is able to detect the optical heterogeneity as small as 1 mm as indicated in the contrast-detail curves in Figure 17.7. MR guidance also enables NIRS to recover the objects of lower contrast relative to NIRS alone (Pogue et al., 2006).

17.4.2 Phantom Results

Validation of new imaging systems usually begins with phantom studies. Although phantoms are incomplete representations of tissue, they can be used to determine the overall performance of the hardware. Optical phantoms are composed of absorbing and scattering materials that can be adjusted separately to control their optical properties. Absorbers such as blood or India ink are often mixed with a scattering liquid like intralipid and suspended in a gelatin or resin medium. Typically, whole blood is used to evaluate the changes in total hemoglobin, but the approach does not provide control over water content or scattering parameters. As a result, instrumentation is evaluated by recovering linearly varying concentrations and relative contrast (Poplack et al., 2007).

17.4.2.1 Stand-Alone NIRS Optical Phantoms

Since phantom validation is critical to the introduction of a new imaging modality, optical phantoms were first imaged using NIRS alone. Because photon propagation through tissue is inherently three-dimensional, the emphasis here will be placed on 3D phantom imaging. Hebden et al. investigated contrast recovery in 3D using a system of three concentric rings to image a cone-shaped phantom (Hebden et al., 2001). The investigators showed that despite the radius of a particular set of detectors, recovered optical properties were uniform throughout the phantom except in the embedded heterogeneities. These results agree with earlier studies by Pogue et al. demonstrating that the logarithm of optical absorption was linear with source–detector path length (Pogue et al., 2000). In the experiments reported, only 10% contrast (rather than the 100% that existed) was recovered due to limited mesh resolution and distortion of spatial information. With no spatial priors to guide the image reconstruction, the inclusion's finite contrast was distributed over a larger area and the value of μ_a was lower than expected.

A 2003 study by Dehghani et al. found similar results in a cylindrical phantom with an absorbing sphere. The authors were able to show good localization and property separation, but they were not able to accurately quantify the optical properties (Dehghani et al., 2003). The results suggested that more sophisticated regularization methods and *a priori* information could improve the image quantification.

17.4.2.2 MRI/NIRS Optical Phantoms

As Dehghani et al. indicated, *a priori* information can improve spatial resolution, and therefore, contrast recovery. The first examples were demonstrated in simulation by Ntziachristos et al. for single wavelengths and by Brooksy et al. a year later for multiple wavelengths (Ntziachristos et al., 2002b, Brooksby et al., 2003). Phantom results obtained with prior information

are far superior to those without these constraints. An initial estimate of the spatial distribution of absorbers allows the algorithm to focus on finding their concentrations. Once suitable algorithms were developed and *a priori* information was available, Brooksby et al. demonstrated that both spatial resolution and contrast recovery were enhanced by spatial priors (Brooksby et al., 2004, 2005a). In these studies, qualitative improvements resulting from spatial priors were shown relative to stand-alone optical image reconstructions. Chromophores were recovered more accurately both quantitatively and spatially and spatial fluctuations in the images were reduced. A demonstration of the image quality that can be obtained from these techniques in phantoms is provided in Figure 17.8.

Later phantom studies by Carpenter et al. and Mastanduno et al. indicate that MR-guided NIRS is capable of recovering optical properties very accurately using algorithms incorporating both the spectral and spatial priors (Carpenter et al., 2008, Mastanduno et al., 2011). These experiments showed contrast is recovered more accurately relative to absolute quantities. Specifically, phantom contrast was experimentally recovered to nearly the expected value as long as the inclusion's optical properties were less than approximately two times the background. NIRS contrast recovery suffers at higher contrasts and MRI/

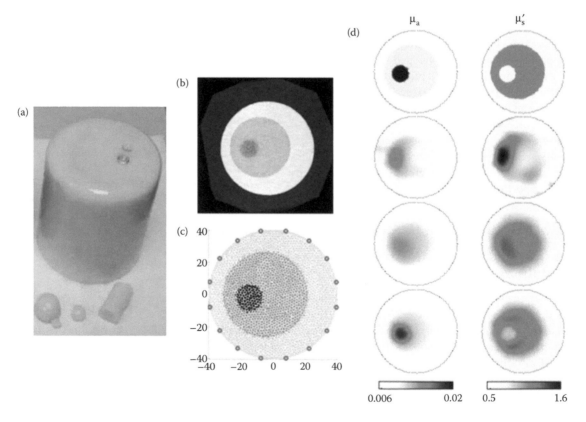

FIGURE 17.8 (a) Gelatin phantom with gelatin inclusions used to evaluate the performance of incorporating spatial priors. (b) MRI image of the phantom. (c) FEM mesh derived from the MR image in (b). (d) Images of the true optical properties (top row), without spatial priors (2nd row), with the outermost two regions of spatial priors (3rd row) and with all three regions of spatial priors (bottom row). (From Brooksby, B. et al. 2005a. *J Biomed Opt*, 10, 051504.)

NIRS is no exception (Pogue et al., 2006). Since hemoglobin contrast in malignant breast tissue is typically below twofold, this limitation is not particularly significant from a clinical standpoint.

17.4.3 *In Vivo* Human Imaging Results

While nearly a decade has passed since the first MRI/NIRS systems were reported, the number of *in vivo* breast imaging studies using MRI/NIRS remains small, in part because only a few research groups have combined these methods for human breast imaging applications. Fewer women undergo MR breast exams relative to other modalities; hence, the number of potential subjects for combined MRI/NIRS studies is more modest, unlike when evaluating other multimodality breast imaging platforms such as NIRS combined with x-ray tomosynthesis (Fang et al., 2011).

17.4.3.1 Healthy Volunteers

The majority of MRI-guided NIRS studies reported in the literature have been conducted on healthy volunteers to characterize the optical properties in clearly separated adipose and glandular tissues. One study by Brooksby et al. involved 11 volunteers and a spectrally unconstrained image reconstruction algorithm to produce images and statistics on healthy breast tissue (Brooksby et al., 2006). Representative images are presented in Figure 17.9. Modest variation in tissue composition was found between subjects with radiologically scattered and dense breasts. While these results are important for furthering our understanding of breast tissues, the authors acknowledged that many questions remained regarding the

spatial resolution at which tumor tissue can be characterized with MRI/NIRS.

17.4.3.2 Cancer Patients

Ntziachristos et al. first used MRI/NIRS to image breast lesions *in vivo* after the injection of indocyanine green (ICG). While the study emphasized the comparison of NIRS-monitored uptake of ICG with MRI-monitored uptake of Gd, it was the first utilization of combined MRI/NIRS in patients with cancer (Ntziachristos et al., 2000). An image from this landmark paper can be found in Figure 17.10, where a ductal carcinoma is differentiated from the background tissue. The largest study of MRI/NIRS reports results from 14 patients: nine benign lesions and five malignant tumors. Published in 2002 by Ntziachristos et al., the aim of the work was to understand the optical properties of different cancers and demonstrate the technique's performance in a patient population with breast abnormalities (Ntziachristos et al., 2002b). The authors were able to distinguish the malignant lesions from their benign counterparts based on intrinsic contrast, but the paper did not report the sensitivity and specificity of the combined imaging system. As a result, MRI/NIRS is currently viewed as an add-on to clinical MRI rather than a separate imaging tool. Ntziachristos et al. showed the imaging of breast patients was possible and established the idea of using optics to add the specificity and/or sensitivity to DCE-MRI, which is in need of improvement (Orel and Schnall, 2001, Kuhl, 2007).

Since these initial studies, several efforts on algorithm development have appeared, specifically focused on the inclusion of prior information during image reconstruction (Brooksby et al., 2005c, Pogue et al., 2011). Unfortunately, a definitive patient study to characterize these new algorithms and

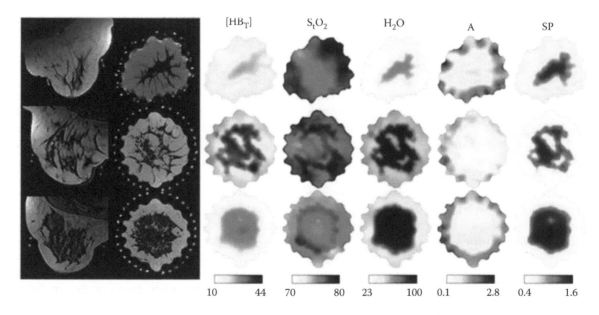

FIGURE 17.9 Examples of T_1-weighted breast MRI and NIRS image reconstructions in the breasts of three healthy women with radiographic density varying from scattered to heterogeneously dense. The study separately characterized the optical properties of glandular and adipose tissues with regard to radiographic density in a healthy population. (From Brooksby, B. et al. 2005a. *J Biomed Opt*, 10, 051504.)

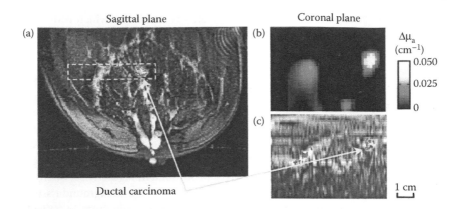

FIGURE 17.10 Example of MRI/NIRS results in a cancer patient with ductal carcinoma first published by the UPenn group in the year 2000. Spatial priors were used to define the suspicious regions. (From Ntziachristos, V. et al. 2000. *Proc Natl Acad Sci USA*, 97, 2767–2772.)

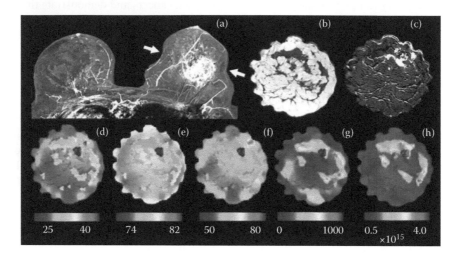

FIGURE 17.11 (**See color insert.**) Example from Carpenter et al. of using MRI/NIRS to differentiate malignant from benign lesions. (a) MRI maximum intensity projection image. (b) T_1-weighted anatomically coronal cross section in the plane of the optical fibers. (c) Same as (b) but after Gd contrast injection. (d) NIRS total hemoglobin image. (e) NIRS oxygen saturation image. (f) NIRS water content image. (g) NIRS scatter amplitude image. (h) NIRS scatter power image. (From Carpenter, C. M. et al. 2007. *Opt Lett*, 32, 933–935.)

hardware advances in MRI/NIRS has not been undertaken to date. Some interesting case reports have suggested that MRI/NIRS can be used to characterize tumor physiology (Srinivasan et al., 2006), create volumetric tissue maps (Carpenter et al., 2008), and even separate malignant tissues from benign conditions (Carpenter et al., 2007). Figure 17.11 shows a nice demonstration of the synergistic qualities of MRI/NIRS in a malignant case that appeared benign. Investigations of larger numbers of patients with the latest techniques are currently under way and hopefully researchers will have the evidence that NIRS boosts the specificity of DCE-MRI within a few years.

17.4.4 Small-Animal Imaging Using NIRS-MRI

In addition to clinically focused breast imaging research, volumetric fluorescence imaging of small animals is being used to study the underlying molecular processes of disease. Fluorescence molecular tomography (FMT), also known as fluorescence diffuse optical tomography (FDOT), has been used to assess the blood kinetics with nonspecific fluorophores (Corlu et al., 2007), enzymatic activity-targeted fluorophores (Weissleder et al., 1999, Ntziachristos et al., 2002a, McCann et al., 2009), deoxyglucose uptake (Li et al., 2009), and transmembrane cellular receptor status with antibody, ligand, or other targeted fluorophores (Ntziachristos et al., 2004, Davis et al., 2010b,c). In most applications, FMT has been deployed as an extension of NIRS, and thus, it relies on diffusion-based assumptions of photon transport. FMT is subject to the same image-degrading photon scattering as absorption-based NIRS; hence, the rationale for including additional information about internal tissue structure from a secondary imaging modality applies to both approaches. Incorporating spatial priors from MRI in FMT is accomplished in much the same manner as for NIRS in clinical geometries. After simultaneous MRI-optical imaging, the MR images are segmented into broadly defined tissue regions, and are used as an anatomical template to guide the optical image reconstruction. The extent to which the algorithm encodes the MRI template to constrain the optical image recovery varies depending on the type of spatial priors, and care must be exercised to avoid overconstraining the problem. These approaches have been shown to improve image quality and fluorescence activity quantification (Davis et al., 2007, Lin et al., 2007, Hyde et al., 2009, Kepshire et al., 2009, McCann et al., 2009).

To date, much of the research on multimodality FMT has involved coupling with small-animal CT scanners. While CT is faster, less expensive, and less restrictive for integrating optical imaging, MRI provides unmatched soft-tissue contrast and the ability to emphasize a wide range of physiological features in an image volume of interest. The first demonstration of spatially guided MRI/FMT was published by Davis et al. and used a multichannel spectrometer-based optical FMT system integrated with a clinical 3T MRI scanner (Davis et al., 2007, 2008). This system was later used to quantify the receptor status in human gliomas implanted in mice (Davis et al., 2010b) and to investigate the different methods of applying spatial priors in the FMT imaging algorithm (Davis et al., 2010a). Meanwhile, Lin et al. developed an MRI-coupled small animal imaging system with PMT (photomultiplier tube) detectors (Lin et al., 2011). The use of PMTs facilitates the acquisition of time-harmonic measurements in the frequency domain, allowing the recovery of the fluorescence lifetime of the fluorophore, which could be an important quantity in tissue diagnosis. Studies using these systems demonstrate various methods of incorporating internal tissue structures from the MR images to guide the optical imaging algorithms. Other studies that have reported combined MRI/FMT imaging have shown MRI images for postprocessing and MRI/FMT image overlays, but have not included the MR data as priors in the optical imaging algorithms (Unlu et al., 2008, McCann et al., 2009, Stuker et al., 2011).

The hardware requirements for coupling FMT systems into MRI scanners present unique challenges. Using a finite number of optical fibers to couple light between the tissue in the magnet bore and the optical system, which is often positioned outside the scanner room, limits the size and geometry of the tissue volume to be imaged. Additionally, the relatively restrictive space in MRI scanners requires careful design of the tissue interface; specialized RF coils are often needed. While surmountable, these constraints have limited the widespread adoption of MRI/FMT. New systems are under development to address some of the challenges and streamline the imaging process by providing noncontact raster-scanning light excitation (Stuker et al., 2011) and MRI-compatible detector systems (Seetamraju et al., 2011). Until these advances are fully validated, the current generation of MRI/FMT imaging systems remains as capable research tools which provide unique insight into tissue pathology *in vivo*.

17.5 Conclusions/Future Directions

The field of optical breast imaging is diverse and imaging systems are under evaluation for use in several stages of the disease management process. Since visible light is highly attenuated, it is primarily used for surface imaging and spectroscopy, and has the greatest potential for application in therapy monitoring or biopsy. Near-infrared light is highly scattered but only weakly absorbed; hence, it offers significant potential in tomography applications. The integration of spectral tomography with conventional imaging (such as x-ray and MRI) is emerging as a major area of interest, where the synergy of the combined information may prove valuable for the diagnosis of complex diseases such as breast cancer.

Where MRI/NIRS will fit into clinical practice is one question that remains. MRI is currently being recommended as a screening tool for women at high risk for breast cancer and women with radiographically dense breasts. Since integrated NIRS imaging requires no additional time and adds relatively little expense to MR scans, it may be able to increase the sensitivity and specificity of breast MR alone. Because MRI/NIRS is too expensive for general breast cancer screening (due to the cost of MRI), researchers are also integrating NIRS with both ultrasound and x-ray tomosynthesis.

Alternatively, MRI/NIRS may be well suited for exams prescribed after a patient has had an abnormal mammogram. In the diagnostic setting, fewer patients are involved and the target cost per scan is higher (~$1000) than in screening. Alternative imaging modalities such as additional mammographic views, ultrasound, CT, and MRI are commonly used before biopsy to gain as much information as possible about the suspicious region of tissue. Each of these modalities exploits a different contrast mechanism, and, therefore, may be able to yield the diagnostic information before biopsy or about sites that are not going to be biopsied. MRI/NIRS would likely follow this same path as the data is a functionally rich and spatially resolved indicator of aggressive cancer.

Another application where MRI/NIRS might be beneficial is monitoring the neoadjuvant chemotherapy. Here, cancer patients are benefited and the cost of the exam can be much higher (exceeding $2000 per scan). MRI is currently used to monitor the therapy, but is slow to show response since breast tumors tend to change the size long after other functional parameters have been altered such as hemoglobin content which can be imaged with NIRS techniques. As an adjunct to MRI for monitoring therapy, combined MRI/NIRS could help differentiate nonresponders and allow medical oncologists to stop chemotherapy earlier or change treatment strategies based on the intermediate imaging results.

MRI, in particular, can benefit from the addition of NIRS by providing functional information in conjunction with the highly resolved spatial maps that MR produces. Like other multimodality imaging platforms, synergy exists between MRI and NIRS whereby anatomical information from MR is obtained prior to optical image reconstruction and used to spatially constrain the NIRS image estimation process. This approach improves the contrast resolution of NIRS relative to NIRS alone, improves spatial resolution to approximately 1 mm, and is expected to improve the ROC curves for diagnosis. Coregistered image volumes of tissue function at the molecular level on a millimeter scale could aid clinicians in making the diagnostic decisions and eliminate the unnecessary biopsy procedures caused by MRI's high sensitivity but low specificity. MRI/NIRS is an emerging imaging modality that could help to inform clinical decisions and improve the quality of life of those affected with breast cancer.

Acknowledgments

The authors acknowledge and appreciate the substantial contributions of former graduate students, specifically Ben Brooksby, PhD, and Colin Carpenter, PhD. Collaborations with radiologists, in particular, Steven P. Poplack, MD, and Roberta M. diFlorio-Alexander, MD, are also appreciated and have been critical to the clinical evolution of the MRI/NIRS technique. The willingness of volunteers and patients to participate in our clinical breast exams is recognized as an essential component of moving this new technology forward toward clinical acceptance. The research described in this chapter has been supported in part by NIH grants R01-CA69544 and P01-CA80139 awarded by the National Cancer Institute.

References

Arridge, S. R. 1995. Photon-measurement density functions. Part I: Analytical forms. *Appl Opt,* 34, 7395–409.

Arridge, S. R. and Hebden, J. C. 1997. Optical imaging in medicine: II. Modelling and reconstruction. *Phys Med Biol,* 42, 841–53.

Brooksby, B., Dehghani, H., Pogue, B. W., and Paulsen, K. D. 2003. Near infrared (NIR) tomography breast image reconstruction with a priori structural information from MRI: Algorithm development for reconstructing heterogeneities. *IEEE J STQE,* 9, 199–209.

Brooksby, B., Jiang, S., Dehghani, H., Pogue, B. W., Paulsen, K. D., Kogel, C., Doyley, M., Weaver, J. B., and Poplack, S. P. 2004. Magnetic Resonance-Guided Near-infrared tomography of the breast. *Rev Sci Instr,* 75, 5262–5270.

Brooksby, B., Jiang, S., Dehghani, H., Pogue, B. W., Paulsen, K. D., Weaver, J., Kogel, C., and Poplack, S. P. 2005a. Combining near-infrared tomography and magnetic resonance imaging to study *in vivo* breast tissue: Implementation of a Laplacian-type regularization to incorporate magnetic resonance structure. *J Biomed Opt,* 10, 051504.

Brooksby, B., Pogue, B. W., Jiang, S., Dehghani, H., Srinivasan, S., Kogel, C., Tosteson, T., Weaver, J. B., Poplack, S. P., and Paulsen, K. D. 2006. Imaging breast adipose and fibroglandular tissue molecular signatures using hybrid MRI-guided near-infrared spectral tomography. *Proc Nat Acad Sci USA,* 103, 8828–8833.

Brooksby, B., Srinivasan, S., Jiang, S., Dehghani, H., Pogue, B. W., Paulsen, K. D., Weaver, J., Kogel, C., and Poplack, S. P. 2005b. Spectral priors improve near-infrared diffuse tomography more than spatial priors. *Opt Lett,* 30, 1968–1970.

Brooksby, B., Srinivasan, S., Jiang, S., Dehghani, H., Pogue, B. W., Paulsen, K. D., Weaver, J., Kogel, C., and Poplack, S. P.

2005c. Spectral-prior information improves Near-Infrared diffuse tomography more than spatial-prior. *Opt Lett,* 30, 1968–1970.

Carpenter, C., Srinivasan, S., Pogue, B., and Paulsen, K. 2008. Methodology development for three-dimensional MR-guided near infrared spectroscopy of breast tumors. *Opt Express,* 16, 17903–17914.

Carpenter, C. M., Pogue, B. W., Jiang, S. J., Dehghani, H., Wang, X., Paulsen, K. D., Wells, W. A. et al. 2007. Image-guided spectroscopy provides molecular specific information in vivo: MRI-guided spectroscopy of breast cancer hemoglobin, water, and scatterer size. *Opt Lett,* 32, 933–935.

Chance, B., Nioka, S., Zhang, J., Conant, E. F., Hwang, E., Briest, S., Orel, S. G., Schnall, M. D., and Czerniecki, B. J. 2005. Breast cancer detection based on incremental biochemical and physiological properties of breast cancers: A six-year, two-site study. *Acad Radiol,* 12, 925–933.

Choe, R., Corlu, A., Lee, K., Durduran, T., Konecky, S. D., Grosicka-Koptyra, M., Arridge, S. R. et al. 2005. Diffuse optical tomography of breast cancer during neoadjuvant chemotherapy: A case study with comparison to MRI. *Med Phy,* 32, 1128–1139.

Corlu, A., Choe, R., Durduran, T., Rosen, M. A., Schweiger, M., Arridge, S. R., Schnall, M. D., and Yodh, A. G. 2007. Three-dimensional *in vivo* fluorescence diffuse optical tomography of breast cancer in humans. *Opt Express,* 15, 6696–6716.

Culver, J. P., Choe, R., Holboke, M. J., Zubkov, L., Durduran, T., Slemp, A., Ntziachristos, V., Chance, B., and Yodh, A. G. 2003. Three-dimensional diffuse optical tomography in the parallel plane transmission geometry: Evaluation of a hybrid frequency domain/continuous wave clinical system for breast imaging. *Med Phy,* 30, 235–247.

Davis, S. C., Dehghani, H., Wang, J., Jiang, S., Pogue, B. W., and Paulsen, K. D. 2007. Image-guided diffuse optical fluorescence tomography implemented with Laplacian-type regularization. *Opt Express,* 15, 4066–4082.

Davis, S. C., Pogue, B. W., Springett, R., Leussler, C., Mazurkewitz, P., Tuttle, S. B., Gibbs-Strauss, S. L., Jiang, S. S., Dehghani, H., and Paulsen, K. D. 2008. Magnetic resonance-coupled fluorescence tomography scanner for molecular imaging of tissue. *Rev Sci Instr,* 79, 064302-1–064302-10.

Davis, S. C., Samkoe, K. S., O'hara, J. A., Gibbs-Strauss, S. L., Paulsen, K. D., and Pogue, B. W. 2010a. Comparing implementations of magnetic-resonance-guided fluorescence molecular tomography for diagnostic classification of brain tumors. *J Biomed Opt,* 15, 051602-10.

Davis, S. C., Samkoe, K. S., O'hara, J. A., Gibbs-Strauss, S. L., Payne, H. L., Hoopes, P. J., Paulsen, K. D., and Pogue, B. W. 2010c. MRI-coupled fluorescence tomography quantifies EGFR activity in brain tumors. *Acad Radiol,* 17, 271–276.

Davis, S. C., Samkoe, K. S., O'hara, J. A., Gibbs-Strauss, S. L., Payne, H. L., Hoopes, P. J., Paulsen, K. D., and Pogue, B. W. 2010b. MRI-coupled fluorescence tomography quantifies EGFR activity in brain tumors. *Acad Radiol,* 17, 271–276.

Dehghani, H., Eames, M. E., Yalavarthy, P. K., Davis, S. C., Srinivasan, S., Carpenter, C. M., Pogue, B. W., and Paulsen, K. D. 2008. Near infrared optical tomography using NIRFAST: Algorithm for numerical model and image reconstruction. *Commun Numer Methods Eng*, 25, 711–732.

Dehghani, H., Pogue, B. W., Shudong, J., Brooksby, B., and Paulsen, K. D. 2003. Three-dimensional optical tomography: Resolution in small-object imaging. *Appl Opt*, 42, 3117–28.

Dehghani, H. and Pogue, B.W. 2011. NIRFAST. Dartmouth College.

Fang, Q. Q., Carp, S. A., Selb, J., Boverman, G., Zhang, Q., Kopans, D. B., Moore, R. H., Miller, E. L., Brooks, D. H., and Boas, D. A. 2009. Combined optical imaging and mammography of the healthy breast: Optical contrast derived From breast structure and compression. *IEEE Trans Med Imaging*, 28, 30–42.

Fang, Q., Selb, J., Carp, SA., Boverman, G., Miller, EL., Brooks, DH., Moore, RH., Kopans, DB., and Boas, DA. 2011. Combined optical and x-ray tomosynthesis breast imaging. *Radiology*, 258(1), 89–97.

Hebden, J. C., Veenstra, H., Dehghani, H., Hillman, E. M. C., Schweiger, M., Arridge, S. R., and Delpy, D. T. 2001. Three-dimensional time-resolved optical tomography of a conical breast phantom. *Appl Opt*, 40, 3278–3287.

Hyde, D., Kleine, R. D., Maclaurin, S. A., Miller, E., and Brooks, D. H. 2009. Hybrid FMT–CT imaging of amyloid-ß plaques in a murine Alzheimer's disease model. *Camera*, 44, 1304–1311.

Intes, X. 2005. Time-domain optical mammography SoftScan: Initial results. *Acad Radiol*, 12, 934–47.

Intes, X., Maloux, C., Guven, M., Yazici, B., and Chance, B. 2004. Diffuse optical tomography with physiological and spatial a priori constraints. *Phys Med Biol*, 49, N155–NI63.

Jacques, S. L. and Pogue, B. W. 2008. Tutorial on diffuse light transport. *J Biomed Opt*, 13, 041302.

Jacques, S., Wang, L. 1995. Monte Carlo modeling of light transport in tissues. In: Welch, A. J., van Gemert, M. J. C. (ed.) *Optical-Thermal Response of Laser Irradiated Tissues*. NYC: Springer.

Jiang, S., Pogue, B. W., Mcbride, T. O., and Paulsen, K. D. 2003. Quantitative analysis of near-infrared tomography: Sensitivity to the tissue-simulating precalibration phantom. *J Biomed Opt*, 8, 308–315.

Kepshire, D., Mincu, N., Hutchins, M., Gruber, J., Dehghani, H., Hypnarowski, J., Leblond, F., Khayat, M., and Pogue, B. W. 2009. A microcomputed tomography guided fluorescence tomography system for small animal molecular imaging. *Rev Sci Instrum*, 80, 043701.

Kriege, M., Brekelmans, C. T., Boetes, C., Besnard, P. E., Zonderland, H. M., Obdeijn, I. M., Manoliu, R. A. et al. 2004. Efficacy of MRI and mammography for breast-cancer screening in women with a familial or genetic predisposition. *N Engl J Med*, 351, 427–437.

Kuhl, C. K. 2007. Current status of breast MR imaging—Part 2. Clinical applications. *Radiology*, 244, 672–691.

Kuhl, C. K., Schrading, S., Leutner, C. C., Morakkabati-Spitz, N., Wardelmann, E., Fimmers, R., Kuhn, W., and Schild, H. H. 2005. Mammography, breast ultrasound, and magnetic resonance imaging for surveillance of women at high familial risk for breast cancer. *J Clin Oncol*, 23, 8469–8476.

Kukreti, S., Cerussi, A. E., Tanamai, W., Hsiang, D., Tromberg, B. J., and Gratton, E. 2010. Characterization of metabolic differences between benign and malignant tumors: High-spectral-resolution diffuse optical spectroscopy. *Radiology*, 254, 277–284.

Li, C., Mitchell, G. S., Dutta, J., Ahn, S., Leahy, R. M., and Cherry, S. R. 2009. A three-dimensional multispectral fluorescence optical tomography imaging system for small animals based on a conical mirror design. *Opt Express*, 17, 7571–7585.

Lin, Y., Gao, H., Nalcioglu, O., and Gulsen, G. 2007. Fluorescence diffuse optical tomography with functional and anatomical a priori information: Feasibility study. *Phys Med Biol*, 52, 5569–5585.

Lin, Y., Ghijsen, M. T., Gao, H., Liu, N., Nalcioglu, O., and Gulsen, G. 2011. A photo-multiplier tube-based hybrid MRI and frequency domain fluorescence tomography system for small animal imaging. *Phys Med Biol*, 56, 4731.

Li, D., Meaney, P. M., Tosteson, T. D., Jiang, S., Kerner, T. E., McBride, T. O., Pogue, B. W., Hartov, A., and Paulsen, K. D. 2003. Comparisons of three alternative breast modalities in a common phantom imaging experiment. *Med Phy*, 30, 2194–2205.

Lynch, D. R. 2005. *Numerical Partial Differential Equations for Environmental Scientists and Engineers: A First Practical Course*. New York: Springer.

Mastanduno, M. A., Jiang, S., Diflorio-Alexander, R., Pogue, B. W., and Paulsen, K. D. 2011. Remote positioning optical breast magnetic resonance coil for slice-selection during image-guided near-infrared spectroscopy of breast cancer. *J Biomed Opt*, 16, 066001.

McCann, C. M., Waterman, P., Figueiredo, J. L., Aikowa, E., Weissleder, R., and Chen, J. W. 2009. Combined magnetic resonance and fluorescence imaging of the living mouse brain reveals glioma response to chemotherapy. *Neuroimage*, 45, 360–369.

Moon, J. A., Mahon, R., Duncan, M. D., and Reintjes, J. 1993. Resolution limits for imaging through turbid media with diffuse light. *Opt Lett*, 18, 1591–1593.

Moon, J. A. and Reintjes, J. 1994. Image resolution by use of multiply scattered light. *Opt Lett*, 19, 521–523.

Ntziachristos, V., Schellenberger, E. A., Ripoll, J., Yessayan, D., Graves, E., Bogdanov, A., Jr., Josephson, L., and Weissleder, R. 2004. Visualization of antitumor treatment by means of fluorescence molecular tomography with an annexin V-Cy5.5 conjugate. *Proc Natl Acad Sci USA*, 101, 12294–12299.

Ntziachristos, V., Tung, C.-H., Bremer, C., and Weissleder, R. 2002a. Fluorescence molecular tomography resolves protease activity in vivo. *Nat Med*, 8, 757–760.

Ntziachristos, V., Yodh, A. G., Schnall, M., and Chance, B. 2000. Concurrent MRI and diffuse optical tomography of breast after indocyanine green enhancement. *Proc Natl Acad Sci USA*, 97, 2767–2772.

Ntziachristos, V., Yodh, A. G., Schnall, M. D., and Chance, B. 2002b. MRI-guided diffuse optical spectroscopy of malignant and benign breast lesions. *Neoplasia,* 4, 347–354.

Orel, S. G. and Schnall, M. 2001. MR imaging of the breast for detection, diagnosis, and staging of breast cancer. *Radiology,* 220, 13–30.

Patterson, M. S., Chance, B., and Wilson, B. C. 1989. Time resolved reflectance and transmittance for the non-invasive measurement of tissue optical properties. *Appl Opt,* 28, 2331–2336.

Poellinger, A., Martin, J. C., Ponder, S. L., Freund, T., Hamm, B., Bick, U., and Diekmann, F. 2008. Near-infrared laser computed tomography of the breast first clinical experience. *Acad Radiol,* 15, 1545–1553.

Pogue, B. W., Davis, S. C., Leblond, F., Mastanduno, M. A., Dehghani, H., and Paulsen, K. D. 2011. Implicit and explicit prior information in near-infrared spectral imaging: Accuracy, quantification and diagnostic value. *Philos Transact A Math Phys Eng Sci,* 369, 4531–4557.

Pogue, B. W., Davis, S. C., Song, X., Brooksby, B. A., Dehghani, H., and Paulsen, K. D. 2006. Image analysis methods for diffuse optical tomography. *J Biomed Opt,* 11, 33001.

Pogue, B. W., Geimer, S., Mcbride, T. O., Jiang, S., Osterberg, U. L., and Paulsen, K. D. 2001. Three-dimensional simulation of near-infrared diffusion in tissue: Boundary condition and geometry analysis for finite-element image reconstruction. *Appl Opt,* 40, 588–600.

Pogue, B. W., Jiang, S., Dehghani, H., Kogel, C., Soho, S., Srinivasan, S., Song, X., Tosteson, T. D., Poplack, S. P., and Paulsen, K. D. 2004. Characterization of hemoglobin, water, and NIR scattering in breast tissue: Analysis of intersubject variability and menstrual cycle changes. *J Biomed Opt,* 9, 541–552.

Pogue, B. W., Paulsen, K. D., Abele, C., and Kaufman, H. 2000. Calibration of near-infrared frequency-domain tissue spectroscopy for absolute absorption coefficient quantitation in neonatal head-simulating phantoms. *J Biomed Opt,* 5, 185–193.

Poplack, S. P., Tosteson, T. D., Wells, W. A., Pogue, B. W., Meaney, P. M., Hartov, A., Kogel, C. A., Soho, S. K., Gibson, J. J., and Paulsen, K. D. 2007. Electromagnetic breast imaging: Results of a pilot study in women with abnormal mammograms. *Radiology,* 243, 350–359.

Schmitz, C. H., Klemer, D. P., Hardin, R., Katz, M. S., Pei, Y., Graber, H. L., Levin, M. B. et al. 2005. Design and implementation of dynamic near-infrared optical tomographic imaging instrumentation for simultaneous dual-breast measurements. *Appl Opt,* 44, 2140–2153.

Schmitz, C. H., Locker, M., Lasker, J. M., Hielscher, A. H., and Barbour, R. L. 2002. Instrumentation for fast functional optical tomography. *Rev Sci Instr,* 73, 429–439.

Schneider, B. P. and Miller, K. D. 2005. Angiogenesis of breast cancer. *J Clin Oncol,* 23, 1782–1790.

Schweiger, M. and Arridge, S. R. 2011. *TOAST: Image Reconstruction in Optical Tomography.* University College London, Birmingham, UK. Retrieved: http://web4.cs.ucl.ac.uk/research/vis/toast/index.html

Seetamraju, M., Zhang, X., Davis, S., Gurjar, R., Myers, R., Pogue, B. W., and Entine, G. 2011. Concurrent magnetic resonance and diffuse luminescence imaging for hypoxic tumor characterization. In: *Proc. SPIE 7892, Multimodal Biomedical Imaging VI,* San Francisco, CA: 78920K-10.

Srinivasan, S., Pogue, B. W., Jiang, S., Dehghani, H., Kogel, C., Soho, S., Gibson, J. J., Tosteson, T. D., Poplack, S. P., and Paulsen, K. D. 2006. *In vivo* hemoglobin and water concentrations, oxygen saturation, and scattering estimates from near-infrared breast tomography using spectral reconstruction. *Acad Radiol,* 13, 195–202.

Stuker, F., Baltes, C., Dikaiou, K., Vats, D., Carrara, L., Charbon, E., Ripoll, J., and Rudin, M. 2011. Hybrid small animal imaging system combining magnetic resonance imaging with fluorescence tomography using single photon avalanche diode detectors. *IEEE Trans Med Imaging,* 30, 1265–1273.

Tofts, P. S., Berkowitz, B., and Schnall, M. D. 1995. Quantitative analysis of dynamic Gd-DTPA enhancement in breast tumors using a permeability model. *Magn Reson Med,* 33, 564–568.

Tromberg, B. 2011. *Virtual Photonics Technology Initiative.* Beckman Laser Institute, Irving, CA. Retrieved: http://www.virtualphotonics.org/Pages/Page/Item?slug=welcome-to-the-virtual-photonics-technology-initiative

Tromberg, B. J., Pogue, B. W., Paulsen, K. D., Yodh, A. G., Boas, D. A., and Cerussi, A. E. 2008. Assessing the future of diffuse optical imaging technologies for breast cancer management. *Med Phys,* 35, 2443–2451.

Unlu, M. B., Lin, Y., Birgul, O., Nalcioglu, O., and Gulsen, G. 2008. Simultaneous *in vivo* dynamic magnetic resonance-diffuse optical tomography for small animal imaging. *J Biomed Opt,* 13, 060501-3.

Vander Heiden, M. G. 2011. Targeting cancer metabolism: A therapeutic window opens. *Nat Rev Drug Discov,* 10, 671–684.

Weissleder, R., Tung, C.-H., Mahmood, U., and Bogdanov, A., Jr. 1999. *In vivo* imaging of tumors with protease-activated near-infrared fluorescent probes. *Nat Biotechnol,* 17, 375–378.

Yalavarthy, P. K., Pogue, B. W., Dehghani, H., Carpenter, C. M., Jiang, S., and Paulsen, K. D. 2007a. Structural information within regularization matrices improves near infrared diffuse optical tomography. *Opt Express,* 15, 8043–8058.

Yalavarthy, P. K., Pogue, B. W., Dehghani, H., and Paulsen, K. D. 2007b. Weight-matrix structured regularization provides optical generalized least-squares estimate in diffuse optical tomography. *Med Phys,* 34, 2085–2098.

Three-Dimensional *In Vivo* Microscopy

Photoacoustic Microscopy for Ophthalmic Applications

Hao F. Zhang
Northwestern University

Shuliang Jiao
University of Southern California

18.1 Introduction

Photoacoustic microscopy (PAM) has the unique ability to image the optical absorption contrast in biological tissue with scalable spatial resolution and penetration depth. Recently, researchers have been investigating the use of PAM for both fundamental biomedical research and for possible clinical uses. A promising potential clinical application is for ophthalmic imaging, by integrating PAM with existing clinical modalities, such as optical coherence tomography (OCT) and scanning laser ophthalmoscopy (SLO). In this chapter, we first briefly introduce the principles of image acquisition and formation in PAM, especially optical resolution PAM (OR-PAM). Then, we describe the stepwise technical development of PAM toward retinal imaging. In the end, *in vivo* multimodal retinal imaging of rodent eyes is demonstrated and prospects for future development of ophthalmic PAM are discussed.

18.2 Principles of Photoacoustic Microscopy

All photoacoustic (PA) imaging techniques, including PAM and photoacoustic computed tomography (PAT), share the same physical principle: absorption of optical energy leads to adiabatic thermal-elastic vibration that can be detected using ultrasonic transducers and used to form images. However, the image-formation procedure in PAM is much more straightforward than that in reconstruction-based PAT and no sophisticated reconstruction algorithms are involved.

There are two types of PAM: ultrasonic-resolution PAM (UR-PAM) (Maslov et al., 2005, Zhang et al., 2006b, 2007c) and optical-resolution PAM (OR-PAM) (Maslov et al., 2008, Hu et al., 2009, Hu and Wang, 2010, Ku et al., 2010, Zhang et al., 2010a). UR-PAM is designed for deep-tissue imaging, where laser illumination is diffusively delivered into the tissue and only ultrasonic detection is focused. The laser beam can be illuminated either from the sides of a focused transducer (Song and Wang, 2007) or through a dark-field ring positioned coaxially with a focused transducer (Zhang et al., 2006b, 2007c). The lateral resolution of UR-PAM is determined by the ultrasonic focal size and the axial resolution is determined by the ultrasonic center frequency and bandwidth (Hu and Wang, 2010). At a center frequency of 50 MHz and a bandwidth of 30 MHz, the reported axial and lateral resolutions are 15 and 35 μm, respectively (Zhang et al., 2006b).

In OR-PAM, optical illumination is strongly focused and ultrasonic detection can be either focused (Maslov et al., 2008) or nonfocused (Xie et al., 2009b). In the design with focused ultrasonic detection, the ultrasonic focus spatially overlaps the optical focus to form a confocal geometry in either transmission mode (Li et al., 2009) or reflection mode (Maslov et al., 2008). The goal of OR-PAM is to work within a comparable depth range as existing optical microscopic technologies but to provide optical absorption contrast. The lateral resolution of OR-PAM is, therefore, determined by the optical focal spot and the axial resolution is still determined by the ultrasonic parameters. Typical lateral and axial resolutions in OR-PAM are 5 and 15 μm, respectively, at an ultrasonic center frequency of 75 MHz and a bandwidth of 100 MHz. Using OR-PAM, single capillaries were imaged *in vivo* with high contrast-to-background ratio (Maslov et al., 2008) without resorting to extrinsic labels.

Both UR-PAM and OR-PAM share identical data-acquisition and image-visualization procedures as shown in Figure 18.1. During data acquisition, the optical-ultrasonic assembly is raster-scanned mechanically in the *x–y* plane along a zig-zag route (Figure 18.1a). At each position in the *x–y* plane, a laser pulse is fired and a time-resolved PA signal is detected and converted to a depth-resolved one-dimensional signal along the *z* axis (referred to as A-line) based on the ultrasonic propagating velocity in the tissue (~1.5 mm/μs). A cross-sectional image (B-scan) in the *z–x* plane is formed by aligning all the A-lines with an identical *y* coordinate and a three-dimensional image is further generated by combining all the B-scan images. Here, the *x* axis is considered the fast scanning axis.

PAM images can be visualized at different levels of details. In *in vivo* imaging of subcutaneous microvascular network in a normal rodent, a typical A-line is shown in Figure 18.1b, where

three PA peaks corresponding to the skin surface and two subcutaneous vessels are observed. In the PAM B-scan image (Figure 18.1c), the continuous skin surface and vessel cross sections are observed, which is similar to an ultrasound B-scan image but showing optical absorption contrast rather than ultrasonic impedance mismatch. A more widely used visualization method in PAM is to generate a maximum-amplitude-projection (MAP) image (Figure 18.1d), where the maximum PA value in each A-line is identified and reorganized into a two-dimensional (2D) image according to its *x–y* coordinate (Zhang et al., 2006a, 2007c). The MAP method was previously widely used in radiological angiography and has proved to be effective in visualizing the vascular networks in PAM. The most vivid method is, of course, to visualize the data in three dimensions. Figure 18.1e is a pseudocolored volume rendering of the subcutaneous vessel network using VolView™ (Kitware), where vessels at different depths can be observed. When the subject contains multiple layers of anatomic features, such as skin surface and underlying vessels, a direct MAP may not be able to

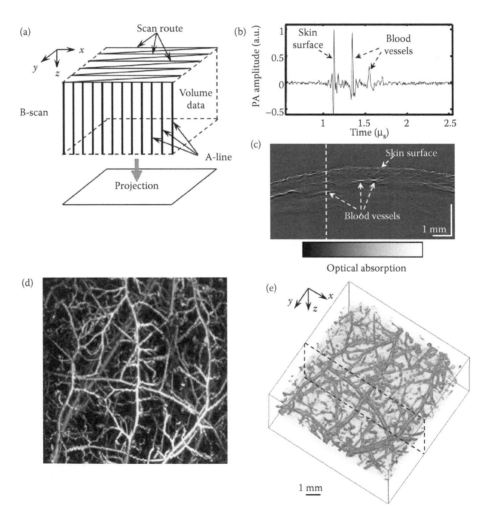

FIGURE 18.1 Image acquisition and visualization in PAM. (a) Illustration of scanning route and formation of a PAM volumetric data; (b) typical PAM A-line; (c) typical PAM B-scan image. The vertical dashed line indicates the position of the A-line shown in b; (d) typical two-dimensional maximum-amplitude-projection image; (e) directly rendered three-dimensional image. The dashed box indicates the position of the B-scan image shown in (c). (Reprinted with permission from Zhang, H. F., Maslov, K. and Wang, L. V. 2007d. *Nat Protoc*, 2, 797–804.)

visualize the vessel network correctly (Zhang et al., 2009). In this case, a boundary detection algorithm becomes necessary to identify and remove the unwanted PA signals from certain layers (Zhang et al., 2009).

18.3 Technical Developments of Photoacoustic Microscopy Toward Retinal Imaging

Before PAM could be applied to ophthalmic imaging, we needed to overcome a few technological obstacles. First, the existing PAM usually needs a long data-acquisition time as mechanically translating all the necessary optical and ultrasonic components can be slow. Because the eye is the fastest-moving organ in humans, a high imaging speed is essential to minimize any possible motion artifact. Second, almost all existing PAM methods need a water bath coupling the ultrasonic detector to the sample and the ultrasonic transducer is usually placed directly above the subject. However, such a detection geometry is not suitable for ophthalmic imaging because the physical contact may damage the cornea, especially when the imaging time is long. Third, the existing PAM is not compatible with the current clinical ophthalmic imaging modalities, such as OCT and SLO. As a result, it will be hard for PAM to gain benefits from the established technologies and to be easily accepted by ophthalmologists. We think that the optimal solution for these obstacles is to design an optical-scanning PAM and to keep the ultrasonic detector stationary during imaging (Xie et al., 2009b). As a result, we can avoid the use of water bath and can achieve both high imaging speed and compatibility with established technologies.

18.3.1 Laser-Scanning Optical-Resolution Photoacoustic Microscopy

The schematic diagram of the original laser-scanning OR-PAM (LSOR-PAM) reported in 2009 (Xie et al., 2009b) is shown in Figure 18.2. A tunable dye laser pumped by a Q-switched Nd:YLF laser system was used as the irradiation source (Figure 18.2a).

The laser system was triggered externally at a pulse repetition rate (PRR) of 1 kHz and the optical wavelength was 580 nm. The output laser light from the dye laser was spatially filtered by an iris and expanded to reach a beam diameter of 8 mm. After being attenuated, the expanded laser beam passed through a two-dimensional galvanometer scanner and a final objective lens.

A polarization controller (PC)-controlled analog-output board synchronized the laser triggering and the optical scanner. The time courses of the analogue (AO) output are shown in Figure 18.2b. The x axis is the fast scanning axis and the y axis is the slow axis. Within an active window, 512 triggering pulses were sent to the pump laser and one roundtrip scan (containing two B-scans) was performed; hence, each B-scan image contains 256 scanning points in LSOR-PAM. After sending one triggering pulse train, the AO board entered a resting period (2 s) allowing data transfer, processing, and display. In case of unexpected mechanical scanning errors, an interlock signal (Figure 18.2b) is sent to the laser system to disable the laser firing. This feature is important for animal and clinical applications as a key laser safety measure, especially in ophthalmic imaging.

The induced PA waves were detected by an unfocused ultrasonic transducer (center frequency: 10 MHz; bandwidth: 80%; active element diameter: 6 mm). The ultrasonic transducer was placed 30 mm from the sample and tilted about 15°. A fast photodiode detected the laser pulses and triggered the data acquisition to avoid the impact of laser jittering. The energy of each laser pulse was also recorded from the photodiode to compensate for the pulse energy instability. The sampling rate of the data acquisition board was 200 MS/s and the PA signals were recorded for 3 μs. No signal averaging was employed during data acquisition.

One major difference between LSOR-PAM and the mechanical scanning PAM is the field of view (FOV). In LSOR-PAM, the FOV is fundamentally limited by the ultrasonic transducer. Because the ultrasonic transducer is kept stationary during imaging, only PA signals generated within a certain ultrasonic sensitivity region can be detected with sufficient signal-to-noise ratio. In comparison, the FOV in mechanical scanning PAM is, in principle, unlimited.

The experimentally measured FOV in LSOR-PAM is shown in Figure 18.3a. The imaged target was a carbon mesh grid printed

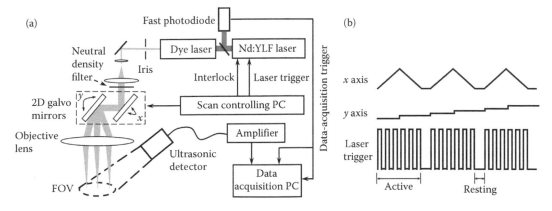

FIGURE 18.2 Experimental setup of LSOR-PAM. (a) Schematic of the experimental setup; (b) time courses of the galvanometer driving voltages and laser triggers. (Reprinted with permission from Xie, Z. et al. 2009a. *Opt Lett*, 34, 1771–3.)

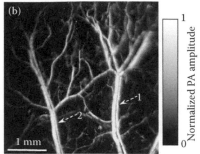

FIGURE 18.3 LSOR-PAM image of (a) a mesh grid to quantify FOV and (b) microvascular network in a mouse ear *in vivo*. (Reprinted with permission from Xie, Z. et al. 2009a. *Opt Lett*, 34, 1771–3.)

on a transparency using a commercial laser printer. The grid size was 865 μm and the optically scanned area in LSOR-PAM was 8 × 8 mm². According to the size of the grid, the FOV was a circular area with a diameter of at least 6 mm, where more than seven grids were observed with a local contrast-to-noise ratio (CNR) greater than 6 dB. This measured result agreed with simulation result obtained using Field II.* The experimentally quantified lateral resolution was 7.8 μm as measured by imaging a USAF 1451 resolution target and the axial resolution was measured to be 50 μm by applying the shift-and-sum of a point spread function along the *z* axis (Xie et al., 2009b).

The *in vivo* LSOR-PAM image of microvasculature in a Swiss Webster mouse ear is shown in Figure 18.3b. As previously demonstrated (Hu et al., 2009), OR-PAM is able to resolve the capillary network with high quality and offers the capability to study the associated physiological changes in the capillaries (Yao et al., 2011b). Although LSOR-PAM has lower spatial resolution and CNR compared to other reported OR-PAM, it shows a similar quality in imaging microvascular structures. Up to seven orders of vessel branching can be observed, and the smallest vessel is imaged by a single pixel in the image. Parallel veins and arteries, as pointed out in Figure 18.3b, were also clearly shown.

As we can learn from the above, the advantages of LSOR-PAM stem from several factors. First, removing the bulky load (all the optical and ultrasonic components) that used to be physically translated in OR-PAM can potentially achieve higher scanning speeds. In the results shown above, the laser PRR was only 1 kHz, which gave a total image acquisition of around 2 min. In our later studies, the PRR was 24 kHz and the image acquisition time was reduced to <3 s (Jiao et al., 2010b), a compelling demonstration of the advantages of optical scanning. Second, optical scanning provides the freedom to implement sophisticated scanning patterns, such as the radial and circular scans (Srinivasan et al., 2008, Cheung et al., 2009) that are often used in ophthalmic imaging.

What is more important is that the optical delivery and scanning mechanism in LSOR-PAM makes it compatible with most existing optical microscopic modalities and lays the foundation

for integrating LSOR-PAM with confocal microscopy (Zhang et al., 2010b), OCT (Jiao et al., 2009b, Liu et al., 2011b), fluorescence microscopy (Zhang et al., 2010c), and adaptive optics (Jiang et al., 2010). The integration of PAM and OCT later proved to be essential for PA retinal imaging because OCT adds not only additional optical contrast to the multimodal imaging, but also provides critical guidance for initial optical alignment.

18.3.2 Integrating Photoacoustic Microscopy with Optical Coherence Tomography

Now, we demonstrate the integration of LSOR-PAM with spectral-domain OCT (SD-OCT) for both anatomic (Jiao et al., 2009a) and functional (Liu et al., 2011a) imaging of microvascular network *in vivo*.

Figure 18.4 shows a schematic diagram of the integrated system. The configuration and data acquisition of the LSOR-PAM subsystem are the same as shown above. The PAM illumination beam was merged with the OCT probing light by a dichroic mirror. The combined light beams were scanned by an *x–y* galvanometer and were focused on the sample by an achromatic objective lens.

The fiber-based SD-OCT subsystem (Jiao et al., 2005) consists of a broadband superluminescent diode (SLD) (center wavelength: 840 nm, full-width-half-maximum bandwidth: 50 nm), a 3 dB fiber coupler, a reference arm, a sample arm coupled with the PAM, and a spectrometer for detecting the interference signals in the spectral domain. The exposure time of the line-scan charge-coupled device (CCD) camera in the spectrometer was 36 μs, which gave an A-line rate of 24 kHz. The calibrated depth resolution was 6 μm in tissue and the lateral resolution was estimated to be 20 μm.

To ensure the automatic registration of the two imaging modalities under all circumstances, the timing of the LSOR-PAM laser triggering and data acquisition, the scanning of the galvanometer scanner, and the OCT data acquisition were controlled by the same time base generated by an analog-output board. As a result, the two subsystems are synchronized despite the difference in their achievable imaging speed. Besides just integrating the two imaging systems for performing naturally registered multimodal imaging, the key improvement is the ability to acquire the images simultaneously using the two different

---

* Field II Simulation Program http://server.oersted.dtu.dk/personal/jaj/field/.

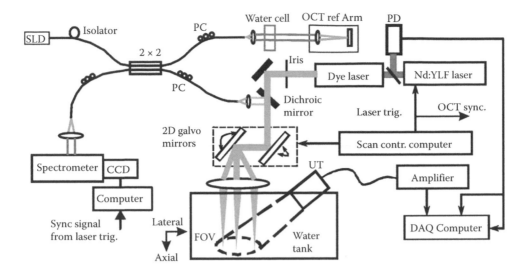

FIGURE 18.4 Schematic diagram of the integrated LSOR-PAM and SD-OCT. SLD: superluminescent diode; PC: polarization controller; PD: photodiode; UT: ultrasonic transducer. (Reprinted with permission from Jiao, S. et al. 2009a. *Opt Lett*, 34, 2961–3.)

modalities, which is critical for ophthalmic imaging to eliminate the image distortion between two sequential acquisitions.

The *in vivo* dual-mode images of a Swiss Webster mouse ear are shown in Figure 18.5. Figure 18.5a is a typical OCT B-scan image consisting of 400×220 (depth × lateral) pixels at the location highlighted in Figure 18.5b. Different anatomical layers, including epidermis, dermis, and the cartilaginous backbone, can be seen (Barker et al., 1989, So et al., 1998). There are also shadows in the images created by either blood vessels or clusters of scatterers, presumably from various glands that blocked the light from probing tissues below them, but we cannot tell the exact locations of the blood vessels. Figure 18.5b shows the LSOR-PAM projection image that was acquired simultaneously. Since the PAM image visualizes the distribution of optical-energy deposition, only the blood vessels appear clearly in the image whereas no information about skin anatomy can be observed, which demonstrates the complementary nature of the contrasts contributing to OCT and PAM.

When PAM and OCT images are fused, both the skin anatomy and microvasculature can be visualized. Figure 18.5c is the fused B-scan image at the location marked in Figure 18.5b. The registration of the two images along the lateral directions was automatic and was guaranteed by the system design and optical alignment. The SD-OCT color map was inverted to highlight the blood vessels and the blood vessels are pseudocolored red. The volume rendering of the final fused image is shown in Figure 18.5d, where the vessels imaged by LSOR-PAM fill in the shadows in the SD-OCT volume.

Moving beyond anatomic imaging, functional information can be acquired by the two respective modalities and can be further combined to measure more sophisticated physiological parameters. Here, we demonstrate the measurement of the metabolic rate of oxygen (MRO_2) in small animals based on multi-wavelength PAM and Doppler OCT without extrinsic contrast agents (Liu et al., 2011a). In future ophthalmic applications, measuring retinal MRO_2 could be significant for early detection and better understating of, for example, diabetic retinopathy (Alder et al., 1997) and retinopathy of prematurity (Stone et al., 1996).

To measure the MRO_2 *in vivo*, we selected an artery–vein pair (Figure 18.6a) imaged from a mouse ear. Then we acquired the

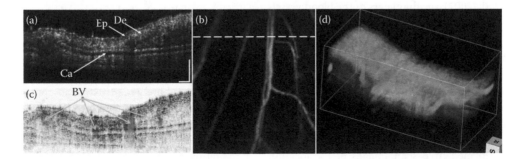

FIGURE 18.5 (**See color insert.**) Simultaneously acquired LSOR-PAM and SD-OCT images of a mouse ear. (a) SD-OCT B-scan at the location in panel b highlighted by the dashed line; (b) LSOR-PAM projecting image; (c) fused LSOR-PAM and SD-OCT B-scan image; and (d) fused LSOR-PAM and SD-OCT volumetric image. Ep: epidermis; De: dermis; Ca: cartilaginous backbone; BV: blood vessels; bar: 200 µm. (Reprinted with permission from Jiao, S. et al. 2009a. *Opt Lett*, 34, 2961–3.)

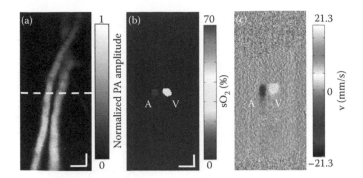

FIGURE 18.6 (**See color insert.**) Multimodal functional imaging using multiwavelength PAM and Doppler OCT. (a) Imaged sO_2 using LSOR-PAM; (b) imaged blood flow velocity using SD-OCT. Bar: 100 μm. A: artery; V: vein. (Reprinted with permission from Liu, T. et al. 2011a. *Biomed Opt Express*, 2, 1359–65.)

following parameters: the hemoglobin oxygen saturation (sO_2) in artery and vein (sO_{2a} and sO_{2v}, dimensionless), the mean total hemoglobin concentration ($\overline{HbT}$, g/L) in blood, the diameters of the artery and vein (d_a and d_v, m), and the mean blood flow velocities in the artery and vein ($\overline{v}_a$ and $\overline{v}_v$, m/s). Assuming that vessels have circular cross sections, that total hemoglobin concentration is constant, and that each hemoglobin molecule carries four oxygen molecules, the MRO_2 (g/min) is given by (Wang, 2008)

$$
\begin{aligned}
MRO_2 &= 60 \cdot \frac{4 \cdot W_{O_2}}{W_{HbO_2}} \cdot \left[\overline{[HbT]} \cdot sO_{2a} \cdot (\frac{\pi}{4} \cdot d_a^2) \cdot \overline{v}_a - \overline{[HbT]} \cdot \right. \\
&\quad \left. sO_{2v} \cdot (\frac{\pi}{4} \cdot d_v^2) \cdot \overline{v}_v \right] \\
&= \frac{60 \cdot \pi \cdot W_{O_2}}{W_{HbO_2}} \cdot \overline{[HbT]} \cdot \left(sO_{2a} \cdot d_a^2 \cdot \overline{v}_a - sO_{2v} \cdot d_v^2 \cdot \overline{v}_v \right),
\end{aligned}
$$

$$(18.1)$$

where W_{O_2} and W_{HbO_2} are the molecular weights of O_2 and oxyhemoglobin, which are 32 and 68,000, respectively. Here, the sO_2 was calculated based on multiwavelength LSOR-PAM measurements (Zhang et al., 2006c, 2007a, b, d, Maslov et al., 2007); the blood flow velocities were measured by Doppler OCT (Wehbe et al., 2007, Wang et al., 2007); the diameters of the blood vessels were measured by LSOR-PAM along the lateral direction; and the $\overline{HbT}$ was measured from blood samples by a commercial veterinary laboratory. Alternatively, $\overline{HbT}$ can potentially be acquired by PAM measured at an isosbestic point in the molar extinction spectrum of hemoglobin (Yao et al., 2011a) after calibration.

The *in vivo* results of sO_2 and blood flow velocity measurements in the artery–vein pair are shown in Figure 18.6b and c, respectively. The quantification of sO_2 and blood flow in these two vessels was performed in B-scan images at the location highlighted by the dashed line. Figure 18.6b shows the sO_2 values calculated from LSOR-PAM B-scan images acquired at the four optical wavelengths. The measured sO_2 distinguished the artery (sO_{2a} = 68.9%) and vein (sO_{2v} = 38.3%) in the vessel pair.

The vessel diameters were measured to be d_a = 42.69 ± 0.09 μm and d_v = 61.13 ± 0.25 μm. Figure 18.6c shows the Doppler OCT results, where the mean velocities were $\overline{v}_a$ = −15.58 ± 0.43 mm/s and $\overline{v}_v$ = 8.11 ± 0.48 mm/s.

We compared the blood flow rates (F_a: flow rate in artery; F_v: flow rate in vein) in this vessel pair. Using the measured blood flow velocities and diameters, we have F_a = 1.34 ± 0.04 μL/min and F_v = 1.43 ± 0.10 μL/min, which are comparable and, thus, show that the measurements are self-consistent. Finally, based on the $\overline{HbT}$ value of 344g/L, the calculated MRO_2 was 242.77 ± 30.50 ng/min, assuming that the vessel pair supports the same tissue volume. The measured MRO_2 was further converted to a more widely used unit as 193.70 ± 24.33 nL/min when considering typical blood temperature (311.15 K), molar mass of oxygen (32 g/mol), and standard atmosphere pressure (101,325 Pa).

18.4 Photoacoustic Ophthalmoscopy

18.4.1 Introduction

Ophthalmic imaging, especially retinal imaging, plays a key role in screening, diagnosis, intraoperative guidance, and treatment evaluation of all major blinding diseases. The existing retinal imaging modalities include fundus photography, SD-OCT, SLO, autofluorescence imaging, and fluorescein angiography (FA), all of which are commercially available and can provide superb image quality. They all rely on the detection of photons either reflected from the retinal tissues or re-emitted from an endogenous chromophore or an extrinsic contrast agent. In other words, existing retinal imaging technologies can only measure optical scattering and fluorescence properties; the capability to quantify the physiological-specific optical absorption properties of the retina is still missing.

The fundamental understanding and early detection of several irreversible blinding diseases can potentially be improved by noninvasive imaging of the optical absorption properties in the retina. For example, if the optical absorption by retinal blood could be imaged, the retinal sO_2 could be quantified with a high precision to potentially benefit the early detection of diabetic retinopathy and to provide better understanding of the molecular pathway of its progress, which is strongly associated with retinal ischemia in its early stage (Mohamed et al., 2007). If the optical absorption of melanin in the retinal pigment epithelium (RPE) could be measured, early detection of age-related macular degeneration (AMD) and better understanding of retinal aging could potentially be achieved because RPE cell dysfunction and AMD (Delori et al., 2001) are both often associated with melanin photobleaching (Burke and Hjelmeland, 2005). As a result, we have developed photoacoustic ophthalmoscopy (PAOM) to fill this technological void, and we have further integrated PAOM with SD-OCT, SLO, and FA for multimodal imaging and cross-modality validation.

18.4.2 Laser Safety

Before *in vivo* experiments, laser safety in PAOM was carefully evaluated and the American National Standards Institute (ANSI) ocular laser safety standard* was strictly followed. The ocular maximum permissible exposure (MPE) of the incident laser depends on the optical wavelength, pulse duration, exposure duration, and exposure aperture. Since no signal averaging will be used in PAOM, each scanning location on the retina will be exposed only to a single laser pulse. As a result, by taking all the experimental parameters into consideration, a pulse energy of 40 nJ is within the ANSI safety limit and should be considered safe. More detailed derivations can be found in Hu et al. (2010). When compared with a clinical ocular ultrasound biomicroscope (UBM), the amplitude of the laser-induced PA signal (~100 Pa) is four orders of magnitude weaker than the ultrasound amplitude of UBM (~1 × 10⁶ Pa).

---

* American National Standard for Safe Use of Lasers ANSI Z136.1-2007 (American National Standards Institute Inc., New York, NY, 2007).

18.4.3 Optical Coherence Tomography-Guided Photoacoustic Ophthalmoscopy

The schematic diagram of integrated SD-OCT-guided PAOM is shown in Figure 18.7a (Jiao et al., 2010a). Besides providing intrinsically registered multimodal images, another key reason for integration is that we can use SD-OCT as an alignment tool to achieve the optimal optical illumination and to minimize the discomfort caused by visible-light illumination in PAOM.

The optical setup of the SD-OCT-guided PAOM is similar to what we have described in Figure 18.4. The major difference is in the last stage of optical delivery, where the collimated SD-OCT and PAOM beam pass through a telescope assembly and then enter the eye through the dilated pupil. In our experiments, the PAOM laser pulse energy was 40 nJ/pulse and the SD-OCT light source power was 0.8 mW. The PAOM illuminating wavelength was 532 nm and the light source in SD-OCT had a center wavelength of 830 nm and bandwidth of 60 nm. To detect the induced PA waves, we used a small-footprint custom-built needle ultrasonic transducer (center frequency: 30 MHz; bandwidth: 50%; active element diameter: 1 mm) and placed it in gentle contact with the eyelid coupled by ultrasound gel (Figure 18.7b). The axial resolutions of PAOM and SD-OCT were 23 and 4 μm, respectively; the lateral resolutions of PAOM and SD-OCT in rat eyes were comparable (~20 μm) and differed only by the ratio of the two optical wavelengths between PAOM and OCT.

The OCT-guided PAOM has two operation modes: alignment mode and acquisition mode. At the beginning of each imaging session, the PAOM worked in the alignment mode and the PAOM laser was blocked. Only the real-time OCT display was activated to provide guidance for locating the retinal region of interest (ROI) and optical focusing. After the ROI was identified, the PAOM laser was turned on and the system was switched to the acquisition mode. In both PAOM and SD-OCT, the A-line rate was 24 kHz; each B-scan image contained 256 A-lines and

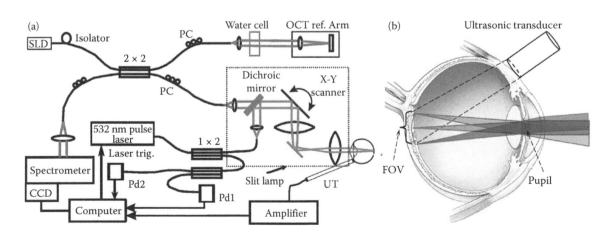

FIGURE 18.7 (**See color insert.**) Experimental system of the OCT-guided PAOM. (a) Schematic diagram of the optical setup; and (b) geometry of the optical illumination and ultrasonic detection in PAOM. SLD: superluminescent diode; PC: polarization controller; Pd: photodiode; FOV: field of view; UT: ultrasonic transducer. (Reprinted with permission from Jiao, S. et al. 2010a. *Opt Express*, 18, 3967–72.)

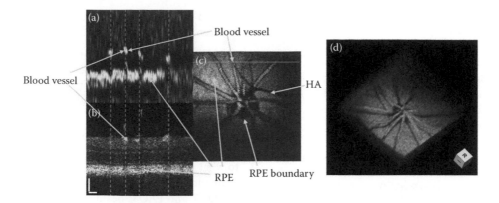

FIGURE 18.8 Comparison of SD-OCT and PAOM images acquired simultaneously *in vivo.* (a) PAOM B-scan image in pseudocolors; (b) OCT B-scan image; (c) MAP image of the PAOM dataset; and (d) volume rendering of the PAOM dataset. Bar: 100 μm. (Reprinted with permission from Jiao, S. et al. 2010a. *Opt Express,* 18, 3967–72.)

the B-scan frame rate was 93 Hz. The simultaneous acquisition of the volumetric PAOM and OCT images (containing 256 B-scans) took 2.7 s.

Figure 18.8 shows the simultaneously acquired PAOM (Figure 18.8a) and OCT (Figure 18.8b) B-scan images of a pigmented rat eye at the position marked in Figure 18.8c. As seen in the PAOM B-scan, the blood vessels appear as clusters of high-amplitude PA signals. Beneath the blood vessels, the continuous line of high PA amplitude is the RPE layer. The vertical lines highlight the corresponding positions of the recognized vessels, which demonstrate the automatic registration of the two modalities. Figure 18.8a and b demonstrates the contrast mechanism of PAOM—the stronger the optical absorption, the stronger the generated ultrasonic signals. We can see that the hemoglobin in the blood vessels and the melanin in the RPE cells are strong absorbers of the illuminating light, while other retinal tissues have very low optical absorption. These differences in optical absorption provide the foundation for functional and anatomical imaging of the blood vessels and the RPE.

Figure 18.8c shows the PAOM projection image. The circular boundary of the RPE at the optic disc can be observed. We can see that inside the optic disc, only blood vessels absorb

the illuminating light; thus, the optic disc appears dark except for the blood vessels. Figure 18.8d demonstrates the volumetric imaging capability of PAOM. The shadows on the RPE are created by blood vessels blocking the illuminating laser light. A movie visualizing the imaged retinal vessels and RPE from different viewing angles can be found from Jiao et al. (2010a).

Owing to the presence of high melanin concentration in the RPE in pigmented rats, PAOM illumination light did not penetrate the RPE and, thus, the choroidal vascular network was not imaged by PAOM. In fact, choroidal vascular circulation is critical to maintaining a healthy physiological environment to support normal vision. The dysfunctions in the choroidal vascular system may lead to, for example, glaucoma and choroidal neovascularization (CNV), and eventually, macular degeneration. Noninvasive imaging can provide insight into the development of choroidal vascular dysfunctions and may lead to better understanding and treatment of several blinding diseases. One advantage of PAOM is that when the melanin concentration in the RPE is minimized, such as in albino eyes, both retinal and choroidal vascular networks can be imaged with comparable quality (Wei et al., 2011). Hence, PAOM can be a useful tool to study the functions and pathologies of choroidal circulation in animal models.

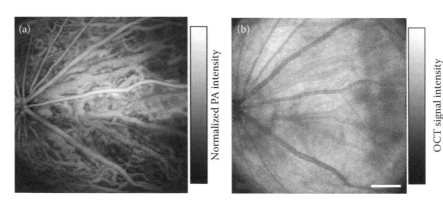

FIGURE 18.9 Comparison of chorioretinal vascular network imaged by (a) PAOM and (b) SD-OCT in an albino rat. Bar: 500 μm. (Reprinted with permission from Wei, Q. et al. 2011. *J Mod Opt,* 58, 1997–2001.)

Figure 18.9a is the PAOM projection image of both the retinal and choroidal vessels from an albino Sprague Dawley rat. There are dual circulation systems in the rodent eye: the retinal vessels supply the nerve fiber layer and the external part of the inner nuclear layer; and the choroidal vessels support the rest of the retina.* In rodent eyes, the optic disk is usually located at the center of the posterior orbit. In Figure 18.9a, the optic disk was intentionally placed off-center within the FOV in order to emphasize the choroidal vascular network in regions where the retinal vessel density is low. As the retinal arteries come out from the optic disk and the retinal veins return to the optic disk, the retinal vessels demonstrate a star-shaped architecture (Paques et al., 2006, Seeliger et al., 2005). In comparison, choroidal vessels show a densely packed network without noticeable features. Note that choroidal vessels with different sizes, ranging from capillaries, medium vessels, and large vessels in both inner Sattler's and outer Haller's layers (Nickla and Wallman, 2010), can be observed in the PAOM fundus image, but not in the SD-OCT fundus image (Figure 18.9b) due to lack of contrast.

18.4.4 A More Comprehensive Multimodal Imaging Platform

Based on the compatibility of PAOM with other established retinal imaging modalities, we further integrated PAOM with SLO and FA.† Such a multimodal system has the potential to combine the merits of SLO/FA, SD-OCT, and PAOM for a more comprehensive evaluation of the retina based on their complementary contrast mechanisms. By combining SD-OCT with PAOM, a high-resolution, cross-sectional retinal layer structure image can be obtained together with a high-contrast vascular image from PAOM. More importantly, the future combination of Doppler SD-OCT, which measures retinal blood flow, with spectroscopic PAOM, which provides retinal sO_2, will allow for evaluation of oxygen metabolism in the retina. By combining SLO with PAOM, the PAOM light source will be shared with the SLO without increasing retinal light exposure. While PAOM detects PA signals from the absorbed portion of the illumination photons, the SLO will generate a scattering-based image by collecting the reflected portion of the illumination photons, which would otherwise be discarded. By adding a 488 nm laser excitation and detecting the corresponding fluorescence emissions, integration of PAOM, SD-OCT, and FA-SLO becomes possible. In this section, we demonstrate the integration of PAOM with SD-OCT, SLO, and FA for its capabilities of comprehensive anatomic imaging and cross-modality validation.

Figure 18.10 shows the imaging results using the integrated PAOM, SD-OCT, and FA-SLO, where each modality resolves different and complementary anatomic features. Although both the PAOM fundus (Figure 18.10d) and the SD-OCT fundus (Figure 18.10f) images have bright backgrounds when imaging pigmented eyes, the background signals come from different sources. PAOM is an optical absorption-based technology; as a result, the background may contain contributions from RPE melanin, the choroidal vascular network, and choroidal melanin. Owing to the relatively poor axial resolution of PAOM (32 μm), it cannot identify the contribution from every possible source. Although there is cell-to-cell variation in RPE melanin content at the microscopic level (Burke and Hjelmeland, 2005), the current PAOM cannot resolve such variations due to limited lateral resolution. SD-OCT is an optical scattering-based technology and its near-infrared probing light is capable of penetrating deeper into the retina than the visible light used in PAOM. Thus, the "streaky background" in Figure 18.10f is likely to be caused by major choroidal vessels (Bhutto and Amemiya, 2001, Srienc et al., 2010).

A more interesting phenomenon in comparing PAOM with FA-SLO is highlighted by the arrows in the dashed boxes in Figure 18.10d and e. The dark region in the PAOM image can potentially be attributed to a lower melanin concentration in the RPE, which is supported by the FA-SLO, where the dark region in the PAOM image becomes bright. The reason is that when the melanin concentration is high in the RPE, the fluorescence excitation light is significantly attenuated and the fluorescence emission from the choroid is blocked and cannot be detected; however, in RPE regions with low melanin concentration, more fluorescence emission can be generated and detected, which leads to high signal amplitudes in the FA-SLO as seen in Figure 18.10e.

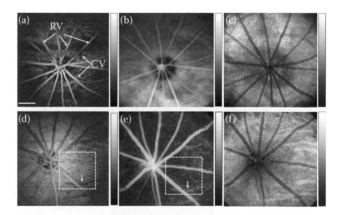

FIGURE 18.10 *In vivo* multimodal retinal imaging of an albino rat (top row) and a pigmented rat (bottom row). Panels (a) and (d) are PAOM fundus images; panels (b) and (e) are FA-SLO images; and panels (c) and (f) are *en face* SD-OCT images. RV: retinal vessels; CV: choroidal vessels. Bar: 500 μm. (Reprinted with permission from Burke, J. M. and Hjelmeland, L. M. 2005. *Mol Interv*, 5, 241–9.)

* Smith R. S. ed, *Systematic Evaluation of the Mouse Eye Anatomy Pathology, and Biomethods*. CRC Press, 2002.

† W. Song, Q. Wei, T. Liu, D. Kuai, J. M. Burke, S. Jiao, and H. F. Zhang. 2012. Integrating photoacoustic ophthalmoscopy with scanning laser ophthalmoscopy, optical coherence tomography, and fluorescein angiography for a multimodal retinal imaging platform, *Journal of Biomedical Optics*, 17, 061206.

18.4.5 Toward Functional Imaging of RPE Using Photoacoustic Ophthalmoscopy and Autofluorescence Imaging

Melanin and lipofuscin are two major pigments in the RPE. Melanin protects RPE cells from oxidative damage by acting as an antioxidant (Ostrovsky et al., 1987). Unfortunately, melanin may lose its protective function with aging and can also be irreversibly photobleached as a result of longtime light exposure (Sarna et al., 2003). On the other hand, lipofuscin is a by-product of phagocytosis of the photoreceptor outer segments, which accumulates with aging of the retina. Excessive levels of lipofuscin accumulation could compromise essential RPE functions and contribute to the pathogenesis of AMD (Delori et al., 2001). Hence, *in vivo* imaging of melanin (by PAOM) and lipofuscin (by autofluorescence imaging) (Zhang et al., 2011) can provide important aging information of the retina, which can be significant for AMD research and clinical diagnosis.

Figure 18.11a and b shows the autofluorescence images of the retina in pigmented rats at 10 weeks and 18 weeks after birth, respectively. Figure 18.11c and d are the corresponding PAOM images. The autofluorescence images represent the lipofuscin distribution in the RPE: the stronger the autofluorescence signal, the higher the lipofuscin concentration. Statistical analysis of the pixel intensity in autofluorescence images showed that lipofuscin concentration increased with age (Zhang et al., 2011). In a separate study involving imaging of albino rats, autofluorescence images showed higher intensity in albino rats than in pigmented rats from comparable age groups and also found that the increase of autofluorescence intensity with age is more obvious in albino rats than in pigmented rats (Zhang et al., 2011).

18.4.6 Prospects of Photoacoustic Ophthalmoscopy

Some of the demonstrated advantages of PAOM are briefly introduced above. In the near future, technical developments should achieve (1) better spatial resolution in retinal imaging; (2) capability of imaging large eyes toward patient imaging, and (3) functional imaging of retinal physiological parameters.

The lateral resolution of PAOM is determined by the smallest achievable illumination spot size on the retina. In *in vivo* retinal imaging, for large pupil diameters, monochromatic aberration of the eye is the major factor that limits the achievable resolution of retinal images. Although the depth resolution of PAOM could be improved by increasing the bandwidth of the ultrasonic detecting system, the lateral resolution of PAOM is fundamentally limited by ocular optics. One possible approach for improving the lateral resolution of PAOM is to use adaptive optics to compensate for the ocular aberrations. Adaptive optics has been applied to ophthalmic imaging with great success, making it possible to image single photoreceptors (Morgan et al., 2009). Adaptive optics has also been successfully integrated in OCT (Zawadzki et al., 2005) and SLO (Pircher et al., 2008). A feasibility study was conducted on the integration of adaptive optics with PAM for *ex vivo* imaging of ocular tissues (Jiang et al., 2010), which serves as the technical preparation for the future adaptive optics—PAOM for *in vivo* retinal imaging.

The current PAOM system is only suited for small-animal eyes due to its limited FOV and small transducer footprint. Compared with human eyes, whose typical diameter is more than 20 mm, the rodent eyes are usually <10 mm. The current PAOM is not compatible with human eye imaging for two reasons: the high ultrasonic center frequency (35 MHz) and small ultrasonic footprint (1 mm²). To accommodate larger eyes, lower ultrasonic center frequency is needed due to the frequency-dependent ultrasonic attenuation. It is estimated that a human PAOM will need to have an ultrasonic center frequency of around 10–15 MHz. A larger FOV is also necessary to image large eyes. Because there is a physical limit on the transducer size, novel transducers (Xie et al., 2011, Chen et al., 2011) should be employed to reach this goal.

PAM has long been used to study the sO₂ in microvasculature. It is anticipated that PAOM will be able to measure the sO₂ in retinal vessels accurately. Multiwavelength PAOM measurements are necessary for calculating the retinal sO₂. After being integrated with SD-OCT, the blood flow can be measured by Doppler OCT; as a result, the retinal MRO₂ can be derived based on the measured sO₂, blood flow, and vessel diameters. Compared to reported retinal oximetry (Harris et al., 2003, Hardarson et al., 2006), PAOM directly measures optical absorption of hemoglobin and is expected to be insensitive to the photon propagation paths and RPE melanin concentration; hence, the precision of PAOM is expected to be higher than reflection-based retinal oximetry.

  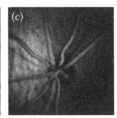 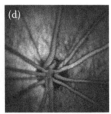

FIGURE 18.11 Autofluorescence (a) and (b) and PAOM (c) and (d) images of pigmented rat eyes. (a and c): 10 weeks old; (b and d): 18 weeks old. (Reprinted with permission from Zhang, X. et al. 2011. *J Biomed Opt*, 16, 080504.)

Acknowledgment

We believe that the ophthalmic application of PAM has a bright future. This is an ongoing project and there is still a long way before reaching our final clinical goals. During the technical development process, we have received enormous help and support from colleagues, lab members, and funding agents and we want to express our sincere thanks to all of them.

References

Alder, V. A., Su, E. N., Yu, D. Y., Cringle, S. J., and Yu, P. K. 1997. Diabetic retinopathy: Early functional changes. *Clin Exp Pharmacol Physiol*, 24, 785–8.

Barker, J. H., Hammersen, F., Bondar, I., Uhl, E., Galla, T. J., Menger, M. D., and Messmer, K. 1989. The hairless mouse ear for *in vivo* studies of skin microcirculation. *Plast Reconstr Surg*, 83, 948–59.

Bhutto, I. A. and Amemiya, T. 2001. Microvascular architecture of the rat choroid: Corrosion cast study. *Anat Rec*, 264, 63–71.

Burke, J. M. and Hjelmeland, L. M. 2005. Mosaicism of the retinal pigment epithelium: Seeing the small picture. *Mol Interv*, 5, 241–9.

Chen, S. L., Ling, T., and Guo, L. J. 2011. Low-noise small-size microring ultrasonic detectors for high-resolution photoacoustic imaging. *J Biomed Opt*, 16, 056001.

Cheung, C. Y. L., Yiu, C. K. F., Weinreb, R. N., Lin, D., Li, H., Yung, A. Y., Pang, C. P., Lam, D. S. C., and Leung, C. K. S. 2009. Effects of scan circle displacement in optical coherence tomography retinal nerve fibre layer thickness measurement: A RNFL modelling study. *Eye*, 23, 1436–41.

Delori, F. C., Goger, D. G., and Dorey, C. K. 2001. Age-related accumulation and spatial distribution of lipofuscin in RPE of normal subjects. *Invest Ophthalmol Vis Sci*, 42, 1855–66.

Hardarson, S. H., Harris, A., Karlsson, R. A., Halldorsson, G. H., Kagemann, L., Rechtman, E., Zoega, G. M., Eysteinsson, T., Benediktsson, J. A., Thorsteinsson, A., Jensen, P. K., Beach, J., and Stefansson, E. 2006. Automatic retinal oximetry. *Invest Ophthalmol Vis Sci*, 47, 5011–6.

Harris, A., Dinn, R. B., Kagemann, L., and Rechtman, E. 2003. A review of methods for human retinal oximetry. *Ophthalmic Surg Lasers Imaging*, 34, 152–64.

Hu, S., Maslov, K., and Wang, L. H. V. 2009. Noninvasive label-free imaging of microhemodynamics by optical-resolution photoacoustic microscopy. *Opt Express*, 17, 7688–93.

Hu, S., Rao, B., Maslov, K., and Wang, L. V. 2010. Label-free photoacoustic ophthalmic angiography. *Opt Lett*, 35, 1–3.

Hu, S. and Wang, L. V. 2010. Photoacoustic imaging and characterization of the microvasculature. *J Biomed Opt*, 15, 011101 (article number).

Jiang, M., Zhang, X., Puliafito, C. A., Zhang, H. F., and Jiao, S. 2010. Adaptive optics photoacoustic microscopy. *Opt Express*, 18, 21770–6.

Jiao, S., Jiang, M., Hu, J., Fawzi, A., Zhou, Q., Shung, K. K., Puliafito, C. A., and Zhang, H. F. 2010a. Photoacoustic ophthalmoscopy for *in vivo* retinal imaging. *Opt Express*, 18, 3967–72.

Jiao, S. L., Jiang, M. S., Hu, J. M., Fawzi, A., Zhou, Q. F., Shung, K. K., Puliafito, C. A., and Zhang, H. F. 2010b. Photoacoustic ophthalmoscopy for *in vivo* retinal imaging. *Opt Express*, 18, 3967–72.

Jiao, S., Knighton, R., Huang, X., Gregori, G., and Puliafito, C. 2005. Simultaneous acquisition of sectional and fundus ophthalmic images with spectral-domain optical coherence tomography. *Opt Express*, 13, 444–52.

Jiao, S., Xie, Z., Zhang, H. F., and Puliafito, C. A. 2009a. Simultaneous multimodal imaging with integrated photoacoustic microscopy and optical coherence tomography. *Opt Lett*, 34, 2961–3.

Jiao, S. L., Xie, Z. X., Zhang, H. F., and Puliafito, C. A. 2009b. Simultaneous multimodal imaging with integrated photoacoustic microscopy and optical coherence tomography. *Opt Lett* 34, 2961–3.

Kitware Inc. http://www.kitware.com/products/volview.html.

Ku, G., Maslov, K., Li, L., and Wang, L. H. V. 2010. Photoacoustic microscopy with 2-µm transverse resolution. *J Biomed Opt*, 15, 021302 (article number).

Li, L., Maslov, K., Ku, G., and Wang, L. V. 2009. Three-dimensional combined photoacoustic and optical coherence microscopy for *in vivo* microcirculation studies. *Opt Express*, 17, 16450–5.

Liu, T., Wei, Q., Wang, J., Jiao, S., and Zhang, H. F. 2011a. Combined photoacoustic microscopy and optical coherence tomography can measure metabolic rate of oxygen. *Biomed Opt Express*, 2, 1359–65.

Liu, T., Wei, Q., Wang, J., Jiao, S., and Zhang, H. F. 2011b. Combined photoacoustic microscopy and optical coherence tomography can measure metabolic rate of oxygen. *Biomed Opt Express*, 2, 1359–65.

Maslov, K., Stoica, G., and Wang, L. H. V. 2005. *In vivo* dark-field reflection-mode photoacoustic microscopy. *Opt Lett*, 30, 625–7.

Maslov, K., Zhang, H. F., Hu, S., and Wang, L. V. 2008. Optical-resolution photoacoustic microscopy for *in vivo* imaging of single capillaries. *Opt Lett*, 33, 929–31.

Maslov, K., Zhang, H. F., and Wang, L. V. 2007. Effects of wavelength-dependent fluence attenuation on the noninvasive photoacoustic imaging of hemoglobin oxygen saturation in subcutaneous vasculature in vivo. *Inverse Probl*, 23, S113–S122.

Mohamed, Q., Gillies, M. C., and Wong, T. Y. 2007. Management of diabetic retinopathy: A systematic review. *JAMA*, 298, 902–16.

Morgan, J. I., Dubra, A., Wolfe, R., Merigan, W. H., and Williams, D. R. 2009. *In vivo* autofluorescence imaging of the human and macaque retinal pigment epithelial cell mosaic. *Invest Ophthalmol Vis Sci*, 50, 1350–9.

Nickla, D. L. and Wallman, J. 2010. The multifunctional choroid. *Prog Retin Eye Res*, 29, 144–68.

Ostrovsky, M. A., Sakina, N. L., and Dontsov, A. E. 1987. An antioxidative role of ocular screening pigments. *Vis Res*, 27, 893–9.

Paques, M., Simonutti, M., Roux, M. J., Picaud, S., Levavasseur, E., Bellman, C., and Sahel, J. A. 2006. High resolution fundus imaging by confocal scanning laser ophthalmoscopy in the mouse. *Vis Res*, 46, 1336–45.

Pircher, M., Zawadzki, R. J., Evans, J. W., Werner, J. S., and Hitzenberger, C. K. 2008. Simultaneous imaging of human cone mosaic with adaptive optics enhanced scanning laser ophthalmoscopy and high-speed transversal scanning optical coherence tomography. *Opt Lett*, 33, 22–4.

Sarna, T., Burke, J. M., Korytowski, W., Rozanowska, M., Skumatz, C. M., Zareba, A., and Zareba, M. 2003. Loss of melanin from human RPE with aging: Possible role of melanin photooxidation. *Exp Eye Res*, 76, 89–98.

Seeliger, M. W., Beck, S. C., Pereyra-Munoz, N., Dangel, S., Tsai, J. Y., Luhmann, U. F., van de Pavert, S. A. et al. 2005. *In vivo* confocal imaging of the retina in animal models using scanning laser ophthalmoscopy. *Vis Res*, 45, 3512–9.

So, P., Kim, H. and Kochevar, I. 1998. Two-photon deep tissue ex vivo imaging of mouse dermal and subcutaneous structures. *Opt Express*, 3, 339–50.

Song, K. H. and Wang, L. V. 2007. Deep reflection-mode photoacoustic imaging of biological tissue. *J Biomed Opt*, 12, 060503 (article number).

Srienc, A. I., Kurth-Nelson, Z. L., and Newman, E. A. 2010. Imaging retinal blood flow with laser speckle flowmetry. *Front Neuroenergetics*, 2, 128 (article number).

Srinivasan, V. J., Monson, B. K., Wojtkowski, M., Bilonick, R. A., Gorczynska, I., Chen, R., Duker, J. S., Schuman, J. S., and Fujimoto, J. G. 2008. Characterization of outer retinal morphology with high-speed, ultrahigh-resolution optical coherence tomography. *Invest Ophthalmol Vis Sci*, 49, 1571–9.

Stone, J., Chan-Ling, T., Pe'er, J., Itin, A., Gnessin, H., and Keshet, E. 1996. Roles of vascular endothelial growth factor and astrocyte degeneration in the genesis of retinopathy of prematurity. *Invest Ophthalmol Vis Sci*, 37, 290–9.

Wang, L. V. 2008. Prospects of photoacoustic tomography. *Med Phys*, 35, 5758–67.

Wang, Y., Bower, B. A., Izatt, J. A., Tan, O., and Huang, D. 2007. *In vivo* total retinal blood flow measurement by Fourier domain Doppler optical coherence tomography. *J Biomed Opt*, 12, 041215.

Wehbe, H., Ruggeri, M., Jiao, S., Gregori, G., Puliafito, C. A., and Zhao, W. 2007. Automatic retinal blood flow calculation using spectral domain optical coherence tomography. *Opt Express*, 15, 15193–206.

Wei, Q., Liu, T., Jiao, S. L., and Zhang, H. F. 2011. Image chorioretinal vasculature in albino rats using photoacoustic ophthalmoscopy. *J Mod Opt*, 58, 1997–2001.

Xie, Z., Chen, S. L., Ling, T., Guo, L. J., Carson, P. L., and Wang, X. 2011. Pure optical photoacoustic microscopy. *Opt Express*, 19, 9027–34.

Xie, Z., Jiao, S., Zhang, H. F., and Puliafito, C. A. 2009a. Laser-scanning optical-resolution photoacoustic microscopy. *Opt Lett*, 34, 1771–3.

Xie, Z. X., Jiao, S. L., Zhang, H. F., and Puliafito, C. A. 2009b. Laser-scanning optical-resolution photoacoustic microscopy. *Opt Lett*, 34, 1771–3.

Yao, J., Maslov, K. I., Zhang, Y., Xia, Y., and Wang, L. V. 2011a. Label-free oxygen-metabolic photoacoustic microscopy in vivo. *J Biomed Opt*, 16, 076003.

Yao, J. J., Maslov, K. I., Zhang, Y., Xia, Y. N., and Wang, L. V. 2011b. Label-free oxygen-metabolic photoacoustic microscopy in vivo. *J Biomed Opt*, 16, 067003 (article number).

Zawadzki, R. J., Jones, S. M., Olivier, S. S., Zhao, M., Bower, B. A., Izatt, J. A., Choi, S., Laut, S., and Werner, J. S. 2005. Adaptive-optics optical coherence tomography for high-resolution and high-speed 3D retinal *in vivo* imaging. *Opt Express*, 13, 8532–46.

Zhang, X., Jiang, M., Fawzi, A. A., Li, X., Shung, K. K., Puliafito, C. A., Zhang, H. F., and Jiao, S. 2010c. Simultaneous dual molecular contrasts provided by the absorbed photons in photoacoustic microscopy. *Opt Lett*, 35, 4018–20.

Zhang, H. F., Maslov, K., Li, M. L., Stoica, G., and Wang, L. H. V. 2006a. *In vivo* volumetric imaging of subcutaneous microvasculature by photoacoustic microscopy. *Opt Express*, 14, 9317–23.

Zhang, H. F., Maslov, K., Sivaramakrishnan, M., Stoica, G., and Wang, L. H. V. 2007a. Imaging of hemoglobin oxygen saturation variations in single vessels *in vivo* using photoacoustic microscopy. *Appl Phys Lett*, 90, 3.

Zhang, H. F., Maslov, K., Sivaramakrishnan, M., Stoica, G., and Wang, L. H. V. 2007b. Imaging of hemoglobin oxygen saturation variations in single vessels *in vivo* using photoacoustic microscopy. *Appl Phys Lett*, 90, article number: 053901.

Zhang, H. F., Maslov, K., Stoica, G., and Wang, L. H. V. 2006b. Functional photoacoustic microscopy for high-resolution and noninvasive *in vivo* imaging. *Nat Biotechnol*, 24, 848–51.

Zhang, H. F., Maslov, K., Stoica, G., and Wang, L. V. 2006c. Functional photoacoustic microscopy for high-resolution and noninvasive *in vivo* imaging. *Nat Biotechnol*, 24, 848–51.

Zhang, H. F., Maslov, K., and Wang, L. H. V. 2007c. *In vivo* imaging of subcutaneous structures using functional photoacoustic microscopy. *Nat Protoc*, 2, 797–804.

Zhang, H. F., Maslov, K., and Wang, L. V. 2007d. *In vivo* imaging of subcutaneous structures using functional photoacoustic microscopy. *Nat Protoc*, 2, 797–804.

Zhang, H. F., Maslov, K., and Wang, L. H. V. 2009. Automatic algorithm for skin profile detection in photoacoustic microscopy. *J Biomed Opt*, 14, 024050 (article number).

Zhang, C., Maslov, K., and Wang, L. H. V. 2010a. Subwavelength-resolution label-free photoacoustic microscopy of optical absorption in vivo. *Opt Lett*, 35, 3195–97.

Zhang, H. F., Wang, J., Wei, Q., Liu, T., Jiao, S. L., and Puliafito, C. A. 2010b. Collecting back-reflected photons in photoacoustic microscopy. *Opt Express*, 18, 1278–82.

Zhang, X., Zhang, H. F., Puliafito, C. A., and Jiao, S. 2011. Simultaneous *in vivo* imaging of melanin and lipofuscin in the retina with photoacoustic ophthalmoscopy and autofluorescence imaging. *J Biomed Opt*, 16, 080504.

Interferometric Synthetic Aperture Microscopy

P. Scott Carney
University of Illinois at Urbana-Champaign

Steven Adie
University of Illinois at Urbana-Champaign

Stephen A. Boppart
University of Illinois at Urbana-Champaign

19.1 Background

Optical coherence tomography (OCT) has grown from a new research tool into a modality with widespread biomedical applications ever since its introduction in the early 1990s [1–5]. OCT is the optical analog of ultrasound imaging, permitting tomographic imaging with micrometer-scale resolution, and with penetration depth of 1–3 mm in biological tissues. The success of OCT can be attributed in large part to the ideal match between imaging technology and the clinical requirements for imaging in the human eye. An excellent early review of the basic methods may be found in Ref. [6]. Within 15 years, OCT became an established imaging modality in ophthalmology [7,8], and to date has made significant advances toward applications in cardiology [9], dermatology [10], and oncology [11].

OCT is based on low-coherence interferometry in order to measure the "echo time" of light that is backscattered from tissue structures. A broadband (and therefore low-coherence) optical source is utilized to localize these structures to within the source coherence length, enabling the micrometer-scale axial resolution [12]. The interferometer of a typical OCT system is fiber-based, facilitating compact instrumentation, as well as endoscopic catheter [13,14] and needle-based [15] imaging geometries capable of imaging deep within the human body. A typical OCT system performs scanned imaging acquisition, as in confocal microscopy, but only over the transverse scan axes. This provides important advantages for imaging in scattering tissue, including confocal rejection of out-of-focus light and reduced cross talk (that is otherwise present in full-field holographic imaging with spatially coherent sources).

Transverse resolution in OCT depends on the numerical aperture (NA) of the focused imaging beam. In a typical OCT system, the NA is low in order to maintain a large depth-of-field (DOF). Higher NA imaging in optical coherence microscopy (OCM) suffers from the compromise between transverse resolution and DOF [16,17]. Various methods have been proposed to address this compromise, including the fusion of datasets acquired at different focus depths [18], and the use of Bessel beams that have a large DOF [18]. However, the need to combine multiple datasets results in a reduced overall acquisition speed, while the second approach produces a point-spread function (PSF) with sidelobes and requires a more complicated sample arm design to improve the light collection.

Interferometric synthetic aperture microscopy (ISAM) is a computed imaging approach for broadband interferometric tomography that overcomes the DOF limitations of higher NA imaging [19–22]. The technique is so named because the method of aperture synthesis, over the multiple illumination angles present in a focused optical beam, is performed via a Fourier space resampling scheme similar to that previously utilized in synthetic aperture radar (SAR) [23]. This resampling relationship, known as the Stolt mapping in SAR, removes defocus for all depths, resulting in a 3D reconstruction of the sample with spatially invariant resolution. In principle, ISAM can provide spatially invariant cellular resolution throughout a 3D volume without having to scan the focus in depth.

In this chapter, we provide an algebraic approach to the underlying theory of ISAM, and describe the reconstruction strategies employed to achieve real-time reconstructions [24]. The main theoretical results are supported by experimental results in tissue phantoms and biological tissue.

19.2 Interferometric Measurements

OCT tomograms were first acquired using time-domain acquisition, in which the broadband interference signal was measured as a function of reference arm length. In time-domain OCT (TD–OCT), interference between sample and reference beams occurs when the reference path length is matched (to within the coherence length) to the path length to a given reflection or scattering event in the sample arm. The development of spectral-domain (SD–OCT) [25,26] and later swept-source OCT, where the raw interferometric signal is acquired as an optical spectrum, provided the advantage of a static reference arm, as well as improved signal-to-noise ratio [27–29]. These Fourier-domain data were then resampled to be a linear function of wavenumber, to correct dispersion broadening of the axial PSF [30,31]. Utilizing the result of the cross-correlation theorem, each spatial-domain A-scan can be computed from the Fourier-domain data via the inverse Fourier transform.

Since the Fourier-domain data provide a measurement of the real part of the interferometric signal, the corresponding spatial-domain data is Hermitian symmetric about zero optical path delay. This symmetry can produce overlap in the tomogram corresponding to regions of the sample that lie on opposite sides of zero optical path delay [32]. While several techniques have been developed to measure the complex interferometric signal [32–34], a common method is to compute the imaginary part of the Fourier-domain data via the Hilbert transform. This provides access to the complex analytic signal that is free of overlap artifacts when the entire sample is positioned on one side of the zero optical path delay.

In the analysis below, we assume the complex analytic signal is accessible. We do not deal with the interferometry problem, nor the methods for obtaining the complex signal from the real one.

19.3 The Forward Problem

The first step in the solution of any inverse problem is the construction or identification of a model for the forward problem. The system shown in Figure 19.1 can be analyzed using a modular approach. To that end, we describe each component in the optical system with an operator. Since the system is linear, it is convenient to work in a notation designed for linear algebra, namely the Dirac notation. Throughout the manuscript, vectors belonging to different Hilbert spaces, or drawn from different sets within the same Hilbert space, are differentiated only by the symbols used for labels. For instance, $|\mathbf{r}_\parallel\rangle$ is a basis element for the square integrable functions L_2, on $\mathbb{R}^2$ (together with its closure) while $|\mathbf{k}_\parallel\rangle$ is also a basis element for the same space drawn from a different set. Vectors describing the field in a plane at constant axial coordinate z are written $|U_z\rangle$ with Hermitian adjoint $\langle U_z|$. The plane wave component of the field at spatial frequency $\mathbf{k}_\parallel$ is written $\tilde{U}(\mathbf{k}_\parallel) = \langle \mathbf{k}_\parallel \mid U_z\rangle$ and the value of the field at a point $\mathbf{r}_\parallel$ in the plane of constant z is given by $U_z(\mathbf{r}_\parallel) = \langle \mathbf{r}_\parallel \mid U_z\rangle$.

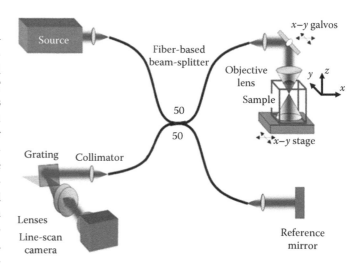

FIGURE 19.1 Schematic of the fiber-based SD–OCT system.

The measured signal in all cases is a function of frequency and some parameters describing the configuration of the system. For instance, in Cartesian scanning ISAM, a beam is projected into the sample normal to the scanning plane, centered on points $\mathbf{r}_0$. The collected signal is denoted $|S\rangle$ and the values of the signal at a particular frequency and position is $\langle \mathbf{r}_0, k \mid S\rangle$.

19.3.1 Illumination

We assume the field originates at some source and enters the system through an aperture. The source field is denoted $|U_0\rangle$, a member of a Hilbert space H_0. As an example, the source may be completely described by a complex amplitude, in which case, $H_0 = L_2(\mathbb{R})$ and the field amplitude as a function of wavenumber k is given by $U_0(k) = \langle k \mid U_0\rangle$. Alternatively, the field might be multimode and the source characterized by the amplitude in each of n modes or coherent modes so that $H_0 = L_2(\mathbb{R}) \otimes \mathbb{R}^n$.

The action of the aperture on the field is denoted by the operator $\mathbf{A}: H_0 \to H_1$ so that the field emerging from the aperture is $|U_{zA}\rangle = \mathbf{A}|U_0\rangle$. Typically, the aperture acts as a multiplicative screen and so $H_1 = H_1' \otimes H_0$ where, for a simple scalar field and a planar screen $H_1' = L_2(\mathbb{R}^2)$. Finally, we are interested in the measurements made with a set of illuminating fields, each datum collected with the aperture translated by some displacement. The fields obtained for the whole set of aperture positions are obtained by acting with an operator $\mathbf{T}: H_1 \to H_2$, where $H_2 = H_0 \otimes L_2(\mathbb{R}^2 \otimes \mathbb{R}^2)$. With a single-mode source, the value of the illuminating field at position $\mathbf{r}_\parallel$ in the plane $z = z_A$ is

$$\langle \mathbf{r}_{0a}, \mathbf{r}_\parallel, k \mid U_{z_A}\rangle = \langle \mathbf{r}_{0a}, \mathbf{r}_\parallel, k \mid \mathbf{TA}|U_0\rangle$$
$$= \langle \mathbf{r}_\parallel - \mathbf{r}_{0a}, k \mid \mathbf{A}|U_0\rangle, \tag{19.1}$$

where k is the wavenumber and $\mathbf{r}_{0a}$ is the translation of the aperture mask. For continuous planar Cartesian scanning

$$\langle \mathbf{r}_{0a}, \mathbf{r}_\parallel', k' \mid \mathbf{T}|\mathbf{r}_\parallel, k\rangle = \delta(k - k')\delta(\mathbf{r}_\parallel' + \mathbf{r}_{0a} - \mathbf{r}_\parallel). \tag{19.2}$$

19.3.2 Propagation of Fields

The field is propagated between parallel planes, along an axis we conventionally label z, by an operator $\mathbf{K}_d$, where d is the distance between planes. The field is generally a vector in H_2 but we again assume separability and unless explicitly needed, only treat the field as an element of H_1 or H_1'. Thus, $\mathbf{K}_d: H_1 \to H_1$ so that elements of the field in plane $z = z_1$, $\langle \mathbf{r}_\parallel, k \mid U_{z_1} \rangle$ to elements of the field in plane $z = z_2$, $\langle \mathbf{r}_\parallel', k \mid U_{z_2} \rangle$. This is accomplished with an integral operator whose kernel is $\langle \mathbf{r}_\parallel', k \mid \mathbf{K}_d \mid \mathbf{r}_\parallel, k \rangle$ so that

$$\langle \mathbf{r}_\parallel', k \mid U_{z_2} \rangle = \int d^2 r_\parallel \langle \mathbf{r}_\parallel', k \mid \mathbf{K}_{|z_1 - z_2|} \mid \mathbf{r}_\parallel, k \rangle \langle \mathbf{r}_\parallel, k \mid U_{z_1} \rangle. \quad (19.3)$$

The dependence of the elements of the states on $\mathbf{r}_{0a}$ and k is suppressed below for compactness. Noting that $\int d^2 r_\parallel |\mathbf{r}_\parallel\rangle\langle \mathbf{r}_\parallel|$ is the identity

$$\left| U_{z_2} \right\rangle = \mathbf{K}_d \left| U_{z_1} \right\rangle. \quad (19.4)$$

We assume that the propagation is unidirectional, that is, the sources of the field being acted upon are entirely to one side of both planes. In the coordinate space representation, $\mathbf{K}_d$ is translationally invariant. It is thus advantageous to work in the two-dimensional Fourier basis whose elements are denoted $|\mathbf{k}_\parallel\rangle$ and

$$\langle \mathbf{r}_\parallel \mid \mathbf{k}_\parallel \rangle = \frac{1}{2\pi} e^{i\mathbf{r}_\parallel \cdot \mathbf{k}_\parallel}. \quad (19.5)$$

The matrix elements of $\mathbf{K}_d$ are given by

$$\langle \mathbf{k}_{1\parallel} \mid \mathbf{K}_d \mid \mathbf{k}_{2\parallel} \rangle = \delta(\mathbf{k}_{2\parallel} - \mathbf{k}_{1\parallel}) e^{ik_z(\mathbf{k}_{1\parallel})d}, \quad (19.6)$$

where δ is the Dirac delta function. The function $k_z(\mathbf{k}_\parallel) = \sqrt{k^2 - k_\parallel^2}$ ensures that the field obeys the Helmholtz equation. The positive root is taken by definition. Alternatively, the operator may be constructed:

$$\mathbf{K}_d = \int d^2k \left| \mathbf{k}_\parallel \right\rangle e^{ik_z(\mathbf{k}_\parallel)d} \left\langle \mathbf{k}_\parallel \right|. \quad (19.7)$$

In the coordinate domain, we have

$$\langle \mathbf{r}_\parallel' \mid \mathbf{K}_d \mid \mathbf{r}_\parallel'' \rangle = \int d^2k \, e^{i(\mathbf{r}_\parallel' - \mathbf{r}_\parallel'') \cdot \mathbf{k}_\parallel} e^{ik_z(\mathbf{k}_\parallel)d}. \quad (19.8)$$

Some care must be taken here. For $z > z'$, the propagation from z' to z is accomplished with $\mathbf{K}_d$, the distance being $d = |z' - z| = z - z'$, so in constructing the forward operator, an appropriate d must be chosen. Alternatively, we could have defined $\mathbf{K}_d$ with $d \to |d|$, thus ensuring that the phase always accumulates with the correct sign. However, the definition we chose, allowing negative values of d, has the advantage that

$\mathbf{K}_d \mathbf{K}_{d'} = \mathbf{K}_{d+d'}$ for all d and d'. Thus, given a source at plane $z = z_0$, it is possible to compute a virtual field distribution in the plane $z = z_1 < z_0$ that would produce at a third plane, $z = z_2$ the field produced by the source.

It might also be noted that $\mathbf{K}_d^{-1} = \mathbf{K}_{-d}$ and that when evanescent waves are disallowed, that is, when $\mathbf{K}_d$ is restricted to the subspace spanned by $|\mathbf{k}_\parallel\rangle$, with $k_\parallel < k$, $\mathbf{K}_d^{-1} = \mathbf{K}_d^* = \mathbf{K}_{-d}$, where the asterisk denotes the Hermitian conjugate. Thus, restricted to propagating fields, $\mathbf{K}_d$ is unitary.

19.3.3 The Optical System

As illustrated in Figure 19.1, the field propagates to an optical system of some sort, most simply a lens, represented by an operator $\mathbf{L}$. The field in an arbitrary plane $z = z_\ell$ is then given by

$$\left| U_z \right\rangle = \mathbf{K}_{z-z_1} \mathbf{L} \mathbf{K}_{z_1-z_A} \mathbf{TA} \left| U_0 \right\rangle. \quad (19.9)$$

The details of the elements of $\mathbf{L}$ are not necessarily important. Rather, it is often the case that the operator $\mathbf{K}_{z-z_1} \mathbf{L} \mathbf{K}_{z_1-z_A} \mathbf{TA}$ may be known for a particular value of z and the kernel elements may be determined for other planes through the appropriate use of the propagation operator $\mathbf{K}_{z'-z}$.

19.3.4 Interaction with the Sample

The object is described by a susceptibility function $\eta(\mathbf{r}) = \langle \mathbf{r} \mid \eta \rangle$, where $\langle \mathbf{r} \mid$ is composed $\langle \mathbf{r} \mid = \langle \mathbf{r}_\parallel \mid \langle z \mid$. The field excites a secondary source in the object proportional to this susceptibility. That is, the secondary source is $\eta(\mathbf{r})U(\mathbf{r})$. This secondary source radiates the scattered field

$$U^{scatt}(\mathbf{r}') = \int d^3r \, G(\mathbf{r}', \mathbf{r})\eta(\mathbf{r})U(\mathbf{r}), \quad (19.10)$$

where G is the Green function for the system. The field appearing in the integral on the right-hand side of Equation 19.10 is the total field $U = U^{scatt} + U^{inc}$, the sum of the incident and scattered fields. In general, then U^{scatt} is a nonlinear function of the susceptibility η. However, in the near-infrared in biological samples, it is often possible to linearize this equation. This is the basic assumption made implicitly in OCT. Keeping only terms linear in η, that is, within the first Born approximation, the scattered field in a plane $z = z'$ is given by

$$\left| U_{z'} \right\rangle = \int d^2 r_\parallel dz \, \mathbf{G}_{z'} \left| \mathbf{r} \right\rangle \langle \mathbf{r} \mid \eta \rangle \langle \mathbf{r}_\parallel \mid U_z \rangle. \quad (19.11)$$

The operator $\mathbf{G}_{z'}$ has matrix elements given by the Green function $\langle \mathbf{r}_\parallel' \mid \mathbf{G}_{z'} \mid \mathbf{r} \rangle = G(\mathbf{r}', \mathbf{r})$. Explicitly, for a scalar field in free-space

$$\mathbf{G}_{z'} = \int d^2k \, dz \left| \mathbf{k}_\parallel \right\rangle \frac{2\pi i e^{ik_z(\mathbf{k}_\parallel)|z-z'|}}{k_z(\mathbf{k}_\parallel)} \langle \mathbf{k}_\parallel \mid \langle z \mid. \quad (19.12)$$

19.3.5 Detection

The scattered field is propagated back to the optical system. It is possible to detect the scattered field either in a direct-backscattering (monostatic) geometry or in a bistatic arrangement. In the more general bistatic case, the field propagates back through the lens, through a beam splitter, to an aperture that couples the field into a fiber to be detected interferometrically. The coupling of the field into the interferometer is described by an operator $\mathbf{B}^\dagger$. Thus, the detected signal is given by

$$|S\rangle = \mathbf{B}^\dagger \mathbf{T} \mathbf{K}_{z_1-z_B} \mathbf{L}^\dagger \mathbf{K}_{z'-z_1} |U_{z'}\rangle. \tag{19.13}$$

Explicitly

$$\langle k \mid S\rangle = \int d^2 r_\| \langle k \mid \mathbf{B}^\dagger |\mathbf{r}_\|\rangle\langle \mathbf{r}_\| \mid \mathbf{T} |U_{zB}\rangle. \tag{19.14}$$

For backscattering experiments, $\mathbf{B} = \mathbf{A}$. In principle, the set of translations of the apertures $\{\mathbf{r}_{0a}\}$ and $\{\mathbf{r}_{0b}\}$ may be distinct and so the operator $\mathbf{T}$ appearing in Equation 19.14 should be labeled $\mathbf{T}_B$, and similarly $\mathbf{T}_A$ in Equation 19.9 though that distinction is not made explicitly below.

19.3.6 Complete Model

The field exits the illumination aperture ($\mathbf{TA}$), propagates to the lens ($\mathbf{K}_{z_1-z_A}$), interacts with the lens ($\mathbf{L}$), propagates to the sample ($\mathbf{K}_{z-z_1}$), interacts with the sample, propagates back to the lens ($\mathbf{K}_{z-z_1}$), interacts again with the lens ($\mathbf{L}^\dagger$), propagates back to the aperture ($\mathbf{K}_{z_1-z_B}$), and is coupled back into the collection aperture ($\mathbf{B}^\dagger \mathbf{T}$). Thus the measured field is

$$|S\rangle = \mathbf{B}^\dagger \mathbf{T} \mathbf{K}_{z_1-z_B} \mathbf{L}^\dagger \mathbf{K}_{z'-z_1} \int d^2 r_\| dz\, \mathbf{G}_{z'} |\mathbf{r}\rangle\langle \mathbf{r}_\| \mid \mathbf{K}_{z-z_1} \mathbf{L} \mathbf{K}_{z_1-z_A} \mathbf{TA} |U_0\rangle$$
$$\times \langle \mathbf{r} \mid \eta\rangle. \tag{19.15}$$

The forward operator is thus

$$\mathbf{P} = \mathbf{B}^\dagger \mathbf{T} \mathbf{K}_{z_1-z_B} \mathbf{L}^\dagger \mathbf{K}_{z'-z_1} \int d^2 r_\| dz\, \mathbf{G}_{z'} |\mathbf{r}\rangle\langle \mathbf{r}_\| \mid \mathbf{K}_{z-z_1} \mathbf{L} \mathbf{K}_{z_1-z_A} \mathbf{TA} |U_0\rangle\langle \mathbf{r}|. \tag{19.16}$$

The elements of the data should formally be indexed by the settings for each aperture separately and the wavenumber, that is, $\langle \mathbf{r}_{0a}, \mathbf{r}_{0b}, k \mid S\rangle = \langle \mathbf{r}_{0a} \mid \langle \mathbf{r}_{0b} \mid \langle k \mid S\rangle$.

19.3.7 The Inverse Problem

Formally, the object $|\eta\rangle$ may be recovered by acting on the data with the inverse of $\mathbf{P}$, $|\eta\rangle = \mathbf{P}^{-1} \mid S\rangle$. In general, $\mathbf{P}^{-1}$ does not exist and so as a substitute, the pseudoinverse $\mathbf{P}^+$ is used. The

pseudoinverse is the inverse on the orthogonal complement of the null-space of $\mathbf{P}$. Alternatively, it gives the minimum-norm solution of the variational problem

$$\mathbf{P}^+ |S\rangle = |\hat{\eta}\rangle = \underset{\hat{\eta}}{\arg\min} \parallel \mathbf{P} |\hat{\eta}\rangle - |S\rangle \parallel^2. \tag{19.17}$$

The pseudoinverse may be constructed

$$\mathbf{P}^+ = [\mathbf{P}^\dagger \mathbf{P}]^+ \mathbf{P}^\dagger. \tag{19.18}$$

19.3.8 Gaussian Beams

In many cases of interest, the illuminating field is well described by a scalar Gaussian beam. For the cases of arbitrary illumination and detection, see Ref. [21]. An ideal focused Gaussian beam achieves a minimal width in the so-called waist-plane. We denote this plane z_{0a} or z_{0b} for the illumination and detection focal planes, respectively. The normalized beam-shape function in the waist-plane we denote

$$g_a(\mathbf{r}, k)\delta(k-k') = \langle \mathbf{r}, k \mid \mathbf{K}_{z_{0a}-z_1} \mathbf{L} \mathbf{K}_{z_1-z_A} \mathbf{A} |k'\rangle. \tag{19.19}$$

The field in the waist-plane is real, therefore

$$g_a(\mathbf{r}, k)\delta(k-k') = \langle k' \mid \mathbf{A}^\dagger \mathbf{K}_{z_1-z_A} \mathbf{L}^\dagger \mathbf{K}_{z_{0a}-z_1} |\mathbf{r}, k\rangle. \tag{19.20}$$

For detection, g_b is similarly defined.
In the Fourier basis

$$\tilde{g}_a(\mathbf{k}_\|, k)\delta(k-k') = \langle \mathbf{k}_\|, k \mid \mathbf{K}_{z_{0a}-z_1} \mathbf{L} \mathbf{K}_{z_1-z_A} \mathbf{A} |k'\rangle. \tag{19.21}$$

Including translations in a mixed Fourier-coordinate basis

$$e^{-i\mathbf{k}_\|\cdot\mathbf{r}_{0a}} \tilde{g}_a(\mathbf{k}_\|, k)\delta(k-k') = \langle \mathbf{r}_{0a}, \mathbf{k}_\|, k \mid \mathbf{K}_{z_{0a}-z_1} \mathbf{L} \mathbf{K}_{z_1-z_A} \mathbf{TA} |k'\rangle, \tag{19.22}$$

or entirely in a Fourier basis

$$2\pi\delta(\mathbf{k}_\| + \mathbf{k}'_\|) \tilde{g}_a(\mathbf{k}_\|, k)\delta(k-k') = \langle \mathbf{k}'_\|, \mathbf{k}_\|, k \mid \mathbf{K}_{z_{0a}-z_1} \mathbf{L} \mathbf{K}_{z_1-z_A} \mathbf{TA} |k'\rangle, \tag{19.23}$$

or finally

$$e^{-i\mathbf{k}'_\|\cdot\mathbf{r}_\|} \tilde{g}_a(\mathbf{k}_\|, k)\delta(k-k') = \langle \mathbf{k}'_\|, \mathbf{r}_\|, k \mid \mathbf{K}_{z_{0a}-z_1} \mathbf{L} \mathbf{K}_{z_1-z_A} \mathbf{TA} |k'\rangle. \tag{19.24}$$

The signal can be written

$$|S\rangle = \int d^2 k''_\| d^2 k'_\| d^2 r_\|\, dz\, \mathbf{B}^\dagger \mathbf{T} \mathbf{K}_{z_1-z_B} \mathbf{L}^\dagger \mathbf{K}_{z_{0b}-z_1} |\mathbf{k}''_\|\rangle\langle \mathbf{k}''_\| \mid \mathbf{K}_{z'-z_{0b}}$$
$$\times \mathbf{G}_{z'} |\mathbf{r}\rangle\langle \mathbf{r}_\| \mid \mathbf{K}_{z_{0a}-z} |\mathbf{k}'_\|\rangle\langle \mathbf{k}'_\| \mid \mathbf{K}_{z_{0a}-z_1} \mathbf{L} \mathbf{K}_{z_1-z_A} \mathbf{TA} |U_0\rangle\langle \mathbf{r} \mid \eta\rangle, \tag{19.25}$$

so that

$$\langle \mathbf{r}_{0a}, \mathbf{r}_{0b}, k \mid S \rangle = \int d^2 k_{\parallel}'' \int d^2 k_{\parallel}' \int d^2 r_{\parallel} \, dz \, e^{i \mathbf{k}_{\parallel}'' \cdot \mathbf{r}_{0b}} \tilde{g}_b(\mathbf{k}'', k) \langle \mathbf{k}_{\parallel}'' \mid$$
$$\times \, \mathbf{G}_{z_{0b}} \mid \mathbf{r} \rangle \langle \mathbf{r}_{\parallel} \mid \mathbf{K}_{z_{0a}-z} \mid \mathbf{k}_{\parallel}' \rangle e^{-i \mathbf{k}_{\parallel}' \cdot \mathbf{r}_{0a}} \tilde{g}_a(\mathbf{k}', k) \langle k \mid U_0 \rangle \langle \mathbf{r} \mid \eta \rangle.$$

(19.26)

In the backscattering geometry commonly encountered, $g_a = g_b = g$ and we enforce $\mathbf{r}_{0a} = \mathbf{r}_{0b} = \mathbf{r}_0$ so that

$$\langle \mathbf{r}_0, k \mid S \rangle = \int d^2 k_{\parallel}'' \int d^2 k_{\parallel}' \int d^2 r_{\parallel} \, dz \, e^{i(\mathbf{k}_{\parallel}'' - \mathbf{k}_{\parallel}') \cdot \mathbf{r}_0} \tilde{g}(\mathbf{k}'', k) \langle \mathbf{k}_{\parallel}'' \mid$$
$$\times \, \mathbf{G}_{z_0} \mid \mathbf{r} \rangle \langle \mathbf{r}_{\parallel} \mid \mathbf{K}_{z_0-z} \mid \mathbf{k}_{\parallel}' \rangle \tilde{g}(\mathbf{k}', k) \langle k \mid U_0 \rangle \langle \mathbf{r} \mid \eta \rangle,$$

(19.27)

which becomes somewhat simpler in the Fourier basis

$$\langle \mathbf{k}_{\parallel}, k \mid S \rangle = 2\pi \int d^2 k_{\parallel}' \int d^2 r_{\parallel} \, dz \, \tilde{g}(\mathbf{k}', k) \langle \mathbf{k}_{\parallel}' \mid \mathbf{G}_{z_0} \mid \mathbf{r} \rangle \langle \mathbf{r}_{\parallel} \mid$$
$$\times \, \mathbf{K}_{z_0-z} \mid \mathbf{k}_{\parallel}' - \mathbf{k}_{\parallel} \rangle \tilde{g}(\mathbf{k}_{\parallel}' - \mathbf{k}_{\parallel}, k) \langle k \mid U_0 \rangle \langle \mathbf{r} \mid \eta \rangle$$

(19.28)

and can be expressed

$$\langle \mathbf{k}_{\parallel}, k \mid S \rangle = 2\pi \langle k \mid U_0 \rangle \int d^2 k_{\parallel}' \int dz \, \tilde{g}(\mathbf{k}_{\parallel}', k) \frac{ie^{\left[k_z(\mathbf{k}_{\parallel}') + k_z(\mathbf{k}_{\parallel}' - \mathbf{k}_{\parallel}) \right](z-z_0)}}{k_z(\mathbf{k}_{\parallel}')}$$
$$\times \, \tilde{g}(\mathbf{k}_{\parallel}' - \mathbf{k}_{\parallel}, k) \langle \mathbf{k}_{\parallel} \mid \langle z \mid \eta \rangle.$$

(19.29)

This expression gives a Fredholm integral of the first kind relating the signal and the object structure, η. Despite at first glance appearing to be a three-folded integral equation, this is in fact only a one-dimensional problem, an integral over z. The kernel of this integral is represented as an integral over $\mathbf{k}_{\parallel}'$, but this integral does not explicitly involve the unknown η. The solution may be found for η using standard techniques; however, an alternative approach follows.

19.3.9 Asymptotics

The result in Equation 19.29 provides a connection between the data and the object structure. In principle, this integral equation can be solved for the object structure. However, as shown below, the kernel of the integral in Equation 19.29 takes on particularly simple form in two asymptotic regimes.

19.3.9.1 Far from Focus

For samples with structure only far from the focal plane, the integral over $\mathbf{k}_{\parallel}'$ may be evaluated by the method of stationary phase. The stationary point is at $\mathbf{k}_{\parallel}' = \mathbf{k}_{\parallel}/2$, so that

$$\langle \mathbf{k}_{\parallel}, k \mid S \rangle \sim 2\pi^2 \langle k \mid U_0 \rangle \tilde{g}^2(\mathbf{k}_{\parallel}/2, k) \int dz \, e^{2ik_z(\mathbf{k}_{\parallel}/2)(z-z_0)}$$
$$\langle \mathbf{k}_{\parallel} \mid \langle z \mid\mid z - z_0 \mid^{-1} \mid \eta \rangle,$$

(19.30)

where we have made use of that fact that $\tilde{g}$ and k_z are even functions.

$$\langle \mathbf{k}_{\parallel}, k \mid S \rangle = H_F \langle \mathbf{k}_{\parallel} \mid \int dz \langle -2k_z(\mathbf{k}_{\parallel}/2) \mid z \rangle \mid z - z_0 \mid^{-1} \langle z \mid \eta \rangle,$$

(19.31)

where $H_F = 4\pi^3 \langle k \mid U_0 \rangle \tilde{g}^2(\mathbf{k}_{\parallel}/2, k) e^{-2ik_z(\mathbf{k}_{\parallel}/2)z_0}$. This equation is easily inverted to obtain the object structure in the coordinate basis

$$\langle \mathbf{r} \mid \eta \rangle = \mid z - z_0 \mid \int d\beta \langle z \mid -\beta \rangle \int d^2 k_{\parallel} \langle \mathbf{r}_{\parallel} \mid \mathbf{k}_{\parallel} \rangle H_F^{-1}$$
$$\left\langle \mathbf{k}_{\parallel}, \frac{1}{2}\sqrt{\beta^2 + k_{\parallel}^2} \mid S \right\rangle.$$

(19.32)

The pseudoinverse is thus approximately

$$\mathbf{P}^+ = \int dz \mid z \rangle \mid z - z_0 \mid \int d\beta \langle z \mid -\beta \rangle \int d^2 k_{\parallel} \mid \mathbf{k}_{\parallel} \rangle H_F^{-1} \langle \mathbf{k}_{\parallel}, \frac{1}{2}\sqrt{\beta^2 + k_{\parallel}^2} \mid.$$

(19.33)

To implement this solution, data $\langle \mathbf{r}_{\parallel}, k \mid S \rangle$ are acquired on a regular Cartesian grid and a Fourier transform is taken over the transverse scanning coordinate to obtain $\langle \mathbf{k}_{\parallel}, k \mid S \rangle$. The data are needed on a new set of points, $k = \frac{1}{2}\sqrt{\beta^2 + k_{\parallel}^2}$ and may be obtained by interpolation. The result $\left\langle \mathbf{k}_{\parallel}, \frac{1}{2}\sqrt{\beta^2 + k_{\parallel}^2} \mid S \right\rangle$ is multiplied by H_F^{-1} and a three-dimensional inverse Fourier transform is taken. Finally, the coordinate domain image is multiplied by the distance from focus.

In this whole procedure, the critical step is the resampling onto the new grid. The resampling has the effect of bringing the entire volume into focus in one step. Usually, the factor of H_F^{-1} may be left out without much change to the image. The scaling with distance from the focus corrects for the fall-off in the holographic signal. It also multiplies any noise in the image and may also be left out for certain applications. Any real system will only provide data on some finite range of k and so the object is only reconstructed up to a band-limited version of the original.

19.3.9.2 Near Focus

For sample structure close to the focus, the specific form of the field in the beam waist is needed. We take as an example a Gaussian beam with width that increases with wavelength to ensure that even at high numerical aperture (NA), the focal spot does not violate the diffraction limit

$$\tilde{g}(\mathbf{k}_{\parallel}, k) = e^{-\alpha^2 k_{\parallel}^2 / 2k^2},$$

(19.34)

where $\alpha = \pi/NA$ and NA is the numerical aperture of the system.

Near the focus, the method of stationary phase is not applicable to Equation 19.29. However, the factors of the $\tilde{g}$ limit the effective range of the integral over $\mathbf{k}_\parallel'$. A good approximation may be found by expanding the rest of the integrand around the peak of the product $\tilde{g}(\mathbf{k}_\parallel' - \mathbf{k}_\parallel, k)\tilde{g}(\mathbf{k}_\parallel', k)$, which for this specific beam is the point $\mathbf{k}_\parallel' = \mathbf{k}_\parallel/2$ just as in the far-field case [19]. The signal is then given by

$$\langle \mathbf{k}_\parallel, k \mid S \rangle = H_N \langle \mathbf{k}_\parallel \mid \int dz \langle -2k_z(\mathbf{k}_\parallel / 2) \mid z \rangle \langle z \mid \eta \rangle, \quad (19.35)$$

where

$$H_N = \frac{2ik^3}{\pi^2 \alpha^2 k_z^2(k_\parallel^2)} H_F.$$

This equation can be inverted to obtain an expression much like Equation 19.32

$$\langle \mathbf{r} \mid \eta \rangle = \int d\beta \langle z \mid -\beta \rangle \int d^2 k_\parallel \langle \mathbf{r}_\parallel \mid \mathbf{k}_\parallel \rangle H_N^{-1} \langle \mathbf{k}_\parallel, \frac{1}{2}\sqrt{\beta^2 + k_\parallel^2} \mid S \rangle, \quad (19.36)$$

with the corresponding pseudoinverse

$$\mathbf{P}^+ = \mid -\beta \rangle \int d^2 k_\parallel \mid \mathbf{k}_\parallel \rangle H_N^{-1} \langle \mathbf{k}_\parallel, \frac{1}{2}\sqrt{\beta^2 + k_\parallel^2} \mid. \quad (19.37)$$

19.4 Implementation and Results

The near- and far-field asymptotics produce inversion schemes that are very similar, differing only in normalization H_F or H_N and the weighting factor $|z - z_0|$. A single approach to reconstructing the sample in all regions is desirable for ease of implementation and computational efficiency. Except at very high NA, the factors of H_F or H_N may be ignored. The transition from near- to far-field happens around one Rayleigh range from the focus. The Rayleigh range is $z_R = 2\alpha^2/k\pi^2$. It thus makes sense to implement a reconstruction scheme given in Equation 19.38 and then multiply the result by $|z - z_0| + z_R$. A more sophisticated hybrid approach with better performance at high NA is given in Ref. 21.

19.4.1 Two- and Three-Dimensional Implementations

The algorithm may be applied to a single slice or so-called B-scan, or may be applied along just one axis of the system. For instance, the 2D version of the algorithm, implemented along the *y*-axis is

$$\langle \mathbf{r} \mid \eta \rangle = \left(| z - z_0 | + z_R \right) \int d\beta \langle z \mid -\beta \rangle \int dk_y \langle y \mid k_y \rangle$$
$$\times \left\langle x, k_y, \frac{1}{2}\sqrt{\beta^2 + k_y^2} \mid S \right\rangle. \quad (19.38)$$

The appearance of the image in any fixed *x*-plane is greatly improved by this algorithm. The features appear to be in-focus everywhere. However, the reconstruction still suffers from artifacts associated with scatterers out of the image plane. Implementing ISAM along one axis brings the PSF back to its focal-plane width in the image plane, but leaves the PSF out-of-plane at the same width as in OCT.

Two- and three-dimensional ISAM are demonstrated in Figure 19.2 with a phantom consisting of 1 μm TiO$_2$ particles embedded in silicone. It may be noted that the image looks qualitatively improved in the 2D version (center panel) but that the full 3D implementation produces further changes still. This effect becomes worse farther from the original focal plane and may be understood to result from the fact that the PSF grows with distance from the focal plane. This narrowing of the PSF along one axis at a time may be seen clearly in Figure 19.3. In Figure 19.4, we demonstrate the same effect in a biological sample consisting of mouse adipose tissue.

19.4.2 Cross-Validation

The fact that the image appears to improve is self-evident. In Figure 19.5, we demonstrate that the ISAM reconstruction produces the correct image by coregistration with an in-focus OCT image [22]. That is, data are acquired and a reconstruction is performed and a single *en face* image far from focus is compared to an image acquired by moving the focus to the plane of interest. It may be seen that the reconstructed image agrees well with the measurement.

19.5 Applications and Outlook

ISAM can be applied anywhere OCT has been applied and well beyond. ISAM eliminates the need to choose between a high resolution or a large DOF. Moreover, ISAM allows for the construction of fixed-focal-plane instruments. These instruments are smaller, more robust, and more economical. The ability to design compact, hand-held systems opens new vistas of applications of OCT. A small-hand-held probe can go into the operating suite or can be kept in the lab for use among several experiments. Extremely sensitive samples unsuitable for examination with OCT and its focused illumination could be inspected with ISAM with the focus positioned just outside the sample.

The emergence of ISAM at this point in history is hardly accidental. The modern GPU provides unprecedented power in parallel processing that makes possible ISAM processing at speeds comparable to the speed of data acquisition. The faster data can be acquired, the easier ISAM is to implement. Recent developments in swept-source lasers and fast detectors for spectral domain OCT make A-scan acquisition rates above 100 KHz feasible. Finally, advances in hardware for beam steering now provide opportunities for smaller and more robust beam scanning. Progress on all three of these fronts is still expanding and should only make ISAM more powerful and useful.

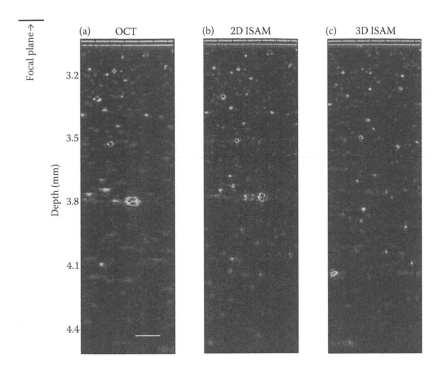

FIGURE 19.2 (**See color insert.**) ISAM in a silicone phantom with 1 μm TiO₂ particles. OCT cross-sectional image of a phantom (a), in-plane 2D ISAM (b), and 3D ISAM (c). The data were acquired at an A-scan rate of 8 kHz on a 1300 nm swept-source OCT system with NA = 0.07. (Courtesy of Diagnostic Photonics, Inc.)

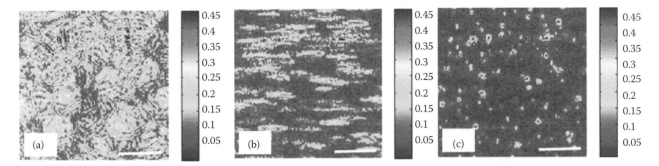

FIGURE 19.3 (**See color insert.**) ISAM in a silicone phantom with 1 μm TiO₂ particles. *En face* planes 17 Rayleigh ranges above the focus with (a) OCT processing, (b) 2D ISAM processing, and (c) 3D ISAM processing. The scale bar denotes 100 μm. Gamma correction (signal $S_{Corr} = S^{\gamma}$, $\gamma = 0.3$) was utilized for dynamic range compression. The data were acquired at an A-scan rate of 92 kHz on a 1300 nm spectral-domain OCT system with NA = 0.1.

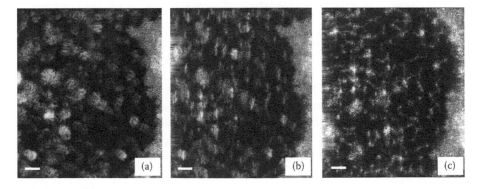

FIGURE 19.4 ISAM of mouse adipose tissue. *En face* planes are 990 μm (18 Rayleigh ranges) above the focus with (a) OCT processing, (b) 2D ISAM processing along the fast (horizontal) axis, and (c) 3D ISAM processing. The scale bar denotes 100 μm. The data were acquired at an A-scan rate of 92 kHz on a 1300 nm spectral-domain OCT system with NA = 0.1.

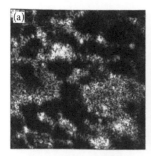
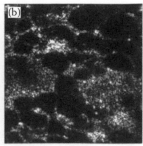
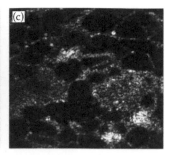

FIGURE 19.5 ISAM and OCT in *ex vivo* rat adipose tissue. (a) *En face* OCT of a plane 388 μm (optical depth) above the focal plane (and 55 μm below the tissue surface). (b) ISAM reconstruction of the same *en face* plane. (c) *En face* OCT with the focal plane moved up to bring the plane of interest in (a) into focus. The field of view in each panel is 500 × 500 μm. All images have gamma correction, with γ = 0.8. Data were acquired with an 800 nm OCT system with NA = 0.01.

This research was supported in part by grants from the National Institutes of Health (NIBIB, R01 EB012479, and R01 EB013723, S.A.B.) and the National Science Foundation (CBET 10-33906, S.A.B.). This animals used in this research were handled and cared for under protocols approved by the Institutional Animal Care and Use Committee at the University of Illinois at Urbana-Champaign.

References

1. D. Huang, E. A. Swanson, C. P. Lin, J. S. Schuman, W. G. Stinson, W. Chang, M. R. Hee et al. Optical coherence tomography. *Science*, 254(5035):1178–1181, 1991.

2. J. G. Fujimoto, M. E. Brezinski, G. J. Tearney, S. A. Boppart, B. Bouma, M. R. Hee, J. F. Southern, and E. A. Swanson. Optical biopsy and imaging using optical coherence tomography. *Nature Medicine*, 1(9):970–972, 1995.

3. G. J. Tearney, M. E. Brezinski, B. E. Bouma, S. A. Boppart, C. Pitris, J. F. Southern, and J. G. Fujimoto. *In vivo* endoscopic optical biopsy with optical coherence tomography. *Science*, 276(5321):2037–2039, 1997.

4. A. F. Fercher, W. Drexler, C. K. Hitzenberger, and T. Lasser. Optical coherence tomography—Principles and applications. *Reports on Progress in Physics*, 66(2):239–303, 2003.

5. J. G. Fujimoto. Optical coherence tomography for ultrahigh resolution *in vivo* imaging. *Nature Biotechnology*, 21(11):1361–1367, 2003.

6. J. M. Schmitt. Optical coherence tomography (OCT): A review. *IEEE Journal of Selected Topics in Quantum Electronics*, 5(4):1205–1215, 1999.

7. W. Drexler, U. Morgner, R. K. Ghanta, F. X. Kartner, J. S. Schuman, and J. G. Fujimoto. Ultrahigh-resolution ophthalmic optical coherence tomography. *Nature Medicine*, 7(4):502–507, 2001.

8. J. S. Schuman, C. A. Puliafito, and J. G. Fujimoto. *Optical Coherence Tomography of Ocular Diseases*. SLACK Inc., Thorofare, NJ, 2004.

9. B. E. Bouma, G. J. Tearney, H. Yabushita, M. Shishkov, C. R. Kauffman, D. D. Gauthier, B. D. MacNeill et al. Evaluation of intracoronary stenting by intravascular optical coherence tomography. *Heart*, 89(3):317–320, 2003.

10. T. Gambichler, G. Moussa, M. Sand, D. Sand, P. Altmeyer, and K. Hoffmann. Applications of optical coherence tomography in dermatology. *Journal of Dermatological Science*, 40(2):85–94, 2005.

11. S. A. Boppart, W. Luo, D. L. Marks, and K. W. Singletary. Optical coherence tomography: Feasibility for basic research and image-guided surgery of breast cancer. *Breast Cancer Research and Treatment*, 84(2):85–97, 2004.

12. B. Povazay, K. Bizheva, A. Unterhuber, B. Hermann, H. Sattmann, A. F. Fercher, W. Drexler et al. Submicrometer axial resolution optical coherence tomography. *Optics Letters*, 27(20):1800–1802, 2002.

13. S. A. Boppart, B. E. Bouma, C. Pitris, G. J. Tearney, J. G. Fujimoto, and M. E. Brezinski. Forward-imaging instruments for optical coherence tomography. *Optics Letters*, 22(21):1618–1620, 1997.

14. J.M. Zara and C.A. Lingley-Papadopoulos. Endoscopic OCT approaches toward cancer diagnosis. *IEEE Journal of Selected Topics in Quantum Electronics*, 14(1):70–81, 2008.

15. D. Lorenser, X. Yang, R. W. Kirk, B. C. Quirk, R. A. McLaughlin, and D. D. Sampson. Ultrathin side-viewing needle probe for optical coherence tomography. *Optics Letters*, 36(19):3894–3896, 2011.

16. J. A. Izatt, M. R. Hee, E. A. Swanson, C. P. Lin, D. Huang, J. S. Schuman, C. A. Puliafito, and J. G. Fujimoto. Micrometer-scale resolution imaging of the anterior eye *in-vivo* with optical coherence tomography. *Archives of Ophthalmology*, 112(12):1584–1589, 1994.

17. S. A. Boppart, B. E. Bouma, C. Pitris, J. F. Southern, M. E. Brezinski, and J. G. Fujimoto. *In vivo* cellular optical coherence tomography imaging. *Nature Medicine*, 4(7):861–865, 1998.

18. R. A. Leitgeb, M. Villiger, A. H. Bachmann, L. Steinmann, and T. Lasser. Extended focus depth for Fourier domain optical coherence microscopy. *Optics Letters*, 31(16):2450–2452, 2006.

19. T. S. Ralston, D. L. Marks, S. A. Boppart, and P. S. Carney. Inverse scattering for high-resolution interferometric microscopy. *Optics Letters*, 31:3585–3587, 2006.

20. T. S. Ralston, D. L. Marks, P. S. Carney, and S. A. Boppart. Interferometric synthetic aperture microscopy. *Nature Physics*, 3(2):129–134, 2007.

21. B. J. Davis, S. C. Schlachter, D. L. Marks, T. S. Ralston, S. A. Boppart, and P. S. Carney. Nonparaxial vector-field modeling of optical coherence tomography and interferometric synthetic aperture microscopy. *Journal of the Optical Society of America A—Optics Image Science and Vision*, 24(9):2527–2542, 2007.

22. T. S. Ralston, S. G. Adie, D. L. Marks, S. A. Boppart, and P. S. Carney. Cross-validation of interferometric synthetic aperture microscopy and optical coherence tomography. *Optics Letters*, 35(10):1683–1685, 2010.

23. B. J. Davis, D. L. Marks, T. S. Ralston, P. S. Carney, and S. A. Boppart. Interferometric synthetic aperture microscopy: Computed imaging for scanned coherent microscopy. *Sensors*, 8(6):3903–3931, 2008.

24. T. S. Ralston, D. L. Marks, P. S. Carney, and S. A. Boppart. Real-time interferometric synthetic aperture microscopy. *Optics Express*, 16(4):2555–2569, 2008.

25. A. F. Fercher, C. K. Hitzenberger, G. Kamp, and S. Y. Elzaiat. Measurement of intraocular distances by backscattering spectral interferometry. *Optics Communications*, 117(1–2):43–48, 1995.

26. G. Häusler and M. W. Lindner. "Coherence radar" and "spectral radar" new tools for dermatological diagnosis. *Journal of Biomedical Optics*, 3(21):21–31, 1998.

27. J. F. de Boer, B. Cense, B. H. Park, M. C. Pierce, G. J. Tearney, and B. E. Bouma. Improved signal-to-noise ratio in spectral-domain compared with time-domain optical coherence tomography. *Optics Letters*, 28(21):2067–2069, 2003.

28. R. Leitgeb, C. K. Hitzenberger, and A. F. Fercher. Performance of Fourier domain vs. time domain optical coherence tomography. *Optics Express*, 11(8):889–894, 2003.

29. M. A. Choma, M. V. Sarunic, C. H. Yang, and J. A. Izatt. Sensitivity advantage of swept source and Fourier domain optical coherence tomography. *Optics Express*, 11(18):2183–2189, 2003.

30. M. Wojtkowski, V. J. Srinivasan, T. H. Ko, J. G. Fujimoto, A. Kowalczyk, and J. S. Duker. Ultrahigh-resolution, high-speed, Fourier domain optical coherence tomography and methods for dispersion compensation. *Optics Express*, 12(11):2404–2422, 2004.

31. D. L. Marks, A. L. Oldenburg, J. J. Reynolds, and S. A. Boppart. Autofocus algorithm for dispersion correction in optical coherence tomography. *Applied Optics*, 42(16):3038–3046, 2003.

32. M. Wojtkowski, A. Kowalczyk, R. Leitgeb, and A. F. Fercher. Full range complex spectral optical coherence tomography technique in eye imaging. *Optics Letters*, 27(16):1415–1417, 2002.

33. M. V. Sarunic, M. A. Choma, C. H. Yang, and J. A. Izatt. Instantaneous complex conjugate resolved spectral domain and swept-source OCT using 3x3 fiber couplers. *Optics Express*, 13(3):957–967, 2005.

34. L. An and R. K. Wang. Use of a scanner to modulate spatial interferograms for *in vivo* full-range Fourier-domain optical coherence tomography. *Optics Letters*, 32(23):3423–3425, 2007.

20

Fourier Transform Infrared Spectroscopic Imaging: An Emerging Label-Free Approach for Molecular Imaging

Kevin Yeh
University of Illinois at Urbana-Champaign

Rohith Reddy
University of Illinois at Urbana-Champaign

Rohit Bhargava
University of Illinois at Urbana-Champaign

20.1 Introduction

Fourier transform infrared (FT-IR) spectroscopy is a widely used technique for measuring the chemical composition of a sample. Different functional groups in molecules have resonant frequencies in the mid-IR spectral range and will absorb light corresponding to their resonant frequencies. The absorption spectrum of a molecule, hence, acts as its chemical signature and will contain contributions from each of the different functional groups constituting the molecule. Even when a mixture of complex composition is used, the chemically specific spectrum is often termed a "molecular fingerprint" of the material. While the practice of FT-IR spectroscopy is over 100 years old, there has been a recent renaissance in theoretical understanding, instrumentation, and applications that has generated considerable interest in scientific study and applications in this area. In particular, the ability to couple IR spectroscopy with microscopy has been advanced significantly with the development of

array detectors. This combination of spectroscopy, microscopy, and multichannel detection is often referred to as FT-IR spectroscopic imaging (Bhargava and Levin 2001) or FT-IR imaging. Since the primary contrast mechanism in FT-IR imaging is intrinsic, molecular, and quantitative, it alleviates the need for external contrast agents or chemicals. Consequently, FT-IR imaging has found a wide variety of applications in forensics, cancer histopathology, polymer science, drug diffusion studies, art restoration, and tissue engineering. The fundamentals of instrumentation, theory, and applications will be presented in this work.

20.2 History

The theoretical discoveries enabling modern FT-IR spectroscopy and imaging were developed primarily over the past 40 years. In the prior dispersive spectroscopic technology, a prism was used to separate the frequencies in the light source and a small grating would select a specific wavelength of light. The resulting monochromatic beam was shone through a sample and the amount of light absorbed was measured. Then, the grating would be shifted and the measurement repeated for each individual wavelength. In 1948, P. Jacquinot discovered that the energy throughput using an interferometer is much higher than what could be realized in the dispersive spectrometer, where the light intensity is restricted by the slits of the optical grating (Saptari 2004). This is called the Jacquinot or throughput advantage. By replacing the grating with an interferometer, FT-IR spectrometers have an increased sensitivity of roughly 60-fold for the same resolving power and instrument size (Saptari 2004). In 1949, a discovery by P.B. Fellgett, now referred to as the Felgett or multiplex advantage, showed that the signal intensity at multiple wavelengths could be simultaneously measured when the interferogram is deconstructed using a Fourier transform (Saptari 2004). However, a Fourier transform of a spectrum consisting of tens of thousands of data points was a prohibitively slow process at that time.

The fast Fourier transform (FFT) algorithm developed by Cooley and Tukey in 1965 dramatically decreased the computational requirements for performing a Fourier transform, making the modern-day FT-IR spectroscopy practical. The combination of the aforementioned three insights allowed an interferometer-based spectrometer to achieve the signal-to-noise ratio (SNR) comparable to a dispersive instrument while significantly reducing the measurement time. In the ensuing years, the interferometers being developed typically involved various configurations of stationary or scanning paradigms. Most stationary interferometers are based on the Sagnac design (Griffiths and De Haseth 2007). The beam is split and made to travel in different directions along a common closed loop path. In one example developed by Okamoto et al., the angles of the mirrors along the common path are arranged to disperse the beam such that they form an interferogram upon recombination at the beam splitter (Okamoto et al. 1985). This interferogram is then focused onto a linear array detector. Although this design has the advantage of having no moving parts, it is not as stable as scanning interferometers since the generation of the interferogram is extremely sensitive to the angle of the mirrors along the path. Furthermore, the resolution of the interferogram is limited by the number of detector elements in the array.

Most scanning interferometers used for spectroscopy are based on the basic design by Michelson (details presented in the next section). The vast majority of commercial instruments are based on this design. Cube-corner interferometers were designed to compensate for mirror tilt effects by ensuring that the reflected beam is always parallel to the incident beam and are used in some commercial implementations. Their performance is nearly identical to a well-aligned basic Michelson interferometer. There are also other designs that may have specific advantages but are not routinely commercially available. Refractively scanned interferometers use prisms in place of plane mirrors in order to increase the optical path difference (OPD). Lamellar grating interferometers reflect light off of a surface consisting of adjustable parallel interleaved mirrors, where the OPD is dependent on the

shift between the mirror facets. Interferometers have also been designed to take advantage of polarization, where an OPD is introduced by rotating a rocking mirror. The reliability of many of these designs depends heavily on the accuracy of movement of the reference mirror. Other designs have been proposed to be stable or to scan rapidly. Precision electromotors and advanced control via computers, however, have made Michelson interferometers the preferred design in modern FT-IR spectroscopy.

While IR spectroscopy technology was developing, significant efforts were also expended in developing IR microscopy. The combination of a microscope and IR spectrometer was reported more than 60 years ago (1957). The instrument, however, relied on dispersion of light and collected spectral elements one by one for each spatial position. To acquire a spectrum at every point, the spectrum at every spatial position had to be serially recorded and the sample had to be raster scanned across many positions, making data acquisition very slow and impractical for most uses. Microscopy is challenging as light throughput for a small spatial region is necessarily small. As opposed to many other optical imaging techniques, mid-IR lasers for the broadband spectral regions were not developed. Hence, only weak thermal sources were available. Since the spectral region was not driven by either consumer demand or by visible imaging needs, detector development was also slow. Hence, microscopy did not make much progress until the advent of FT-IR spectroscopy, better detectors, and microcomputer control and processing. Coupling a microscope to an FT-IR spectrometer resulted in significant speed advantages but was commercially available only in the late 1880s. The sample was still raster scanned to collect the spatial data. The late 1990s saw the introduction of wide-field detection with focal plane array (FPA) detectors that were declassified from military applications (Levin and Bhargava 2005). The tremendous multichannel advantage imparted by array detectors provided a speed advantage over the previous point-by-point mapping technology (Bhargava et al. 2000). To distinguish the multichannel detection-based instruments from the point-scanning instruments, the latter are referred to as "imaging" techniques, while the former are termed "mapping" techniques. A second generation of detectors—both linear array and revamped FPA —provided further advances in the early 2000s. The period since then has been dominated by various applications and data-analysis methods. Lately, a focus on quantitative and fundamental understanding has driven both instrumentation and applications. For example, the use of high numerical aperture optics coupled with optimized data sampling has enabled high-definition FT-IR spectroscopic imaging (Mohlenhoff et al. 2005, Bassan et al. 2009, Lee et al. 2007, Boiana 2000). These developments have fueled and are reenergizing the use of FT-IR spectroscopic imaging in a wide array of fields including polymeric materials (Nishikawa et al. 2012), forensics (Elkins 2011), art restoration (Navas et al. 2008), and tissue histopathology (Holton et al. 2011).

20.3 Theory

IR microscopy is an unusually rich area of research because the spectral diversity and unique light–matter interactions in the mid-IR spectral range give rise to interesting theoretical questions and instrument-design trade-offs. In this section, we first present the basic theory of a Michelson interferometer, which helps us to understand the trade-offs in spectroscopic acquisition and analysis. Next, we discuss the optical theory of IR microscopy, which helps understand the spectral response of heterogeneous systems and spectral recording when using a microscope. Combined, the two aspects afford a full understanding of the data acquisition and aid in the interpretation of recorded data.

20.3.1 Interferogram

When two waves propagate along the same path, they interfere with each other, either constructively or destructively depending on their frequency, magnitude, and phase. An interferogram is the signal obtained by superimposing several such waves. In a Michelson interferometer, light from an IR source is split into two arms using a beam splitter and reflected back to the beam splitter by mirrors in each arm. A schematic diagram of such a setup is demonstrated in Figure 20.1. As opposed to other uses of interferometry, as used in depth ranging, for example, this approach is the predominant one used for spectral encoding in mid-IR. The common path of the beam provides for an exceptionally stable signal in which noise sources, including source fluctuations and phase errors among others, that often dominate other applications of interferometry are naturally eliminated. The length of path traversed by light in each arm can be varied by changing the relative positions of the mirrors and the difference in path length is called retardation and denoted by δ. The intensity of light at the detector is recorded as a function of δ and is called an interferogram. The intensity, I, as a function of retardation, δ, for a Michelson interferometer is given by

$$I'(\delta) = 0.5I(\tilde{v}_o)(1 + \cos(2\pi\tilde{v}_o\delta)), \qquad (20.1)$$

where $\tilde{v}_o$ is the wavenumber in cm^{-1}. Typically, only the oscillating component (AC signal) is useful for spectrometry and the constant term (average DC signal) is subtracted either in hardware or software. With this simplification, Equation 20.1 can be written as

$$I(\delta) = 0.5I(\tilde{v}_o)\cos 2\pi\tilde{v}_o\delta. \qquad (20.2)$$

The instrument parameters including the beamsplitter efficiency, detector response, and amplifier characteristics modify the signal and are included into the measurement via a multiplicative term $G(\tilde{v}_o)$ (Griffiths and De Haseth 2007). The signal, $S(\delta)$, from the amplifier is

$$S(\delta) = 0.5G(\tilde{v}_o)I(\tilde{v}_o)\cos(2\pi\tilde{v}_o\delta). \qquad (20.3)$$

All the multiplicative factors can be simplified into the beam spectral intensity, $B(\tilde{v}_o)$ resulting in

$$S(\delta) = B(\tilde{v}_o)\cos(2\pi\tilde{v}_o\delta). \qquad (20.4)$$

$S(\delta)$ is the Fourier cosine transform of $B(\tilde{v}_o)$. Consequently, the spectrum of the signal can be calculated by computing the inverse Fourier cosine transform of $S(\delta)$. The interferogram and spectrum form a Fourier cosine transform pair and can be computed from one another as shown below:

$$S(\delta) = \int_{-\infty}^{+\infty} B(\tilde{v}_o)\cos(2\pi\tilde{v}_o\delta)d\tilde{v}_o, \qquad (20.5)$$

and

$$B(\tilde{v}_o) = \int_{-\infty}^{+\infty} S(\delta)\cos(2\pi\tilde{v}_o\delta)d\delta. \qquad (20.6)$$

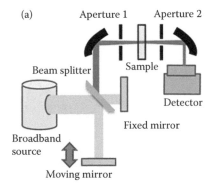
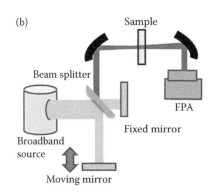

FIGURE 20.1 Schematic diagram of (a) FT-IR point spectroscopy using far-field apertures and (b) FT-IR imaging with a multielement FPA detector. Note that the interferometer, consisting of the source, beam splitter, and mirror with their attendant structural, electrical, and control elements, is common between instruments and the commonly available FT-IR spectrometers. The output of the interferometer is directed to an IR microscope and the detectors can be monolithic or FPA.

20.3.2 Resolution

Consider a spectrum that consists of features, at two distinct wavenumbers, $\tilde{\nu}_1$ and $\tilde{\nu}_2$, but with identical magnitudes. The resolution of an interferometer is the minimum difference in these positions, $\Delta\tilde{\nu}$ such that the corresponding spectral features can still be resolved. In the time domain, when these two beams with distinct wavenumbers are superimposed, the interference pattern will repeat every multiple of $(\Delta\tilde{\nu})^{-1}$ where the beams are in phase. Therefore, in order to just resolve the two wavenumbers, the maximum travel distance of the movable mirror, Δ_{max}, has to be such that

$$\Delta\tilde{\nu} = \left(\Delta_{max}\right)^{-1}. \qquad (20.7)$$

Obviously, the longer the distance an interferometer scans, the finer the resolution. Longer scanning will also require longer times and results in an overall lower SNR per resolution element. Hence, a resolution that is sufficient for the problem at hand is usually recommended. A finite resolution implies a finite maximum retardation. The effect of this finite maximum retardation is to introduce a truncation function, $D(\delta)$, called the instrument line shape to the beam, where

$$D(\delta) = \begin{cases} 1, & -\Delta \le \delta \le +\Delta \\ 0, & |\delta| > |\Delta| \end{cases} \qquad (20.8)$$

and

$$B(\tilde{\nu}_o) = \int_{-\infty}^{+\infty} S(\delta)D(\delta)\cos(2\pi\tilde{\nu}_o\delta)d\delta. \qquad (20.9)$$

The resulting spectrum is a convolution of the original spectrum with the Fourier transform of $D(\delta)$. Choosing an appropriate truncation function is important to obtain the useful data. For the boxcar truncation function above, the FT is a $sinc$ function. The $sinc$ function, however, is not a useful line shape because of its large oscillating amplitudes away from the center. The side lobes of a $sinc$ function can reach negative values with a magnitude 22% of the primary peak and small features that fall in these negative regions will appear omitted from the computed spectrum. Therefore, apodization is used to control the effect of the finite retardation by modulating the instrument line shape in order to minimize these secondary effects.

20.3.3 Apodization

The molecular fidelity of the spectrum and resolution of the interferogram can be improved by controlling appropriately the instrument line shape, $D(\delta)$. When the boxcar apodization function is used to modulate the instrument line shape, a spectrum consisting of two adjacent spectral lines of equal intensity as shown in Figure 20.2, the two features cannot be resolved when the separation is $<0.5/\Delta$. Analogous to the definition of resolution for optical microscopy, the resolution for spectrometers is often defined using the Rayleigh criterion and the full width at half-height criterion (Griffiths and De Haseth 2007). For the Rayleigh criterion, we can consider these lines distinguishable if the resulting curve has a dip of approximately 20% of the maximum intensity, which occurs at $0.73/\Delta$ separation, as shown in Figure 20.3. The lines are considered fully resolved if the separation dips down to the baseline.

Several apodization techniques have been proposed to replace the boxcar function including trapezoidal, triangular, triangular-squared, Bessel, $sinc$-squared, and Gaussian functions (Griffiths and De Haseth 2007). All these functions have smaller oscillatory side lobes. The following example illustrates the use of a triangular function (defined later) in apodization. The FT of a triangular function is a $sinc^2$ function. Using the same separation ($0.5/\Delta$) as Figure 20.2a earlier, the triangular apodization function allows the distinct components of the spectrum to be resolved:

$$T(\delta) = \begin{cases} 1 - \left|\dfrac{\delta}{\Delta}\right|, & \Delta \le \delta \le +\Delta \\ 0, & \delta > |-\Delta| \end{cases}. \qquad (20.10)$$

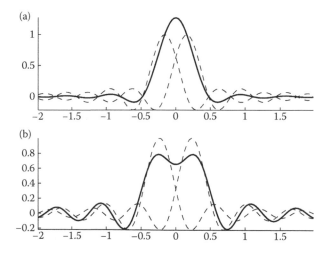

(a)

(b)

FIGURE 20.2 $Sinc$ functions of a spectrum consisting of two distinct wavenumbers spaced (a) $0.5/\Delta$ and (b) $0.73/\Delta$ apart. The features in the first scenario cannot be resolved.

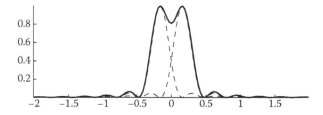

FIGURE 20.3 $Sinc^2$ function of a spectrum consisting of two distinct wavenumbers spaced $0.5/\Delta$ apart. Unlike the **sinc** function, the triangular apodization function clearly allows the closely spaced features in this spectrum to be resolved.

20.3.4 Sampling

The analog signal that arrives at the detector eventually undergoes a digitization process. While an extremely high sampling rate can better emulate a continuous signal, this is rarely done for two reasons. First, this is not practical because the FFT calculation will take a prohibitively long time. Second, the source, optical components, or detector may either be spectrally limited in their response or the throughput in the system can be limited by the use of appropriate optical filters. Hence, the combination of a useful bandpass for the desired spectral analysis and minimal computation can lead to optimal spectral recording and processing. Once the spectral and optical parameters are set, the appropriate sampling rate so as to retain all the information from a band-limited signal is given by the Shannon–Nyquist sampling theorem. The theorem states that the sampling rate must be at least twice the bandpass of the system in order to accurately represent the original analog signal. Care must be taken to ensure that a signal being sampled at a rate $2F$ does not have frequencies above F. The presence of frequencies above F results in aliasing and can potentially affect the entire signal. Figure 20.4a shows the FT of a signal with bandwidth F being sampled at a frequency, $f_s < 2F$. This causes a corruption of the data in the (overlap) frequency range below frequency F. To prevent an overlap of frequencies, that is, aliasing, the data must be sampled at not $<2F$ as shown in Figure 20.4b.

The analog optical signals are digitized using an analog-to-digital converter (ADC). Modern ADCs typically have a resolution between 14 and 18 bits, which allows the ability to distinguish between 2^{14} and 2^{18}, or respectively 16,384 and 262,144, separate voltage levels. However, at least 1 bit is typically reserved for noise to prevent truncation errors. For a spectrum with a resolution of $\Delta \tilde{v}$ and a bandpass between $\tilde{v}_{\max}$ and $\tilde{v}_{\min}$, the number of digital levels required to record its interferogram is shown in Equation 20.37 and the dynamic range is approximated by its square root.

$$M = \frac{\tilde{v}_{\max} - \tilde{v}_{\min}}{\Delta \tilde{v}}. \tag{20.11}$$

A common mid-IR spectrum with a bandpass of 3600 cm⁻¹ at a resolution of 4 cm⁻¹ will require 900 digital levels. The dynamic range of this spectrum is the number of digital levels of the ADC divided by the dynamic range of the interferogram, which in this case is roughly 1000:1 for a 16-bit ADC. Commercial FT-IR instruments can even exceed this spectral dynamic range and use higher-resolution ADCs.

20.3.5 Empirical Errors

The prior theoretical analysis assumes a perfectly collimated beam, which is difficult to achieve experimentally. Smaller sources are more collimated but have a weaker signal at the detector. A beam traveling through the interferometer with the largest solid angle possible without degrading the resolution will yield the highest signal. A monochromatic beam passing through a Michelson interferometer will create interference fringes, called Hädinger fringes, on the image plane at the detector. These fringes will oscillate sinusoidally when the retardation increases. If the entire beam was measured, an interferogram could not be generated because there would be no average change in intensity. An aperture, called a Jacquinot stop, limits the beam such that only the central peak of the fringe pattern passes through the detector. The maximum solid angle allowed for the beam is given by

$$\Omega_{\max} = 2\pi \frac{\Delta \tilde{v}}{\tilde{v}_{\max}}. \tag{20.12}$$

Alignment of the fixed mirror, moving mirror, and beam splitter relative to one another will affect the interferogram. The mirror drive mechanism must be capable of keeping the plane of the moving mirror at a constant angle to the beam path at all points during the scan. A small angle discrepancy of any mirror during the scan will result in a misaligned detector causing a significant reduction of signal intensity. Moreover, fringes can be formed at the detector plane resulting in poor data quality.

20.3.6 Signal-to-Noise Ratio

The SNR of an acquired spectrum is one of the most critical quantities for obtaining the best analytical results. Owing to the abovementioned considerations, the noise in FT-IR spectrometers is usually dominated by the noise at the detector. Since the noise is random and uncorrelated with the source intensity, signal averaging offers significant advantages. Often, extensive signal averaging is required, especially in microscopy formats. The SNR scales as $t^{1/2}$, ultimately providing a diminishing return and yielding to the effects of other noise sources, including medium-term drift and source fluctuations. When all parameters (range, source intensity, and resolution) have been optimized for the

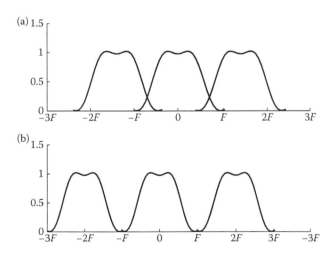

FIGURE 20.4 A signal sampled at (a) $f_s < 2F$ results in aliasing, which can be avoided by sampling at (b) $f_s \geq 2F$.

experiment and the best possible hardware is chosen, signal averaging is the best recourse for improving SNR.

20.3.7 FT-IR Spectroscopic Imaging

FT-IR spectroscopic imaging couples a Michelson interferometer to an IR microscope. There have been significant developments in the theory of FT-IR spectroscopic imaging in the past 3 years. Since its inception, FT-IR imaging data has been treated as a direct extension of FT-IR spectroscopy, with the additional trivial consideration that data are spatially resolved. Other than reflecting species concentration, the spatial structure of the sample was not thought to play a major role on the recorded data. However, there are fundamental differences in the data recorded between homogeneous and heterogeneous samples that arise because of the microstructure of samples on the scale of the wavelength. Recent work has shown that the recorded spectra can be significantly different in important ways from FT-IR spectroscopy and these differences arise due to the shape and distribution of different chemical species (Davis et al. 2010a, b, Reddy et al. 2011) as well as the effects of focusing optics, which are considerably less impactful in non-imaging systems. Rigorous electromagnetic models analyzing the interaction of light with heterogeneous samples have quantified the nature of these spectral differences. A systematic and robust difference is indeed observed experimentally, validating the theory.

The theory of interaction of IR light with layered samples has shown that the recorded data can be distorted from bulk spectra because of sample thickness. If the sample thickness is known *a priori*, iterative techniques can be used to estimate the bulk spectra from FT-IR imaging data. In the absence of thickness information, coestimating both geometry and spectra is difficult and convergence to the correct result is not always guaranteed. Extending this model to include both lateral and transverse heterogeneities has also been proposed (Davis et al. 2010b). Using these models, it is possible to predict the collected spectrum when the IR optical properties and the geometry of the sample are known. The spectral differences are typically more prominent in transflection FT-IR imaging data relative to transmission data. These spectral differences fundamentally arise due to scattering of light at optical boundaries and the nature of these distortions can be quantified. This work has demonstrated that distortions are largest when the point of focus of an IR light is directly at the edge and they diminish as we move away from the edge. Simultaneously recovering both the geometry and spectra of samples using IR imaging is an open problem and the subject of current research. In the next section, we present a rigorous theoretical model for understanding data from FT-IR spectroscopic imaging instruments.

20.3.8 Light–Sample Interaction

Here, we present a rigorous electromagnetic model to describe light–sample interaction and its effect on recorded data. The work presented here borrows heavily from work from our laboratory, especially Elkins (2011), Navas et al. (2008), and Bhargava and Kong (2008). The general optical setup for the spectroscopic imaging system working in transmission mode is shown in Figure 20.5. Light is focused on to the sample using Cassegrain 1 and is collected by Cassegrain 2 after interacting with the sample. The sample is assumed to be a linear system consisting of multiple layers of varying (complex) refractive indices. We separate the problem of light–sample interaction into the transition between pairs of homogeneous layers and transition between homogeneous and heterogeneous layers.

Owing to the linear nature of the system, for interfaces between homogeneous layers, the electromagnetic field can be described as consisting of a linear combination of plane waves each of which satisfies Maxwell's equations at the boundaries. We identify the response of the system to a single plane wave and finally obtain the total response as a sum of each of these individual responses. The electric E and magnetic H fields at a position $r = (x, y, z)^T$ are represented by their complex amplitudes. The permittivity and permeability of free space are denoted by ϵ_0 and μ_0, respectively, and the real and imaginary parts of the refractive index of a single layer are denoted by $n(\upsilon)$ and $k(\upsilon)$, respectively. For homogeneous layers, the electric and magnetic fields can be written as

$$E(r,\tilde{v},t) = E_0 e^{i2\pi\tilde{v}s\hat{v}r}, \tag{20.13}$$

and

$$H(r,\tilde{v},t) = \sqrt{\frac{\epsilon_0}{\mu_0}}(s \times E_0)e^{i2\pi\tilde{v}s\hat{v}r}, \tag{20.14}$$

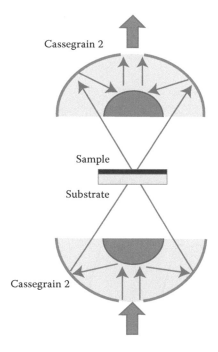

FIGURE 20.5   General setup for a transmission mode FT-IR imaging.

where $s = (s_x, s_y, s_z)^T$ and $|s| = \epsilon(\upsilon) = [n(\upsilon) + ik(\upsilon)]^2$. The temporal dependence of the fields is not explicitly stated for notational simplicity. The field in the lth layer (between z^{l-1} and z^l) is

$$
E^{(l)}(x, y, z, \tilde{\nu}) = \tilde{\nu} \iint_{\mathbb{R}^2} \left\{ \begin{array}{l} B^{(l)}(s_x, s_y, \tilde{\nu}) e^{\left[i2\pi\tilde{\nu}s_z^{(l)}(z - z^{(l-1)})\right]} \\ + \hat{B}^{(l)}(s_x, s_y, \tilde{\nu}) e^{\left[-i2\pi\tilde{\nu}s_z^{(l)}(z - z^{(l)})\right]} \end{array} \right\}
$$
$$
e^{\left[i2\pi\tilde{\nu}(s_x x + s_y y)\right]} ds_x ds_y, \qquad (20.15)
$$

where B is the angular spectrum in a transverse plane. Next, we set up boundary conditions at each layer-interface. Using Gauss' law for electricity, we have

$$
s_x B_x^{(l)}(s_x, s_y, \tilde{\nu}) + s_y B_y^{(l)}(s_x, s_y, \tilde{\nu}) + s_z^{(l)} B_z^{(l)}(s_x, s_y, \tilde{\nu}) = 0, \qquad (20.16)
$$

and

$$
s_x \hat{B}_x^{(l)}(s_x, s_y, \tilde{\nu}) + s_y B_y^{(l)}(s_x, s_y, \tilde{\nu}) - s_z^{(l)} \hat{B}_z^{(l)}(s_x, s_y, \tilde{\nu}) = 0. \qquad (20.17)
$$

Faraday's law and Ampere's law require that the transverse components of the electric and magnetic fields be continuous. Therefore

$$
B_x^{(l)} e^{\left[i2\pi\tilde{\nu}s_z^{(l)}(z^{(l)} - z^{(l-1)})\right]} + \hat{B}_x^{(l)} = B_x^{(l+1)} + \hat{B}_x^{(l+1)} e^{\left[-i2\pi\tilde{\nu}s_z^{(l+1)}(z^{(l)} - z^{(l+1)})\right]}, \qquad (20.18)
$$

$$
B_y^{(l)} e^{\left[i2\pi\tilde{\nu}s_z^{(l)}(z^{(l)} - z^{(l-1)})\right]} + \hat{B}_y^{(l)} = B_y^{(l+1)} + \hat{B}_y^{(l+1)} e^{\left[-i2\pi\tilde{\nu}s_z^{(l+1)}(z^{(l)} - z^{(l+1)})\right]}, \qquad (20.19)
$$

and

$$
(s_y B_z^{(l)} - s_z^{(l)} B_y^{(l)}) e^{\left[i2\pi\tilde{\nu}s_z^{(l)}(z^{(l)} - z^{(l-1)})\right]} + (s_y \hat{B}_z^{(l)} + s_z^{(l)} \hat{B}_y^{(l)})
$$
$$
= (s_y B_z^{(l+1)} - s_z^{(l+1)} B_y^{(l+1)})
$$
$$
+ (s_y \hat{B}_z^{(l+1)} + s_z^{(l+1)} \hat{B}_y^{(l+1)}) e^{\left[-i2\pi\tilde{\nu}s_z^{(l+1)}(z^{(l)} - z^{(l+1)})\right]} \qquad (20.20)
$$

$$
(s_z^{(l)} B_x^{(l)} - s_x B_z^{(l)}) e^{\left[i2\pi\tilde{\nu}s_z^{(l)}(z^{(l)} - z^{(l-1)})\right]} + (-s_z^{(l)} \hat{B}_x^{(l)} - s_x \hat{B}_z^{(l)}) = (s_z^{(l+1)} B_x^{(l+1)}
$$
$$
- s_x B_z^{(l+1)}) - (s_z^{(l+1)} \hat{B}_x^{(l+1)} + s_x \hat{B}_z^{(l+1)}) e^{\left[-i2\pi\tilde{\nu}s_z^{(l+1)}(z^{(l)} - z^{(l+1)})\right]}. \qquad (20.21)
$$

Finally, note that there is no incoming light from the last boundary (L) and therefore,

$$
\hat{B}^{(L)}(s_x, s_y, \tilde{\nu}) = 0. \qquad (20.22)
$$

The above equations provide the complete set of boundary conditions for analyzing a multilayer sample consisting of homogeneous layers. Knowledge of the refractive indices and thicknesses of all layers would allow us to completely solve the system

of equations. Therefore, the electric and magnetic fields at every point in the sample can be completely determined. The field on a detector is a magnified version of the field exiting Cassegrain 2 and the total light intensity at the detector ($I = |E|^2$) can be computed. This detector intensity is the raw data recorded from the instrument (before processing). Therefore, the recorded data can be obtained from the model, provided that the source and sample properties are known.

Next, we consider the case when some layers are heterogeneous. In an imaging system, the region of transverse heterogeneity can always be assumed to be of finite area and therefore we can approximate the structure of the object by its two-dimensional Fourier series. Each heterogeneous layer is characterized by a transversely varying complex refractive index or equivalently by its permittivity ε. Truncating the Fourier series of ε and ε^{-1} to a finite number of coefficients gives

$$
\epsilon(x, y, \tilde{\nu}) \approx \sum_{p=-N_U}^{N_U - 1} \sum_{q=-N_w}^{N_W - 1} \phi_{p,q}(\tilde{\nu}) e^{[i(pUx + qWy)]}, \qquad (20.23)
$$

and

$$
[\epsilon(x, y, \tilde{\nu})]^{-1} \approx \sum_{p=-N_U}^{N_U - 1} \sum_{q=-N_w}^{N_W - 1} \psi_{p,q}(\tilde{\nu}) e^{[i(pUx + qWy)]}. \qquad (20.24)
$$

Owing to the periodic nature of the sample, an incident plane wave $e^{[i(\delta x + \sigma y)]}$ is shifted only by integer frequencies integer multiples of spatial frequencies U and W, resulting in plane waves with spatial frequency components given by

$$
u_p = pU + \delta, \qquad (20.25)
$$

and

$$
w_q = qW + \sigma. \qquad (20.26)
$$

The electric and magnetic fields can be represented using plane wave decompositions as

$$
E^{(\Delta)}(x, y, z, \tilde{\nu}) = \sum_p \sum_q \begin{bmatrix} X_{p,q}(z, \tilde{\nu}) \\ Y_{p,q}(z, \tilde{\nu}) \\ Z_{p,q}(z, \tilde{\nu}) \end{bmatrix} e^{[i(u_p x + w_q y)]}, \qquad (20.27)
$$

and

$$
H^{(\Delta)}(x, y, z, \tilde{\nu}) = \sqrt{\frac{\varepsilon_0}{\mu_0}} \sum_p \sum_q \begin{bmatrix} I_{p,q}(z, \tilde{\nu}) \\ J_{p,q}(z, \tilde{\nu}) \\ K_{p,q}(z, \tilde{\nu}) \end{bmatrix} e^{[i(u_p x + w_q y)]}. \qquad (20.28)
$$

Note that the electric and magnetic field are not necessarily orthogonal in the presence of transverse heterogeneity, but can be related using Faraday's law and Ampere's law, resulting in

$$\nabla \times E(\boldsymbol{r}, \tilde{\boldsymbol{v}}) = i2\pi\tilde{v}\sqrt{\frac{\mu_0}{\epsilon_0}}H(\boldsymbol{r}, \tilde{\boldsymbol{v}}), \qquad (20.29)$$

and

$$\nabla \times H(\boldsymbol{r}, \tilde{\boldsymbol{v}}) = -i2\pi\tilde{v}\epsilon(\boldsymbol{r}, \tilde{\boldsymbol{v}})\sqrt{\frac{\epsilon_0}{\mu_0}}E(\boldsymbol{r}, \tilde{\boldsymbol{v}}). \qquad (20.30)$$

Expanding the above equations gives

$$\frac{dX_{p,q}(z, \tilde{\boldsymbol{v}})}{dz} = i2\pi\tilde{v}J_{p,q}(z, \tilde{\boldsymbol{v}}) + iu_p Z_{p,q}(z, \tilde{\boldsymbol{v}}), \qquad (20.31)$$

$$\frac{dY_{p,q}(z, \tilde{\boldsymbol{v}})}{dz} = -i2\pi\tilde{v}I_{p,q}(z, \tilde{\boldsymbol{v}}) + iw_q Z_{p,q}(z, \tilde{\boldsymbol{v}}), \qquad (20.32)$$

$$K_{p,q}(z, \tilde{\boldsymbol{v}}) = \frac{1}{2\pi\tilde{v}}[u_p Y_{p,q}(z, \tilde{\boldsymbol{v}}) - w_q X_{p,q}(z, \tilde{\boldsymbol{v}})], \qquad (20.33)$$

$$\frac{dI_{p,q}(z, \tilde{\boldsymbol{v}})}{dz} = -i2\pi\tilde{v}\sum_{p''}\sum_{q''}\phi_{p-p'',q-q''}(\tilde{v})Y_{p'',q''}(z, \tilde{\boldsymbol{v}}) + iu_p K_{p,q}(z, \tilde{\boldsymbol{v}})$$

$$(20.34)$$

and

$$\frac{dJ_{p,q}(z, \tilde{\boldsymbol{v}})}{dz} = i2\pi\tilde{v}\sum_{p''}\sum_{q''}\phi_{p-p'',q-q''}(\tilde{v})X_{p'',q''}(z, \tilde{\boldsymbol{v}}) + iw_q K_{p,q}(z, \tilde{\boldsymbol{v}}).$$

$$(20.35)$$

These equations can be reorganized into a matrix differential equation of the form

$$\begin{bmatrix} \dfrac{dX(z,\tilde{v})}{dz} \\ \dfrac{dY(z,\tilde{v})}{dz} \\ \dfrac{dI(z,\tilde{v})}{dz} \\ \dfrac{dJ(z,\tilde{v})}{dz} \end{bmatrix} = i2\pi\tilde{v}\Phi(\tilde{v})\begin{bmatrix} X(z,\tilde{v}) \\ Y(z,\tilde{v}) \\ I(z,\tilde{v}) \\ J(z,\tilde{v}) \end{bmatrix}. \qquad (20.36)$$

The above matrix equation couples the electric and magnetic fields and it is possible to decouple these equations by computing the eigenvalues and eigenvectors of Φ. We can now find X, Y, I, and J in terms of the eigenvalues and eigenvectors. The boundary conditions for each component are along the same lines as those for homogeneous layers given above. Therefore, we can solve the matrix equation and find the electric (E) and magnetic (H) fields in every layer. The field on the detector can again be computed by propagating light via Cassegrain 2 to the detector. Finally, the detector intensity I, and therefore the recorded raw data, can be calculated given the source and sample properties.

Spectra from such a calculation for a heterogeneous layer are shown in Figure 20.6. Note that the green spectrum, which corresponds to the light focused at an edge, is distorted relative to the black spectrum, which is away from the edge.

20.3.9 Noise Reduction via Postprocessing

While noise can be reduced experimentally via signal averaging, this approach requires a significant increase in data collection time. Alternatively, computational noise reduction techniques can be used following the data acquisition to reduce

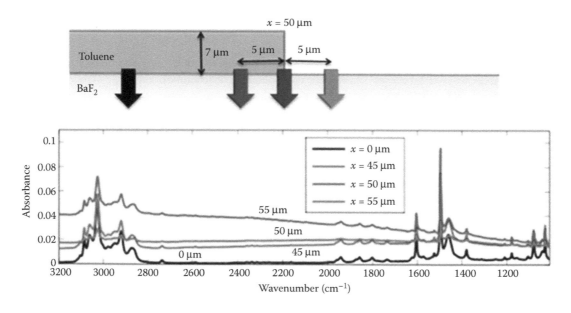

FIGURE 20.6   Recorded spectra when light is focused at different positions on the sample (indicated by arrows).

the noise and they can provide high SNR data from rapidly obtained, low-SNR data without requiring any modifications to the FT-IR instrument setup. Several different computation methods are available. These techniques have the same mathematical framework. Typically, the data is transformed into a new domain where noise is uncorrelated and the spectra, that is, the signal, have a higher degree of correlation (Vogt et al. 2005, Vogt 2006). Common data transformation methods include principal component analysis, singular value decomposition, wavelet transform, and the minimum noise fraction transform. In the transformed space, the signal can be represented by a smaller number of coefficients or eigenvalues, whereas the noise components are spread across all eigenvalues. Noise can then be removed from the signal by keeping only the subset containing highly correlated eigenvalues chosen by various methods (Vogt et al. 2005, Cattell 1966, Wold 1978). Their corresponding eigenimages will contain high signal content and then the inverse transform will yield a signal with high noise rejection. However, retaining too few eigenimages will result in a loss of spectral detail. Conversely, selecting a span of eigenimages too wide will result in less noise rejection. Methods for optimally choosing appropriate eigenimages have been proposed recently. An order of magnitude improvement in SNR is routinely possible using these noise reduction techniques.

20.4 Instrumentation

20.4.1 Hardware

The optical system in an FT-IR spectrometer primarily consists of a Michelson interferometer. This is a device that creates interference between a beam of light and its time-delayed counterpart. A general design is depicted by the schematic in Figure 20.7. Light from an IR source is directed toward a beam splitter, which reflects about half the incident beam while transmitting the remaining half. The first half of the beam, having reflected off the splitter causing a half-wavelength phase shift, then reflects off a fixed mirror causing another half-wavelength shift, and travels straight back through the splitter to the detector. Concurrently, the remaining half, after first passing through the splitter, reflects off a moving mirror, and then travels back to the beam splitter, which

reflects the beam toward the detector resulting in an identical net phase shift relative to the first half. The moving mirror is used to change the distance traveled by the beam. When both path lengths are equal, at zero OPD or retardation, the beams recombine constructively. If the movable mirror is displaced by a quarter wavelength, resulting in an optical retardation of a half wavelength, the two beams are now 180° out of phase, so would be complete destructive interference when recombined at the beam splitter. A signal (interferogram) is obtained as a function of retardation (Griffiths and De Haseth 2007).

The ability of the mirror drive motor to keep the movable mirror in constant alignment and speed is critical for interferogram quality. Furthermore, as the only moving part in the interferogram, the motor assembly must be resilient to wear. The drive in the modern interferometers consists of a brushless direct current (DC) servomotor mounted on air bearings for frictionless motion. These servomotors contain an electromagnetic transducer with a fixed magnet. The force produced by the electromotor is a function of the length of the coil wire, the current passing through it, and the field strength of the fixed magnet. The maximum force is typically around 1–2 N and can move a mirror assembly with a mass between 50 and 500 g. The advanced modern precision linear motor stages can provide a single nanometer resolution with 5 nm guaranteed repeatability. These stages contain laser position encoders that have subnanometer resolution and are capable of providing positional feedback useful for spectroscopy. Their use has not been reported in interferometers for IR spectroscopy but one has been implemented in our laboratory.

20.4.2 Continuous Scan Interferometer

This mirror drive is currently the most common implementation (Griffiths and De Haseth 2007, Snively et al. 1999, Snively and Koenig 1999) of FT-IR spectrometers. The mirror travels at a near constant velocity, usually faster than 0.1 mm s⁻¹ during each scan. To obtain accurate phase data, the motor must accelerate and decelerate before and after the data collection region so the mirror can maintain a constant velocity. Usually, multiple passes are made and the signal is summed. This improves the SNR since noise is random and only increases with the square root of the number of scans. This signal averaging process is performed as many times as necessary in order to obtain the desired SNR. Depending on the quality of the motor, it may not have the necessary accuracy to generate the consistent interferograms. Alignment of the interferogram center burst may be necessary prior to averaging. Since the retardation is dependent on the scanning velocity V' over t seconds

$$\delta = 2V't. \tag{20.37}$$

Equation 20.4 can then be modified as

$$S(\delta) = B(\tilde{v}_0)\cos(4\pi\tilde{v}_0 V't). \tag{20.38}$$

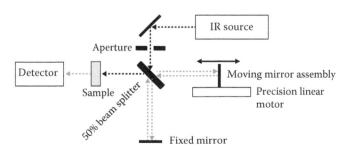

FIGURE 20.7 Schematic diagram of a Michelson interferometer.

20.4.3 Step-Scan Interferometer

In a step-scan interferometer, the moving mirror is stopped at discrete retardation until the measurement is completed. The interferogram is built by discrete retardation points rather than discrete time points. Since time is decoupled from the mirror position, a desired retardation can be held and measured for any amount of time. However, the step-scan interferometer suffers several limitations: (1) it is slower due to the time required for the mirror to stabilize, (2) it is less efficient in data acquisition due to these delays, (3) it may suffer from additional noise sources due to positioning and limited filtering capability, and (4) it requires more expensive and complicated hardware. This method was only more favorable in early FT-IR instruments due to slower detector speeds (Levin and Bhargava 2005). Faster detectors and electronics have made the rapid-scan interferometer more popular for many applications. Step-scan interferometers are now primarily used to record the evolution of transient molecular events over a period of time (Levin and Bhargava 2005). At each retardation setting, the molecular event is retriggered in order to take a new measurement point.

20.5 Detectors

20.5.1 Thermal and Quantum Detectors

The two types of IR detectors available are of thermal or quantum design. The sensing element of a thermal detector responds to changes in temperature. A thermocouple generates an electromotive force, bolometers or semiconductors change resistance, and pneumatic detectors contain a diaphragm that moves due to the expansion of a gas. Most of these have become obsolete for modern FT-IR spectroscopy due to slow response times. The only mid-IR thermal detector that has sufficient speed, linearity, low cost, and ability to operate near room temperatures is the pyroelectric bolometer. The sensing element of pyroelectric detectors initially used in most rapid-scanning FT-IR spectrometers was triglycine sulfate (TGS). In the 1980s, deuterated triglycine sulfate (DTGS) was found to have a higher Curie temperature, the point of failure, so it replaced TGS in FT-IR spectrometer designs. Most recently, deuterated L-alanine-doped triglycine sulfate (DLATGS) has been preferred.

Quantum detectors contain either a photomultiplier tube (PMT) or a semiconductor sensing element. In a PMT, when a photon strikes the photoemissive surface, electrons are produced due to the photoelectric effect. A focusing electrode directs these electrons toward electron multipliers causing a cascading effect. Eventually, enough of these secondary emissions are generated in order to produce a detectable electric current. PMTs require high energies in order to produce a current. Therefore, they can only be used in near-IR regions lower than 1 μm. Semiconductor-based detectors are used for the near to mid-IR region. In the semiconductor element of a quantum detector, the electrons in the valence band become excited to the higher energy level conduction bands when interacting with IR radiation. The level of

excitation depends on the energy of each photon, which is proportional to its wavenumber. However, the wavenumber can only increase until a cutoff where the detector response drops suddenly. Semiconductors are often used for their relatively small bandgap between the valence and conduction bands. Adjusting the doping level of the material by introducing small impurities is often used to adjust the electrical properties of the semiconductor. A common semiconductor used in IR detectors is mercury cadmium telluride (MCT). The ratio of the HgTe and CdTe used to create the MCT determines the bandpass frequency response of the detector and is closely related to its sensitivity, which is characterized by its specific detectivity.

20.5.2 Single Point Detectors

FT-IR spectroscopy can be used to raster scan a sample to create an image, where each pixel contains the IR spectrum of the specimen at that specific point (Reddy et al. 2011). The beam from the interferometer is directed at an IR opaque aperture that restricts light to the prescribed sample area as shown in Figure 20.8. The basic apertures are typically manufactured from carbon black-coated metal. However, more advanced microscope designs use apertures formed from IR-absorbing glasses in order to allow visible images to be obtained while recording IR spectra (Levin and Bhargava 2005). However, light passing through the aperture will form a diffraction pattern that allows unwanted light to reach the detector. The problem due to these diffraction artifacts is resolved by using a second aperture on the opposite side of the sample in front of the detector to serve as a rejection mask (Levin and Bhargava 2005). Since the point spread function narrows, we can achieve a higher resolution but with the drawback of a lower SNR (Levin and Bhargava 2005). Longer acquisition times and multiple passes are necessary to compensate.

The sensitivity of a detector is usually described in terms of the noise equivalent power (NEP), which is the optical power required to give a signal equal to the noise level (Griffiths and De Haseth 2007). NEP is proportional to the square of the detector area A_D and the specific detectivity D^*. The noise power, N' in Watts, of a measurement over t seconds is given (Reddy et al. 2011) by

$$N' = \frac{\text{NEP}}{t^{1/2}}, \quad \text{NEP} = \frac{A_D^{1/2}}{D^*}. \qquad (20.39)$$

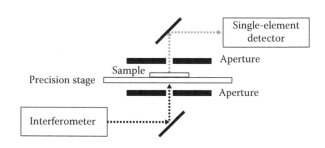

FIGURE 20.8 Design of an FT-IR instrument incorporating a single-element detector. Dual apertures on either side of the sample minimize error due to diffraction artifacts.

Given the spectral brightness $U_v(T)$ derived from the Planck equation, using an interferometer at resolution $\Delta\tilde{v}$, throughput Θ, and efficiency ξ, the signal power, S', is given (Reddy et al. 2011) by

$$S' = U_v(T)\Theta\Delta\tilde{v}\xi. \qquad (20.40)$$

Therefore, the SNR is

$$\text{SNR} = \frac{S'}{N'} = \frac{U_v(T)\Theta\Delta\tilde{v}\xi t^{1/2}}{\text{NEP}} = \frac{U_v(T)\Theta\Delta\tilde{v}\xi t^{1/2}D^*}{A_{\text{D}}^{1/2}}. \qquad (20.41)$$

FT-IR instruments are limited by trade-offs between SNR, resolution, and acquisition time. This property is known as the "spectroscopic trading rules" (Griffiths and De Haseth 2007). Under the criteria of constant throughput and constant mirror velocity, $U_v(T)$, Θ, ξ, and D^* are all constant (Griffiths and De Haseth 2007). To measure at twice the resolution, in other words, half the $\Delta\tilde{v}$, over the same time period t, SNR will be halved and noise doubles (Griffiths and De Haseth 2007). The relation between SNR and time is shown in Equation 20.42 (Griffiths and De Haseth 2007). But, if the throughput is variable, then Θ is also halved so the SNR actually decreases by a factor of four (Griffiths and De Haseth 2007). The same principle applies when changing the velocity of the movable mirror, where the result is identical to varying the scan time.

$$\text{SNR} \propto t^{1/2}. \qquad (20.42)$$

The time can also be impacted by the interferometer resolution $\Delta\tilde{v}$. If $\Delta\tilde{v}$ is smaller than the full width at half-max (FWHM) of the smallest features of the spectrum, lowering $\Delta\tilde{v}$ will have little impact on the features' intensity. However, it will increase noise by a factor depending on the constant or variable throughput criterion described previously (Griffiths and De Haseth 2007). If $\Delta\tilde{v}$ is much larger than the FWHM of the features, then the feature will appear narrower but with a higher amplitude. If the spectrum is primarily composed of weak narrow lines, lowering the resolution has no effect on the SNR of the spectrum (Griffiths and De Haseth 2007). Therefore, measuring a spectrum that has broad features at excessively high resolution will decrease the potential SNR.

Apodization will affect the SNR of a spectrum. When changing from the boxcar apodization function shown in Figure 20.2 to the triangular function shown in Figure 20.3, the SNR is decreased by a factor of $\sqrt{3}$ on average (Griffiths and De Haseth 2007). Apodization functions are mostly identical at retardations near zero, so changing the apodization will not affect low spatial frequency noise. It will, however, affect the degree at which high spatial frequency noise is attenuated. Apodization functions that decay more sharply will smoothen the spectrum more (Griffiths and De Haseth 2007).

FT-IR instruments incorporating single point detectors also have their speed impacted by individual point mapping.

A programmable microscope stage is required to be capable of performing precise movements at the micron scale and then return accurate positional feedback. The stage is used to move the sample such that the beam scans across the desired collection region. Incrementing over thousands of micron-sized points across an entire image can take days to weeks to complete. While FT-IR allows measurements at high fidelity, the acquisition time of raster scanning each point of an image individually limits its applications in time-critical environments (Reddy and Bhargava 2010a).

20.5.3 Multielement Detectors and Imaging

Detectors can be arrayed in a linear format or as an FPA. These multielement detectors, which consist of many detector channels in parallel, allow for faster data collection and are commonly used for imaging purposes (Levin and Bhargava 2005, Bhargava 2007a, 2010, Lewis et al. 1995). The beam of IR light covers the entire detector surface, allowing many points to be captured simultaneously at each position of the raster scan. The readout time of multiple channels will be slower than reading from a single element detector. The additional delay depends on system electronics. Yet, this delay is still much less than the time it takes for the stage to move. Imaging with a multielement detector, in general, will be faster by a factor of the number of channels in that array. For instance, a 256×256 pixel detector will have a speed advantage of 65,536-fold.

Additionally, higher SNR can be obtained because it eliminates the need for apertures (Levin and Bhargava 2005). The dimensions of pixels in a modern detector array are in the tens of microns, compared to single element detectors which are 100–250 μm in size (Levin and Bhargava 2005). Since SNR is proportional to the inverse square root of detector size, the smaller multielement detectors allow for considerable SNR increase. From the single-element SNR analysis summarized by Equation 20.41, the theoretical SNR of an FPA detector can be derived (Levin and Bhargava 2005) as

$$\text{SNR} = \frac{0.12\pi A\left(1 - \sqrt{1 - (\text{NA})^2}\right)U_v(T)\Theta\Delta\tilde{v}t^{1/2}D^*}{A_{\text{D}}^{1/2}}, \qquad (20.43)$$

where NA is the numeric aperture and A is the area of the sample collected per pixel.

The number of pixels on a multielement detector has been increasing steadily and is expected to increase in the foreseeable future. The speed advantage of parallel data acquisition using multielement detectors is a key driving force behind the broad deployment of FT-IR imaging. These improvements in hardware, in conjunction with the dramatic increase in computing power due to faster processors, have made FT-IR imaging a powerful tool for molecular analysis. The development of robust signal processing and data-analysis algorithms in the recent past has also been important. These developments are making real-time analysis of samples using FT-IR spectroscopic imaging practicable in the near future.

20.6 High-Definition Imaging

20.6.1 Synchrotron FT-IR Spectroscopic Imaging

Synchrotron FT-IR spectroscopic imaging combines IR light from multiple synchrotron beams with wide-field detection using an IR microscope (Nasse et al. 2011). Compared to IR thermal sources, synchrotron sources are stable, broadband, and emit high-brightness radiation. The resolution of any FT-IR imaging system is limited by the diffraction limit of the mid-IR light and the numeric aperture of the objective. For the shortest wavelength of interest, 2.5 μm, and a numerical aperture of 0.65, a pixel spacing of around 0.96 μm or smaller may be achieved. However, commercial instruments are typically limited to a spatial resolution of 5.5 μm in order to ensure a sufficient SNR from the weak thermal source. Using a synchrotron source with the same setup, hence, alleviates the lack of throughput and enables smaller pixel sizes at the sample plane to be recorded. One example of such a device is the synchrotron at the University of Madison–Wisconsin that combines 12 high-flux IR beams through a 74× objective to completely illuminate an FPA detector (Nasse et al. 2011). The use of multichannel 128 × 128 pixel FPA detectors removes the need for apertures while allowing for significantly lower acquisition times. However, these synchrotron FT-IR imaging systems have limited applications due to the size and accessibility of the source.

FT-IR synchrotron instruments have allowed for IR diffraction-limited imaging measurements of human breast and prostate tissue (Nasse et al. 2011). As shown in Figure 20.9, synchrotron FT-IR instruments are capable of a dramatic increase in spatial resolution. This study identified the chemical characterization of cell types and tissue structures such as the basement membrane in human prostate tissue. The ability to image at fine detail permits the identification of tissues that are key components of cancer diagnosis.

20.6.2 Attenuated Total Reflectance FT-IR Imaging

An attenuated total reflectance (ATR) FT-IR imaging relies on the total internal reflection of the incident beam using a solid immersion lens that creates an evanescent electric wave that interacts with the sample. Figure 20.10 demonstrates the use of an ATR crystal in an FT-IR imaging system. The evanescent waves penetrate 1–4 μm into the sample. The lens must have a higher refractive index than the sample, so it is typically made from germanium, diamond, or zinc selenide crystal. Furthermore, the sample must be in complete contact with the lens to prevent the presence of air bubbles, where the change in refractive index would systematically distort the data. The physical contact may cause damage to the sample as well as changes in conformation. In most cases, liquid samples can be imaged in their natural state with little or no preparation required. ATR can also be an excellent technique for measuring the composition of solids which often absorb too much of energy to be measured by IR transmission.

The spatial resolution of the instrument increases at higher angles of incident light. Since ATR and FT-IR imaging is a

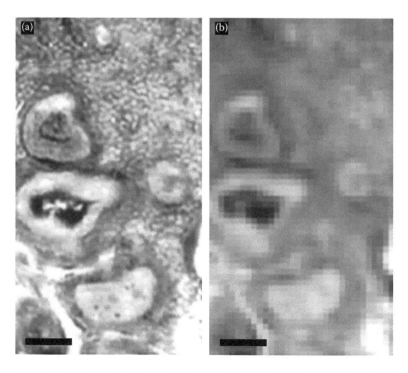

FIGURE 20.9 Cancerous prostate tissue (a) imaged with a multibeam synchrotron FPA (0.54 μm pixel size) compared to (b) an image from a thermal source linear array (6.25 μm pixel size) (50 μm scale bar). (Reprinted with permission from Nasse, M. J. et al. 2011. *Nature Methods* 8:413–416.)

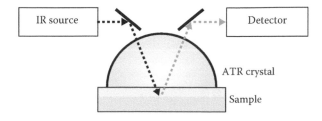

FIGURE 20.10 Schematic of an ATR FT-IR imaging system.

near-field technique, there is no theoretical diffraction limit. However, there are practical limitations on spatial resolution due to SNR. At higher incident angles, there is a trade-off between increased spatial resolution and decreased wave penetration depth. IR radiation can only interact with the sample up to a depth (dp) depending on the wavelength λ, incident angle θ, and refractive indices n_1 and n_2 of the materials.

$$dp = \frac{\lambda}{2\pi\sqrt{n_1^2\sin\theta^2 - n_2^2}}. \tag{20.44}$$

ATR FT-IR instruments have been shown to be capable of imaging large samples at nearly 1.5 by 2 mm in size, at spatial resolutions at up to 1.25 µm, using linear and FPA detectors with dimensions up to 256×256 pixels (Patterson et al. 2007). Studies have demonstrated that this can identify important components in tissues and yield new insights into disease processes and their associated chemical changes (Walsh et al.).

20.6.3 Bench-Top High-Definition Imaging

Bench-top high-definition FT-IR spectroscopic imaging has been a relatively recent development. Rigorous electromagnetic models for light propagation through an FT-IR imaging instrument have been proposed recently. Creating a complete model for the propagation of light through the instrument has led to key insights and improved instrument design. An appropriate choice of pixel size, Cassegrain magnification, numerical aperture, and wavelength has been used to optimize the image quality. Consequently, previously obscured spatial information can now be obtained using the new instruments. Data from table-top FT-IR spectroscopic imaging instruments are now capable of spatial resolution and data quality comparable to that from a system using a gigantic synchrotron source, although the SNR from the table-top instrument is significantly lower for a single scan.

The advent of bench-top high-definition imaging instruments has important implications for the application of FT-IR imaging. In a wide array of applications, molecular information coupled to high spatial detail can be a powerful investigative tool. In the study of diseases like breast cancer, for example, high spatial detail can be important in diagnosis. The IR spectrum from FT-IR imaging provides information that helps us to distinguish

epithelial tissue from stroma and this is an important parameter in disease diagnosis. However, the spatial resolution of current FT-IR imaging instruments is insufficient for distinguishing interlobular stroma which can be potentially diagnostic. High-definition imaging has solved this problem by providing significantly higher spatial resolution and image quality. New instruments incorporating the capability of high-definition imaging are likely to be adopted widely in the near future.

As noted in the section on FT-IR imaging theory, one has to be careful in the interpretation of spectra when observing chemically heterogeneous samples. Differences between spectra of homogeneous and heterogeneous samples need careful attention to avoid the incorrect interpretation. Spectral differences can be utilized in characterization and identification of chemical contents of a sample. However, one has to be careful not to assign the chemical significance to every variation in the spectrum. A robust identification of chemicals from the distorted spectra is a subject of current research.

20.7 Applications

FT-IR spectroscopic imaging has a wide range of applications today. Many ongoing studies deal with investigating fundamental processes involved in disease initiation and progression. Imaging also has significant implications for the design of methodologies for disease diagnosis and prognosis in a clinical environment. In addition, the ability to characterize various natural and artificial materials at a chemical level is a desirable tool used across many disciplines. There are numerous studies and applications that can be found in several recent reviews from our group and others. Rather than an exhaustive review of the past work, here, we present examples of some of the exciting recent developments primarily to provide an illustration of the possibilities.

20.7.1 Cancer Histopathology

The preliminary screening of cancer tends to be sensitive to the disease but at a cost of specificity. Further histologic analysis via biopsies is the gold standard for cancer diagnosis. These histopathology tests typically involve hematoxylin and eosin (H and E) staining of the sectioned tissue. A pathologist trained to recognize the morphology of cell types must inspect the patterns in each tissue sample and determine if there are alterations, if these alterations indicate a disease, and, if so, characterize its severity. The vast majority of biopsies, however, are actually benign and this consumes excessive healthcare resources (Nakhleh 2008). Additionally, the human factor is often not accurate when dealing with indeterminate cases. FT-IR imaging is developing to become a high-throughput, automatic, and accurate tool capable of making a clinical impact by improving the efficiency of cancer screening (Walsh and Bhargava 2010, Walsh et al. 2007, Reddy and Bhargava 2010b).

Automated histopathologic methods for classifying FT-IR imaging data using Bayesian classification have been developed

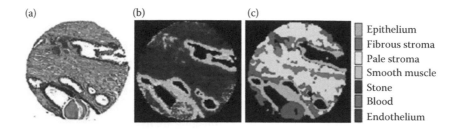

FIGURE 20.11 Prostate tissue section hematoxylin and eosin (H&E) stained (a) compared to its FT-IR absorption data at 1,080 cm^{-1} (b). The FT-IR spectrum can be analyzed for unique spectral features for each cell type (c) thus providing more information than the conventional H&E stain. The diameter of the sample is approximately 500 microns. (Reprinted with permission from Bhargava, R. 2007b. *Analytical and Bioanalytical Chemistry* 389:1155–1169).

recently (Reddy and Bhargava 2010b, Bhargava 2007b) (Figure 20.11). The prediction algorithm is trained to high confidence levels using hundreds of tissue samples in microarrays. Only an optimal subset of metrics that has a diagnostic potential is incorporated into the algorithm. The conditional probabilities for the metrics are used to build a discriminant function that returns probabilistic recognitions of cellular states. The model is then applied on FT-IR image data. It is first segmented into regions of similar cell types by applying pattern recognition techniques. Furthermore, spectra and spatial information from the local neighborhood of each data point are used to determine the presence of a disease and its severity. A supervised classification is performed using chemically stained data segmented by an expert pathologist as a gold standard data (Bhargava 2007b). Validation of the resulting algorithm is performed against an independent data set and receiver operating characteristic (ROC) curves are used as a measure of classification accuracy. The information from the classification of FT-IR imaging data can be combined with H&E-stained visible images (Kwak et al. 2011) to provide further improvements in segmentation. First, the images are aligned and overlaid. This is done by using a smoothing filter to identify rough shapes that can be transformed with a least-squares error minimization approach. Segmentation and feature extraction algorithms then process the combined data sets. This technique has been shown to yield highly accurate classification results (Kwak et al. 2011, Bhargava and Kong 2008, Fernandez et al. 2005, Keith and Bhargava 2006, Mackanos and Contag 2009, Kelly et al. 2010).

20.7.2 Cell Culture

In vitro cell cultures can serve as a model for many environments in the human body. Imaging cell cultures with FT-IR techniques allows us to research the fundamental processes involved in disease initiation and progression (Holman et al. 2000, Hammiche et al. 2005, Boydston-White et al. 1999, 2006). During carcinoma progression, the tumor microenvironment transforms both chemically and morphologically. ATR FT-IR spectroscopic imaging has been used to examine these transitions prevalent in early cancer development (Holton et al. 2011). The increased spatial resolution allowed the location of the transformations to be identified. The changes in the spectra were found to be associated with protein modifications in the cell cytoplasm rather than events occurring in the nucleus (Holton et al. 2011).

20.7.3 Tissue Characterization and Engineering

Morphologic, histologic, and chemical analyses using FT-IR spectroscopy have been performed characterizing the porcine skin to determine if it can serve as an accurate model for human skin (Kong and Bhargava 2011). Porcine skin is a convenient choice for laboratory environments because it can be cryo-prepared and remains chemically stable over a prolonged time. In this study, a marker molecule such as dimethyl sulfoxide was tracked over the depth of the skin. This molecule has a unique absorbance spectrum which allows its location to be accurately monitored. The diffusion concentration profile indicated that porcine skin was very similar to human skin both structurally and chemically as shown in Figure 20.12 (Kong and Bhargava 2011).

An active area of study is the use of three-dimensional (3D) tissue-engineered model systems. Many complex interactions found in 3D tissue cultures are often not found in two-dimensional (2D) models and the engineered skin is able to represent the human skin because it has a simple laminar structure. The use of FT-IR imaging systems on 3D tissue samples is a recent development that has not yet been well characterized. The various spectral changes associated with different cell types within engineered skin versus human skin have been studied using FT-IR image analysis (Kong et al. 2010). Investigations have used FT-IR imaging to monitor the chemical changes of stromal cells surrounding the malignant melanocytes. 3D cultures of engineered skin were designed to model melanoma (Kong et al. 2010, Yamada and Cukierman 2007, Yang et al. 2005). FT-IR analysis characterized the chemical changes of the stromal cells surrounding the malignant melanocytes (Kong et al. 2010). Furthermore, this research led to novel insights that may allow for improved quality control and standardization in tissue engineering (Kong et al. 2010).

20.7.4 Drug Diffusion

FT-IR spectroscopy is a valuable tool for drug diffusion studies. There are many advantages to transdermal drug delivery in handling, safety, controlled release, and avoiding the gastrointestinal

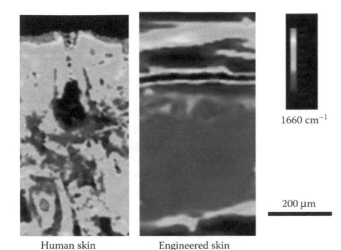

1660 cm^{-1}

200 μm

Human skin   Engineered skin

FIGURE 20.12 FT-IR images of human skin versus 3D tissue engineered skin. (Reprinted with permission from Kong, R. et. al. 2010. *Analyst* 135:1569–1578.)

tract. Some FTIR-ATR designs can allow for noninvasive real-time measurements of drug diffusion (Hartmann et al. 2004, Boncheva et al. 2008, Andanson et al. 2009). The most important advantage of the ATR technique is that it is nondestructive to the sample and provides a profile of drug concentration over tissue depth. In this setup, a membrane acting as a drug acceptor is placed between the ATR crystal and the drug donor. The drug diffuses through the acceptor membrane and builds up at the ATR crystal. The concentration of the drug at the membrane-crystal interface is monitored by the appearance and increase of drug-specific IR bands over time.

20.7.5 Polymers

IR spectroscopy is often used to characterize the properties of polymers and IR imaging was widely used to characterize heterogeneous polymer systems and dynamics therein. Most of these studies focused on fundamental polymer science and less on medically relevant applications. Recently, however, there have been several reports of the analysis of biomaterials, analysis of materials used in biological applications, and biologically derived polymers. For instance, many applications in tissue engineering and medical devices rely on hydrated polymer coatings on polymer substrates. The thickness of the coating has consequences for tissue response as well as tissue mechanical properties. FT-IR spectroscopy can be used to determine the thickness measurements relative to initial calibration with atomic force microscopy (Kane et al. 2009). The diffusion of low-molecular-weight species into polymers also has implications for designing drug release devices (Ribar et al. 2000). FT-IR images taken periodically generated spatially resolved spectral data and absorbance profiles of the polymer (Ribar et al. 2000). Another study used FT-IR spectroscopy to determine the lower critical solution temperature of hydrogels commonly used in medical devices (Percot et al. 2000). This technique can be applied to both linear and cross-linked polymers. The spectral changes, especially in the C–H stretching region, showed conformational changes when heated above the critical temperature (Percot et al. 2000).

20.8 Summary

IR spectroscopic imaging, combining the advantages of optical microscopy and molecular spectroscopy, is a powerful tool for analysis in several physical sciences and biomedical disciplines. By combining interferometry and wide-field detection in microscopy, in particular, FT-IR imaging allows the spatial examination of the spectroscopic fingerprints of the chemical constituents of a sample. Recent developments in FT-IR instrumentation, data processing, and theoretical analyses have led to FT-IR imaging becoming a rapid, robust, and reliable tool for molecular analysis with a clear understanding of the underlying theory and origin of signals. While significant progress has been reported in using the approach for biomedical imaging, the translation of these emerging capabilities to cell and tissue analysis is now being pursued and will likely result in many new applications or ones with significantly enhanced capability in both spatial and chemical analyses.

References

Andanson, J. M., K. L. Chan, and S. G. Kazarian. 2009. High-throughput spectroscopic imaging applied to permeation through the skin. *Applied Spectroscopy* 63:512–517.

Bassan, P., H. J. Byrne, F. Bonnier et. al. 2009. Resonant Mie scattering in infrared spectroscopy of biological materials—Understanding the "dispersion artefact". *Analyst* 134:1586–1593.

Bhargava, R. and G. S. Mintz 2007a. Fourier transform-infrared spectroscopic imaging: The emerging evolution from a microscopy tool to a cancer imaging modality. *Spectroscopy* 22:40–51.

Bhargava, R. 2007b. Towards a practical Fourier transform infrared chemical imaging protocol for cancer histopathology. *Analytical and Bioanalytical Chemistry* 389:1155–1169.

Bhargava, R. and R. Kong. 2008. *Structural and Biochemical Characterization of Engineered Tissue Using FTIR Spectroscopic Imaging: Melanoma Progression as an Example*. R. J. Nordstrom, ed. SPIE, San Jose, CA. pp. 687004–687010.

Bhargava, R. and I. W. Levin. 2001. Fourier transform infrared imaging: Theory and practice. *Analytical Chemistry* 73:5157–5167.

Bhargava, R., B. G. Wall, and J. L. Koenig. 2000. Comparison of the FT-IR mapping and imaging techniques applied to polymeric systems. *Applied Spectroscopy* 54:470–479.

Boiana O. B. 2000. Minimization of optical non-linearities in Fourier transform-infrared microspectroscopic imaging. *Vibrational Spectroscopy* 24:37–45.

Boncheva, M., F. H. Tay, and S. G. Kazarian. 2008. Application of attenuated total reflection Fourier transform infrared imaging and tape-stripping to investigate the three-dimensional distribution of exogenous chemicals and the molecular organization in stratum corneum. *Journal of Biomedical Optics* 13:064009.

Boydston-White, S., T. Gopen, S. Houser, J. Bargonetti, and M. Diem. 1999. Infrared spectroscopy of human tissue. V. Infrared spectroscopic studies of myeloid leukemia (ML-1) cells at different phases of the cell cycle. *Biospectroscopy* 5:219–227.

Boydston-White, S., M. Romeo, T. Chernenko et al. 2006. Cell-cycle-dependent variations in FTIR micro-spectra of single proliferating HeLa cells: Principal component and artificial neural network analysis. *Biochimica et Biophysica Acta* 1758:908–914.

Cattell, R. 1966. The screen test for the number of factors. *Multivariate Behavioral Research* 1:245–276.

Davis, B. J., P. S. Carney, and R. Bhargava. 2010a. Theory of midinfrared absorption microspectroscopy: I. Homogeneous samples. *Analytical Chemistry* 82:3474–3486.

Davis, B. J., P. S. Carney, and R. Bhargava. 2010b. Theory of mid-infrared absorption microspectroscopy: II. Heterogeneous samples. *Analytical Chemistry* 82:3487–3499.

Elkins, K. M. 2011. Rapid presumptive "fingerprinting" of body fluids and materials by ATR FT-IR spectroscopy. *Journal of Forensic Sciences* 56:1580–1587.

Fernandez, D. C., R. Bhargava, S. M. Hewitt, and I. W. Levin. 2005. Infrared spectroscopic imaging for histopathologic recognition. *Nature Biotechnology* 23:469–474.

Griffiths, P. R. and J. A. De Haseth. 2007. *Fourier Transform Infrared Spectrometry*. Wiley-Interscience, Hoboken, NJ.

Hammiche, A., M. J. German, R. Hewitt, H. M. Pollock, and F. L. Martin. 2005. Monitoring cell cycle distributions in MCF-7 cells using near-field photothermal microspectroscopy. *Biophysical Journal* 88:3699–3706.

Hartmann, M., B. D. Hanh, H. Podhaisky et al. 2004. A new FTIR-ATR cell for drug diffusion studies. *Analyst* 129:902–905.

Holman, H. Y., M. C. Martin, E. A. Blakely, K. Bjornstad, and W. R. McKinney. 2000. IR spectroscopic characteristics of cell cycle and cell death probed by synchrotron radiation based Fourier transform IR spectromicroscopy. *Biopolymers* 57:329–335.

Holton, S. E., M. J. Walsh, and R. Bhargava. 2011. Subcellular localization of early biochemical transformations in cancer-activated fibroblasts using infrared spectroscopic imaging. *Analyst* 136:2953–2958.

Kane, S. R., P. D. Ashby, and L. A. Pruitt. 2009. ATR-FTIR as a thickness measurement technique for hydrated polymer-on-polymer coatings. *Journal of Biomedical Materials Research. Part B, Applied Biomaterials* 91:613–620.

Keith, F. A. and Bhargava, R. 2006. Data processing for tissue histopathology using Fourier transform infrared spectral data. *Signals, Systems and Computers*, ACSSC '06:71–75.

Kelly, J. G., P. P. Angelov, J. Trevisan et al. 2010. Robust classification of low-grade cervical cytology following analysis with ATR-FTIR spectroscopy and subsequent application of self-learning classifier eClass. *Analytical and Bioanalytical Chemistry* 398:2191–2201.

Kong, R., R. Bhargava. 2011. Characterization of porcine skin as a model for human transdermal diffusion. *Analyst* 136:2359–2366.

Kong, R., R. K. Reddy, and R. Bhargava. 2010. Characterization of tumor progression in engineered tissue using infrared spectroscopic imaging. *Analyst* 135:1569–1578.

Kwak, J. T., S. M. Hewitt, S. Sinha, and R. Bhargava. 2011. Multimodal microscopy for automated histologic analysis of prostate cancer. *BMC Cancer* 11:62.

Lee, J., E. Gazi, J. Dwyer et al. 2007. Optical artefacts in transflection mode FTIR microspectroscopic images of single cells on a biological support: The effect of back-scattering into collection optics. *Analyst* 132:750–755.

Levin, I. W. and R. Bhargava. 2005. Fourier transform infrared vibrational spectroscopic imaging: Integrating microscopy and molecular recognition. *Annual Review of Physical Chemistry* 56:429–474.

Lewis, E. N., P. J. Treado, R. C. Reeder et al. 1995. Fourier transform spectroscopic imaging using an infrared focal-plane array detector. *Analytical Chemistry* 67:3377–3381.

Mackanos, M. A. and C. H. Contag. 2009. FTIR microspectroscopy for improved prostate cancer diagnosis. *Trends in Biotechnology* 27:661–663.

Mohlenhoff, B., M. Romeo, M. Diem, and B. R. Wood. 2005. Mie-type scattering and non-Beer-Lambert absorption behavior of human cells in infrared microspectroscopy. *Biophysical Journal* 88:3635–3640.

Nakhleh, R. E. 2008. Patient safety and error reduction in surgical pathology. *Archives of Pathology & Laboratory Medicine* 132:181–185.

Nasse, M. J., M. J. Walsh, E. C. Mattson et al. 2011. High-resolution Fourier-transform infrared chemical imaging with multiple synchrotron beams. *Nature Methods* 8:413–416.

Navas, N., J. Romero-Pastor, E. Manzano, and C. Cardell. 2008. Benefits of applying combined diffuse reflectance FTIR spectroscopy and principal component analysis for the study of blue tempera historical painting. *Analytica Chimica Acta* 630:141–149.

Nishikawa, Y., T. Nakano, and I. Noda. 2012. Detection of reversible nonlinear dynamic responses of polymer films by using time-resolved soft-pulse compression attenuated total reflection step-scan Fourier transform infrared spectroscopy. *Applied Spectroscopy* 66:312–318.

Okamoto, T., S. Kawata, and S. Minami. 1985. Optical method for resolution enhancement in photodiode array Fourier transform spectroscopy. *Applied Optics* 24:4221.

Patterson, B. M., G. J. Havrilla, C. Marcott, and G. M. Story. 2007. Infrared microspectroscopic imaging using a large radius germanium internal reflection element and a focal plane array detector. *Applied Spectroscopy* 61:1147–1152.

Percot, A., X. X. Zhu, and M. Lafleur. 2000. A simple FTIR spectroscopic method for the determination of the lower critical solution temperature of N-isopropylacrylamide copolymers and related hydrogels. *Journal of Polymer Science Part B: Polymer Physics* 38:907–915.

Reddy, R. K. and R. Bhargava. 2010a. Accurate histopathology from low signal-to-noise ratio spectroscopic imaging data. *Analyst* 135:2818–2825.

Reddy, R. K. and R. Bhargava. 2010b. Chemometric methods for biomedical Raman spectroscopy and imaging emerging. P. Matousek, and M. D. Morris, eds. *Raman Applications and Techniques in Biomedical and Pharmaceutical Fields*. Springer, Berlin, Heidelberg. pp. 179–213.

Reddy, R., B. Davis, P. S. Carney, and R. Bhargava. 2011. Modeling Fourier transform infrared spectroscopic imaging of Prostate and Breast Cancer tissue specimens. In Biomedical Imaging: From Nano to Macro, IEEE International Symposium, Hyatt Regency McCormick Place, Chicago, pp. 738–741.

Ribar, T., R. Bhargava, and J. L. Koenig. 2000. FT-IR imaging of polymer dissolution by solvent mixtures. 1. Solvents. *Macromolecules* 33:8842–8849.

Saptari, V. 2004. *Fourier Transform Spectroscopy Instrumentation Engineering*. SPIE Optical Engineering Press, Bellingham, WA.

Snively, C. M. and J. L. Koenig. 1999. Characterizing the performance of a fast FT-IR imaging spectrometer. *Applied Spectroscopy* 53:170–177.

Snively, C. M., S. Katzenberger, G. Oskarsdottir, and J. Lauterbach. 1999. Fourier-transform infrared imaging using a rapid-scan spectrometer. *Optics Letters* 24:1841–1843.

Vogt, F. 2006. Trends in remote spectroscopic sensing and imaging—Experimental techniques and chemometric concepts. *Current Analytical Chemistry* 2:107–127.

Vogt, F., J. Cramer, and K. Booksh. 2005. Introducing multidimensional "hybrid wavelets" for enhanced evaluation of hyperspectral image cubes and multi-way data sets. *Journal of Chemometrics* 19:510–520.

Walsh, M.J., Bhargava, R. 2010. Infrared spectroscopic imaging: An integrative approach to pathology. G. Popescu, ed. *Nanobiophotonics*, McGraw-Hill, New York.

Walsh, M. J., M. J. German, M. Singh et al. 2007. IR microspectroscopy: Potential applications in cervical cancer screening. *Cancer Letters* 246:1–11.

Walsh, M. J., S. E. Holton, A. Kajdacsy-Balla, and R. Bhargava. 2012. Attenuated total reflectance Fourier-transform infrared spectroscopic imaging for breast histopathology. *Vibrational Spectroscopy* 60:23–28.

Wold, S. 1978. Cross-validatory estimation of the number of components in factor and principal components models. *Technometrics* 20:397–405.

Yamada, K. M. and E. Cukierman. 2007. Modeling tissue morphogenesis and cancer in 3D. *Cell* 130:601–610.

Yang, Y., J. Sule-Suso, G. D. Sockalingum et al. 2005. Study of tumor cell invasion by Fourier transform infrared microspectroscopy. *Biopolymers* 78:311–317.

Index

9 7 8 0 3 6 7 4 4 5 8 9 8

T - #0564 - 071024 - C12 - 279/216/16 - PB - 9780367445898 - Gloss Lamination